HUMAN GEOGRAPHY

SECOND EDITION

HUMAN GEOGRAPHY

SECOND EDITION

WILLIAM NORTON

OXFORD UNIVERSITY PRESS
TORONTO   NEW YORK   OXFORD
1995

Oxford University Press
70 Wynford Drive, Don Mills, Ontario M3C 1J9

Oxford New York
Athens Auckland Bangkok Bombay
Calcutta Cape Town Dar es Salaam Delhi
Florence Hong Kong Istanbul Karachi
Kuala Lumpur Madras Madrid Melbourne
Mexico City Nairobi Paris Singapore
Taipei Tokyo Toronto

and associated companies in
Berlin Ibadan

Oxford is a trade mark of Oxford University Press

This book is printed on permanent (acid-free) paper. ∞

Canadian Cataloguing in Publication Data

Norton, William, 1944–
    Human geography

2nd ed.
Includes bibliographical references and index.
ISBN 0-19-541083-1

1. Human geography.   I. Title.

GF41.N67 1995   304.2   C95-930221-2

Editor: Valerie Ahwee
Design: Heather Delfino
Cover painting: 'Bridges, Looking West.' 1992 by Jack Darcus. Copyright Jack Darcus.
(Transparency courtesy of Diane Farris Gallery)
Illustrations: Marjorie Halmarson and Ed Pachanuk
Compositor: Indelible Ink

Printed and bound in Canada by Friesen Printers

2 3 4 5 99 98 97 96

Contents

List of figures / vi
List of tables / ix
Preface / xi
To the student / xv

Introduction / 1
Chapter 1 What is human geography? / 10
Chapter 2 Studying human geography / 29
Chapter 3 The earth: A human environment / 56
Chapter 4 The earth: A fragile home /76
Chapter 5 The human population: History and concepts / 99
Chapter 6 The human population: An unequal world / 122
Chapter 7 Cultures: The evolution and regionalization of landscape / 144
Chapter 8 Cultures: Symbolic and social landscapes / 172
Chapter 9 The political world / 192
Chapter 10 Agriculture / 218
Chapter 11 Settlement / 242
Chapter 12 Industry and regional development / 267
Chapter 13 The geography of movement / 288
Conclusion: The human geography of the future ... and the future of human geography / 306

Glossary / 312
References / 320
Index /329

List of figures

Intro.1 World political divisions, 1994 / 3
Intro.2 World political divisions, 1938 / 3
1.1 The world according to Eratosthenes / 11
1.2 The world according to Ptolemy / 12
1.3 An example of a T-O map / 12
1.4 An example of a Portolano chart / 13
2.1 The scientific method / 31
2.2 The site of Winnipeg / 37
2.3 The situation of Winnipeg within North America / 38
2.4 Clustered, random, and uniform point patterns / 41
2.5 A characteristic distance decay curve / 41
2.6 Impact of spatial scale on descriptions of point patterns / 43
2.7 Urban centres in Manitoba / 43
2.8 A characteristic S-shaped growth curve / 43
2.9 Spatial concepts of North America in 1763 / 45
2.10 Mapping at a scale of 1:250,000 / 46
2.11 Mapping at a scale of 1:50,000 / 47
2.12 Schematic representation of a dot map / 48
2.13 Schematic representation of a choropleth map / 48
2.14 Schematic representation of an isoline map / 48
2.15 Mercator projection / 48
2.16 Mollweide projection / 48
2.17 Robinson projection / 48
2.18 Two topographic maps of the Love Canal area, Niagara Falls, New York / 50
3.1 Revolution of the earth around the sun / 57
3.2 Moving continents / 58
3.3 Major ocean currents / 62
3.4 Global cordilleran belts / 62
3.5 Global distribution of soil types / 64
3.6 Global distribution of natural vegetation / 64

3.7   Global distribution of climate / 65
3.8   Generalized global environments / 66
4.1   Chemical cycling and energy flows / 77
4.2   Some consequences of human-induced vegetation change / 84
4.3   Location of tropical rain forests / 85
4.4   The Western Australian rabbit-proof fence / 88
4.5   The global water cycle / 91
4.6   Impact of sea-level change on Bangladesh / 94
5.1   World distribution of crude birth rates, 1993 / 104
5.2   Generalized J-shaped curve of death rates and age / 106
5.3   World distribution of crude death rates, 1993 / 107
5.4   World distribution of life expectancy, 1993 / 107
5.5   World distribution of rates of natural increase, 1993 / 109
5.6   Age and sex structure in China, 1990 / 110
5.7   Age structure of populations / 111
5.8   Age-sex structure in Brazil: 1950, 1970, 1987 / 112
5.9   Age-sex structure in Canada: 1861, 1921, 1981, 2036 / 113
5.10   World population growth / 115
5.11   The demographic transition theory / 117
6.1   World population distribution and density, 1993 / 123
6.2   Mental maps / 128
6.3   The push-pull concept and relevant obstacles / 129
6.4   Refugee numbers, 1960–93 / 132
6.5   The Horn of Africa / 134
7.1   Civilizations in the ancient world / 150
7.2   Cultural regions of the world / 151
7.3   Regions of North America / 152
7.4   Cultural regions of the United States / 152
7.5   Regions of Canada / 153
7.6   Europe defined / 154
7.7   Core, domain, and sphere / 156
7.8   World distribution of language families / 160
7.9   Initial diffusion of Indo-European languages / 160
7.10   Diffusion of Indo-European languages into England / 161
7.11   Four official languages in Switzerland / 161
7.12   Dutch and French in Belgium / 161
7.13   French and English in Canada / 162
7.14   Origin areas and diffusion of four major religions / 165
7.15   World distribution of major religions / 167
8.1   North American vernacular regions / 173
8.2   Components of place as historically contingent process / 177
8.3   Apartheid on the national scale in 1975 / 182
9.1   The British empire in the late nineteenth century / 195
9.2   Principal elements in the process of exploration / 195
9.3   United States' territorial expansion / 196
9.4   Jones field theory / 197
9.5   Mackinder heartland theory / 199
9.6   African ethnic regions / 201
9.7   African political areas in the sixteenth, eighteenth, and nineteenth centuries / 202
9.8   Areas of dissent in Europe / 202
9.9   The former Yugoslavia in 1994 / 204
9.10   The former USSR in 1994 / 205
9.11   European ethnic regions / 208
9.12   World distribution of types of states / 210

9.13   The original gerrymander / 211

9.14   Congressional districts in 1960s Mississippi / 211

10.1   Relationship between the US mean annual rainfall and the wheat yield, 1909 / 221

10.2   Crop and livestock associations, US–Canadian border / 224

10.3   Supply and demand curves / 225

10.4   Rent-paying abilities of selected land uses / 225

10.5   Economic rent lines for three crops and related zones of land use / 225

10.6   Agricultural land use in the isolated state / 226

10.7   A navigable waterway / 227

10.8   Multiple markets and a transport network / 227

10.9   Scatter graphs, best-fit lines, and r values / 228

10.10   Agricultural land use in Uruguay / 229

10.11   World agricultural regions / 232

11.1   Growth of urban population relative to growth of world population / 243

11.2   Township survey in the Canadian prairies / 244

11.3   Examples of nucleated rural settlement patterns / 245

11.4   Approved urban areas in the regional municipality of Niagara, 1989 / 247

11.5   Location of British new towns / 251

11.6   A triangular lattice / 255

11.7   Theoretical trading areas / 255

11.8   A central place system — the marketing principle / 256

11.9   A central place system — the transportation principle / 257

11.10   A central place system — the administration principle / 257

11.11   Urban land values / 259

11.12   Three models of the internal structure of urban areas / 259

12.1   A locational triangle / 269

12.2   A simple isotim map / 270

12.3   An isodapane map / 270

12.4   Transport cost and distance / 271

12.5   Stepped transport costs / 271

12.6   Simplified relationship between economic growth and distribution of employment / 281

13.1   A non-straight line, shortest distance route / 289

13.2   A one-way system — impact on the distance travelled / 289

13.3   Time distance in Edmonton / 290

13.4   Toronto in physical space and time space / 291

13.5   Logarithmic transformation of distances from Asby, Sweden / 292

13.6   The London underground system / 293

13.7   Distribution of innovativeness / 297

13.8   Effects of preemption on the adoption curve / 298

13.9   Diffusion of transport innovations in Britain, 1650–1930 / 299

13.10   Transport network evolution / 300

13.11   The road network of Martinique / 301

List of Tables

Intro. 1 Selected demographic data / 6
3.1 Basic chronology of life on earth / 69
3.2 Basic chronology of human evolution / 71
4.1 Shallow and deep ecology compared / 82
5.1 Contraceptive use by regions, 1993 / 102
5.2 Population data, Canada, 1993 / 106
5.3 Countries with highest rates of natural increase (3.5 and above), 1993 / 108
5.4 Countries with lowest rates of natural increase (0.1 and below), 1993 / 108
5.5 Major epidemics, 1500–1700 / 114
6.1 Population by major world regions, 1983 and 1993 / 123
6.2 The ten most populous countries, 1993 / 123
6.3 Population densities of the ten most populous countries, 1993 / 124
6.4 Refugees and asylum seekers by region of asylum, end of 1991 / 132
6.5 Countries with more than 250,000 refugees and asylum seekers in need of protection and assistance, end of 1991 / 132
6.6 Principal sources of the world's refugees and asylum seekers, end of 1991 / 133
6.7 UNHCR-assisted cases by country of resettlement, 1990–2 / 135
6.8 Population densities, selected countries, 1993 / 138
6.9 Natural events and human disasters, regional data, 1947–67 and 1969–89 / 139
6.10 Extremes of human development, 1992 / 140
7.1 A unilinear evolutionary model of culture / 149
7.2 Basic chronology of early civilizations / 150
7.3 An overview of language data, 1987 / 157
7.4 An overview of religious data, 1992 / 166
7.5 Distribution of Christian populations, 1992 / 166
8.1 Gender-sensitive measures of human development: Selected countries, 1992 / 186
9.1 Ethnic groups in the former Yugoslavia / 204
9.2 Ethnic groups in the former USSR / 205
9.3 Disputes and conflicts involving the UN, 1945–90 / 212

10.1   Contrasting farming types in Illinois, 1985 / 222

10.2   Average distances from London, England, to regions of import derivation (miles) / 230

10.3   Determinants of peasant-herder conflicts in northern Ivory Coast / 239

11.1   Some definitions of urban centres / 242

11.2   Percentage change in rural population by region, Canada, 1981–6 / 246

11.3   Factors influential in the decision to move to exurban locations around Woodstock, southwestern Ontario / 247

11.4   The ten largest urban areas, 1992, and projections for 2000 (population in millions) / 248

11.5   Seven stages of premodern urban development / 252

Conclu.1 Disappearing peoples, 1989 / 308

Preface

TEXTBOOK AUTHORS typically feel a need to justify their exertions. I am no exception and the justification is hardly atypical. Simply put, I feel that this volume fills a substantial gap in the introductory human geography textbook market. Over the years university and college textbooks for all subjects have become increasingly lengthy and more detailed. This is not necessarily a desirable trend. Textbooks are tools for instructors to use and do not need to cover all relevant material. For introductory human geography, certainly, there is a need for a textbook that provides the organizational framework, the intellectual skeleton upon which the flesh of human geography can be added by the instructor. This volume is intended for that purpose. It is therefore shorter and somewhat less detailed than most other introductory human geography textbooks.

Notwithstanding its relative brevity, this volume aims to introduce all human geography. Students are able to read about the diverse subject matter of human geography, about the long and exciting history of the discipline, about its close links with other disciplines, and about the reasons why human geographers do what they do. Let me explain.

Human geography is presented as a wide-ranging discipline, focusing on the concepts of space and place, with interests in cultural and economic issues. No one methodology dominates — a true reflection of this eclectic discipline. The discussion of disciplinary history obliges students to appreciate that human geography is an academic discipline that arose in response to society's needs and that changes in response to changing needs. Acknowledgement that human geography is closely linked to other disciplines — social sciences, humanities, and physical sciences — encourages learning and ensures that human geographic learning is applied in other academic contexts. Considering why we ask the questions that we ask and why we answer in the way that we do facilitates students' comprehension that academic disciplines encompass the search for new knowledge and the understanding and explanation of existing knowledge. Emphasizing disciplinary history gives students an appreciation of the longevity of human geography; emphasizing philosophies and concepts helps students integrate human geography with other disciplines.

This second edition includes a number of substantial revisions in addition to the usual correction of errors and updating of information. An attempt is made to link more clearly the short chapter on physical issues to the text as a whole by moving the chapter on environmental change so that it follows the physical chapter. Material included in the two chapters dealing with cultural and social geography is substantially rearranged to reflect more accurately the current concerns of practising human geographers. The addition of a glossary will aid students in identifying, understanding, and using key terms. Terms included in the glossary are highlighted in the text (either when first mentioned or when first discussed substantively if that is not the occasion of first mention), and defined at the end of the book. More generally, there are additional examples throughout the text, particularly those relating to issues of contemporary human geography. Many of these revisions are prompted by the insightful comments of a number of reviewers of the first edition and it is a pleasure to thank the following fellow human geographers for their suggestions: Quentin Chiotti, Pierre Deslauriers, Michael Fox, Frank Innes, Ian MacLachlan, Rod B. McNaughton, Ben Moffat, and Robert Stock.

A good atlas is an invaluable accompaniment to this textbook. Current demographic and political data and overall global change data are useful adjunct sources of information. Instructors adopting this textbook will have little difficulty in using their own examples to complement the content. I have attempted to produce a volume that is logically sequenced, and recommend a course organization that at least approximately parallels the chapters.

Textbook authors owe a debt of thanks to the discipline as a whole and to academics who comprise that discipline both past and present. There is little that is original here. I hope that this volume is a fair reflection of a most exciting discipline and that, once introduced to human geography, students — young geographers — will have the correct background for more advanced study.

It is a pleasure to acknowledge my thanks to Marjorie Halmarson and Ed Pachanuk who have combined to produce all of the graphics. The staff at Oxford University Press, particularly Valerie Ahwee and Phyllis Wilson, have always been most supportive. My greatest debt is to Pauline, Philippa, and Mark.

The following publishers have granted permission to reproduce previously published material:

Academic Press Ltd: Figure 9.2 is adapted from 'A Theory of Exploration' by J.D. Overton from *Journal of Historical Geography* 7 (1981):57, by permission of Academic Press Ltd.

Blackwell Publishers: Figure 4.2 is adapted from *The Human Impact: Man's Role in Environmental Change* by A. Goudie, Blackwell, 1981, by permission of Blackwell Publishers.

Figure 7.7 is adapted from 'The Mormon Culture Region: Strategies and Patterns in the Geography of the American West, 1847–1964' by D.W. Meinig from *The Annals of the Association of American Geographers* 55 (1965):214. Reprinted by permission of Blackwell Publishers.

Figure 8.1 is reprinted from W. Zelinsky, 'North American Vernacular Regions' from *The Annals of the Association of American Geographers* 70 (1980):14. Reprinted by permission of Blackwell Publishers.

Figure 9.14 is adapted from 'The Identification and Evolution of Racial Gerrymandering' by J. O'Loughlin from *The Annals of the Association of American Geographers* 72 1982):180. Reprinted by permission of Blackwell Publishers.

Figure 10.2 is adapted from H.J. Reitsma, 'Crop and Livestock Production in the Vicinity of the US–Canadian Border', *The Professional Geographer* 23, pp. 217, 219, 221, by permission of Blackwell Publishers.

Table 10.3 is reprinted from 'The Political Ecology of Peasant-Herder Conflicts in the Northern Ivory Coast' by T.J. Bassett from *The Annals of the Association of American Geographers* 78 (1988):456. Reprinted by permission of Blackwell Publishers.

Canadian Association of Geographers: Figure 2.9 is reprinted from R.I. Ruggles, 'Spacial Concepts of North America in 1763' from 'The West of Canada: Imagination and Reality' from *The Canadian Geographer*. Permission has been granted by the Canadian Association of Geographers.

Table 11.2 is adapted from P.D. Keddie and A.E. Joseph, 'The Turnaround of the Turnaround? Rural Population Change in Canada, 1976 to 1986' from *The Canadian Geographer*. Permission has been granted by the Canadian Association of Geographers.

Table 11.3 is reprinted from S. Davies and M.H. Yeates, 'Exurbanization as a Component of Migration: A Case Study in Oxford County, Ontario' in *The Canadian Geographer*. Permission has been granted by the Canadian Association of Geographers.

Figure 13.3 is adapted from J.C. Muller, 'The Mapping of Travel Time in Alberta, Canada' from *The Canadian Geographer*. Permission has been granted by the Canadian Association of Geographers.

Christopher Davies (Publishers) Limited: Figure 9.11 is reprinted from L. Kohr, *The Breakdown of Nations* (Swansea: Christopher Davies, 1957). Reprinted by permission.

Clark University: Figure 10.1 is adapted from O.E. Baker, 'The Potential Supply of Wheat' from *Economic Geography*, published by Clark University.

Encyclopaedia Britannica, Inc.: Tables 7.4 and 7.5 are reprinted from *1993 Encyclopedia Britannica Book of the Year*. Table 'Adherents of All Religions by Seven Continental Areas, Mid-1992'. Reprinted with permission from *1993 Britannica Book of the Year* © 1993 by Encyclopaedia Britannica, Inc.

Geographical Magazine: Table Conclu.1 is adapted from 'An Uphill Struggle for Survival' by R. Handbury-Tenison and C. Shankey, *Geographical Magazine*, 1989. Reprinted by permission.

Gordon Ewing and Pion Limited, London: Figure 13.4 is reprinted from G. Ewing and R. Wolfe, 'Toronto in Physical Space and Time Space' from 'Surface Feature Interpolation on Two-Dimensional Time-Space Maps' from *Environment and Planning* A9 (1977):429–37, figures 3 and 6. Reprinted by permission of Gordon Ewing and Pion Limited, London.

A.L. Grove: Excerpt from F.P. Grove, *Settlers of the Marsh*. Reprinted by permission of A.L. Grove, Toronto.

HarperCollins Publishers: Figure 7.6 is reprinted from *The European Culture Area* by T.G. Jordan. Copyright © 1973 by Terry G. Jordan. Reprinted by permission of HarperCollins Publishers Inc.

Talon Books Ltd: Figure 7.5 is adapted from *Concepts and Themes in the Regional Geography of Canada*, Revised Edition. Copyright © 1989, J. Lewis Robinson. Reprinted by permission of Talon Books Ltd, Vancouver, Canada.

UN High Commissioner for Refugees: Table 6.7 is reprinted from R. Colville, 'UNHCR-Assisted Cases by Country, 1990–92' in 'Resettlement: Still Vital After All These Years' from *Refugees 94* (1993), UN High Commissioner for Refugees.
United Nations: Figure 5.8 is reprinted from *United Nations Demographic Yearbook* (years) by permission of the United Nations.

United States Committee for Refugees: Table 6.6 is reprinted from *World Refugee Survey 1992* by permission of the US Committee for Refugees.

University of Chicago Press: Figure 2.18 is reprinted from 'Two Topographic Maps of the Love Canal Area, Niagara Falls, New York' from *How to Lie* with Maps by M. Monmonier, University of Chicago Press, 1993, pp. 121–2. Reprinted by permission of the University of Chicago Press.

Wadsworth Publishing Co.: Table 4.1 is reprinted from G.T. Miller, Jr, 'Shallow and Deep Ecology Compared' from *Living in the Environment*, 4th Edition. Copyright © 1982. Reprinted by permission of Wadsworth Publishing Co.

West Publishing Company: Figure 3.4 is reprinted by permission from p. 295 of *Physical Geography* by R.C. Scott; copyright © 1989 by West Publishing Company. All rights reserved.
Figure 3.5 is reprinted by permission from pp. 276–7 of *Physical Geography* by R.C. Scott; copyright © 1989 by West Publishing Company. All rights reserved.
Figure 3.6 is reprinted by permission from pp. 232–3 of *Physical Geography* by R.C. Scott; copyright © 1989 by West Publishing Company. All rights reserved.
Figure 3.7 is reprinted by permission from pp. 158–9 of *Physical Geography* by R.C. Scott; copyright © 1989 by West Publishing Company. All rights reserved.

V.H. Winston & Son, Inc.: Excerpt from B.J.L. Berry, review of 'Geography's Inner Worlds: Pervasive Themes in Contemporary American Geography', *Urban Geography* 13:490–4. Reprinted by permission of V.H. Winston & Son, Inc.

To the student

IN AN IMPORTANT SENSE any textbook is only as good as the use made of it. This textbook aims to satisfy five general but basic goals:

- to introduce the discipline of human geography
- to provide you with an awareness of the value of human geography as a discipline that aids understanding of our complex and ever-changing world
- to provide you with a solid foundation for further courses in human geography
- to encourage you to read widely and critically, always questioning what you read (including this textbook)
- to encourage you to think logically and critically

You are most likely to achieve these goals if you employ sound time management, studying, and learning strategies. Now is the right time, before you commence your journey into the world of human geography, to offer some advice on how to make the best use of this textbook as a tool to assist your comprehension of this exciting discipline. Although we all study in our own distinctive ways, this general advice may be helpful to all students.

First, organize your time. This is not always as simple as it sounds. Most of us find it all too easy to be distracted from our important tasks. The key to sound time management is planning and scheduling all activities on a daily basis, including academic and social activities. Allocate the appropriate amount of time for studying human geography and stick to your schedule. Study human geography when you are scheduled to study it.

Second, pursue efficient textbook reading and studying habits. How do you achieve this goal? Many students find it helpful to read and study a text chapter in five relatively discrete stages.

- Survey: Acquire an overview of the chapter; use the headings, subheadings, chapter summaries, and glossary to aid you.
- Question: Look over the chapter again, thinking about the facts and ideas presented. Question as you read; do not be a passive reader.
- Read: This is the third stage, but one that many students mistakenly do first. You already have a sound general feel for the chapter, so now is the right time to read in detail. Again, do not be passive — take notes and underline or highlight.
- Recite: Read out the key ideas, ask yourself questions as you read, and be involved.
- Review: This final stage requires you to go over the chapter content one more time to confirm your grasp of the issues raised and facts presented.

This textbook is most likely to fulfil its goals if you work hard and employ the suggested time management practices and studying habits. Enjoy.

Introduction

YOU ALREADY KNOW that this book is an introductory human geography textbook, but do you have a clear idea as to what human geography is and therefore what is being introduced? Note down in a few key words your perception of human geography. It should be interesting to compare this perception with your understanding after you have read these introductory remarks.

Consider the word 'geography'. Its basis is Greek: *geo* means 'the world', *graphei* means 'to write'. Literally, then, geographers are involved in writing about the world. A demanding task! Its sheer breadth necessitates dividing the task into two relatively distinct disciplines. *Physical geography* is concerned with the physical world, and *human geography* is concerned with the human world.

Human geographers, then, write about the human world. But how do they approach their work? What methods do they employ? Not surprisingly, human geographers have diverse approaches and methods from which to choose, a diversity that the later chapters of this textbook reflect. Nevertheless, three recurring themes continue to be central to the study of our general subject matter.

THREE RECURRING THEMES
HUMANS AND LAND

Our human world is not in any sense preordained. It is not the result of a single cause such as climate, physiography, religion, or culture. Rather, our human world is the ever-changing result of human actions, as individuals and as group members, working within physical and institutional frameworks. Thus, human geographers often focus on the evolution of the human world with reference to people, their cultures, and physical environments. It is usual to describe the human world as a **landscape**. There are two closely related aspects of landscape. First, it is what is there: the human addition to and modification of physical geography. As such, landscape includes crops, buildings, lines of communication, and other visible, material features. Second, landscape has significant symbolic content — it has meaning and identity. It is hardly possible, for instance, to view a church without appreciating that it is much more than a mere landscape feature. As human geographers, we are concerned with landscape for what it is and what it means to live in it: landscape interpreted as the outcome of particular human and land relationships.

REGIONAL STUDIES

As one component of writing about the world, human geographers often divide large areas into smaller areas that exhibit a degree of unity. Such divisions are **regions**. A regionalizing process is one of classifying and may employ one or more variables. The fact that we are able to regionalize tells us that human landscapes make sense — they are not random assemblages of features. Groups of people occupying particular areas over a period of time create regions: human landscapes that reflect their occupancy and that differ from other landscapes. Much contemporary regional study emphasizes this social organization of space and the related impact of region creation on social and economic life.

SPATIAL ANALYSIS

Understanding the human world requires that we explain **location**: why things are where they are. A spatial analytic approach tackles this question with a typical focus on theory construction, models, hypothesis testing, and quantitative methods. For example, we may find that towns in an area are spaced at relatively equal distances apart, or that industrial plants are located close to one another. Spatial analysis is concerned with explaining such locational regularities. Thus, we have already identified two types of spatial organization; first, the one that is reflected in the creation of regions and, second, the one that is reflected in distinct patterns of locations.

A COMMON THREAD

These three recurring themes reflect three separate but overlapping traditions. One common thread concerns the fact that the human world is ever changing and hence human geography, regardless of the specific theme adopted, typically incorporates a time dimension. Change can be in response to internal or external factors. Some changes are rapid while others are gradual. Landscapes change in content and meaning; regions increase or lose their distinctiveness, and locations adjust to changing circumstances.

Identifying three principal themes related to our general subject matter — the human world — explicitly acknowledges that multiple approaches are legitimate. It is useful to note also that the manner in which these three have been applied has changed over time; this is inevitable and necessary in any dynamic discipline.

WHY HUMAN GEOGRAPHY?

Although geography has often striven to be a truly integrating field, one that combines physical and human components, human geography and physical geography are relatively distinct disciplines. Contemporary human geographers do not deny the relevance of physical geography to their studies, but they do not insist on always employing physical content. Human geography studies human beings and, as such, has close ties with such academic disciplines as history, economics, anthropology, sociology, psychology, and political studies. One of the particular strengths of human geography is that it considers our human world in multivariate terms, incorporating various physical and human factors as necessary. This introductory textbook, therefore, includes both physical and human content as needed. We need to know about global climates to understand world population distributions, just as we need to know about human perception to understand these same distributions. Thus human geography is a discipline separate from but related to physical geography in much the same way as it is a discipline separate from but related to the other disciplines that study human beings.

The central subject matter of human geography is *human behaviour* as it affects the earth's surface. Expressed in this way, the subject matter of human geography is very similar to that of the various other social sciences, all of which are centrally concerned with human behaviour in some specific context.

THE GOAL OF HUMAN GEOGRAPHY

Writing about the human world to increase our understanding of it has been the goal of human geography since at least Greek times. Two twentieth-century statements exemplify our continuing commitment to this goal and help explain why this commitment remains so important:

> The function of geography is to train future citizens to imagine accurately the conditions of the great world stage and so help them to think sanely about political and social problems in the world around (Fairgrieve 1926:18).

> Geography's *raison d'être* should be to develop appreciation of the great variety of cultures that comprise the contemporary world, and to show how in each society these have evolved — and are evolving — as specific responses to environment, to place and to people (Johnston 1985:334).

Although written some sixty years apart, these two statements express essentially the same view. Human geography is a practical and socially relevant discipline that has a great deal to teach us about the world we live in and how we live in the world. The goal of this textbook is to provide a basis for comprehending the human world as it is today and as it has evolved.

If this goal is achieved, then a contribution will have been made to the more general goal of advancing a just global society. Human geography is in an enviable position to accomplish this more general goal. It is a holistic discipline and is not restricted to a narrowly defined subject matter or a single theme that might limit our ability to appreciate general contexts or what we might call the interrelatedness of things. Traditional links with physical geography clearly play a key role, but human geographers do not hesitate to use material and ideas from many other disciplines. Our textbook explicitly acknowledges the controlled diversity of human geography.

THE HUMAN GEOGRAPHER AT WORK

What do human geographers actually do? What questions do they ask and what problems do they strive to solve? These questions raise others. What is the nature of the world we live in? How do we live in the world? The following four brief vignettes offer preliminary insights into the

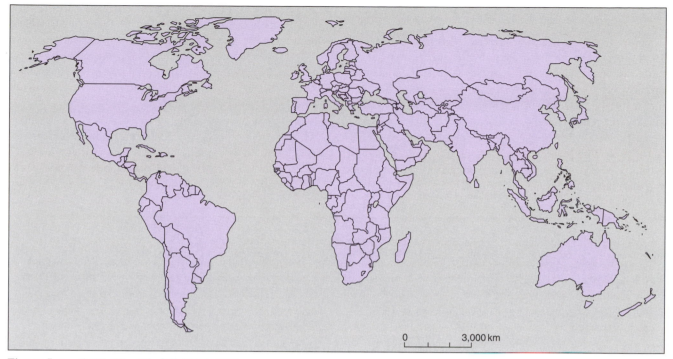

Figure Intro.1 World political divisions, 1994.

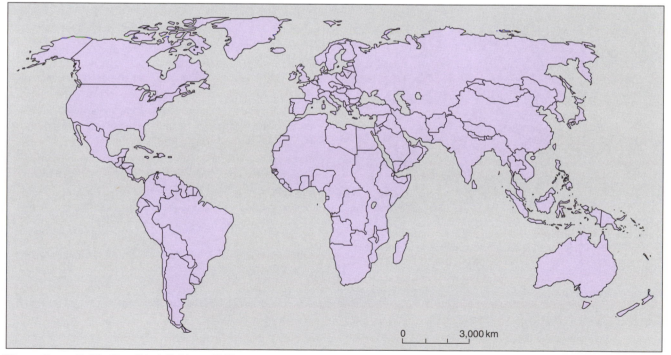

Figure Intro.2 World political divisions, 1938.

concerns and activities of human geographers, insights that are further expanded in later chapters.

A WORLD DIVIDED — 1: POLITICAL DIVISIONS

Figure Intro.1, which shows a simple division of the land surface of the earth into countries, is one example of the diversity of human life in our world. Many questions are raised.

Why is the map so divided? Has it always appeared this way? Is it changing today, between the time I write and the time you read these words? Is the total number of countries increasing or decreasing? What do these political divisions reflect — population distribution, cultures, development? Is it possible to aggregate countries? Are there meaningful groupings? Do basic aggregations such as north versus south or east versus west have value? Human geographers ask all of these questions.

Figure Intro.2 is like Figure Intro.1 except that it depicts 1938 political land divisions. Clearly, the two maps are quite different. This is an elementary yet crucial point. The human world does indeed change — substantially. The major changes evident in a comparison of figures Intro.1 and Intro.2 relate to the decolonization of Africa and Asia, to the political consequences of the Second World War, and to the post-1989 political transformation of Europe and the former USSR (Box Intro.1). The most general consequence is a significant increase in the number of countries, from about seventy in 1938 to over 180 in 1994. There is little reason to believe that a political world map in fifty years' time will closely resemble that of today. The contemporary political world is characterized by two principal movements: first, a tendency for regions to begin a process of integration — the European Economic Community is one example; second, a tendency for portions of countries to seek a separate political identity — Quebec in Canada is one example.

Human geographers are interested in and able to contribute to an understanding of these issues for at least three general reasons. First, political partitioning of the world is an example of one of our recurring themes, namely

---

Box Intro.1: Germany: A New Country

The new Germany was born on 3 October 1990 as a result of the unification of West Germany and East Germany; this unification was achieved peacefully, a remarkable state of affairs given the political tensions in Europe since 1945. German unification was one result of the sweeping changes that occurred throughout eastern Europe beginning in 1989.

Intro.1 The Berlin Wall separated the east and west sectors of the city between 1961 and 1988. It was built to prevent emigration from east to west. (Bundesbildstelle Bonn 108 742)

Germany first appeared as a country on the map of Europe in 1871 following the Franco-Prussian war. The specific territory changed in accord with a series of military activities (most notably, of course, the two world wars) and in 1945, Germany was occupied by the four allied powers and divided into four zones — American, British, French, and Soviet. In 1949, the three western zones linked to create West Germany. The Soviets chose not to cooperate and, in turn, created East Germany. Gradually, during the 1950s, links between them diminished and the 'Iron Curtain', separating not only East and West Germany but also eastern (communist) and western (democratic) Europe was in place. The most notorious indicator of this division was the Berlin Wall, initially set up in 1961, and dividing the then international city of Berlin, which, although located in East Germany, was a microcosm of Germany that was divided into American, British, French, and Soviet sectors. Surprisingly, the division of Germany began to crumble in 1989.

The changes in eastern Europe began in Poland and spread quickly, removing communist governments in favour of democratic political systems. For Germany, the remarkable outcome was a decision, agreed to in February 1990 at a meeting in Ottawa, to unify Germany. Technically, East Germany disappeared some eight months later, absorbed into West Germany, which is now renamed Germany.

Germany is again a major European power with a population of 81.1 million (as of 1993) and a new capital in Berlin, but there are numerous problems to be faced. Economic problems in the former East Germany include the need to make industry more competitive and modernize agriculture. In the former West Germany, recession and inexpensive imports are prompting an economic and social crisis in the major industrial region of the Ruhr. A major environmental problem is the high level of pollution and related industrial hazards in the east. It is estimated, for example, that large areas of forest have died because of sulphur dioxide pollution and that there is also substantial groundwater pollution. There are serious social problems, especially in the urban areas, resulting from the differences between the former states and the immigration of labour from elsewhere. The collapse of communism in the east caused mass unemployment, and in 1993 there were 3.5 million Germans unemployed (about 10 per cent of the labour force). This has prompted a tightening of the rules for admitting refugees. Right-wing nationalists exploit social discontent and direct racist violence against foreign workers, especially Turks and Jews. In 1992, eighty Jewish cemeteries were desecrated; about the same number as in the five-year period from 1926–31. Despite all of these problems, the new Germany is clearly poised to play a key role in the Europe of the twenty-first century.

Intro.2 Urban street scene, Toronto, Ontario, Canada. (Ontario Ministry of Transportation)

regionalizing or organizing space. Second, political identity and our human way of life are closely related with the political unit often being a major determinant of such matters as the numbers of births and deaths and availability of paid employment. Third, many of the specific reasons for political change relate to cultural variations on the surface of the earth such as those exhibited by language, religion, and ethnicity; human geographers are able to identify relationships between culture and political change, often with the help of maps.

A WORLD DIVIDED — 2: TWO WORLDS TODAY

The world is not only divided into political units but also into larger areas to which such labels 'as more developed'

and 'less developed' are often applied. Study photographs Intro.2 and Intro.3. Our human world is full of diversity — languages, religions, ethnic identities, standards of living — all vary from place to place. There are differences between individuals and also more general differences between groups of individuals. Similarly, landscapes vary. Each location is, of course, unique, but, more usefully, there are differences between regions. Photographs Intro.2 and Intro.3 suggest a basic distinction in our contemporary world. In photograph Intro.2 we see middle-class people in a prosperous area of a modern western urban centre. It is probable that these people are employed, own a home, and have surplus income for recreational or other similar purposes. In photograph Intro.3 we see a poor rural setting in a part of

Intro.3 Squatter settlement, Lusaka, Zambia. (A.M. O'Connor)

what is often called the least developed world. The people who live there are probably subsistence agriculturalists without disposable income. We might suggest that the human world we are studying is at least two human worlds.

It is important to recognize at this time that such major differences are evident not only between major world regions as suggested but also at other spatial scales such as regions inside countries (Box Intro.2). Consider also the obvious differences in quality of life between the outer suburbs and the skid row area of the typical North American city.

A WORLD TRANSFORMED — 1:
THE IMPACT OF MODERNIZATION

Table Intro.1 provides basic sample information on population totals, births and deaths, and life expectancy for various years from 1700 to 1900. Again, you will notice that things have changed. Population totals have increased, the birth rate

has declined, the death rate has declined, and life expectancy has increased. Why have these changes taken place? What do these changes mean in terms of human well-being? A process that we might label 'modernization' has been taking place. Birth rates decline primarily because of changing cultural considerations; death rates decline and life expectancy increases primarily because of technological advances related to sanitation and health. Population totals increase because of differences between birth and death rates. Clearly, however, as already noted, this process of modernization has not had similar effects throughout all areas of the world.

Modernization involves much more than these demographic changes; cultural and economic aspects of landscape and life have also been transformed in this time period and human geographers employ their distinctive perspectives (those suggested in our three recurring themes) to describe and explain these changes.

TABLE INTRO.1 *Selected demographic data*

(a) World population totals (millions)		(b) Birth rates and death rates (per 1,000), Finland			(c) Life expectancy at birth (number of years), France	
Year	Total	Year	Birth rate	Death rate	Year	Life expectancy
1700	680	1755	45.3	28.6	1750	27
1800	954	1805	38.4	24.7	1825	41
1900	1,600	1909	31.0	17.7	1905	49
1993	5,500	1990	12.0	12.0	1990	77

Notes:
1. Finland and France are selected because both countries have better than average records.
2. Birth rate refers to the total number of live births in one year for every 1,000 people living.
3. Death rate refers to the total number of deaths in one year for every 1,000 people living.

A WORLD TRANSFORMED — 2:
AN EMERGING POSTINDUSTRIAL SOCIETY?

One component of the larger set of modernization processes that is especially evident today in advanced economies and that also concerns changes in landscape and way of life is the rise of postindustrial society. This process comprises a series of changes, including the increasing importance of information technologies, an economic transformation from a goods-producing to a service economy, the rise of professional and technical workers as a new middle class, and the related demise of the working class. It is possible that this ongoing change will be as dramatic as that resulting from the earlier Industrial Revolution in terms of our social and economic identities. In North American and European cities, for example, there is evidence of the decline of traditional industries, the rise of a professional workforce, and the acceleration of decentralization. Such postindustrial cities include large numbers of disadvantaged people, the unemployed, and the often part-time working poor.

There is a new geography of employment, a new geography of housing, and a new geography of neighbourhoods. Once again, human geographers utilize their particular approaches, as suggested by our three recurring themes, to identify the reasons for and the character of these ongoing trends, and to analyse the impacts on landscapes and ways of life.

ABOUT THIS BOOK

As human geographers, we recognize the legitimate diversity of human geography, and so have much subject matter and many methods to introduce. Our introduction is followed by two chapters that answer two basic questions. In Chapter 1 we ask ourselves about the nature of human geography. A historical perspective is appropriate and you will discover that human geography has a long and distinguished academic pedigree involving many noteworthy individuals. In Chapter 2 we focus on the many ways in which human geography is studied. This is a concern that invokes philosophical and technical issues. Although much of the content of these two chapters may seem somewhat abstract at first reading, you will soon appreciate that the ideas introduced are frequently central to the human geography discussed in later chapters.

Chapter 3 provides basic factual information concerning the earth as a physical environment and the home of humans. This material is an essential prologue to subsequent chapters that discuss how we have chosen to live in this home, but is especially relevant to Chapter 4. Chapter 4 is an overview of humans' global impact on and adjustment to the physical environment. Using our knowledge of the physical environment, we can survey humans' use of the earth through time. Detailed issues of specific activities are covered in subsequent chapters while Chapter 4 considers resource use, resource abuse, and related environmental issues that are often at the core of our contemporary lifestyles.

Chapters 5 and 6 deal with population issues, specifically with questions of growth through time and around the world. Chapters 7 and 8 acknowledge that much human geography is best studied on a group scale and consider the importance of such cultural factors as language, religion, and ethnicity, as well as social factors such as gender, ethnicity,

---

Box Intro.2: Less developed Canada

Although Canada is a part of the developed world, as measured by various cultural and economic criteria, this is not to imply that the benefits are distributed equally among all Canadians. Native Canadians throughout Canada live different lives compared to most other Canadians. Native Canadians are characterized by more youthful populations, a shorter life expectancy, and higher birth rates that result in a high rate of natural increase (about double that of the national average and comparable to rates in many less developed countries). More generally, Native Canadians are a disadvantaged people; in comparison to other Canadians, they are more likely to be unemployed, to rely on social assistance, to live in poor quality housing, to have limited access to medical services, and to have little formal education.

The situation on reserves in the north is particularly deplorable. A survey of housing on northern Manitoban reserves estimated that 1,800 (29 per cent) of the 7,200 units were in need of major repair, and that an additional 1,400 units were required. Further, many of the communities lack piped water and sewage systems, a state of affairs that has clear implications for health. For residents of these reserves, the rate of tuberculosis is eight times the provincial average. Frustration, hopelessness, depression, and apathy are common with high rates of violent deaths, family breakdowns, and substance abuse. All of this is in relatively close proximity to the affluence of most of Canada and yet for many Canadians, it is a scenario to be avoided and ignored. Native Canadians are typically a marginalized group, spatially and socially.

Explaining this state of affairs is far from easy. Writing with general reference to the Canadian north, Bone (1992:197) noted that 'the notion of "ethnically blocked mobility" arises; that is, Native Canadians may value educational and occupational achievements less than other Canadians because of cultural differences such as the sharing ethic and because of perceived or experienced discrimination.' There are also reasons that relate to the historical details of contact between Native and European Canadians and to the character of the social and economic system of capitalism.

and class. Political groupings of people and space are the concern of Chapter 9, and we learn that many of the tensions in our contemporary world result from a discordance between political and cultural spatial organization.

Four uses of land are covered in chapters 10, 11, 12, and 13. Agriculture is considered in Chapter 10, settlement in Chapter 11, industry in Chapter 12, and transportation and trade in Chapter 13. Each of these four topics is discussed in economic and cultural terms and both landscape and life are covered.

The conclusion discusses the future of human geography and the human geography of the future — albeit somewhat tentatively!

Throughout this textbook, there is a consistent thrust. We need to understand the world we live in, and how we live in the world.

ABOUT BEING A HUMAN GEOGRAPHER

Human geography, as you are beginning to discover, is a fascinating subject that is worth studying not only for the importance of the subject matter itself and the approaches employed but also for the career training that is offered. A substantial number of graduating students are able to use their geography in selecting a career because this discipline, especially in conjunction with physical geography, leads to diverse employment opportunities in such areas as education, business, and government. Geography students are employed as cartographers, geographic information specialists, demographers, land-use and environmental consultants, social service advisers, and industrial and transportation planners, to list but a few of the opportunities. There are two distinctive features of an education in geography that contribute to the often high demand for geographers. The first of these is the emphasis on physical and human subject matter that provide the geographer with a particular perspective and knowledge base, and the second is the inclusion of technical content in data collection and analysis.

Human geography is, then, both a scholarly enterprise for understanding our world and a practical area of study that leads to varied career opportunities.

SUMMARY

Geography
Literally, writing about the world.

Three recurring themes
1. Humans and land: The human world, a landscape, is continuously changing because of human actions within institutional and physical frameworks.
2. Regional studies: Describing the earth and dividing the whole into parts — regions — that exhibit some uniformity of one or more variables.

3. Spatial analysis: Explaining the locations of geographic phenomena using abstract arguments and quantitative procedures.

Because the human world is ever changing, human geography typically includes a time dimension.

Human geography/physical geography
Related disciplines for traditional and logical reasons. Best introduced separately at this level because physical geography is a physical science while human geography is a human science.

Our subject matter
Human behaviour as it affects the earth's surface.

Goal of the textbook
To provide a basis for comprehending the human world as it is today and as it has evolved.

This book is about
The world we live in, and how we live in the world.

About being a human geographer
As a human geographer, you have an understanding of our world and a sound training for a variety of careers.

WRITINGS TO PERUSE

DEMKO, G.J., J. AGEL, and E. BOE. 1992. *Why in the World: Adventures in Geography*. Toronto: Anchor Books.
Aimed at the popular market, this entertaining volume asks and answers a multitude of questions to demonstrate the practical relevance of geography.

FIELDING, G.J. 1974. *Geography as Social Science*. New York: Harper and Row.
A human geography textbook in the spatial analytic tradition (the third of our three recurring themes).

JACKSON, W.A.D. 1985. *The Shaping of Our World: A Human and Cultural Geography*. New York: Wiley.
A human geography textbook with an especially strong emphasis on landscape, culture, and change (the first of our three recurring themes).

JORDAN, T.G., L. ROWNTREE, and M. DOMOSH. 1994. *The Human Mosaic: A Thematic Introduction to Cultural Geography*, 6th ed. New York: Harper Collins.
This textbook deals with all human geography from a primarily cultural rather than economic perspective.

JOHNSTON, R.J. 1988. 'There's a Place for Us'. *New Zealand Geographer* 44:8–13.
An article, written by one of the more perceptive commentators on contemporary human geography, that provides an enlightening perspective on what human geographers do and why they do it.

————, ed. 1993. *The Challenge for Geography: A Changing World, A Changing Discipline.* Oxford: Blackwell.

This book demonstrates that the world is changing rapidly, argues that the discipline of human geography needs to respond accordingly, and asks whether human geographers should influence the direction of change.

ROGERS, A., H. VILES, and A. GOUDIE. 1992. *The Student's Companion to Geography.* Oxford: Blackwell.

A clear and comprehensive student guide to both physical and human geography comprising over fifty entries on such diverse topics as the history of geography, the art of interviewing, world libraries, and careers for geographers.

WHEELER, JR, J.H., and J.T. KOSTBADE. 1993. *Essentials of World Regional Geography.* Toronto: Saunders College Publishing.

An example of a geography textbook organized on a regional basis (the second of our three recurring themes).

1

What is human geography?

I would have her instructed in geometry that she may know something of the contagious countries.'

— Mrs Malaprop, *The Rivals* (Richard Brinsley Sheridan 1775)

MRS MALAPROP was, as usual, confusing her words, but there is something particularly appealing about her confusion. After all, geographers themselves frequently debate the nature of geography. Fortunately, however, geographers are not similarly confused about their discipline and their activities! But it is important that we define human geography so that we can understand the contemporary discipline. A discussion of the development of geography makes it clear that the discipline changes through time. Indeed, there would be cause for great concern if such change was not occurring. Geography, like most other academic disciplines, functions to serve society. As society and societal requirements change, so does geography. Geographers have always had a consistent purpose — describing and explaining the world. It is the manner in which they approach this task that changes — not the task itself.

For example, for many years geographers were principally involved in discovering, describing, and explaining an increasingly better-known world. As unknown areas became known, geographers worked feverishly to absorb new facts into established knowledge or to propose radically new knowledge bases. Their most important tool was the map. More recently, since the nineteenth century, geographers have reoriented their activities. Once basic global descriptions were in place, the emphasis switched to providing a better explanation and understanding of geographic facts.

The history of geography is far from dry and tedious. It is long and exciting, full of drama and intriguing individuals. Furthermore, the past is indeed a prologue — an appreciation of the evolution of geography is a major factor in our understanding of contemporary geography.

PRECLASSICAL GEOGRAPHY

It seems likely that the earliest geographic descriptions were in the form of maps — a simple but effective means of communicating spatial information. Indeed, maps must surely have been created, temporarily at least, by all human groups. Maps roughly sketched with a stick in sand or soil, or maps scratched on rock or wood, must often have been used to illustrate the location of water, game, or a hostile group. Geography is indeed about maps and, in this sense, humans have always been geographers. Lack of hard evidence need not prevent us from noting the centrality of geography to our human existence. Our ancestors could no more cope without maps than we can today.

The world's first civilization emerged in Mesopotamia, now southern Iraq, some time after 4000 BC, and there are preserved maps drawn on clay tablets that date from this time. Such maps were typically of local areas, reflecting limited knowledge of areas beyond the immediate environment. This is probably likely for all the civilizations that predate classical Greece, namely those in the Nile Valley, the Indus Valley, China, Crete, the Greek mainland (Minoan and Mycenaean), and southeast Mexico (Olmec).

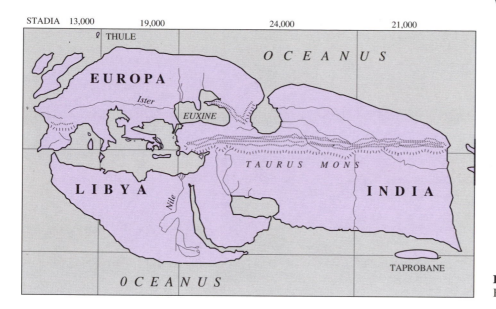

STADIA   13,000   19,000   24,000   21,000

Figure 1.1 The world according to Eratosthenes.

CLASSICAL GEOGRAPHY

With the emergence of classical Greece shortly after 1000 BC, geographic understanding, and hence maps, began to cover more than the local area because the Greeks were the first civilization to experience geographic mobility and the growth of colonies. Greek scholars initiated two major geographic traditions.

The first tradition is *literary*, involving written descriptions of the known world. Many Greek scholars contributed to this tradition, although relatively few of them were geographers. Herodotus (484–*c*.425 BC), for example, is best known as the first great historian, but he was also an accomplished geographer. The descriptions of lands and peoples written by Herodotus make it clear that he saw geography as a necessary background to history and vice versa. These descriptions were based on his observations during a series of extensive travels to such areas as Egypt, Ukraine, and Italy. Possible relationships between latitude, climate, and population density were noted by Aristotle (384–322 BC) who also speculated about the ideal locations for cities and the conflicts between rich and poor groups. Eratosthenes (*c*.273–*c*.192 BC), whom we often designate as the father of geography because he coined the word, contributed to this tradition by writing a book describing the known world. He also mapped the known world at the time (Figure 1.1): the Mediterranean region and adjacent areas in Europe, Africa, and Asia. The literary tradition begun by the Greeks was summarized by Strabo (64 BC–AD 20) in a work entitled *Geographia*, an encyclopedic description of the world known to the Greeks. Unlike many of the earlier descriptive works (that by Eratosthenes, for example), Strabo's still exists. It is a comprehensive work totalling two introductory books, eight books on Europe, six on Asia, and one on Africa. Much of what we know today about Greek geographers

before Strabo comes from the detailed summaries of their work that Strabo provides.

The second tradition is *mathematical*. Thales (*c*.625–*c*.547 BC), one of the originators of this endeavour, viewed the world scientifically and was able to predict an eclipse of the sun in 585 BC. By the fifth century BC, the Greeks knew that the earth was a sphere and in the second century BC, Eratosthenes calculated the circumference of the earth. Shortly after, Hipparchus devised a system of imaginary lines on the surface of the earth based on the poles and the equator — these, of course, are the lines of longitude and latitude. The first development of this grid system was imprecise, but encouraged fairly accurate mapping. Determining the exact position of any location remained difficult. Latitude was relatively easy to calculate by observing the angle of the sun's shadow using an early version of a sundial, but longitude continued to require estimation as there was no way to measure time precisely, especially at sea. Hipparchus focused attention on a second crucial geographic problem, that of mapping a curved earth surface on a flat surface; he was the first person to tackle a problem that remains with us today. Much of this mathematical tradition was summarized by the Alexandrian Ptolemy (AD 90–168) in his eight-volume *Guide to Geography*. Using the grid system devised by Hipparchus, Ptolemy produced the first index of places, or gazetteer, of the world. Unfortunately, this gazetteer includes numerous errors because latitude was not carefully measured and longitude was necessarily estimated.

One outcome of all these endeavours is the world map produced by Ptolemy (Figure 1.2). A grid system is included and the details are generally a clear improvement over Eratosthenes's map (Figure 1.1).

The written works of Strabo and Ptolemy and the world map of Ptolemy combine to provide a useful indication of

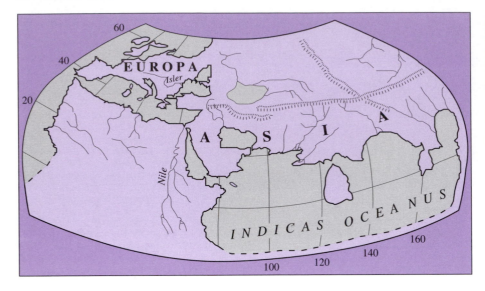

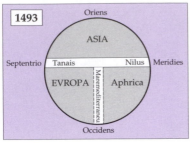

Figure 1.2 *(left)* The world according to Ptolemy.

Figure 1.3 *(above)* An example of a T-O map.

the achievements of Greek civilization in geography. It is apparent that these achievements were but one component — albeit a crucial one — of larger developments, and that the emergence and evolution of geography was in response to larger societal requirements. Specifically, the Greeks moved towards a deliberate assumption of national identity — the growth of city states leading to the concept of a nation state. Competition between city states prompted expansion and colony creation, which, in turn, demanded accurate maps and increases in geographic knowledge.

Roman civilization added very little to geographic knowledge despite the fact that Rome changed from being a nation state to an empire state. There was little development of either the literary or mathematical traditions begun by the Greeks. Some geographic guidebooks were published in response to exploratory and expansionist needs. Although Roman civilization continued until the fifth century AD, classical geography effectively ended with Ptolemy. There seems to be little doubt that Roman leaders relied on Ptolemy, but Roman society did not expand on his work.

Thus, the classical geography of the Greeks is the true beginning of our contemporary discipline. Geography in the classical world was concerned with locations and interconnections of places on the surface of the earth. In addition to the word 'geography', the Greeks also introduced 'chorography' and 'topography'. **Chorography** refers to places smaller than the world, such as countries, while **topography** refers to local areas within countries. Greek geography thus recognized the value of studying places at various scales. As we have seen, the Greeks also produced relatively accurate maps using a sophisticated grid system. What is perhaps most significant is that the mapping procedures devised by Ptolemy have persisted to the present. The basic style and language of our contemporary maps stem from Ptolemy.

THE MIDDLE AGES

The period from the fifth to the fifteenth century was one of sporadic and limited geographic work in Europe. Elsewhere, however, especially in the Islamic world and China, geography flowered. Again, what we find is a close relationship between geography on the one hand and societies experiencing expansion, and hence a thirst for geographic knowledge, on the other.

THE EUROPEAN DECLINE

Medieval Europeans knew little beyond their immediate domain as geographic horizons retreated and mapping deteriorated. Many of the advances made by the Greeks were lost and it was only in monasteries that serious geographic work continued. During the Middle Ages, geography as such no longer existed; the word 'geography' did not enter the English language until the sixteenth century. The clearest evidence of decline relates to maps.

Greek maps were drawn by scholars with expertise in astronomy, geometry, and mathematics. Medieval map makers, although scholarly, are less easily described. Their maps were symbolic, not geographic. They stylized geographic reality to arrive at a predetermined desired structure. As Figure 1.3 shows, these maps are less detailed and accurate than maps produced some 1,500 years earlier. The best example is the twelfth to the fifteenth century's T-O maps, which are a T drawn within an O. T-O maps show the world as a circle divided by a T-shaped body of water. East is at the top of the map. Above the T is Asia, below left is Europe, and below right is Africa. The cross of the T is the Danube–Nile axis and the perpendicular part of the T is the Mediterranean. These maps are ones of scriptural dogma; what was drawn was what Christians were expected to believe. Symbols triumphed over facts. A second type of medieval map divided the world into climatic zones, largely

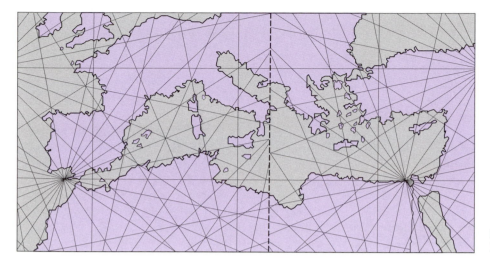

Figure 1.4 An example of a Portolano chart.

hypothetical, on either side of the equator. Other medieval maps were replete with decoration; the Ebstorf *Mappemonde* (*c.*1284) had as background a picture of the crucifixion, while the Hereford map (*c.*1300) is really an encyclopedia. Overall, medieval maps reflect medieval European geography — which, in turn, reflects medieval scholarship. Possibly the exceptions are the maps known as Portolano charts, which date from about 1300. These maps were used at sea and utilized a series of radiating lines. The lines did not serve to locate positions on the map, nor did the maps use any projection. Nevertheless these maps often succeeded in locating coastlines accurately (Figure 1.4).

Medieval Europeans made very few additions to geographic knowledge. Norsemen sailed to Greenland and North America, but no books resulted. Christian Europeans indulged in a series of crusades and military invasions to the Holy Land, but the result, as far as geographic knowledge was concerned, was minimal. The most significant exploratory journey was that of Marco Polo (1254–1323), a Venetian who visited China and wrote a description of places he visited. However, Marco Polo was unable to add to Greek knowledge because he was largely unaware of it. The distinction is not always an easy one, but Marco Polo was an explorer, not a geographer.

It was outside Europe that the major geographic advances took place during the period after the Greeks and prior to the fifteenth century. Two contributions need to be noted.

GEOGRAPHY IN CHINA

In China, a great civilization, clearly the parent of contemporary China, developed before 2000 BC. The longest-lasting civilization in the world, China inevitably made major contributions to geographic knowledge. Chinese writings describing their known world date back to at least the fifth century BC. The Chinese also explored and described areas beyond China; in 128 BC, for example, Chang Chi'en discovered the Mediterranean region, described his travels, and initiated a trade route. Other Chinese geographers reached India, central Asia, Rome, and Paris. Indeed, Chinese travellers reached Europe prior to the travels of Marco Polo.

There is one important aspect in which early Chinese geography differed from the European equivalent. It is a difference of geographic perspective, that is, a different way of looking at the world. Traditionally, Chinese culture views the individual as *a part of* nature whereas Greek and subsequent European culture typically view the individual as *apart from* nature. This distinction is closely tied to the differing attitudes incorporated in Confucianism (which dates from about the sixth century BC) and Christianity. Given the concern with humans and land as one, it is evident that Chinese descriptive geographies often focused on an integrated human and physical description.

Maps were also central to geography in China. There is evidence of a grid system used during the Han dynasty (third century BC to third century AD). It appears that the Chinese map makers began as civil servants whose job it was to draw and revise maps. Viewed in this light, the map maker was an important individual in the service of the state. Maps were symbolic statements, asserting ownership of some territory.

GEOGRAPHY IN THE ISLAMIC WORLD

The second contribution to geography outside Europe came from the Islamic world. The religion of Islam began in the seventh century AD — the prophet Muhammad died in 632 — and very quickly proved to be a powerful unifying force of previously disparate tribes. Consequently, at much the same time as Europe was immersed in the Dark Ages, civilization flowered in Arabia. Conquests beyond the Arab region commenced, increasing the geographic knowledge base to include North Africa, the Iberian Peninsula, and India.

By the ninth century, Islamic geographers were recalculating the circumference of the earth. From then until the fifteenth century, a wealth of geographic writings and maps were produced based on earlier Greek work and Islamic travels. Geographic descriptions included that by Edrisi (1099–1180), whose book on world geography corrected many of the errors contained in the work of Ptolemy. Perhaps the best-known traveller was ibn-Batata (1304–c.1368) who journeyed extensively in Europe, Asia, and Africa. A third major addition to geography came from ibn-Khaldun (1332–1406), a historian who wrote extensively about the relations between humans and the environment. Maps produced by Islamic geographers, including Edrisi, centred on Arabia.

Chinese and Islamic geography prior to the fifteenth century were roughly comparable to Greek geography. Mathematical and literary traditions were evident and map making was central to most geographic endeavours. Both descriptions and maps had horizons that corresponded to the knowledge and needs of particular societies.

THE AGE OF EUROPEAN EXPLORATION AND DISCOVERY

The geography that we have noted so far has been limited in terms of world knowledge. The known world was but a portion of the real world. In the fifteenth century, this truth began to be appreciated by European scholars, and the impact on geography was dramatic and significant. All three components of the geography discussed so far — mathematical, literary, and cartographic — underwent rapid change. Between the fifteenth and seventeenth centuries, there was an outburst of European exploratory activity. Coincidentally, the beginning of this European activity was roughly contemporaneous with a decline in Chinese and Islamic exploratory activity.

Many factors, among them the desire to spread Christianity and the establishment of trade routes, contributed to the European outburst, but there are two interesting additional factors. Printing, which was first applied to maps by the Chinese in 1155, was first used for map production in Europe in 1472. In 1410, editions of the work of Ptolemy were printed; editions including maps appeared in 1477. Printing allowed information, including that contained on maps, to diffuse rapidly. Furthermore, Ptolemy's maps were redrawn using new projections. The second additional factor is related to the thirst for knowledge and involved the deliberate creation of what might be called centres of geographic analysis. The first such establishment was initiated by Prince Henry in 1418 in Portugal and focused on a multitude of key geographic questions, such as the size of the earth and the suitability of tropical environments for habitation. Techniques of navigation were taught and Prince Henry prompted a series of explorations along the western coast of Africa. Interestingly, resultant maps did not always reflect new discoveries. A 1459 map by Fra Mauro presumably deliberately concealed new information to maintain secrecy.

EXPLORATION

Exploration is not, of course, geography, but it was exploration that furnished new facts and provided the basis for new maps and new descriptive geographies. The geography-exploration relationship is a symbiotic one with maps and books encouraging exploration, and exploration generating new maps and books. The major explorations were those by Dias around southern Africa (1486–7); by Columbus to North America (1492–1504); by da Gama to India (1497–9), and by Magellan, who reached Asia by sailing west (1519–22). The last burst of exploratory activity was by Cook, who made three voyages into the Pacific (1769–80). Indeed, it was Cook who corrected one of Ptolemy's most significant errors — the supposed existence of a southern continent. By 1780, the basic outlines of the world map were established. It is not difficult to appreciate that these explorations added considerably to knowledge of the world and thus to the geographic task (Box 1.1).

Exploratory activities were made easier as several basic hurdles were overcome. Most important was the need, previously recognized by the Greeks, to have an accurate method of determining location. Establishing latitude was no problem, but a method for establishing longitude at sea was not available until 1761 and was not used on a major voyage until Cook's second voyage in 1772–5.

Box 1.1: Exploration or invasion?

The overseas expansion of Europe that began in the fifteenth century resulted in much more than new knowledge and new descriptive and mapping tasks for the armchair geographer. Although expansion has characteristically been interpreted as discoveries and explorations, another less innocent and Eurocentric interpretation is possible. As a consequence of expansion, Europeans were able to diffuse their power and authority throughout much of the world, which was a principal cause of the unequal world that currently prevails. Indeed, alternative words to exploration and discovery are invasion and conquest.

The various European activities involved in expansion gave birth and eventual credence to the idea that the European world was distinctive from other worlds; Europe was seen as superior, powerful, enlightened, and civilized, while the other world areas were seen as inferior, powerless, ignorant, and barbaric. It was against such an intellectual backdrop that the discipline of human geography and the various other social sciences began to emerge.

MAPPING

Maps of the world and of specific regions during this period continued the process begun by the Greeks — a process where facts triumphed over imagination. Unlike medieval maps, which were a mixture of fantasy and dogma — a good reflection of medieval geography — the typical map from the fifteenth century onwards was functional. Map making flourished because of the demands of sea travellers. Latitude and longitude returned to replace T-O maps centred on Jerusalem. One of the pioneers in these endeavours was the Polish scientist, Copernicus (1473–1543), who produced maps of Poland that were used to settle a variety of boundary disputes.

Demand for geographic knowledge, and hence for maps, combined with the technique of printing to create new developments. Mercator (1512–94) was undoubtedly the most influential of the new map makers. He tackled the crucial problem of projection — how does one best represent a sphere on a flat surface? His answer was the famous 1569 Mercator projection, still used extensively today. This projection, which showed the earth as a flat rectangle with a grid of latitude and longitude lines, was enormously useful to sea travellers. By the early seventeenth century, Mercator's map had replaced all earlier maps, including the Portolanos, used at sea.

Another map maker, Ortelius (1527–98), produced the first modern atlas in 1570, an atlas that ran into forty-one editions by 1612, such was the demand for geographic knowledge. From Mercator onwards, map making has steadily improved and consistently reflected new geographic knowledge.

GEOGRAPHIC DESCRIPTION

The awakening of Europe and the burst of exploratory activity had an impact not only on map making but also on geographic description. European scholars had to make sense of the new world. Once again, geography had to respond to the demands of society by providing answers to a multitude of questions concerning the shape of the earth, the location of places, physical processes, and human lifestyles. Just as earlier Greek, Chinese, and Islamic geographers described their worlds, so the European geographers of the sixteenth century onwards needed to describe their ever-expanding and ever-changing world. Geography was thus faced with an enormous task: writing about all aspects of the entire world.

The resultant flurry of geographic writing was of mixed quality. Some works of fiction were often regarded as factual.

Box 1.2: The southern continent

The Greeks were the first to suggest the existence of a large southern land mass to counterbalance the large known areas of land in the Northern hemisphere. Along with so many other provocative ideas, this notion was lost during the medieval period; the medieval belief in a flat world did not require any such symmetry. Knowledge of the basic distribution of land and sea in the Southern hemisphere began to unfold with the beginnings of European overseas exploration in the fifteenth century. Early voyages were often prompted by incorrect information and succeeded in discovering fiction as well as fact. For example, in the early sixteenth century, Gonneville claimed to have discovered a tropical paradise in the southern area that was inhabited by people living an easygoing, contented life. All records of this voyage were lost, but the claim clearly influenced French exploration for the next 200 years. Subsequent French and other explorers sought to rediscover this idyllic Gonneville land.

Magellan's voyage of 1518 included the discovery of the Straits of Magellan, separating South America and the island of Tierra del Fuego, and prompted the idea that this island was but the beginning of the large southern continent. Cartographers, such as Orontius in 1531, responded with maps that included the continent. By 1578, however, Drake had shown that Tierra del Fuego was not a large continent. Exploration then turned east with the discovery of the islands of South Georgia by de la Roche in 1675. The French continued to search for Gonneville land, and in 1772 Kerguelen discovered land and reported that it promised all the resources of the mother country — wood, minerals, diamonds, and rubies. He even called his discovery South France — undoubtedly, many thought, the discovery of the long-sought Gonneville land. Locate Kerguelen Island in your atlas to see just how wrong Kerguelen was — the island named after him is in an isolated, far from tropical location in the south Indian Ocean. There is little doubt that Kerguelen discovered what he wanted to discover, just as Gonneville had almost 200 years earlier.

The distribution of land and sea in the Southern hemisphere became clearer in a series of late eighteenth-century voyages at much the same time as Dalrymple was unequivocally asserting that almost all of the area between the equator and 50°S was land. In 1773, Cook spent much time searching for and failing to find the southern continent; Cook sailed further south than any previous explorer and discovered pack ice, not land. During his 1775 voyage, Cook discovered numerous small islands, all inhospitable environments. The coastline of Antarctica was finally mapped during the nineteenth century.

This intriguing example is not necessarily atypical. Once a false idea is established, considerable effort may be needed before the true facts are known. False ideas can and do influence human behaviour.

Imaginative descriptions by an author of uncertain identity, Sir John Mandeville, were widely circulated, with three reprintings of his works in 1530. Other descriptions perpetuated an error begun by the Greeks concerning the presence of a great southern continent. As late as 1767, a Scottish geographer, Sir Alexander Dalrymple (1737–1808), wrote that almost all of the unknown areas between the equator and 50°S were land (Box 1.2).

Other geographers produced works that largely reflected available knowledge and excluded content that was essentially surmise. Their works along with developments in map making provide a clear indication of the progress of geography and geographic understanding.

Apian (1495–1552) was a map maker and writer. In 1524, he produced a geographic work that divided the earth into five zones (one torrid, two temperate, and two frigid), provided notes on each continent, and listed major towns. Munster (1488–1552), a contemporary of Apian, produced *Cosmography* in 1544, the first major work following the initial European expansion activities that included descriptions of the major regions of the earth. The book is a careful compilation of existing knowledge, but is relatively weak for areas other than Europe. It remained the standard geographic work for at least 100 years and was printed in forty-six editions and six languages.

The period from the early fifteenth to the mid-seventeenth century was one of enormous expansion of geographic fact and fiction. Geography itself continued in the tradition begun by the Greeks. The process of making sense of the world — discovering ourselves and our world — involved a questioning of old ideas and substantial exploration. For geography, the results were a drastic increase in factual content. Map makers and writers strove to keep pace with and further encourage these questionings and explorations. By the mid-seventeenth century, geography's main concerns were still mapping and world description.

GEOGRAPHY RETHOUGHT: VARENIUS AND KANT
VARENIUS

Contemporary geography continues to be concerned with map making and description, but also with many additional concerns. The first indication of some of these concerns appeared in 1650 with the publication of *Geographia Generalis* by Varenius (1622–50) (Box 1.3). This work remained the standard geographic text for at least 100 years; the last English language edition appeared in 1765. What was it that was so distinctive and important about this work? Varenius provided an explicit definition of geography:

> Geography is that part of *mixed Mathematics*, which explains the State of the Earth, and of its Parts, depending on Quantity, viz. its Figure, Place, Magnitude, and Motion, with the Celestial Appearances, etc. By some it is taken in too limited a Sense, for a bare description of the several Countries; and by others too extensively, who along with such a Description would have their Political Constitution.

Clearly, Varenius was concerned about establishing an appropriate place for geography in the system of knowledge as a whole. The definition above is further clarified:

> We call that *Universal Geography* which considers the whole Earth in general, and explains its Properties without regard to particular Countries: But *Special* or *Particular Geography*, describes the Constitution and Situation of each single country by itself which is twofold, viz. *Chorographical*, which describes Countries of a considerable Extent; or *Topographical*, which gives a View of some place or small Tract of the Earth.

It is worth noting that the universal/special distinction noted by Varenius was a major intellectual issue for all branches of knowledge at the time. Universal, more typically

---

Box 1.3: Varenius

Bernhardus Varenius was born near Hamburg, north Germany, in 1622. Following an education in philosophy, mathematics, and physics at the university in Hamburg, he moved to Königsberg and then to Leiden before accepting employment as a private tutor in Amsterdam. In this stimulating environment, Varenius first applied his studies in a geographic context by responding to the unquenchable thirst for knowledge about overseas areas. A book published in 1649 provided Amsterdam merchants with information about Japan and southeast Asia, both areas with which the merchants were doing business. This is practical geography responding to the needs of society.

It appears that Varenius was aware of the limitations of such descriptive works as contributions to academic life.

Accordingly, *Geographia Generalis* was published in 1650 after only six months of writing. The volume was a landmark contribution to geography at least partly because of its explicit acknowledgement that regional descriptions needed to be placed in a larger conceptual context if they were to be regarded as contributions to knowledge. *Geographia Generalis* was translated into several languages and was revised by the great mathematician and physicist, Isaac Newton, in 1672 and 1681 for students at the University of Cambridge. Varenius died in 1650, leaving much planned geographic writing undone, yet he had rethought the nature of geography and began the process that culminated in the late nineteenth century with the formal recognition of geography as a university discipline.

1.1 Champlain's map of New France, 1632. (National Archives of Canada 15661)

called *general geography* was seen as crucial. It was not adequate merely to indulge in detailed descriptions of particular places: it was also necessary to develop an understanding of laws that applied to all places. In geography as in any other subjects, this is a basic distinction. Varenius saw the two as mutually supportive. Interestingly, twentieth-century geographers sometimes failed to appreciate the need for both types of study, arguing for one at the expense of the other.

Varenius confirmed geography as the study of the earth's surface, both physical and human. Detailed description — special geography — is essential, as is the development of generalizations — general geography. There is little that is really new here — the Greeks, for example, contributed to both of these areas. Varenius explicitly recognized two different approaches, a recognition that was prompted by the remarkable increases in knowledge and the prevailing intellectual viewpoint.

BROADENING VISTAS: 1650–1800

The one and a half centuries, 1650 to 1800, following the publication of Varenius's major work witnessed major advances in a wide variety of geographic issues. Exploratory activity continued apace, as did map making. But geographic questions were asked that, while often prompted by exploration, had little to do with the acquisition of new facts and the mapping of newly discovered places. Questions were asked about such fundamental issues as the role of the physical environment as a cause of the growth of civilization, the unity of the human race, and the relationship between population density and productivity.

These were broad issues that generated a variety of views. Most of the individuals who tackled these were not geographers as such — indeed, knowledge was not yet divided into the disciplines that we know today. Nevertheless, the views of Bodin (*c.*1528–96) and Montesquieu (1689–1755) on the role of the physical environment, of Buffon (1707–88) on the unity of the human race, and of Malthus (1760–1834) on population and food supply were all eminently geographic. Each of these issues will be considered later in this textbook. For now, it suffices to note the raising of these questions prior to 1800.

In addition to these broadening vistas, much geographic writing followed the prevailing trends. Thus, several important geographies were published after 1650 that stressed geography as an exercise in encyclopedic description. Most such studies concentrated on describing political units. Both Büsching (1724–93), in a six-volume description of Europe published in 1792, and Gatterer (1727–99), who saw geography as earth description and description of humans using the earth, focused almost entirely on geography as regional description.

KANT

Immanuel Kant lived in Königsberg, Germany, from his birth in 1724 until his death in 1804. For most of his adult life, he taught at the University of Königsberg, becoming a professor of logic and metaphysics in 1770. Through his writings, Kant has had a substantial influence over subsequent philosophy, but he was also interested in the natural sciences. His early publications were in astronomy and geophysics and he lectured in physical geography (broadly interpreted to include much human geography) for forty years beginning in 1756. He has been described as 'the outstanding example in western thought of a professional philosopher concerned with geography' (May 1970:3).

To introduce his lectures on physical geography, Kant emphasized that the subject involved a description or classification of facts in their spatial context. It is now generally recognized that this focus was not especially novel, but it does seem clear that Kant has been interpreted by some later geographers, especially Hettner, as an explicit advocate of geography as a regional study. This interpretation arises from Kant's argument that geography was description according to space and history was description according to time. Indeed, in *Physische Geografie*, published in 1802, it is asserted that geography and history combine to comprise all knowledge.

There was nothing especially new about these ideas, but it is possible that Kant contributed heavily to the notion of geography as a descriptive regional science. In an important sense, this can be seen as a departure from Varenius, as Kant saw geography as special, not general. At the same time, there is little doubt that Kant was very much in accord with prevailing late eighteenth-century thought.

UNIVERSAL GEOGRAPHY: 1800–74

The first three-quarters of the nineteenth century witnessed the demise of geography centred on mapping and description and heavily influenced by exploratory activity. This period was one of radical change in the way knowledge was organized. In brief, the academic disciplines that we know today formally emerged at universities. The year 1874 saw the first geography departments established in several Prussian universities. Simply put, geography could not continue to be primarily mapping and description — late nineteenth-century society required a much more academic discipline — and it was appropriate that mid-nineteenth-century scholars laid the groundwork for the formal discipline. The years 1800–74 witnessed the culmination of all that we have discussed so far and the beginnings of much that has happened since.

Following in the established tradition, a Danish geographer, Conrad Malte-Brun (1775–1826), published a major work between 1810 and 1829. This all-embracing work includes both general and special geography, as defined by Varenius. Mathematical, physical, and political principles are discussed as are physical matters including animals and plants, and human matters including race, language, beliefs, and law. All areas of the known world are described. Malte-Brun succeeded in producing a complete geography, a description of the lands and peoples of the earth.

HUMBOLDT AND RITTER

Geography in the first half of the nineteenth century was really dominated by two German scholars, Humboldt (1769–1859) and Ritter (1779–1859) (boxes 1.4 and 1.5). Humboldt's greatest published work is *Cosmos*, a five-volume work published between 1845 and 1862 — the title is that of

---

Box 1.4: Humboldt

Alexander von Humboldt was born in Berlin in 1769 into the land-owning aristocracy. Before attending university, he had the opportunity to mix with a diverse group of liberal intellectuals and this experience, coupled with an enquiring mind, meant that he was interested in a wide range of academic and related topics. Following brief spells at two universities, he commenced studies at the university in Göttingen. It was here that he met George Foster who had recently returned from a world voyage with Cook. It was an encounter that apparently started Humboldt along the geographic path. His studies continued at Freiburg and when he inherited a substantial income, he decided to use it to support travel and related publication.

From 1796 until his death in 1859, Humboldt travelled extensively and published prolifically. His major travels were to Central and South America from 1799 to 1804. His American travels, which culminated with a visit to Washington and a series of meetings with Thomas Jefferson, resulted in a thirty-volume work that was published between 1805 and 1834 and in many languages.

The writing of *Cosmos* occupied Humboldt from the late 1820s until his death. It is not an exaggeration to describe *Cosmos* as a most important and successful piece of scientific writing. Humboldt strove to investigate the interaction of all natural and human forces. In many areas of thought, he was well ahead of his time; he insisted, for example, that all races had a common origin and that no one race was inherently superior to any other.

Although he made no one great discovery, Humboldt was one of the most admired men of the nineteenth century and might justifiably be regarded as the last universal man. Today, as when he lived, Humboldt is acknowledged as a great geographer. His conception of geography as an integrating science and his insistence on seeing humans as a part of, not separate from, nature remain as key ideas to many.

the Greek concept of an orderly universe (as opposed to chaos). As with previous geographies, Humboldt wanted to describe the universe, but there were additional concerns. He wrote, 'my true purpose is to investigate the interaction of all the forces of nature.' Humans were included as part of nature, a shift of emphasis in the European world. In addition to conventional regional descriptions, Humboldt strived to achieve a complete account of the way all things are related. General concepts were carefully blended with precise observation.

Ritter shared the concern with relationships and also argued for coherence in describing the way things are located on the earth's surface. Like Humboldt, he expressed interest in moving from description to description and laws. This, of course, is very much where Varenius had directed our attention. These interests are paramount in *Die Erdkunde*, an only partially complete world geography comprising nineteen volumes published between 1817 and 1859.

Together, Humboldt and Ritter represent a fitting conclusion and an appropriate beginning. Although much of the world was still unknown to Europeans — Africa was largely a void, as was Asia — these two geographers combined to produce two great works of geography. Both recognized the need for complete geographic descriptions and yet neither could truly fulfil that aim. By the year of their deaths, 1859, there was simply too much geography. Complete encyclopedic descriptions could no longer be produced by one scholar. In this sense, they represent a conclusion. In another sense, they represent a beginning. Humboldt and Ritter were the first geographers to pay full attention to concept formulation, that is, the derivation of general statements from the detailed factual information available. Their interest in relations between things and their inclusion of humans as a part of nature established a basis for much twentieth-century

geography. Thus, they represent both an end to the long road begun by the Greeks but no longer a road to be followed, and a beginning to a geography combining regional description and concept formulation.

MAP MAKING, EXPLORATION, AND GEOGRAPHICAL SOCIETIES

Although there was indeed considerable rethinking about geography, there was also a steady, continuing interest in some of the traditional geographic concerns. Map production was considered so important that governments began to assume responsibility for the task (you will recall that in China civil servants were responsible for map making, probably as early as the third century AD) and in England the Ordnance Survey was founded in 1791. Large-scale topographic maps, providing considerable detail of small areas, were made possible following the development of exact survey techniques in eighteenth-century France. Atlases also became much more sophisticated. Exploration continued and was greatly assisted by the founding of geographical societies. The ever-growing interest in overseas areas encouraged prosperous individuals and governments to support the societies established in Paris (1821), Berlin (1828), London (1830), St Petersburg (1845), and New York (1851).

INSTITUTIONALIZATION: 1874–1903

The year 1874 marks the formal beginning of geography as an institutionalized academic discipline. In 1903, the first North American university department of geography was established. This thirty-year period was one of dramatic developments in geography.

First, once geography became legitimized in institutions of higher learning, it was necessary that university geographers

---

Box 1.5: Ritter

Carl Ritter was born in 1779, ten years after Humboldt, and died in 1859, the same year as Humboldt. Although he never became as well known in the larger scientific world, Ritter proved to have a greater influence on German geography.

Ritter had the good fortune to be selected at age five for involvement in a new system of education that encouraged understanding rather than the conventional procedure of rote learning. One of his teachers was a geographer who emphasized that humans were a part of nature, not some separate entity, and this viewpoint was central to Ritter's own writings in later life. Universities attended included those in Halle and Göttingen. He was able to fund his studies by accepting employment as a private tutor. Professor of history at Frankfurt in 1819, in 1820 he obtained the post that he held for the remainder of his life: the first chair of geography in Germany at the University of Berlin.

As early as 1811, Ritter published a two-volume textbook on Europe; the first volume of his major work *Die Erdkunde* was published in 1817. At the time of his death, *Die Erdkunde* had reached nineteen volumes, most of which focused on Asia; he had not even begun to discuss Europe in this series.

For Ritter, geography was clearly an empirical and descriptive science focusing on the unity of humans and nature. Ritter had first met Humboldt in 1807 and the two established an important relationship. There was a clear agreement regarding the importance of unity of all geographic facts. Unlike Humboldt, Ritter did not discover new facts while engaged in exploratory activities, but he was an inspiring lecturer who influenced many subsequent geographers. In both lectures and writings, Ritter employed a conventional regional focus, one that continued to prevail in late nineteenth-century German geography.

clearly define their discipline and delimit it from the various other emerging disciplines. Not surprisingly, not all geographers immediately agreed on the content and methods. Second, the mid- to late nineteenth century was a period of major advance in various areas of knowledge that had direct effects on geography. A mechanistic view of science emphasizing cause and effect prevailed. Third, the publication of Charles Darwin's (1809–82) *Origin of Species* in 1859 had repercussions throughout the scientific world. Fourth, Europe now had a 400-year history of expansion and movement into regions clearly different from the home area. Finally, the rapid growth of physical science encouraged a development of topics that were traditionally geographic in a number of other disciplines, notably in geology.

Thus, at the very time that geography was institutionalized, there was a complex and changing intellectual environment. Not only had geographers to consider the correct place for mapping, description, general and special geography, regions, and human and land relations, all of which were part of the established geographic traditions, but they also had to consider their role in the physical sciences, the social sciences, mechanistic science, the hugely influential Darwinian concepts, and the morality of expansion overseas. In brief, from 1874 onward, geographers had a great deal to accommodate. How did they react?

GERMANY

In 1874, the Prussian government established departments of geography in all Prussian universities. Why? Possibly because the time was ripe; possibly because the value of geography was especially evident following the Franco-Prussian war of 1870–1, possibly because of a belief that geographic knowledge would facilitate political expansion (see Box 1.1). Regardless of specifics, it became the responsibility of these new departments to clarify the identity of geography. Several leading scholars attempted this task.

For Richthofen (1833–1905), geography was regarded once again as the science of the earth's surface. Field methods and observation were basic techniques. The distinction between special and general, first noted by Varenius, was maintained and it was further argued that the two can be combined to form a chorological (regional) approach. The subject matter of geography as **chorology** remained comprehensive, both physical and human, with the physical environment described first and human activities related to the physical environment.

Ratzel (1844–1904) focused on human geography or what he termed 'anthropogeography' (Box 1.6). His major work was published in two volumes (1882 and 1891). Volume 1 had as an overriding theme, the influence of physical geography on humans, while Volume 2 focused on humans using the earth. Interestingly, it was Volume 1 that had the greatest immediate impact, possibly because the inherently simple cause-and-effect logic was in accord with mechanistic views and helped to assert the importance of physical geography. It is appropriate to regard Ratzel as a founder of human geography, for he was probably the first geographer to focus on human-made landscape.

Hettner (1859–1941) was the most influential follower of Richthofen. In his methodological and descriptive work, geography was unequivocally regarded as the chorological science of the earth's surface, a view that was first clearly enunciated by Kant. As a persuasive advocate of regional geography, Hettner was highly influential in the United States. Studies of human and land relations and studies involving time were excluded from the geography of Hettner.

Thus, German geography following 1874 included three rather different interpretations. First, there was geography as

---

Box 1.6: Ratzel

Friedrich Ratzel was born in Karlsruhe in 1844. Following university studies in Heidelberg and Jena, he fought in the Franco-Prussian War (1870–1). After another brief spell at the university in Munich, he travelled in Europe and then in North and Central America in 1874–5. The American travels in particular impressed upon him the importance of humans as makers of landscapes and inspired him to pursue a career in the new field of geography.

He taught geography in Munich from 1875 until 1886 and then became a professor of geography at Leipzig, succeeding Richthofen, where he remained until his death in 1904. The first volume of Ratzel's *Anthropogeographie*, published in 1882, was specifically designed to be a study of the effects of physical landscapes on history. The second volume, published in 1891, focused attention on human and land relations with a primary emphasis on the role of humans. These two volumes together represent a major contribution to human geography. The first volume proved to be especially influential in the United States. Ratzel also made major contributions in political geography, regarding states as spatial organisms.

In many respects, several of Ratzel's ideas were used somewhat indiscriminately by later geographers. Volume 1 of *Anthropogeographie* was seen as a powerful argument for viewing human landscape as secondary to, even caused by, physical landscapes. Its political geographic ideas were used by Nazi geographers in the 1930s to justify spatial aggression. Despite these interpretations, Ratzel remains an important figure largely because he was the first influential geographer to explain clearly the concept of human landscapes as ever-changing additions to original physical landscapes, an idea integral to the thinking of later German, French, and American geographers.

chorology (Richthofen and Hettner); second, there was geography as the influence of physical geography on humans (Ratzel, Volume 1), and, third, there was geography as the study of the human landscape (Ratzel, Volume 2). These are differences that continued well into this century.

FRANCE

Geography in France followed a route independent of German developments, although there is a close accord with the ideas expounded by Ratzel in Volume 2. One influential scholar, Le Play (1806–82), was not a geographer but a sociologist. Le Play focused on human and land relations and on the impact of technology on social groupings. Reclus (1830–1905), along with Le Play, paved the way for later French geography. This geographer and anarchist, who was barred from France and imprisoned at various times, published a descriptive systematic geography of the world and a nineteen-volume universal geography. His work was very much in the Ritter tradition. The key geographer, Vidal de la Blache (1845–1918) (Box 1.7), followed both Ratzel and Le Play and succeeded in introducing a well-articulated geographic method. The overriding concerns are the relations of humans and land, the evolution of human landscapes, and the description of distinctive local regions. Geography for Vidal considers both physical geographic impacts on humans and human modification of physical geography. The parallels with Humboldt, Ritter, and Ratzel (Volume 2) are clear. Vidal and his many followers continue to exert enormous influence on human geography.

BRITAIN

The first British geography department was established in 1900 at Oxford, largely as a result of activities of the Royal Geographical Society. The dominant influence at first was Mackinder (1861–1947). Intriguingly, he planned to be the first European to climb Mount Kenya in Africa so that his geographic views would be favourably received — to be a respected geographer, he felt that one had to be a successful explorer! For Mackinder, geography and history were definitely related, a global geographic perspective was essential, and, following Richthofen, physical geography was a prerequisite for human studies.

There was also a continuing close association in Britain between geography and the traditional pursuits of mapping, description, and exploration: physical rather than human geography.

UNITED STATES

The establishment of the first North American department of geography, at the University of Chicago in 1903, came at a time when American geography was influenced largely by German scholars. Physical geography (physiography) was dominant, possibly because of the powerful personality of Davis (1850–1934), a geologist; the views that physical geography influenced human landscapes and that geography was essentially a regional science were both imported from Germany and promulgated by Davis. Semple (1863–1932) followed the early Ratzel to become a leading proponent of the physical influences school of thought, while a regional

---

Box 1.7: Vidal

Paul Vidal de la Blache was born in 1845, studied history and geography in Paris, and then taught successively at Nancy, Paris, and the Sorbonne until his death in 1918. His influence on French and other geography was enormous, at least partly because his views on geography lacked competition from other geographers in France. It was possible, then, for this one man to lay down ideas that, once accepted and amplified by others, established a dominant French geographic tradition, *la tradition Vidalienne*. This tradition was not, of course, established in an intellectual vacuum as Vidal had been introduced to the great works of Humboldt and Ritter during his schooling and also travelled often to Germany, meeting with such geographers as Richthofen and Ratzel.

The first clear methodological statement by Vidal was in his Sorbonne inaugural address in 1899; other articles appeared in the journal he founded in 1891, *Annales de Géographie*. For Vidal, geography was the study of both physical and human landscapes, not from a viewpoint of physical geography as cause, but rather from a viewpoint of human and land relationships. Central to his blueprint for geography were the ideas of

genre de vie (way of life), *milieu* (local areas), and the regional concept of *pays* (natural regions). Vidal's conception of geography involves chorological analysis, but with a distinctive human and land focus that was not clearly formulated by either Richthofen or Hettner. There are similarities between Vidal and the Ratzel of *Anthropogeographie*, Volume 2, and also between Vidal and the German geographer, Otto Schlüter (1872–1952).

In France, Vidal was a dominant force and his followers included such geographers as Jean Brunhes (1869–1930) and Albert Demangeon (1872–1940). The lack of any alternative French conception of geography did not mean, however, that Vidal's view was unchallenged. There was an ongoing debate between Vidal and the great French sociologist, Emile Durkheim (1858–1917). Simply put, Durkheim was critical of Vidal and geography in general for emphasizing physical landscapes at the expense of, in his view, an adequate emphasis on society. Vidal died in 1918, having founded a distinct school of geographic thought.

focus quickly became a leading interest for many American geographers. At this time there was no evidence of a distinctive school analysing the mutual relations between humans and physical geography. The general legacy of Humboldt and Ritter and the views of Vidal are not reflected, although Marsh (1801–82), an American geographer and congressman, had earlier focused on these same issues.

GEOGRAPHY IN 1903

Once again, we have reached a distinctive juncture. By 1903, geography was a fully fledged academic discipline in many European countries and the United States. Elsewhere geography was institutionalized somewhat later and largely as a consequence of academic contacts with these pioneering countries. In Canada, for example, a partial department of geography was created at the University of British Columbia in 1923 prior to the establishment of a full department at Toronto in 1935. Today Canada, has more than forty geography departments.

In 1903, the general subject matter was clear — there was no real change since Greek times — and a number of different approaches were advocated. Physical geography was, for many, a study in its own right; others focused on physical geography as the cause of human landscapes; others concentrated on regional description, while still others centred their attentions on humans and nature combined. These general threads are maintained through much of the present century. Geography also continued its association with mapping and, to a lesser extent, with exploration.

PRELUDE TO THE PRESENT: 1903–70

The period from 1903 to 1970 was characterized by several alternative approaches to geographic subject matter. Following earlier developments, we find in academic geography three principal emphases. These are physical geography as cause, humans and land, and regional. We also find a fourth, relatively novel, emphasis — spatial analysis.

PHYSICAL GEOGRAPHY AS CAUSE

Much of the attractiveness of this emphasis (most clearly stated in Ratzel's Volume 1) lay in its relative simplicity. Arguing that human landscapes and cultures resulted primarily from physical geography reduced the need to worry about economics, politics, societies, and so forth. Well-known geographers who utilized this approach included Semple (1863–1932), Huntington (1876–1947), and Taylor (1880–1963) (Box 1.8), but it is important principally because it was so all-pervading. Much twentieth-century geography that was not explicitly along these lines was implicitly of this type. The greatest problem with this approach, often called **environmental determinism**, is the unfortunate tendency to use it in isolation. There are, of course, very important and often intimate relations between physical and human geography, but it is never appropriate to view the former as cause and the latter as effect. Fortunately for geography, physical-geography-as-cause, although popular in the early twentieth century and occasionally taken to extremes, was never a major approach and is now a discredited view.

LANDSCAPE GEOGRAPHY

Geography as the study of all things physical and human, specifically as the study of relationships between physical and human facts, has a long tradition. Humboldt, Ritter, and Ratzel (Volume 2) utilized this approach and their specific legacy is evident in much subsequent geography. Such a broad approach necessitated a clarification other than that provided by the physical-geography-as-cause interpretation and this was accomplished by three scholars who combined to produce a sophisticated methodology and specific subject matter. Vidal developed a distinctive French school, *geographie Vidalienne*, beginning about 1899. Schlüter (1872–1952) led

---

Box 1.8: Taylor

Griffith Taylor lived from 1880 to 1963. Born in Britain, he was responsible for introducing geography as a university discipline to Australia and Canada and today, both countries have strong departments of geography in most of their universities. Following a training in geology, Taylor joined the Scott expedition to the Antarctic, 1910–12. This not altogether unusual type of background led Taylor into geography. He first taught geology at the university in Melbourne and was appointed the first professor of geography at the university in Sydney in 1920.

Taylor was a dedicated and effective teacher with strongly held views on the issue of physical geography as an influence over human activities. His views on the settlement potential of the arid interior of Australia, which were diametrically opposed to the propaganda of the Australian government, brought him into considerable disrepute. At least partly as a result of this controversy, he moved to Chicago in 1928 where he taught until 1935. In 1935, he moved to Toronto, where he was responsible for the founding of the first full Canadian department of geography. Taylor remained in Toronto until his retirement in 1951.

Always a stimulating writer, Taylor produced some twenty books and 200 articles. In Canada, as in Australia, he was involved in discussions about settlement of environmentally difficult areas, in this case the Canadian north. Interestingly, when Taylor returned to Australia at the age of seventy, he was treated as a national hero, his earlier views having been vindicated. At the time of his death, Taylor was being considered in Australia for a knighthood.

the growth of a German school of *Landschaftskunde* — landscape science — in 1906. This landscape geography provides a clear definition of subject matter. The third geographer is Sauer (1889–1975), an American who, in 1925, effectively introduced and elaborated upon the various European ideas (Box 1.9). The resultant **landscape school** focuses on the transformation of the physical geographic landscape by human cultural groups over time.

Combined, the ideas of Vidal, Schlüter, and Sauer provide us with a clearly defined approach to our subject matter. Landscape geography explicitly rejects any suggestion of physical geography as cause by virtue of rejecting environmental determinism in favour of **possibilism**.

REGIONAL GEOGRAPHY

Regional geography, or chorology, proved to be the most popular focus during the first half of the nineteenth century. Pioneering German statements, such as that by Richthofen, and the later arguments of Hettner were carried over into English-language geography. In Britain, a view emerged that delimiting regions was the ultimate task of geography. In America, this view was attractive for physical geographers in particular and Davis, for example, produced a map of regions as early as 1899. In 1905, the British geographer, Herbertson (1865–1915), proposed a series of the world's natural regions. These developments culminated with the 1939 publication of *The Nature of Geography* by the American, Hartshorne (1899–1992) (Box 1.10). This substantial contribution to geographic scholarship argued forcefully for geography as the study of regions — or what Hartshorne often called **areal differentiation**.

Geography in the North American world by 1953 was thus characterized by two related but different emphases — landscape geography and regional geography. Both emphases were the products of long geographic traditions and both have continued to change. In 1953, however, a new focus was forcefully introduced.

SPATIAL ANALYSIS

A paper by F.W. Schaefer (1904–53) in 1953 is characteristically seen as the beginning of an emphasis called spatial analysis, an emphasis that, as noted in the Introduction, focuses on explaining the location of geographic facts. Schaefer objected to the regional emphasis and argued that geographers needed to move away from the description characteristic of regional studies to a more explanatory-focused framework based on scientific methods such as the construction of theory and the use of quantitative methods. Once again we find the special/general distinction of Varenius surfacing. Schaefer objected to what he saw as the overly special focus of regional geography. Regardless of the specific merits of Schaefer's arguments, it is clear that his views struck home to many. What may be called a quantitative and possibly a conceptual revolution followed. The decade of the 1960s was characterized by phenomenal growth in detailed analyses that set up hypotheses and tested those by means of quantitative procedures. Regional geography receded in popularity. There is no doubt that scientific spatial analysis was a major and crucially necessary addition to the list of geographic emphases.

By about 1970, this emphasis found a niche along with a continuing concern for somewhat modified versions of both landscape and regional approaches.

CONTEMPORARY GEOGRAPHY

Since 1970, geography has been characterized by a number of important revisions and additions to traditional interests

Box 1.9: Sauer

For students of English-language human geography, Carl Sauer is a major figure. Born in 1889 in Missouri, Sauer obtained a doctorate from the University of Chicago and taught at the University of Michigan from 1915 to 1923, but he is most closely associated with the department of geography at the University of California at Berkeley, where he taught from 1923 to 1957. He died in 1975. The view of geography articulated by Sauer is often labelled the 'Berkeley school'.

Sauer articulated his conception of geography in his inaugural address at Berkeley. An appraisal of this address, entitled 'The Morphology of Landscape', shows that it is a highly original version of earlier European ideas. Sauer saw geography as a chorological discipline, but one largely concerned with the manner in which humans modify physical landscapes. Humboldt, Hettner, Schlüter, and the core French school of Vidal are all evident influences. The idea that physical geography was a cause of human landscapes, a popular view at the time, was explicitly opposed. Sauer employed the term 'landscape' as the object of geographic study in preference to 'region', which he felt was too closely associated with overly detailed descriptions. Although Sauer's view of geography was largely derivative of others, it proved to be a highly influential view that has contributed significantly to the development of historical and cultural emphases.

Sauer succeeded in establishing a landscape geography that continued from the 1920s to the present with relatively little change until the 1980s. This was no mean achievement and reflects the quality of Sauer's methodological and research writings and the legitimacy of his basic arguments. In addition to his initiation of landscape geography, Sauer produced highly original studies of early America and the origins of agriculture.

and approaches. Geography today, while very clearly the product of the earlier endeavours already discussed, thus incorporates some relatively new components. There is the welcome addition of a wide range of new (for the human geographer) philosophies that are prompting major revisions to the traditional landscape and regional approaches. Also, there is the ever-increasing impact of technological advances in data collection and analysis.

To conclude this chapter on the evolution of human geography, and to whet your appetite for the next chapter that provides details on what human geographers are really doing today, this final section briefly summarizes contemporary human geography. Specifically, we identify:

- an increasing separation of the physical and human components of geography
- a revitalized landscape geography
- a revitalized regional geography
- an ongoing spatial analysis
- an increasing concern with applied matters
- an increasing technical content

PHYSICAL AND HUMAN GEOGRAPHY

Our discussion so far has not made any formal attempt to distinguish between physical geography and human geography. As we have seen, geography from the Greeks onwards has physical and human components. Prior to Humboldt and Ritter, however, there was a clear tendency to treat them separately. Descriptions of the earth typically dealt first with physical and second with human matters. Humboldt and Ritter focused very much on an integration of the two. Subsequent developments saw physical geography as paramount (in the physical-geography-as-cause emphasis), the human landscape as the relevant subject matter (in the landscape emphasis), or the two as separate (in the regional emphasis). It was not unusual for geographers to assert a physical-human unity, but there was not much evidence of such a unity. Very few geographers deny the relevance of physical to human geography, but equally few see the one as necessary to the other. Hence, today we tend to teach and research the two separately.

Thus the separation of physical and human geography as a major division is relatively recent and has involved the recognition of human geography as a separate discipline. Of course, some regional accounts and some specific issues require both discussions; Chapter 4 of this book is a clear instance of the value of a geography that is both physical and human. Again, of course, a basic understanding of global physical geography is relevant to much work in human geography. But overall, the two are no longer seen as so closely related that we need to discuss physical and human geography as one. Contemporary human geography is a social science, but one with special and valuable ties to physical sciences, especially physical geography.

CONTEMPORARY LANDSCAPE GEOGRAPHY

At least since the writings of Vidal, Schlüter, and Sauer, landscape geography has maintained a consistent place in human geography along with the regional approach. Variously called social geography (in Britain) and cultural geography (in North America), this emphasis typically involved studies of human ways of life, cultural regions and related landscapes, and relationships between human and physical landscapes. But the landscape approach has recently been enhanced by the inclusion of new conceptual concerns. Since about 1970, concepts such as those associated with humanism and Marxism, along with a generally increased awareness of advances in other social science disciplines, have enriched landscape studies.

Landscape geography, then, is now concerned with visible features such as fields, fences, and buildings and with symbolic features such as meaning and values. Thus much

---

Box 1.10: Hartshorne

Richard Hartshorne was born in 1899, received a doctorate from the University of Chicago, taught at university in Minnesota from 1924 to 1940, and in Wisconsin from 1940 to 1970. In 1939, he published a major book-length statement entitled *The Nature of Geography* and, in 1959, a revision entitled *Perspective on the Nature of Geography*. The first of these is a landmark in the history of writing about geography and was based on extensive research in Europe. It is impossible to do justice to this major methodological statement in a few words, but it is important to note that Hartshorne based his work closely on earlier German writings, such as those by Hettner, that argued for geography as a chorological science. Geography's unique role was the study of 'the world, seeking to describe, and to interpret, the differences among its different parts, as seen at any one time, commonly the present time' (Hartshorne 1939:460).

Although both Sauer and Hartshorne based many of their ideas on earlier German geographic thought, they clearly arrived at rather different conclusions and each became identified with a particular approach to geography—Sauer with the landscape school and Hartshorne with regional studies. Sauer was critical of Hartshorne's conception of geography, especially because of the exclusion of time, but Hartshorne's contention that geography involved the study of physical and human facts in a regional context was very much in accord with the prevailing American view. As such, it was a major reason why regional geographic studies continued to dominate geography until well in the 1950s.

1.2 Rural landscape, Vermont, USA. (Vermont Department of Travel and Tourism)

contemporary landscape geography focuses on the human experience of being in landscape, with landscapes regarded as relevant not simply because of what they are but also because of what we think they are. Further, there is explicit acknowledgement that landscapes, like regions, reflect and affect a variety of cultural, social, political, and economic processes.

This important human and land tradition is now a much more subtle and profound interest than it was previously and is closely tied to conceptual developments in disciplines such as psychology, anthropology, and sociology.

CONTEMPORARY REGIONAL GEOGRAPHY

The rise of spatial analysis was largely at the expense of the regional geography, areal differentiation, articulated by Hartshorne. Since 1970, however, regional geography, like landscape geography, has resurfaced in a rather different guise so that once again it is a central perspective. Most human geographers accept that the earlier regional geography is no longer appropriate; there is a clear need to proceed beyond regional classification and description. At the same time, the regional concept is obviously a valuable and distinctive geographic device. One geographer described regional geography as 'the highest form of the geographer's art' (Hart l982:1).

The contemporary emphasis is on producing meaningful descriptions that help the reader understand what the region is like and what it means to live there. These descriptions are prompted by at least three general concerns, each of which is tied to a particular set of conceptual constructs. First, there is a concern with regions as settings or locales for human activity. Second, there is a concern with uneven economic and social development between regions, including a focus on the changing division of labour. Third, there is a concern with the ways in which regions reflect the characteristics of the occupying society and affect that society. The underlying concepts for each of these three regional geographic concerns are humanism and Marxism; both of these are introduced in Chapter 2. Viewed in this fashion, we can see such regional geography as increasingly satisfying the requirement that human geography serve society by addressing a range of economic and social problems.

The new directions in regional geography and landscape geography are sufficiently similar that these two approaches now share several common interests.

CONTEMPORARY SPATIAL ANALYSIS

The spatial analytic approach that first became a prime interest of human geographers in the 1950s was very influential until about 1970. Since that time it has continued to be an integral part of human geography, but is now best viewed as one of several well-accepted approaches. A central concern is with explaining locations using available theoretical constructs and hence generalizations are typically emphasized at the expense of specifics. The topics studied by spatial analysts are characteristically economic, such as the location of an industrial plant, as opposed to political, cultural, or social. This situation reflects the influence of various economic location theories.

There is a very clear distinction between the spatial analytic approach and the contemporary regional and landscape approaches. Indeed, spatial analysis is criticized by some because of the perception that it is overly concerned with spatial issues, perhaps even with seeing space as a cause

1.3 Aerial photograph, the Red River south of Winnipeg, Canada. (© 1970 Her Majesty the Queen in right of Canada from the National Air Photo Library with permission of Energy, Mines, and Resources Canada.)

of human landscapes, at the expense of acknowledging the full range of human variables as causes of landscapes. Contemporary regional and landscape geography explicitly contend that space is important only when it is analysed in terms of the human use of that space. This is an important philosophical question that is discussed more fully in the next chapter.

APPLIED GEOGRAPHY

Given that geography is an academic discipline that serves society, it is inevitable that there has always been a crucial applied component, using geographical skills to solve problems. Geography as exploration is a prime example of such applied work. In this century, geographers have played a major role in land-use studies. The studies of the British Land Utilization Survey directed by Stamp (1898–1966) in the 1930s proved enormously valuable during the Second World War. American geographers provided major contributions in land use and military matters.

Contemporary human geography responds to the wide variety of social and environmental issues that confront us today with regular contributions on such topics as peace, energy supply and use, food availability, population control, and social inequalities. All of these issues are covered in this textbook for all human geographers recognize that geography (and each individual geographer) has responsibilities to society and the world, of which we are but one part. Much recent geography of this type has used radical arguments, such as those derived from Marx. In this area, as in several of the others noted, there is much evidence of the impact of other disciplines. There are also close links between applied geography and both spatial analysis and technical advances.

TECHNICAL ADVANCES

Human geography today is greatly aided by a variety of technical advances, for example, advances in navigation-assisted exploration. Major technical advances that facilitate data acquisition involve the use of aerial photography and both infrared and satellite imagery. Major technical advances that facilitate mapping and analysing data are primarily related to computer technologies and geographic information systems. These various concerns are an important component of available geographic techniques and receive detailed attention in the next chapter.

POSTSCRIPT

As befits geographers, we have travelled a long way in this chapter. From the earliest map makers to the present, we have witnessed the evolution of geography as an academic discipline. We know that human geography has a rich heritage and exciting, contemporary developments. It is a product of both our past as a discipline and our present needs, but we have not yet travelled far enough. The next chapter provides a fuller discussion of the various contemporary developments that have received only brief mention so far.

Human geography is a responsible social science with the basic aim of advancing knowledge and serving society. In this respect, human geography today is no different than Greek, Chinese, Islamic, or later European geographies. Our subject matter is clear and our methods various. Our key goal — to provide a basis for comprehending the human world as it is today and as it has evolved — is of crucial importance to contemporary world society. As a brochure produced by the Association of American Geographers expresses it, 'Without geography you're nowhere'.

SUMMARY

A *raison d'être* for geography
To serve society.

Preclassical geography
The earliest geographic descriptions were in the form of maps with the earliest preserved maps dating from *c.*4000 BC. Such maps were of local areas.

Classical geography
Because the Greeks experienced geographic mobility, they initiated a descriptive literary tradition, including mapping the known world. Major figures in this tradition include Herodotus, Eratosthenes, and Strabo. Also initiated was a mathematical tradition; Eratosthenes measured the circumference of the earth and Hipparchus devised a grid system for maps. Ptolemy summarized the mathematical tradition. Roman civilization added little to geographic knowledge.

The Middle Ages
From the fifth to the fifteenth centuries, only sporadic and limited geography was accomplished in Europe. Christian dogma prevailed in the production of T-O maps. Only Portolano charts were a development in mapping. Thus, geography largely disappeared. In China from *c.*2000 BC onwards, and in the Islamic world from the seventh century AD onwards, geography flourished; exploration, mapping, description, and mathematical work all proceeded apace.

A European resurgence
From the fifteenth century onwards, Europe expanded overseas; cartography was transformed with new map projections, especially that of Mercator; geographic description reappeared with major works by Apian and Munster. By the mid-seventeenth century, geographers had added significantly to our knowledge of ourselves and our world; geography itself continued to be a concern with mapping and describing the earth.

Rethinking geography
Varenius was the first scholar to formally define geography and distinguish between general and special geography (in 1650). Kant contributed further to the understanding of geography by identifying it as the science that describes or classifies things in terms of area (*c.*1765). Approximately between 1650 and 1800, geography broadened to include discussion of such issues as physical geography as cause, the growth of civilization, the unity of the human race, and population density and productivity relationships. Descriptive geography and mapping continued.

Universal geography
The heyday of geography as mapping and description was in 1859. By this time, geographic knowledge was significant and specialist sciences emerged. Major nineteenth-century geographies include those by Malte-Brun, Humboldt, and Ritter. The latter two made novel contributions by discussing human and land relations in addition to comprehensive descriptions; both considered geography as both general and special.

Institutionalization
Geography was firmly established as a university discipline in

Prussia in 1874. Other countries rapidly followed suit. Programmatic statements about geography were plentiful: Richthofen and Hettner formulated geography as chorology; Ratzel introduced anthropogeography and Vidal followed his lead; Mackinder focused on a world view, and Davis pioneered physiography. The first North American department of geography was established in Chicago in 1903.

Prelude to the present

Four principal emphases have been evident between 1903 and 1970. Physical geography as cause flowered initially. Sophisticated human and land views developed under Vidal, Schlüter, and Sauer and have persisted to the present. Regional geography dominated North American work until the 1950s. Spatial analysis appeared in the 1950s, prospered in the 1960s, and is now one of several emphases.

Human geography today

Contemporary human geography has close ties to the long and illustrious history of geography and is affected by several relatively new inputs. Today, human geography is easily distinguished from physical geography, has revitalized regional and landscape emphases, has a continuing interest in spatial analysis, has a continuing and expanding concern with applied issues, and has a continually expanding technical content.

Conclusion

Knowledge of geography is essential to the education of all humans. It continues to be a discipline designed to serve the society of which it is a part.

WRITINGS TO PERUSE

DICKINSON, R.E. 1969. *The Makers of Modern Geography*. Boston: Routledge and Kegan Paul.

A traditional history of geography that emphasizes German and French scholars and contains much detailed information.

HARTSHORNE, R. 1939. *The Nature of Geography: A Critical Survey of Current Thought in the Light of the Past*. Lancaster: Association of American Geographers.

The leading methodological statement arguing for regional geography; worth perusing, but difficult to read in depth.

HARVEY, D.W. 1969. *Explanation in Geography*. London: Arnold.

A comprehensive and pioneering statement about geography as a science that, like the Hartshorne book, is not an easy read for the beginning geographer, but rewards even a cursory inspection.

HOLT-JENSEN, A. 1980. *Geography: Its History and Concepts*. Totowa: Barnes and Noble.

A stimulating evaluation of geography as a discipline that is especially readable for the new geographer.

JOHNSTON, R.J. 1991. *Geography and Geographers: Anglo-American Human Geography Since 1945*, 4th ed. London: Arnold.

A detailed overview of English language geography since 1945. Stimulating debate of the role of geography in society.

LIVINGSTONE, D.N. 1992. *The Geographical Tradition*. Oxford: Blackwell.

A major and particularly fascinating account of the last 500 years of geography; this book comprises a series of discussions of intellectual history, the first of which is titled 'Should the History of Geography be X-Rated?'

MARTIN, G.J., and P.E. JAMES. 1993. *All Possible Worlds: A History of Geographical Ideas*, 3rd ed. New York: Wiley.

The best comprehensive history of geography with details of geographic developments and geographers from the earliest times.

MINSHULL, R. 1967. *Regional Geography: Theory and Practice*. London: Hutchinson.

A clear account of the regional emphasis, focusing on current issues in method and practice.

MORRILL, R.L. 1984. 'The Responsibility of Geography'. *Annals, Association of American Geography* 74:1–8.

A thoughtful assessment of the obligations of geographers, including our obligations to society and humanity.

SAUER, C.O. 1925. 'The Morphology of Landscape'. *University of California Publications in Geography* 2:19–53.

The landmark article introducing the landscape concept to North American geography.

SEMPLE, E. 1911. *Influences of Geographic Environment*. New York: Henry Holt.

A classic example of the physical-geography-as-cause interpretation of human geography; very readable today.

TAAFE, E.J. 1974. 'The Spatial View in Context'. *Annals, Association of American Geographers* 64:1–16.

A brief and effective summary of major geographic emphases prior to the 1970s.

UNWIN, T. 1992. *The Place of Geography*. Harlow, Essex: Longman.

A readable and critical account that places particular emphasis on the consequences of the current division between human and physical geography.

WOOLDRIDGE, G.W., and W.G. EAST. 1951. *The Spirit and Purpose of Geography*. London: Hutchinson.

An overview co-written by a physical geographer and a human geographer arguing for geography as a link between physical and human sciences.

2

Studying human geography

THE BROAD SUBJECT MATTER of human geography has remained relatively constant over a period of several thousand years and in a variety of different cultural settings. Approaches to this subject matter have changed, however, in accord with the specific needs of time and place. Human geography has evolved into an increasingly sophisticated concern. The contemporary discipline comprises a wide (some might say bewildering) array of questions posed, general concepts, and techniques of analysis. How are we to make sense of all this?

This chapter addresses these matters using a simple tripartite division. First, we consider why geographers ask the questions we do. The answer may seem self-evident, but it is not. Different geographers ask different questions and we need to know why. Second, we consider what general concepts human geographers use to assist our research activities. Once again, we find a number of core concepts that are central to almost any piece of geographic research and a number of other concepts that tend to be associated with particular types of questions. Third, we consider what techniques of analysis are available to help us answer our questions. Not surprisingly, these are several and varied. Some appeal to all geographers, while others tend to relate to the type of question asked — our first consideration.

Although there are three parts to this chapter, we find that the three are intimately related. In turn we consider philosophical, conceptual, and analytical matters.

Why are these philosophically driven matters important to the beginning geographer? Our discussion of the evolution of human geography taught us that the specific content is essentially a product of the preinstitutional phase, of the

need to articulate a specific content at the time of institutionalization, and of recent developments. We can only fully appreciate our content — the way we analyse and conceptualize problems — if we first understand our philosophies. It is our philosophical perspectives that explain our specific content, concepts, and techniques of analysis.

Human geography, in common with other disciplines, has facts, concepts, and techniques that are logically interrelated. It is absolutely crucial that we do not simply accept that these are related. We must know why. This means we must understand the philosophical viewpoints that serve as the 'glue'. Much (but not all) geographic work is guided by a particular philosophical viewpoint. Before we embark on our geographic work, we need to learn about our philosophical options to appreciate why we ask the questions we do, why we conceptualize as we do, and why we utilize particular techniques. Philosophy is at the heart of not only this chapter but much of this book. We begin with an overview of our philosophical options.

PHILOSOPHICAL OPTIONS

Before proceeding to identify and discuss our philosophical options, it is useful to acknowledge that we have already been discussing such matters, albeit indirectly, in the two preceding chapters. This point can be clarified by referring once again to the term 'environmental determinism'.

This term, as we have already seen, implies that the physical environment is the principal determinant of human matters, including human landscapes. This is the anthropo-geography introduced in Ratzel's Volume 1 and subsequently used by many others, notably by Semple, Huntington, and Taylor. But, as we have also already seen, there are other views. Possibilism implies that the physical environment does not determine human matters; rather, it offers a number of possibilities. This is the cultural landscape view variously introduced by Vidal, Schlüter, and Sauer. Immediately, then, we find that what were previously simply two alternative views about human geography are, in actuality, the tip of a massive and difficult philosophical iceberg: determinism versus free will.

Determinism is a philosophical concept postulating that all events, including human actions, are predetermined. Free will, on the other hand, postulates that humans can act in accord with their will. Environmental determinism is a specific version of the larger determinism concept, while possibilism is a specific partial acceptance of the larger free will concept.

It is not difficult to understand why environmental determinism proved attractive to geography.

1. Prevailing late nineteenth-century view of science influenced explanations in terms of cause and effect. Thus, a geography centred on environmental determinism was a geography with scientific credibility. Specifically, such a geography was philosophically akin to the basic format of that landmark work, Darwin's *Origin of Species*, published in 1859.
2. There was a long history of environmental determinist thought beginning with the Greeks and including such philosophers as Bodin and Montesquieu. We might argue that Ratzel, in 1882, was merely introducing to geography a long-established and appropriate concept.
3. When geography was in its institutional infancy, any approach that explicitly asserted the importance of geography to human affairs was one that gave the new discipline an enhanced status.
4. Perhaps environmental determinism proved popular simply because it was superficially so attractive. Taylor once wrote, 'as young people we were thrilled with the idea that there was a pattern anywhere, so we were enthusiasts for determinism' (cited in Spate 1952:425).

These points make it quite clear that geographers did not enter into any substantial determinism-versus-free will argument. Rather, those geographers following Ratzel's Volume 1 simply accepted the logic of the emphasis. It was not until the 1950s that essentially philosophical cases for environmental determinism and possibilism were made (Box 2.1).

Possibilism emerged at much the same time as the environmental determinist emphasis for these reasons:

1. Geographers were able to turn to a substantial literature centring on humans and land and allowing humans to make decisions not explicitly caused by the physical environment. (Humboldt and Ritter are its two most notable authors.)

2. Geographers had no shortage of examples where different human landscapes were evident in essentially similar physical landscapes.
3. Possibilism was philosophically attractive to some because it accorded with a view, popular in history, to the effect that every event results from individual human decision making.

This brief discussion of environmentalism and possibilism is a valuable reminder that when we ask geographic questions, we are using some philosophical underpinnings. It seems fair to characterize geography as essentially aphilosophical only in terms of such matters as exploration and mapping. Our concern now is to identify the major philosophies that are used, implicitly or explicitly, in contemporary geography — empiricism, positivism, humanism, and Marxism.

Remember: as human geographers we study the human face of the earth, but we are not the only discipline that focuses on this topic. Our distinguishing features result from the questions we ask.

EMPIRICISM

Much human geography may appear to have been and continue to be aphilosophical. By and large, prior to the mid-1950s, human geographers simply ignored philosophical matters and conducted research that was considered appropriate given the historical development of the discipline. Such work, particularly in regional and cultural studies, made no claim to have a philosophical base; human geographers were simply doing what human geographers did, but it can be argued that there was an implicit philosophy in such regional and cultural work. This is the philosophy of **empiricism**, which contends that we know through experience and that we only experience those things that actually exist. Empiricism typically sees knowledge acquisition as a gradual process endlessly correcting or verifying factual statements. In practical terms, an empiricist approach allows the facts to speak for themselves. By definition, empiricism is sceptical of any philosophy that purports to be an all-embracing system.

Empiricism is a fundamental assumption of positivism, but is rejected by most other philosophies.

POSITIVISM

This is a very attractive philosophy because it is rigorous and formal. In principle a clear and straightforward philosophy for human geography, **positivism** argues for the following characteristics.

1. Human geography needs to be objective, in that any personal beliefs of the geographer should not influence research activity. Do you agree? If you do agree in principle, do you believe that it is possible to research human geography without being affected by your personal

beliefs? We will soon discover that according to the following two philosophical options, objectivity is not only undesirable but is also not a feasible goal.

2. Human geography can be studied in a manner comparable to any other science. For the positivist, there is really no such thing as a separate geographic method; all sciences utilize the same method. Specifically, positivism first found favour in the physical sciences and applications in human geography reflect the belief that humans can be treated in a similar fashion as can physical objects. Once again, we will discover that our following two philosophical options regard this assumption as a dehumanization of human geography.

3. The specific method that positivism argues as appropriate for all sciences, physical and human, is known as the **scientific method** (Figure 2.1). Research begins with facts in accord with the empiricist character of the philosophy; a **theory** is derived from facts and from any

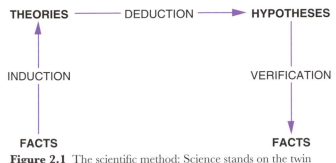

Figure 2.1 The scientific method: Science stands on the twin legs of facts and theory; theoretical work depends on and stimulates factual work.
Source: Adapted from J.G. Kemeny, *A Philosopher Looks at Science* (Englewood Cliffs: Prentice Hall, 1959):81.

available laws or appropriate assumptions; a **hypothesis** (or a set of hypotheses) is derived from the theory, and that hypothesis becomes a **law** when verified by the real world of facts. This scientific method thus involves the

---

Box 2.1: Environmental determinism

'In the early part of this century, a major stumbling block to meaningful objective research in American geography was the assumption that the physical environment largely determined the cultural landscape' (Dohrs and Sommers 1967:121).

There is no denying the long scholarly pedigree of this view in western thought. Thus, both Plato and Aristotle regarded Greece as having an ideal climate for government; Montesquieu contended that humans progressed especially in areas of high winds and frequent storms, and the French philosopher Cousin stated:

> Yes, gentlemen, give me the map of a country, its configuration, its climate, its waters, its winds and all its physical geography; give me its natural productions, its flora, its zoology, and I pledge myself to tell you, a priori, what the man of this country will be, and what part this country will play in history, not by accident but of necessity, not at one epoch but at all epochs (quoted in Febvre 1925:10).

Nor is there any denying that in the late nineteenth century, environmental determinism was both in accord with prevailing views of science and seen by some to offer a distinct and needed basis for the new discipline.

Nevertheless, in retrospect it is somewhat surprising that so many geographers accepted the view in a relatively uncritical manner. Thus, Davis considered any statement to be geographic if it linked physical factors to some human response, while Semple (1911:1) wrote, 'Man is a product of the earth's surface.' Most significantly, until the 1950s environmental determinism was implicitly acknowledged in much geography, including much regional geography.

Many of the apparently extreme statements, such as those quoted from Cousin and Semple above, are easily misinterpreted out of context. A careful reading of the original suggests that such dramatic statements are not substantiated in the larger work — see, for example, Lewthwaite (1966:9). Indeed, Huntington, a human geographer usually regarded as an archetypal determinist, wrote, 'Physical environment never compels man to do anything: the compulsion lies in his own nature. But the environment does say that some courses of conduct are permissible and others impossible' (Huntington 1927:vi).

Debates about the merits of determinism *vis-à-vis* alternative interpretations were evident by the late 1940s and flowered in the 1950s. Environmental determinists posited that it was essential to talk in terms of cause and effect, and therefore it was necessary to be a determinist. The basic counter to this assertion was that any environmental determinist statement was not a statement that was capable of being tested; rather it was a working assumption that was only useful for suggesting research directions. According to this logic, the fundamental error of much environmental determinism is that a certain cause and effect is simply assumed to be true when in actuality, there is no basis for any such assumption.

Many other social scientists regarded geography with some dismay because of what they perceived as an overemphasis on the role of physical factors in discussions of human behaviour. Durkheim even objected to the inclusion of physical factors in the French school of geography initiated by Vidal, a school that was itself antienvironmental determinist.

Regardless of where one stands on the question of geography as an integrated physical and human discipline, there is no doubt that it is inappropriate to talk about physical geography as a cause or an influence.

study of facts, the construction of theory, the derivation of hypotheses, and the related recognition of laws (Box 2.2). For the positivist, any science rests on the twin pillars of facts and theory, and a disciplinary focus on one at the expense of the other is wrong.

Positivist philosophy has roots in the work of the French social philosopher, Comte (1798–1857), and was introduced into human geography relatively late (1953). The principal impetus for the initial proposal of positivism by Comte was to distinguish between science on the one hand and metaphysics and religion on the other. In human geography it was closely associated with quantitative methods and theory development during the 1960s. Positivism is an integral part of what we described earlier as the spatial analytic approach. This philosophy was introduced to geography in a painful fashion because it directly challenged the regional approach dominant at the time. Box 2.3 provides you with some idea of the controversy.

HUMANISM

A third philosophy that we need to consider is **humanism**. Compared to positivism, this is a loosely structured set of ideas with several significantly different versions. Contemporary humanistic geographers argue for intellectual origins in earlier geographic work, such as that of Vidal and Sauer, but it is fair to point out that this earlier work was essentially aphilosophical, whereas the contemporary work is typically rooted in some philosophical ground.

Humanistic geography developed from about 1970 onwards, initially in strong opposition to positivism. Its focus is on humans as individual decision makers, on the way humans perceive their world, and on subjective matters in general. Thus, there are substantial differences between the two philosophies. Positivists argue for an objective world that can be studied, while humanists reject such an objective world. Positivists focus on generalizations, while humanists study specific matters. Positivists contend that geographic research can be conducted without being affected by the personal qualities of the researcher; humanists disagree and see research as irrevocably related to the individual researcher. As part of their research, positivists use theories, test hypotheses, and develop laws; humanists do not have a comparable package of methods. Rather, humanistic research is essentially the personal product of the individual geographer involving intuition and interpretation.

Box 2.2 Positivistic human geography

It is useful at this time to summarize one example of positivistic human geography as a way of demonstrating the value of working within this philosophical framework. The example is that of an analysis of the retailing and wholesaling activities in urban centres and the research proceeds through five conventional stages (Yeates 1968:99–107).

1. *Statement of problem:* The purpose of the study is to determine the relationships between the numbers of retailing and wholesaling establishments and the size of urban centres in a part of southern Ontario. This problem is suggested by earlier work, both factual and theoretical, that suggests that urban land uses and activities may be explained in terms of variables such as size.
2. *Hypotheses:* Three specific hypotheses are identified on the basis of available human geographic theory (the theory in question is known as central place theory and is discussed in Chapter 12). These are: (1) the number of retail establishments and the population size of an urban centre are directly related; (2) the number of retail and service functions in an urban centre increases with the size of that centre, but at a decreasing rate; (3) the number of wholesale establishments and population size of an urban centre are directly related.
3. *Data collection:* Population, retail, and wholesale data are collected for all urban centres of 4,000 or more population

(1966). In addition to fieldwork, the data sources used include city directories, telephone directories, and censuses. Note that the use of a minimum population of 4,000 is an example of an operational definition (a way of describing how an individual object can be identified as belonging to a general set).

4. *Data analysis:* This is the stage in the research when information is subjected to appropriate analysis to formally test the stated hypotheses (the verification stage). In this study, the data are graphed, for example, the number of establishments against the population of urban centres, and a statistical procedure known as correlation analysis is used to calculate the extent of the relationship between the two sets of data.
5. *Stating conclusions:* In standard positivistic research, the formal testing of a hypothesis results in a statement concerning the validity of the hypothesis. In this study all three of the stated hypotheses are confirmed by the statistical analysis, thus increasing our confidence in the theory that generated the hypotheses and also providing insights into the specific landscape investigated.

After reading through this example, it is not difficult to appreciate what is meant by stating that positivism is both a rigorous and formal philosophy.

It is not difficult to appreciate that positivistic and humanistic geographic research are vastly different. Not surprising also is the fact that geographers utilizing these different approaches tend to focus on different types of research problems. Humanists tend to focus on humans and land, landscapes, and regions, whereas positivists address spatial analytic topics. These differences will become apparent in later chapters.

A discussion of humanism can easily become very complicated for there are actually several humanistic philosophies, among them pragmatism and phenomenology.

1. The philosophical approach known as **pragmatism** developed in the United States at the end of the nineteenth century and was especially influential before 1945. Pragmatism is centred on the idea that our human actions are structured by our subjective interpretations of the world we live in. Moreover, our interpretations have a marked practical — pragmatic — content. The central argument is that every perception of the world and every action that occurs in the world is tied to self-evident conditions and successful habits. As such, pragmatism is closely associated with empiricism. There is little evidence to date of explicit uses of this philosophy by human geographers, although it is basic to some of our qualitative research techniques.

Unlike some other humanistic philosophies, pragmatism is strongly group or society (as opposed to individual) in focus. One influential offshoot of pragmatism sees individual behaviour as comprehensible only by reference to the behaviour of the appropriate larger groups to which the individual belongs. This offshoot is known as interactionism.

2. The philosophy of **phenomenology** is based on the key idea that all knowledge is subjective. The goal is to reconstruct the worlds of individuals, the ultimate aim being to understand human behaviour. There are many types of phenomenology, but all are in accord with the above generalization. One popular component of this philosophy is the idea that researchers need to demonstrate **verstehen,** or sympathetic understanding of the issue being researched. The distinction with positivism is clear. Phenomenology strives to understand individual

---

Box 2.3: The Schaefer-Hartshorne debate

Overviews of the evolution of an academic discipline and, specifically, of philosophical and related changes can rarely do justice to the individuals involved and to the events in which they participate, as there is a need to remain relatively dispassionate. On this one occasion, however, we will permit ourselves a brief departure.

By the mid-1950s, geography seemed ripe for change. The combination of regional geography and environmental determinism was seen by many as being of less and less value, while the landscape school led by Sauer appealed to a minority, albeit a substantial minority. The transition to spatial analysis was not, however, an easy one. One aspect of the transition was the Schaefer-Hartshorne debate.

An orderly account of this debate has recently been rendered by Martin (1989). In 1952, Fred Schaefer submitted a manuscript titled *Exceptionalism in Geography: A Methodological Examination* to a major American geographical periodical, the *Annals of the Association of American Geographers*. The article was published shortly after Schaefer's death in 1953 (Schaefer 1953). The major purpose of the article was to object to the perceived claim that geography was necessarily a regional discipline, and to advocate the alternative philosophy and related methods of positivism. Schaefer objected specifically to the argument that he saw as implicit in Richard Hartshorne's work, to the effect that geography studied unique places or regions and was therefore unlike other disciplines.

Hartshorne corresponded with the *Annals* editor in 1953 to complain that Schaefer's article was 'a palpable fraud, consisting of falsehoods, distortions and obvious omissions' (quoted in Martin 1989:73) and subsequently published a series of substantial rebuttals (Hartshorne 1955, 1959).

One geographer has chosen to interpret the debate in a non-academic as well as an academic context. Bunge (1979:128) described the early life of the socialist Schaefer in Germany as Hitler rose to power: 'Schaefer and people like him were immediately subjected to terrorism ... He was under constant police surveillance.' Schaefer became a political refugee in England and subsequently moved to Iowa City where he became a member of the university's geography department. Bunge further described the harassment that Schaefer faced in the early 1950s at the time of 'McCarthyism', a term that refers to the seeking out and possible punishment of suspected communists by a Senate committee headed by Senator Joseph McCarthy. For Bunge, the production of the 1953 article has to be seen in the light of these earlier life experiences. The production of the article, his 'achievement' in life, was not without strong opposition, which further drained his energies. Hartshorne, in a commanding position in geography, turned his fury against Schaefer. With sheer brilliance of argument as his only weapon, Schaefer fought to be heard. Fire engine sirens vividly reminded him of Hitler's terrorism ... McCarthyism repulsed and sickened Schaefer. In the winter of 1952–3, Schaefer suffered a heart attack, and on 6 June 1953 he died of a second one. His article — his 'reason for being' — had not yet been published (Bunge 1968:20–1).

behaviour by using the terms and ideas of individuals, whereas positivism seeks to explain by reference to ideas that the researcher defines and imposes. One especially thoughtful geographer has produced a series of books and articles that are phenomenological in focus (Box 2.4).

Several other humanistic philosophies, such as **existentialism** and **idealism**, have been advocated by geographers, but they have not yet proved to be highly influential.

Our brief discussion of humanism has identified two general issues. First, it is useful to characterize the positivism/ humanism distinction as one of objectivism versus subjectivism. Positivism contends that the study of human phenomena can be objective; humanism says not. The contention is an acknowledged, long-established, and unresolved issue in the social sciences. These queries are relevant here:

- Is there an interaction between the researcher and the researched that invalidates the information collected?
- Does a researcher have a personal background that effectively influences choice of problem, methods, and interpretation of results?
- If we view humans objectively, does this mean we are viewing them as objects? If so, can this be seen as dehumanization?

Answering 'no' to these three questions means that you have positivistic tendencies; 'yes' means you have humanistic

tendencies. The most likely answer is probably 'don't know', which is most understandable at this time.

The second issue that is becoming apparent is that of social scale. Do we as human geographers study individuals or groups of people? Traditionally we have studied and continue to study groups, but you will have noted that the classic humanistic philosophies place some emphasis on individuals. For many human geographers, an emphasis on individuals is inappropriate and much geographic humanistic work has operated on a group scale. Characteristic groups are those defined by culture, religion, language, ethnicity, or class.

MARXISM

This third philosophy is most easily introduced by using the terminology of structuralism, a set of methods arguing that understanding the observed world first requires an understanding of underlying structures and systems. There are three levels of analysis identified in structuralism, two of which are relevant here: the **superstructure** is the level of appearances; the **infrastructure (base)** is the level of processes. Marxism is a version of structuralism as it concerns itself with these two levels. A primary concern of Marxism is indicating the crucial role of the infrastructure or base, in this case the economic structure of capitalist society, as a factor influencing the superstructure of the human geographic world. The human geographer who works within

---

Box 2.4: Humanistic human geography

What questions does a humanistic geographer ask? What types of research issues are addressed? What types of answers are sought? Notwithstanding the fact that there are various and varied humanistic approaches, one way to think about these matters is to consider the work of one leading practitioner, Yi-Fu Tuan.

Tuan (1930–) was born in China and educated at London and Oxford universities in England and at Berkeley in California. An initial interest in physical geography was soon replaced by a broad focus on questions of humans and nature. Today, Tuan is a leading humanistic geographer and one of the first to make explicit reference to philosophy, although he does not claim to subscribe to any one single approach: 'My point of departure is a simple one, namely, that the quality of human experience in an environment (physical and human) is given by people's capacity — mediated through culture — to feel, think and act ...' (Tuan 1984:ix).

With such an aim in mind, Tuan is thus critical of much geography for ignoring individuals and the relationships that they have with one another. For Tuan, there is a need to be comprehensive, like a novelist: 'comprehensiveness — this attempt at complex, realistic description — is itself of high intellectual value; places and people do exist, and we need to see them as

they are even if the effort to do so requires the sacrifice of logical regour and coherence' (Tuan 1983:72).

Three examples of Tuan's work, all well received within and beyond our discipline, are a discussion of landscapes of fear, recognizing that fear is a component part of some environments and that, in turn, it helps create environments (Tuan 1979); a discussion of gardens as an example of our attempts to dominate environments (Tuan 1984); and a discussion of the links between territory and self (Tuan 1982). This third example contends that the presence of segmented spaces, for instance, the many discrete spaces in any urban area, leads to segmented societies. Humans retreat into segmented social worlds as response to the difficulties of coping in a complex society. There is a clear distinction between a small settlement in the less developed world, which has few human-constructed barriers and few spaces assigned for exclusive activities, and a modern city with numerous physical and social barriers. The human experience in these two environments is very different and this in turn has significant implications for individual and group behaviour.

These three examples provide some insights into humanistic geography. The distinction with positivistic work is striking. There are different problems tackled, different questions raised, and different answers achieved.

Marxism thus analyses superstructures, at any location at any time, as they derive from the relevant infrastructure or base, which is a set of economic processes.

Marx summarized the character of society as follows. There are **forces of production** — these are raw materials, implements, and workers that actually produce goods. There are also **relations of production** — these are the economic structures of a society; in a capitalist society, for example, the economic structure is such that workers are able to sell their labour on the open market. Combined, these forces and relations make up the **mode of production**. Examples of modes of production include slavery, feudalism, capitalism, and socialism. It is helpful to think of mode of production as a stage in a development process. Further, there is the already noted separation into infrastructure and superstructure where the former comprises the relations of production and the latter comprises, for our purposes, the larger human geographic world.

This Marxist perspective is often termed **historical materialism**.

Marxism is a philosophy very different from positivism and humanism. Laws of the superstructure are not sought, as in much of humanism. Human behaviour, and hence the human world, are constrained by economic processes.

But Marxism does have another significant dimension. Working within Marxism typically implies that one is trying to understand the human world and striving to change it (Box 2.5). Ultimately, the long-term aim of such work is to facilitate a social and economic transformation from capital-ism to socialism. In the shorter term, Marxism is an attractive philosophy for many human geographers who feel strongly that their work needs to focus explicitly on social and environmental ills and help solve them. Other philosophies, such as positivism, are seen as merely describing the superstructure and implicitly seeing that superstructure as appropriate. Much geography other than Marxist geography also strives to help solve environmental and social ills, but Marxist work has the most explicit such focus.

There is much more to Marxist philosophy than has been intimated so far. Box 2.6 outlines some of this thought and demonstrates the diversity of the larger philosophy.

PHILOSOPHY AND HUMAN GEOGRAPHY

It was aphilosophical human geography, or empiricism, to which positivism objected in the mid-1950s. In turn, by about 1970, positivism was criticized by humanism and Marxism, philosophies that also object to aspects of each other. Not surprisingly, human geographers continue to welcome further additions to an already varied range of options, some of which we will use in later chapters. You can look forward to encounters with such current developments in social thought as feminism, postmodernism, and contextual theory.

There is one more important distinction to be drawn, namely that between idiographic and nomothetic methods. An **idiographic** approach is concerned with individual phenomena; traditional empiricist regional geography, the areal differentiation of Hartshorne, was idiographic in focus as is much humanistic geography. A **nomothetic** approach,

---

Box 2.5: Marxist human geography

Much contemporary human geography either implicitly or explicitly employs a Marxist philosophical framework. To help clarify what doing Marxist research really means, consider the following discussion of the global spread of a capitalist culture (Peet 1989).

What is culture? How is it formed? Why is capitalist culture able to spread across the globe and destroy previously existing cultures? A Marxist response to these questions begins with the idea of the consciousness of a culture, that is, the general way that humans think about things. Put simply, different human experiences in different regions of the world result in different regional consciousnesses and cultures. Thus, before the introduction of capitalism as a new economic and cultural system (a new mode of production), the world comprised a series of regional cultures, each with its own distinct kind of consciousness and characterized by different means and modes of production. One way to explain these regional variations is to recognize that human culture was closely related to physical geography (this is not an environmental determinist assertion but rather a recognition that humans with limited technology are especially closely related to the environment).

To understand culture change as a result of the rise of capitalism, a Marxist acknowledges that human history is one of domination by forces that are beyond human control. Initially, as we have seen, the dominating force was nature, physical geography, but the rise of capitalism was able to replace nature as the dominating force.

Like nature before it, capitalism is able to cause human wealth or poverty, success or failure. It is a powerful and all-pervading economic and cultural system that can overwhelm and replace earlier regionally based economies and cultures. Thus capitalist culture is now a world culture, while regional non-capitalist cultures in the less developed world have either been or are being dramatically transformed. In explicitly Marxist terminology, this process can be seen as one of change in the mode of production.

Typically, a Marxist regards these developments as unfortunate. The regional cultures that are being lost are those that deal with a wide range of environmental problems while the new culture, with its new ways of life and thought, is unsuited to handling such regional issues.

on the other hand, formulates generalizations or laws. Research in a positivist tradition is nomothetic as is much Marxism.

Contemporary human geography, like any other social science, is philosophically diverse and prompts a varied group of concepts and methods. Philosophical diversity is a healthy state of affairs. Our subject matter, namely human behaviour in a spatial context, requires that we employ a variety of philosophies and hence the many concepts and methods that these imply. Some aspects of the human geography in this textbook are clearly aphilosophical or empiricist, some are positivist, some are humanist, some are Marxist, and some are most closely related to other aspects of current social thought. Some may be difficult to clearly assign, a reflection of the fact that the division of philosophies employed is not adequate to describe all philosophically motivated human geography.

HUMAN GEOGRAPHIC CONCEPTS

As is the case with other academic disciplines, human geography involves two basic endeavours. First, there is a need to establish facts — to know where places are and to be aware of their fundamental characteristics. This is what we might call *geographic literacy*. Certainly, facts are a vital starting point. Second, there is a need to understand and explain the facts — to know why the facts are the way they are. This is what we might call *geographic knowledge*. Understanding and explaining requires that we ask intelligent questions, which, in turn, requires sensible employment of an appropriate

philosophy. But there is much more involved in the acquisition of geographic knowledge. Philosophy guides us, but does not necessarily provide all of the needed tools. We are now concerned with identifying these tools, first in the form of concepts and, in the subsequent section, in the form of techniques. Some of the concepts and techniques are relevant regardless of philosophical bent, while others are philosophy-specific. Where appropriate, concepts are linked to the parent philosophy. Concepts and techniques that are not tied to particular philosophies typically focus on factual descriptions such as stating where things are on the surface of the earth; those tied to a philosophy typically focus on understanding or explaining why things are where they are.

SPACE

We have already needed to use this word on a number of occasions. Indeed, it is not uncommon to describe human geography as a spatial discipline, one that deals primarily with **space**. To better appreciate the meaning of this term, it is helpful to distinguish between absolute and relative space. *Absolute space* is objective, it exists, and refers to the areal relations among phenomena on the earth's surface. This conception of space is at the heart of much geography, notably the making of maps, chorology, and spatial analysis. It is central to the ideas of Kant, who saw the geographer ordering phenomena in space. *Relative space* is perceptual, it is socially produced and therefore, unlike absolute space, subject to continuous change.

Some human geographers, notably those sympathetic to a humanistic philosophy, are critical of spatial analysis for

---

Box 2.6: Marxism

The ideas of Karl Marx need to be taken very seriously. He was not only a revolutionary, he was a scientific revolutionary who, with Friedrich Engels, realized that if one could understand how a society actually worked, then one would know how to change that society as needed.

Unfortunately, Marx is a very difficult writer to summarize briefly. Nowhere are his ideas presented in a clear unequivocal fashion and it is not possible to produce a 'correct' reconstruction. Regardless, most interpreters accept that Marx argued that humans affect their own life circumstances and that the character of human history results from the character of human nature. Further, it seems clear that Marx saw humans not as individuals but as members of a social group. Thus, for Marx, humans have needs that they satisfy using particular social forms; as needs change, so do the social forms.

With these concepts of society in mind, Marx can be interpreted as arguing that the economy (the infrastructure or base) determines the superstructure. This is clearly a form of economic determinism. A reasonable interpretation might

argue that superstructures are determined by bases, but not in a manner that permits firm predictions.

Marx also provided sketches of a history of economic change, specifically seeing a transition from feudalism to capitalism to socialism. Marx was especially critical of the capitalism that flourished in his lifetime because of his belief that one class was able to (and did) exploit the other—owners exploit workers. This occurs because the owners are concerned with profits. Exploitation itself was not new, but Marx recognized that what was previously open (slavery, serfdom) was disguised under capitalism. For this fundamental reason, Marx saw a need for a socialist revolution.

What form did such a revolution need to take? Class struggle can overturn one mode of production and replace it with another and with different classes.

Marx died in 1883, but his ideas, variously interpreted, live on. The writings of Marx are essential reading for any social scientist, especially for those critical of our current societies.

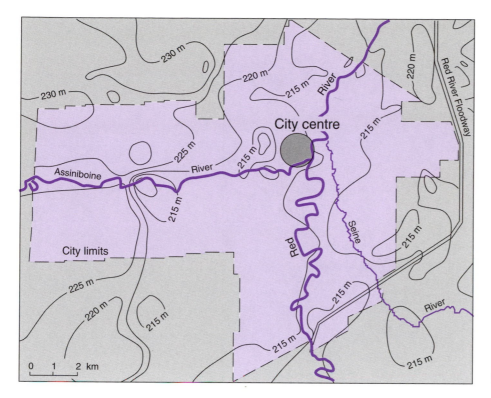

Figure 2.2 The site of Winnipeg.

committing the error of **spatial separatism**, or spatial fetishism, whereby human geography is viewed as a discipline focusing on space as an explanation of human behaviour. Such criticisms are based on the argument that space itself is devoid of content and is only important when given status by humans.

LOCATION

Regardless of philosophical persuasion, there is an initial need to answer the central question: where? Location is the basic concept; it refers to a particular position within space, usually but not necessarily a position on the earth's surface. As we have just noted with regard to space, there are two ways to describe location. *Absolute location* identifies position by reference to an arbitrary mathematical grid system such as latitude and longitude. Winnipeg, Canada, for example, is located at 49°53'N latitude, 97°09'W longitude. Such mathematically precise statements are often essential, but on occasions may not be as meaningful as a statement of *relative location*, that is, the location of one place relative to the location of one or more other places. Absolute locations are unchanging, whereas some relative locations do change. The location of Winnipeg, relative to lines of communication, changes over time.

There are other means by which the location of a geographic fact can be identified. Locations can be described simply by reference to their place name or toponym. Thus, one can be located with an increasing degree of exactness in Canada, Manitoba, Winnipeg, or Portage Avenue. Geographers also make reference to site and situation. Site refers to the local characteristics of a location, whereas situation refers to a location relative to other locations, as is the case with the concept of relative location. Figures 2.2 and 2.3, respectively, illustrate the site and the situation of Winnipeg.

PLACE

The term place has developed a special meaning in human geography. It refers to a location, but specifically to the values that we associate with that location. We all recognize that some locations are in some way distinctive to us or to others. A place, then, is a location that has a particular identity. Our home, the church to which we belong, and the shops we frequent are all examples. Clearly, for one person or another, many locations on the surface of the earth qualify as places. Use of the term in this fashion leads to the idea of a sense of place, a phrase that was popularized in the 1970s by humanistic geographers with philosophical roots in phenomenology. It refers to the attachments that we have to a location such as our home and also to especially memorable or distinctive locations. It is possible for one to be aware that a location evokes a sense of place without necessarily visiting that location; Mecca, Jerusalem, Niagara Falls, and Disney World are all examples.

A closely related concept in sociology, now also utilized by humanistic and other geographers, is that of **sacred space**. This term refers to landscapes that are particularly esteemed by an individual or a group, usually for a religious but possibly also for a political or some comparable reason. For Mormons, a Protestant religious group based in Utah, there are a series of sacred spaces. All their temples are

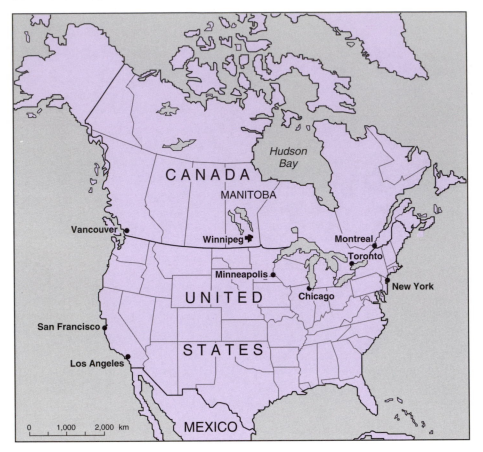

Figure 2.3 The situation of Winnipeg within North America.

sacred because they are open only to Mormons and are considered as locations where communication between earth and heaven occurs; another sacred Mormon space is a location near Palmyra in New York state where their founder received God's message (Jackson and Henrie 1983). Sacred space is different from the concept of mundane space, that which is occupied by humans but has no special quality. Humanistic geographers use the related concept of **placelessness** to identify landscapes that are relatively homogeneous and standardized; examples include many tourist landscapes, urban commercial strips, and suburbs (Relph 1976). There is an interesting parallel here with the arguments in Box 2.5; places may be thought of as especially characteristic of precapitalist cultures, whereas placelessness is more evident in the modern industrial world.

Tuan, the humanistic geographer whose work is discussed in Box 2.4, introduced **topophilia**, literally love of place, to human geography. This term refers to the full range of emotional ties between humans and environments, especially the positive feelings for places. This concept can be contrasted with the less frequently used term 'topophobia', which refers to dislike of a landscape that may prompt feelings of anxiety, fear, or suffering.

Overall, these concepts provide a valuable addition to the more general concept of location. Any human geographic work that aspires to concern itself with people and their

relationship with land is likely enriched by using these ideas. Indeed, we can think of places as emotional anchors for much human activity.

REGION

Region is one of the most useful and yet at the same time one of the most confusing of geographic concepts:

> So much geography is written on a regional basis that the idea of the region and the regional method is as familiar, and as accepted as is Mercator's map in an atlas. Yet as with so many other familiar ideas which we use every day and take for granted, the concept of the region floats away when one tries to grasp it, and disappears when one looks directly at it and tries to focus (Minshull 1967:13).

Our discussions in Chapter 1 identified regional geography as a traditional focus. In the simplest sense, geographers have long recognized that geographic accounts cannot encompass the earth as a whole, hence the logical tendency to subdivide. In similar fashion, historians have traditionally subdivided the span of human history into periods. By the early twentieth century, however, regional geography proceeded to formally define the concept of region as 'a device for selecting and studying areal groupings of the complex

2.1 Maydan-e-Naghsh-e-Jahan, Isfahan, Iran. (Embassy of Pakistan, Interests Section of the Islamic Republic of Iran, Washington, DC)

2.2 Niagara Falls, Canada–United States border. (Bell Aircraft Corporation/Bell Aerospace Textron)

2.3 Settlement and wheat fields in the Canadian prairies. (George Hunter, Mississauga, Ontario)

phenomena found on the earth. Any segment or portion of the earth surface is a region if it is homogeneous in terms of such an areal grouping' (Whittlesey 1954:30). Such a definition accommodated a wide variety of types of region, both physical and human.

Dividing a large area into regions or **regionalization** is a process of classification with each specific location assigned to a region. Human geographers recognize various types of regions, most notably the **formal** (or uniform) **region**, as an area with one or more traits in common, and the **functional** (or nodal) **region**, an area with locations related to each other or to a given location. Thus an area of German-language speakers qualifies as a formal region, while the sales distribution area of a city newspaper qualifies as a functional region. Delimiting formal regions has long been notoriously troublesome. Consider, for example, how you might delimit a wheat-growing area in the Canadian prairies — on the basis of percentage of wheat farmers per township, percentage of acreage under wheat, percentage of farm income from wheat, or one of many other possibilities? Clearly, there is considerable subjectivity involved in the choice of measure. An additional problem inherent in region delimitations is the implication that regions are geographically meaningful. A prairie wheat-growing region may or may not be a significant portion of geographic space.

With the rise of spatial analysis, regions, as used in traditional regional geography, became somewhat discredited during the 1960s. Functional region delimitation increased in popularity as a result of the rise of statistical analyses. Since *c.*1970 humanistic geographers have argued for a revitalized regional geography involving **vernacular regions**, regions perceived to exist either by those within or without. The 'Bible belt' of the American south and midwest is one example. This region is readily identifiable to most Americans as an area with a particularly strong commitment to various conservative Protestant religions. For many of those living in the region, the name is a source of pride; Oklahoma City declares itself as the 'buckle on the Bible belt'.

The regional concept is a core one. We have identified four separate applications. First, regions are a valuable simplifying device, comparable to the period of the historian; the task of classifying may be a very valuable exercise to aid understanding of landscapes. Second, formal region delimitation was central to the chorological approach that dominated during much of the first half of the twentieth century. Third, functional region delimitation was one part of the spatial analytic thrust. Fourth, much contemporary work sees vernacular regions as central to our understanding of human landscapes.

DISTANCE

Our first three concepts — location, place, and region — each refer to portions of the earth's surface and logically we are also interested in the **distance** between locations,

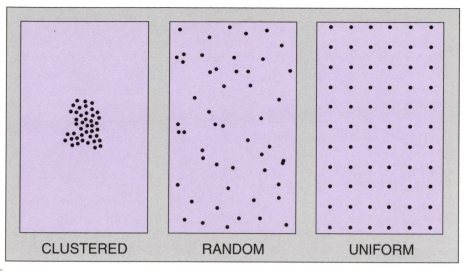

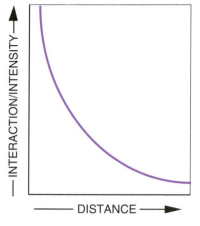

Figure 2.4 Clustered, random, and uniform point patterns. A clustered pattern typically results if geographic facts benefit from close proximity; for example, specialty retail outlets such as jewellery stores. A uniform pattern results if geographic facts benefit from separation because they compete for surrounding space; for example, urban centres in a region. Random patterns may reflect the effect of more than one process or may be a temporary stage during changing circumstances.

Figure 2.5 A characteristic distance decay curve. The specific slope is a function of the particular variable being plotted against distance. The slope usually becomes less steep over time, reflecting decreasing distance friction as a result of constantly improving technology.

places, and between or inside regions. Not surprisingly, because distance is quantifiably measurable, the greatest interest has been shown by spatial analysts. Distances may be measured in many ways, but most measuring systems typically involve some standard unit such as kilometres. Human geographers also measure distances in terms of time and cost or, rather more difficult to measure, in terms of cognitive or social variables.

Regardless of the measure, there is a clear tendency for geographic facts — such as churches or towns — to be located such that a close relationship with some measure of distance is evident. Geographers often talk about distributions, patterns, or forms to refer to the mapped appearance of spatial facts (see Figure 2.4). It is often the case that such maps reveal some sort of order. Specialty retail stores (jewellers and shoe stores, for example) tend to locate near one another. Similarly, financial institutions prefer to be close to one another. It is useful to characterize such distributions as resulting from a clustering process. Urban centres, on the other hand, typically locate apart, as one of their principal reasons for location is serving surrounding rural populations. Urban centres locate at least partly in response to competition. Other geographic facts locate apart because they involve the provision of services without involving competition; hospitals, community centres, and recreation areas are examples.

Much of what has been said about distance so far is evident in what Tobler (1970) referred to as the first law of geography, namely, everything is related to everything else, but near things are more related than distant things. This is the notion of **distance decay**, graphically depicted in Figure 2.5 and sometimes referred to as the effect or friction of distance; typically both time and cost are involved in overcoming distance. This concept lies at the heart of much spatial analytic research (for example, that discussed in Box 2.2) and is subject to criticism from humanistic geographers who interpret the concept as an example of spatial separatism in that distance is seen as having causal powers.

The distance concept is related to several other ideas. **Accessibility** refers to the relative ease with which a given location can be reached from other locations and therefore indicates the relative opportunities for contact and interaction; it is a key concept in the agricultural, settlement, and industrial location theories that we encounter later in this text. **Interaction** refers to the act of movement, trading, or any other means of communication between locations. **Agglomeration** describes the situation where locations, usually of productive activities, are in close proximity, while **deglomeration** refers to those situations where the geographic facts are characterized by separation rather than togetherness.

Distance is conventionally measured as absolute distance using some standard unit. Distance measured in economic, temporal, cognitive, or social terms is relative. Such distances can be and frequently are quite different to corresponding absolute distances. Minimum cost distances between two locations often involve greater absolute distances because of the cheaper cost of water compared to land transport or because of physical or political barriers. Distance may be measured in terms of the time required; times are usually less in developed than in less developed areas and also usually decrease with increasing levels of technology. Cognitive and social distance vary according to individual and group circumstances, often being related to knowledge of the areas between the two locations.

This concept of distance, along with the variety of closely related concepts discussed, often lies at the heart of geographic research. Indeed much of world economic history can be interpreted as involving the gradual overcoming of distance as obstacle. One leading Australian historian viewed Australian development in terms of the 'tyranny of distance' (Blainey 1968).

SCALE

One of the first decisions made in any piece of geographic research relates to the selection of appropriate scales — spatial, temporal, and social. It is possible to study a large area or a small area, a long period of time or a particular moment in time, a great many people or a single individual. The choice of scale is usually determined by the questions posed, but it is important to recognize that different scales can generate different answers. With reference to *spatial scale*, if we ask questions about whether a given set of locations are clustered together (agglomerated) or are dispersed (deglomerated), the answer will vary with the area selected (figures 2.6 and 2.7). Spatial scale also clearly needs to be carefully identified when making any statements of density. In Canada, for example, a single statistic masks enormous variations, such as those between urban centres and large unpopulated areas.

The choice of *temporal scale* is also important. If questions are asked concerning the evolution of landscape, then a temporal perspective is essential; if, alternatively, questions are asked concerning the manner in which a given area functions, then a temporal perspective may be quite unnecessary. Historical and cultural geography typically emphasize time, while chorological and spatial analytic focuses tend to emphasize the present. Contemporary human geography exhibits a willingness to utilize whatever temporal scale is most appropriate.

The relevance of *social scale* has emerged as a major concern with the rise of humanistic interests. Traditionally, the focus has been on groups, typically delimited by reference to culture. Some humanism, however, favours an individual scale of analysis.

Selecting scales that reflect the questions posed and that facilitate the acquisition of correct answers is not simple. Scales need to be selected with care and justified, for it is only with judicious selections that we can be confident of our results.

DIFFUSION

Landscapes, regions, and locations are subject to change — our three recurring themes from Chapter 1. **Diffusion** — the spread of a phenomenon over space and growth through time — is one way change occurs. Migration of people, the movement of an idea (such as a religion), or the expansion of land use (such as wheat cultivation) are all examples of diffusion. Diffusion-centred research has long been central to

cultural geography because of the need to understand landscape evolution. Probably the greatest impetus was a result of the gradual 'diffusion' of the work of a pioneering Swedish geographer, Hagerstrand, who developed a series of diffusion-related concepts in 1953; his work was partially known in the English-language world by the early 1960s, but was not translated *in toto* until 1967. Hagerstrand and the work that he inspired, which was largely positivisitic in character, combined to introduce a series of important ideas.

First, there is *the neighbourhood effect*. This term describes the situation where diffusion is distance-biased with the earliest expansion of the phenomenon to individuals or groups nearest its place of origin. A second and alternative situation involves a *hierarchical effect*. In this case, the phenomenon first diffuses to large centres, then to centres of decreasing size. One geographer noted that the 1832 cholera epidemic in North America diffused in a neighbourhood fashion, while the 1867 epidemic diffused hierarchically. Why the difference? Because North America in 1832 had a limited number of urban centres and a limited communication network, whereas North America in 1867 had many urban centres and a well-developed communication network. A third contribution focused on the well-established fact that most diffusion situations proceed slowly at first, then very rapidly, followed by a final slow stage to produce an *S-shaped curve* (Figure 2.8).

The diffusion concept is perhaps best described as a process that prompts changes in landscapes, regions, and locations.

PERCEPTION

In 1850, Humboldt noted, 'in order to comprehend nature in all its vast sublimity, it would be necessary to present it under a twofold aspect, first objectively, as an actual phenomenon, and next subjectively as it is reflected in the feelings of mankind' (quoted in Saarinen 1974:255–6). Despite this pioneering statement, geographers paid relatively little attention to subjective matters, especially the perceived environment, until the late 1960s. We now recognize that all humans relate not to some real physical or social environment but rather to their **perception** of that environment, a perception that varies with knowledge and is closely related to cultural and social considerations. Humanistic geographers in particular discuss the mental images of places and other people and seek to describe and understand the mental maps that we carry in our heads. These images and maps are important for at least five reasons.

1. Mental **images** of other places and other people are always changing. We all live in a shrinking world. We are becoming increasingly aware that when we pollute a specific location, we are in fact polluting the world. Our actions today can affect others and will certainly affect our children. But is our increasing knowledge of the

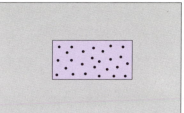

Figure 2.6 Impact of spatial scale on descriptions of point patterns. Whether we describe this point pattern as clustered or dispersed depends upon the area in which it is contained. Using the inner boundary, the pattern is dispersed; using the outer boundary, it is clustered.

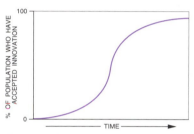

Figure 2.8 A characteristic S-shaped growth curve. Adoption of the new idea begins slowly, then proceeds rapidly, only to slow down as the adoption rate reaches 100 per cent.

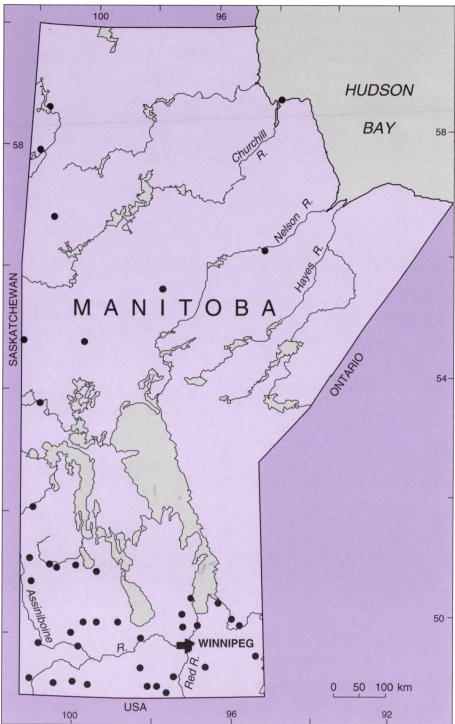

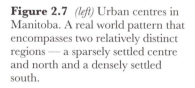

Figure 2.7 *(left)* Urban centres in Manitoba. A real world pattern that encompasses two relatively distinct regions — a sparsely settled centre and north and a densely settled south.

world affecting our behaviour for the better? The evidence here is contradictory. Nations today dump atomic wastes, insecticides end up in people's diets, sewers pollute rivers and seas, and acid rain causes massive environmental damage. In terms of spatial scale, we might suggest that there is a need to think globally but act locally.

2. **Mental map** research demonstrates that humans have varying perceptions of environments. These perceptions help explain population movements. In the less developed world, the most evident migrations are those from rural to urban areas; in the United States, there is a continuing movement to the south.

3. The mental maps of particular individuals are of great importance — for example those people who make decisions about where to locate employment opportunities. The location of an industrial plant, involving perhaps

2.4 Steel mills, Hamilton, Ontario, Canada. (Ontario Ministry of the Environment and Energy)

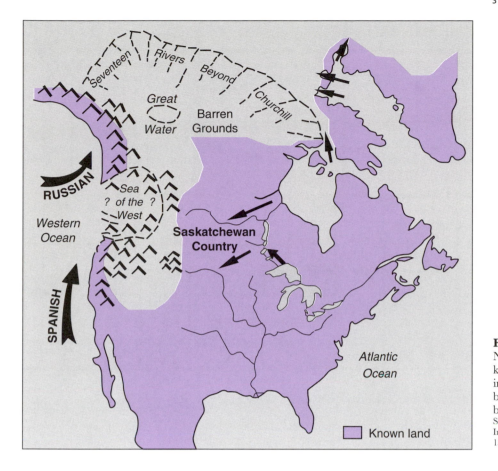

Figure 2.9 Spatial concepts of North America in 1763. Limited knowledge resulted in many inappropriate decisions. As reality became clearer, spatial behaviour became increasingly appropriate.
Source: R.I. Ruggles, 'The West of Canada: Imagination and Reality', *Canadian Geographer* 15 (1971):237.

1,000 new jobs, may often reflect the biases and values of one person or one group. The traditional industrial areas of the developed world are rarely seen favourably by such decision makers. The favoured areas may be those close to recreational amenities and/or large cities. In the United States, the south and west are generally favoured over the north and east; and in England, the London region is favoured over the midlands and north. In Canada, there is an additional factor. Many locational decisions reflect the biases and values of Americans, not Canadians. Mental maps of decision makers are especially important because so many industries today are footloose, that is, not tied to specific raw materials or power sources.

4. It is distressing that groups of individuals — for instance government officials and the military — may often have highly distorted mental maps. When an American president mistakes one country for another, it is far from humorous. Research in the 1980s showed that the mental maps of the typical American military officer completely confused the sizes of areas and relative populations, and tended to see one threatening country only (the former USSR).

5. Mental maps do change. In Brazil, the movement of the capital from Rio de Janeiro to Brasilia completely changed the image of the interior from one of a worthless wasteland to one of huge potential. In Canada, the image of the north changed with the beginning of oil exploration, while the images of Quebec and the west change in response to political separatist movements.

Mental maps of relatively unknown areas are especially subject to error. Figure 2.9 is an interesting composite of the mid-eighteenth century image of North America as held by Europeans.

Clearly we acquire our mental maps from a variety of sources, but one is of particular relevance to us: human geography teaches us about the world, about where things are located, why they are there, and what they really are. Many university students often develop improved mental image maps as a result of studying introductory human geography.

DEVELOPMENT

As one part of their studies of the landscapes created by humans, human geographers recognize that any one area changes through time and that different areas have varied landscapes. Such conditions are often interpreted in terms of **development**. The general meaning of this term involves measures of economic growth, social welfare, and

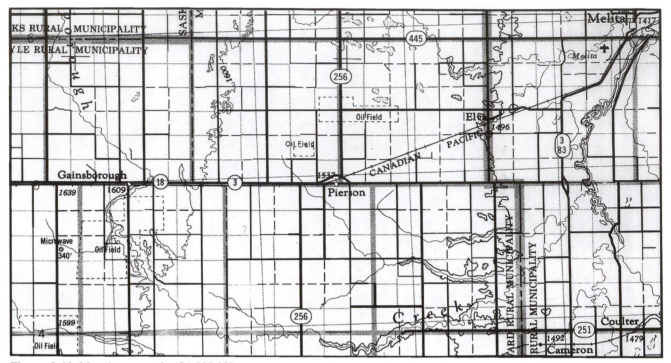

Figure 2.10 Mapping at a scale of 1:250,000.

modernization. It is possible to define, for any given area at any given time, the level or state of development. Following this logic, certain areas might qualify as more developed and others as less developed. This is a distinction that we pursue in some detail in Chapter 6.

Such basic identifications are of very real value, but need to be applied cautiously. It is important that human geographers highlight spatial disparities in economic well-being, but it is also important that we interpret variations with reference to cultural and social considerations. If, for example, we use income levels as a measure of development, then we must not neglect to recognize that income level may reflect both economic success and cultural values.

Contemporary human geographers analyse development while fully cognizant of the dangers of trivializing the concept. A Marxist focus might view underdevelopment as a result of the relatively recent and rapid diffusion of the capitalist economic and social system, arguing that areas brought into the expanding capitalist system become dependent. It can also be argued that a capitalist system tends to create depressed areas within any given country, thus prompting uneven development. Such variations are evident throughout what we generally label the developed world. In Canada, for example, there is an economically advantaged core area, the St Lawrence lowlands of Quebec and Ontario, and a series of relatively disadvantaged peripheries, such as the maritime area and the north.

CONCEPTS: A SUMMARY

This discussion of human geographic concepts has introduced the terms that are central to an appreciation of studying human geography. As it is an academic discipline concerned with how human behaviour affects the earth's surface, human geographers employ such concepts as space, location, place, region, distance, scale, diffusion, perception, development, and a series of additional concepts that can be conveniently subsumed under those headings. In most cases, as we have seen, a concept has multiple interpretations depending on the philosophical emphasis employed; think especially about region, distance, scale, and development. All of the terms introduced in this discussion are employed in subsequent accounts of the substantive work of human geographers; the concepts are to be used.

TECHNIQUES OF ANALYSIS

We move now to a discussion of the techniques that aid the human geographer to collect, display, and analyse data. We have an understanding of why human geographers ask the research questions they ask, our philosophical options, and an understanding of the concepts they use to help structure their research activity. But what techniques do human geographers employ to achieve their research goals?

The first three techniques discussed are cartography, computer-assisted cartography, and geographic information systems. Each of these closely related techniques is inherently geographic, largely developed within geography and involve inputting, storing, analysing, and outputting spatial data. A group of techniques that relate primarily to the collection of data are discussed next, namely remote sensing. Finally, we consider both qualitative and quantitative methods of collecting and analysing data.

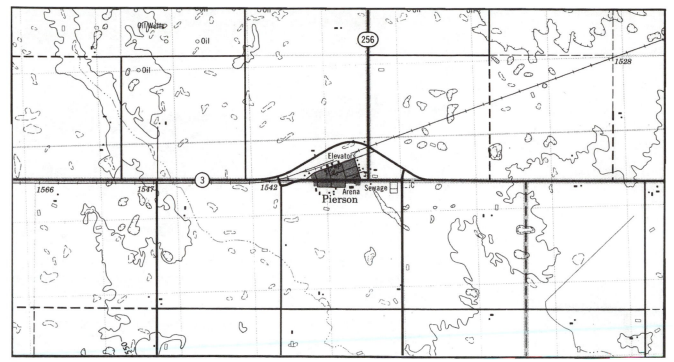

Figure 2.11 Mapping at a scale of 1:50,000.

Although many of these techniques are philosophy-free, this section is not entirely philosophy-free. Data collection by fieldwork clearly includes some procedures that are positivist and others that are humanist; we take advantage of this situation at the conclusion of this section to raise questions about whether our human geography is one of people, for people, or with people.

CARTOGRAPHY

Our discussions in Chapter 1 highlighted the centrality of mapping and maps to the geographic enterprise. The science of map making is known as **cartography**. Until the 1960s, cartography was essentially limited to map production, following data collection by surveyors and preceding analysis by geographers. There was much emphasis on manual skills. The purpose of such maps was to communicate information. Quite simply, maps are an efficient means of portraying and communicating spatial data. Today, however, cartography is less dependent on manual skills and is closely integrated with analysis.

As communication tools, maps describe the location of geographic facts. As analytical tools, maps can be used to clarify questions and suggest research directions. In the production of maps, cartographers need to decide on questions that can significantly affect map appearance and quality: scale, type, and projection.

Map scale relates to the area covered and to the detail presented. Large-scale maps portray small areas in considerable detail, while small-scale maps portray large areas with little detail. A map of a city block is at a large scale and a map of the world at a small scale. Scale is always indicated on a map as a fraction, ratio, a written statement, or a graphic scale. Canada has a National Topographic System with maps at a scale of 1:250,000 covering approximately 15,539 km^2 (6,000 square miles), and with maps at 1:50,000, covering 1,036 km^2 (400 square miles) in greater detail. On a 1:50,000 map, farm buildings are located and contours (lines of equal elevation) are shown at 7.5-m (25-ft) intervals. Figures 2.10 and 2.11 provide examples of each type.

The type of map constructed depends on the information being presented. Dot maps are useful for data such as towns, wheat farming, cemeteries, incidences of disease, and so forth. Typically, each dot represents one occurrence of the phenomenon being mapped (Figure 2.12). A **choropleth map** displays data using tonal shadings that are proportional to the density of the phenomena in each of the defined areal units (Figure 2.13). Choropleth maps sacrifice detail for improved appearance. An **isopleth map** comprises a series of lines, isopleths, that link points having the same value (Figure 2.14). Examples include contour maps, isochrome maps showing lines of equal time, and isotim maps showing lines of equal transport cost.

Small-scale maps (of large areas such as the earth) encounter a fundamental problem; how best to represent a nearly spherical earth on a flat surface, a process of conversion known as a projection. There is not a correct solution to this problem and literally hundreds of projections have been developed (Box 2.7).

Map users need to understand the cartographic design process, including the significance of the chosen scale, the types of symbols, and the projection in order to interpret a map correctly (Box 2.8).

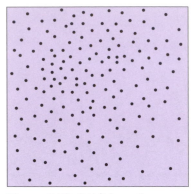

Figure 2.12 *(top)* Schematic representation of a dot map.

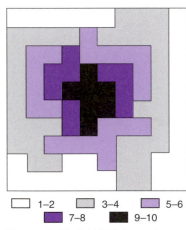

Figure 2.13 *(middle)* Schematic representation of a choropleth map.

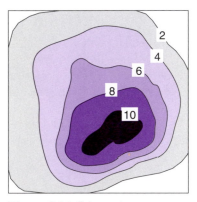

Figure 2.14 Schematic representation of an isoline map.

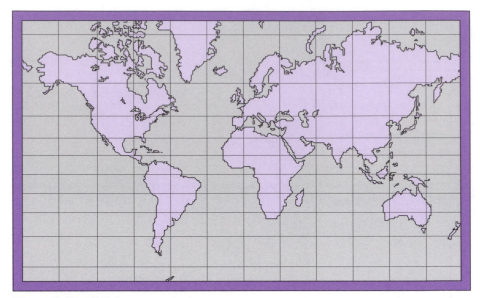

Figure 2.15 *(top)* Mercator projection.

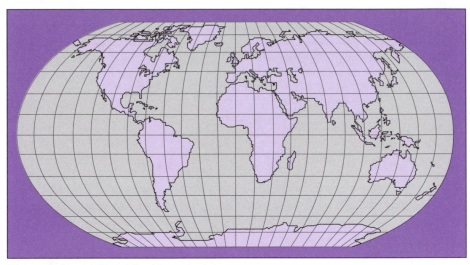

Figure 2.16 *(middle)* Mollweide projection.

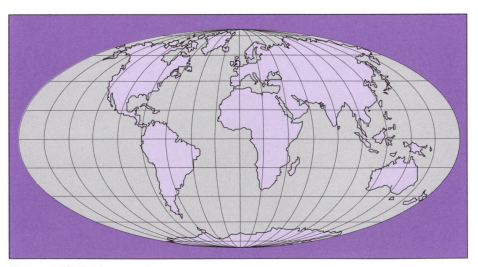

Figure 2.17 Robinson projection.

COMPUTER-ASSISTED CARTOGRAPHY

This technique is covered separately from the previous discussion because computer-assisted cartography, sometimes called digital mapping, is much more than just another production evolution, although technical advances have significantly reduced the demand for manual skills.

Computers allow us to amend maps by incorporating new and revised data; they enable us to produce various versions of the mapped data to create the best version; through the use of mapping packages, they diminish the necessity for artistic skills and allow for map creation in desktop fashion. Computer-assisted cartography has introduced maps and map analysis in a wide range of new arenas, for example, to assist businesses to realign sales and service territories. Computer-generated maps facilitate decision making and are becoming important in academic and applied geography.

As a means of upgrading cartographic method and practice, as opposed to speeding up the production process, computer-assisted cartography is one component of geographic information systems.

GEOGRAPHIC INFORMATION SYSTEMS

A **geographic information system** (GIS) is a computer-based tool that combines the storage, display, analysis, and mapping of spatially referenced data. A GIS includes processing hardware, specialized peripheral hardware, and software. The typical *processing hardware* is a personal computer or workstation, although mainframe computers may be used for especially large applications. *Peripheral hardware* may be used for data input — for example, digitizers and scanners — while printers or plotters produce copies of the output. *Software* production is now a major industry and there are numerous products available for the GIS user; examples include IDRISI, a university-produced package designed primarily for pedagogic purposes, and ARC/INFO, a package developed by the private sector that is widely used by governments, industries, and universities.

The origins of contemporary GIS can be traced to the first developments in computer-assisted cartography and to the Canada GIS of the early 1960s, developments that centred on computer methods of map overlay and area measurement — procedures previously accomplished by hand. Since the early 1980s, however, there has been an explosion of GIS activity related to the increasing need for GIS and the increased availability of personal computers.

The roots of GIS are clearly in cartography, and maps are both the principal input and the principal output. But computers are able to handle only characters and numbers and not spatial objects such as points, lines, and areas. Hence GISs are distinguished according to the method used to translate the spatial data into computer form. There are two principal methods of translation. The **vector** approach describes the spatial data as a series of discrete objects; points are described according to distance along two axes, lines are described by the shortest distance between two points, and areas are described by sets of lines. The **raster** method represents the area mapped as a series of small rectangular cells known as pixels (picture elements); points, lines, and areas are approximated by sets of pixels and the computer maintains a record of which pixels are on or off. Box 2.9 summarizes the characteristic activities involved in using a GIS.

What is the value of a GIS? What applications are possible? The short answer is that GISs have numerous and varied

---

Box 2.7: Map projections

'Not only is it easy to lie with maps, it's essential' (Monmonier 1991:1). Monmonier is referring to the fact that even the best-intentioned of cartographers is obligated to distort reality because it is always necessary to use a map projection, a systematic two-dimensional transformation of the three dimensions of a sphere. This issue is most important when the map is of a large area (small scale).

A projection attempts to retain one or more of the following characteristics of a sphere: shape, size, direction, and distance. The correct projection to use is that which best serves the objectives of the map. The main decision to be made concerns selecting either a *conformal* projection, which depicts shape correctly, or an *equal area* projection, which maintains the relative areas of all parts of the earth.

Figures 2.15, 2.16, and 2.17 show three important projections. The Mercator projection, introduced in 1569, greatly exaggerates the size of high-latitude land masses (Figure 2.15).

Note that Alaska and Brazil appear to be similar sized. Brazil is in fact some six times larger. Mercator is a conformal projection. The great advantage of the Mercator projection is that it can be easily used for navigational purposes; one has only to draw a straight line between two locations on the projection and that line gives the necessary compass direction. The Mollweide projection, introduced in 1805, depicts all regions in correct relative size, but compresses high-latitude regions (Figure 2.16). Mollweide is an equal area projection. The third projection was introduced in 1963 by the geographer, Robinson, and was adopted in 1988 by the National Geographic Society (Figure 2.17). The Robinson projection has distortions both at the equator and at the poles and is only accurate at latitudes 38°N and 38°S. Because this projection provides a good compromise between geographic accuracy and aesthetic, it is used in this text.

applications in any context that may be concerned with spatial data. We are to encounter several applications later in this text; for the moment, it suffices to identify two general applications: measuring and analysing spatially distributed resources and managing spatially distributed facilities. Here are a few specific areas of application: biologists analyse the effect of changing land-use patterns on wildlife, geologists search for mineral deposits, market analysts determine trade areas, and defence analysts select sites for military installations.

The common factor throughout is that the data involved are spatial. Geographers have been handling spatial data for more than 2,000 years and they now have available important new capacities. In brief, a GIS achieves a whole new range of mapping and analytical capabilities — additional ways of handling spatial data.

REMOTE SENSING

The production of maps cannot be accomplished without data. GISs and analytical methods in general also require

---

Box 2.8: The power of maps

'The authoritative appearance of modern maps belies their inherent biases. To use maps intelligently, the viewer must understand their subjective limitations' (Wood 1993:85). While Box 2.7 outlines the unavoidable errors in maps that result from the need to employ a projection, this discussion centres on how maps reflect the assumptions of their creators and how they may be deliberately employed to convey specific misleading messages.

Recall the T-O maps introduced in Chapter 1; these medieval European maps clearly reflect Christian values with Jerusalem placed at the centre of the world. Other maps from that place and time show *terra incognita* to the south and make extensive use of decorative pictures. Nineteenth-century European maps reflect the political and economic concerns of their creators with colonial possessions, for example, prominently displayed. These maps, building on Ptolemaic traditions, established the conventions that most of us take for granted today; north at the top, 0° longitude running through Greenwich, England, and the map centred on either North America or western Europe. Never forget how arbitrary such decisions are and how influential they can be with their implications of a world centre and world outliers.

Other maps may convey a biased message either deliberately or because the map maker has some ulterior motive. As Monmonier (1991) suggests in his book, *How to Lie with Maps*, there are many ways to deceive. Political propaganda maps represent an extreme example. Less extreme, but significant nonetheless, are maps such as those of the Love Canal area in Niagara Falls, New York, that fail to explain the nature of, or in one case even identify, this notorious health hazard contaminated by hazardous waste (Figure 2.18).

The key message here is that 'a single map is but one of an indefinitely large number of maps that might be produced for the same situation or from the same data' (Monmonier 1991:2).

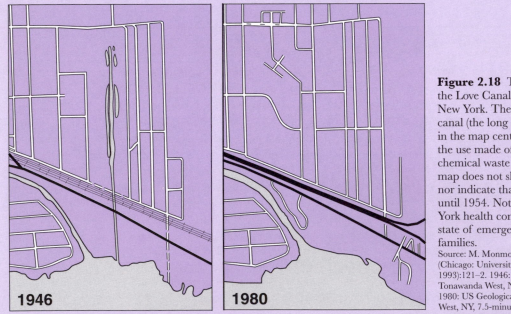

Figure 2.18 Two topographic maps of the Love Canal area, Niagara Falls, New York. The 1946 map shows the canal (the long straight vertical feature in the map centre), but does not show the use made of the canal as a dump for chemical waste since 1942. The 1980 map does not show the filled-in canal nor indicate that dumping continued until 1954. Note that in 1978 the New York health commissioner declared a state of emergency and relocated 239 families.
Source: M. Monmonier, *How to Lie with Maps* (Chicago: University of Chicago Press, 1993):121–2. 1946: US Army Map Service, 1946, Tonawanda West, NY, 7.5-minute quadrangle map. 1980: US Geological Survey, 1980, Tonawanda West, NY, 7.5-minute quadrangle map.

data. Where do these data come from? Early map makers obtained their data from explorers and travellers and, as we saw in Chapter 1, occasionally from imaginative sources. The contemporary geographer collects various data in different ways.

A particularly important method of data collection, actually a group of methods, involves detecting the character of an object from a distance. The term **remote sensing** describes the process of obtaining data using both photographic and non-photographic sensor systems. Actually we all possess remote sensors — our eyes — and one of our continuing aims has been to improve upon these as means of acquiring information. Improving our eyes (using telescopes), or improving our field of vision by gaining altitude (climbing hills or using balloons and aircraft), or improving our recording of what is seen (using cameras) are all advances in remote sensing. Today, most applications of remote sensing rely upon electromagnetic radiation to transfer data from the object of interest to the sensor. Using satellites as far as 36,000 km (22,370 miles) from the earth, we are now able to collect information both for the earth as a whole and what is invisible to our eyes. Electromagnetic radiation occurs naturally at a variety of wavelengths and there are specific sensing technologies for the principal spectral regions (visible, near-infrared, infrared, and microwave).

The conventional camera was the principal sensor used until the introduction of earth orbital satellites in the 1960s. Aerial photography is still used for numerous routine applications, especially in the visible and near-infrared spectral regions. The near-infrared spectral region has proved particularly useful for acquiring environmental data. Today, the emphasis is on satellite imagery, especially since the launching of Landsat (initially called ERTS — Earth Resources Technology Satellite) in 1972 as the first unmanned satellite. Satellite scanners numerically record radiation and transmit numbers to a receiving station; these numbers are used to computer generate pixel-based images. Satellite images are not photographs.

What are the principal advantages of satellite remote sensing? First, repeated coverage of an area facilitates analysis of land-use change. Thus the most recent Landsat satellite provides coverage of most of the earth's surface every sixteen days; these data are homogeneous and comprehensive. Second, because the data collected are in digital format, rapid data transmission and image manipulation are possible. Third, for many parts of the globe, these are the only useful data available. Finally, remote sensing allows the collection of entirely new sets of data; satellite data first alerted us to the changing patterns of atmospheric ozone in high-latitude areas.

For the human geographer, remote sensing is especially valuable in aiding understanding of human use of the earth; much of the discussion in Chapter 4 relies upon remotely sensed data. It is less useful if we are concerned with underlying economic, cultural, or political processes.

Most satellites have been launched by either the United States or by the European Space Agency. Canada anticipates launching a satellite called Radarsat in the mid-1990s.

Box 2.9: Using a GIS

Using a GIS can involve up to eight principal activities, depending on the specific application (Raper and Green 1989).

1. *Data capture:* Vector data are recorded by manual tracing or electronic digitizing of the spatial data. Raster data are recorded by a scanner that uses a light sensor to record the spatial data as 'on' pixels.
2. *Editing:* Once captured, the data are edited to correct errors, for example, relocating misplaced points in vector data or removing stray pixels in raster data.
3. *Structuring:* This involves storing the data in a form suitable for rapid spatial retrieval; a number of technical procedures are available to achieve this goal.
4. *Restructuring:* Changing the level of map detail or converting between vector and raster data types are known as restructuring.
5. *Manipulation:* If necessary, the spatial data can easily be manipulated. Indeed, one of the greatest advantages of a GIS is the ease with which it is possible to make revisions.
6. *Search or retrieval:* Another major benefit of working with a GIS is the ease of identifying and retrieving information if necessary.
7. *Analysis:* The ability to analyse, not merely represent, spatial data is perhaps the critical activity that a GIS is able to perform and is the clearest distinction between GIS and computer-assisted cartography. The quality of GIS software is often judged primarily in terms of this activity.
8. *Integration:* One output possible from a GIS is the integration of two or more maps of an area to form a single map; in the simplest terms, this involves a map overlay procedure that results in the creation of a new composite map.

Several of these activities are technical and are both prompting and responding to technical advances. Not surprisingly, given the difficulty of understanding the internal workings of a GIS, the mapped output must always be carefully interpreted and used.

The Earth Observing System platforms are scheduled to be placed in orbit from 1997 to 2010.

QUALITATIVE METHODS

Human geographers collect and analyse data using a wide range of **qualitative methods**, a term that is in popular use in other social sciences and that refers to examining the social world with a focus on attitudes and behaviour. Qualitative methods are a part of **ethnography**.

Much **fieldwork**, a traditional term for the methods that geographers use to obtain primary data by observation and questionnaires, is qualitative in character. Observation of landscapes was central to much early regional and cultural geography, but decreased in importance about 1960 as geographers tended to focus on secondary data. Recently, however, new types of fieldwork have appeared in association with humanistic concerns and human geographers now employ a range of qualitative methods for collecting and analysing data. Thus early fieldwork was not philosophically motivated, although it was implicitly empiricist because it assumed that reality was present in appearance, whereas contemporary qualitative fieldwork is a response to the humanistic requirement that human geography strive to discern the nature of the social world.

For the humanist, qualitative methods that involve a researcher's observation of and involvement in everyday life are central to an understanding of humans and human landscapes. **Participant observation**, a major method in anthropology and sociology, is now a popular geographic approach. The principal advantage of such a method is the explicit recognition that people and their lives do matter.

To conduct research employing qualitative methods requires considerable skill. A procedure such as participant observation does not provide any means to permit the

Box 2.10: Some qualitative resources

One human geographer used the concept of a qualitative resource, incorporating both methods and data sources, and has identified nine such resources (Holland et al. 1991). Commenting briefly on each of these helps provide an understanding of the breadth of contemporary humanistic research activity.

1. *Talking and listening:* The typical quantitative interview uses a well-structured questionnaire to ensure a careful adherence to specific questions. What such interviews necessarily fail to achieve is an understanding of the human world of the respondents. Interviews ought to be based on the world of the respondent rather than on the world of the academic geographer. At the very least, we ought to devise our questions only after acquiring some understanding of the world we are querying. In summary, talk with and listen to those you are interviewing. This is particularly necessary when the world under consideration is different from your own.

2. *A feminist perspective:* Feminist research centres on the social significance of gender and especially on the fact that power is unequally distributed between women and men. This perspective does influence the questions posed, and the concepts and techniques used to accomplish research.

3. *Literature:* Why do we change landscapes as we do? Some geographers have turned to literature, including poetry, to answer this question. We may find that the accounts of poets add an unexpected and valuable perspective reflecting different cultural attitudes. A comparison of Aboriginal and European settler poetry in any area of European overseas expansion may be enlightening.

4. *Cartoons:* Cartoons may be used as one indication of larger social commentary on many issues. Although, like literature, cartoons are usually the work of an individual, they may reflect a consensus attitude on a political, environmental, or economic issue.

5. *Photographs:* Historians frequently turn to photographs as an indicator of the past and geographers are increasingly acknowledging the value of this resource. As records of past landscapes, they may be invaluable.

6. *Art:* Many artists try to reflect the distinctive character of a particular time and place and in some cases try to reflect a vision of some larger, possibly national, identity. Combining photographs and art might provide clear insights into people and places. What locations are chosen by photographers as opposed to artists and how do their visions differ?

7. *Popular music:* Much music reflects place — blues and the American south, for example. As of the early 1990s, much popular music is a direct reflection of inner-city urban experiences, highlighting both social and environmental ills.

8. *Buildings:* Buildings are a principal feature of human landscapes, indeed a dominant feature in urban areas. Different styles and functions reflect larger past and present geographies.

9. *Monuments, memorials, and cemeteries:* These are important landscape components that, like several of the other resources, are able to reveal much about our past. A careful analysis of these landscape features will inform us about the cultural identities of their creators.

What can we learn from this account of nine of many qualitative resources? Quite simply that our human world can be read by the skilled reader.

researcher to control the relationship between observer and observed. You will recall that this was one of the key issues in our discussion of the differences between humanism and positivism. It may be the case that the researcher, who is often of a higher social status, exhibits **ethnocentrism** or that the researcher identifies with the group being observed and becomes their advocate. Other disadvantages of these procedures include the possibility that the researcher approaches data collection with an inappropriate initial view that subsequently distorts the data, and the possibility that the subjects studied are not representative. You will have an opportunity to assess the quality of some of this work, especially in chapters 8 and 11.

QUANTITATIVE METHODS

Some fieldwork is explicitly quantitative in character, notably the use of a **questionnaire** to survey people. Like the traditional qualitative fieldwork procedure of observation, questionnaires are a part of an empiricist research activity. Unlike qualitative fieldwork, however, a questionnaire asks all individuals the same questions in the same way. The value of these questionnaire results is related to the response rate achieved and to the way potential respondents are selected, namely the **sampling** method. Proper sampling methods, based on statistical sampling theory, allow the results of a sample to be treated as representative of the population, within certain error limits. The most common means of selecting respondents is by using random sampling.

During the 1960s, **quantitative methods** in general developed extensively in association with the spatial analytic school and the general acceptance of a positivist philosophy. The principal methods used were statistical and the purposes were to describe data (by calculating a mean and a standard deviation, for example) and to test hypotheses generated by theory (using correlation tests, for example, as in Box 2.2). We encounter applications of these and other tests in Chapter 10.

The spatial analytic school early recognized that models could play a much increased role in analysing data. A **model** is an idealized, simplified representation of the real world that highlights its key properties and eliminates incidental information. Many of the earliest spatial models employed by human geographers were based on generalizations about the relationships between the distribution of geographic facts and distance. We encounter some of these classic spatial models in chapters 10, 11, 12, and 13.

Geographers use quantitative techniques for a wide variety of purposes, but especially for analysing relationships between spatial patterns and classifying data. Describing relationships is fundamental in producing explanations and is usually broached by proposing a functional relationship such that one variable is dependent on one or more other variables. The relationship specified is, ideally, derived from appropriate theory in accord with the scientific method outlined earlier. Classifying imposes order on data, and a number of techniques facilitate the activity.

A CONCLUDING QUESTION

We conclude this chapter with a provocative, albeit somewhat crude, distinction. Positivism and associated quantitative procedures have a tendency to exclude humans from research because of an emphasis on aggregate data; critics label such work as dehumanized human geography. This may be labelled as a geography *of* people. Humanism and associated qualitative procedures, on the other hand, emphasize the integration of researcher and researched, thus generating a geography *for*, and possibly even *with*, people.

We do not, however, need to debate the relative merits of differing research strategies; we simply need to acknowledge that our contemporary human geography is an exciting and diverse discipline incorporating a variety of some useful philosophies, concepts, and techniques that complement each other well and that combine to offer an enviable range of procedures.

SUMMARY

The importance of philosophy
The subject matter of human geography, the questions asked, and the aids used in obtaining answers are best understood by reference to the appropriate underlying philosophy. Most human geography is explicitly or implicitly guided by a particular philosophical viewpoint.

Philosophical options
Contemporary human geography utilizes four principal philosophies. Empiricism gives priority to facts over generalizations; traditional regional geography, although typically presumed to be aphilosophical, was actually empiricist in character. Positivism argues for use of the scientific method, similar to that applied in physical science, and purports to be objective. The approach known as spatial analysis is prompted by positivistic logic. Humanism is a set of subjective philosophies, including pragmatism and phenomenology, with a focus on humans. Early cultural geography had some humanistic content, but these philosophies emerged as major ideas about 1970. Marxism is a loosely structured philosophy that aims to understand and change the human world. Work inspired by Marxist thought tends to centre on social and environmental problems.

Human geographic concepts
In order to increase geographic knowledge, a great many concepts are used. We constantly refer to locations (where a

geographic fact is present), to place (the quality of a given location), to region (a grouping of similar locations or places), and to distance (the interval, however measured, between locations or places). Geographic research is at specific spatial, temporal, and social scales. Explaining and understanding landscape change is often enhanced by considering both diffusion, the spread and growth of geographic facts, and perception, the way in which landscapes are viewed by individuals and groups. Analyses of landscapes frequently distinguish different degrees of development.

Techniques of analysis

Collecting and handling geographic data can be accomplished by using a variety of techniques. Maps are used to store, display, and analyse data and can be produced manually or by computer. Geographic information systems are computer-based tools that analyse and map data and have a wide range of applications. A major method of collecting data involves remote sensing techniques, such as aerial photography and satellite imagery. Fieldwork, both observation and interviewing, are essentially qualitative procedures for collection and analysis of data and is favoured by humanists. The use of models and quantification is associated with positivistic spatial analysis and involves abstraction and empirical testing of ideas.

WRITINGS TO PERUSE

BIRD, J.H. 1989. *The Changing Worlds of Geography: A Guide to Concepts and Methods*. New York: Oxford University Press.
This is a comprehensive account of concepts and methods often discussed in great detail.

EYLES, J., ed. 1988. *Research in Human Geography: Introductions and Investigations*. Oxford: Blackwell.
Entertaining series of statements by leading geographers concerning research inspirations and practices; includes examples of humanistic and applied work.

GOODCHILD, M.F. 1988. 'Geographic Information Systems'. *Progress in Human Geography* 12:560–6.
A brief history and assessment of geographic information systems that helps place this major growth area in an appropriate perspective; avoids technical terminology.

GOULD, P., and R. WHITE. 1986. *Mental Maps*, 2nd ed. Boston: Allen and Unwin.
Highly readable and well-researched discussion of the topic; the inspiration for this original research is one of the stories told in Eyles (1988).

HAGERSTRAND, T. 1967. *Innovation Diffusion as a Spatial Process*. Translated by A. Pred. Chicago: University of Chicago Press.
A highly influential analysis with an excellent review of diffusion studies by the translator; this is a classic cultural geographic problem approached for the first time in an essentially positivist manner.

JOHNSTON, R.J. 1986. *Philosophy and Human Geography: An Introduction to Contemporary Approaches*, 2nd ed. London: Arnold.
This book is a clear readable statement of our various philosophies; employs a tripartite division into positivism, humanism, and Marxism.

———, D. GREGORY, and D.M. SMITH, eds. 1994. *The Dictionary of Human Geography*, 3rd ed. Oxford: Blackwell.
A remarkably comprehensive, detailed, and clear dictionary with major entries often treated in an almost encyclopedic fashion; highly recommended.

PICKLES, J. 1985. *Phenomenology, Science and Geography*. New York: Cambridge University Press.
A detailed and thoughtful appraisal of phenomenology as applied in geographic research.

QUANI, M. 1982. *Geography and Marxism*. Oxford: Blackwell.
A useful introduction that focuses primarily on European rather than North American work; somewhat dated now, but still a valuable source of information and ideas.

RELPH, E.C. 1976. *Place and Placelessness*. London: Pion.
A good example of writing from a phenomenological perspective.

ROBINSON, A.H., and B.B. PETCHENIK. 1976. *The Nature of Maps: Essays towards Understanding Maps and Mapping*. Chicago: University of Chicago Press.
A highly readable book full of useful ideas. The first author devised the projection used for world maps in this text.

SCHMITT, R. 1987. *Introduction to Marx and Engels: A Critical Reconstruction*. Boulder: Westview Press.
Although not written especially for human geographers, this is an exceptionally clear and comprehensive introduction to this challenging school of thought.

STODDARD, R.H. 1982. *Field Techniques and Research Methods in Geography*. Dubuque: Kendall Hunt.
A comprehensive book covering such topics as data collection and analysis; provides clear guidelines concerning how to conduct positivist research.

STODDART, D.R. 1986. *On Geography*. Oxford: Blackwell.
A highly personal exploration of the origins and character of geography that rejects any division between physical and human geography.

WATSON, J.W. 1955. 'Geography: A Discipline in Distance'. *Scottish Geographical Magazine* 71:1–13.

A landmark statement anticipating the spatial analysis of the 1960s and highlighting the 'distance factor'.

WATSON, M.K. 1978. 'The Scale Problem in Human Geography'. *Geografiska Annaler* 6OB:36–47.

A useful summary of this all-important topic that is a valuable read for all geographers.

3

The earth: A human environment

THE EARTH is a habitable planet for humans — the only such planet known to us. By habitable, we mean that both physical processes and physical environments permitted the emergence and subsequent spread and growth of human life. Indeed, only a series of quite remarkable coincidences have made earth habitable for humans. Our distance from the sun, the axial tilt, the twenty-four-hour rotation and the 'minor' variations in relief all combine to spread the energy received from the sun relatively evenly. The evolution of the earth involved the presence of water to form oceans, an average surface temperature of 15° C, and the presence of an atmosphere. Interestingly, the first life on earth evolved in water and it was the subsequent movement onto land that converted the initial carbon dioxide atmosphere into an oxygen-rich one. This atmospheric transformation included the creation of the **ozone layer** — ozone is a form of oxygen — that prevents the penetration of ultraviolet rays that would sterilize the land surface. It was also the oxygen-rich atmosphere that allowed the emergence of animal life.

From our earthbound perspective, the surface of the earth is tremendously varied. There is an always-changing atmosphere, mountains as high as 8,854 m (29,048 ft) above sea level (Mount Everest), land areas 408 m (1,338 ft) below sea level (the shores of the Dead Sea), and known ocean depths 10,927 m (35,849 ft) below sea level (south of the Mariana Islands in the Pacific Ocean). Our restless planet frequently changes dramatically — volcanic eruptions, earthquakes, and tidal waves are examples. But, on another level, these variations and dramatic changes are minor. Were we able to hold the earth in our hands, it would feel as smooth as a peach.

There are many relationships between physical and human worlds at all spatial scales and at all times since the first human life appeared. To provide an understanding of these important relationships, this chapter gives an overview of the physical earth. There is a discussion of planet earth, an acknowledgement of the range of physical processes — including those that make landscape — and an account of the diversity of contemporary physical environments. This overview of the physical earth is followed by discussions of the origins of life, human origins, the unity of the human **species**, and the mistaken ideas behind racist thinking.

PLANET EARTH

The earth is approximately spherical in shape with an equatorial circumference of 40,067 km (24,897 miles) and polar circumference of 40,000 km (24,855 miles). This shape results from gravity, a force that is also responsible for the tendency of the earth's surface area to be layered; a dense substance, rock, is at the bottom, a less dense substance, water, is often above the rock, and the least dense substance, air, is above the other two. Such an arrangement is crucial for human life, which is thus allowed access to all three essential substances.

Our spherical earth is involved in two principal movements. First, it rotates on its axis, one rotation defined as a day. Second, it orbits around the sun with one revolution defined as a year. Rotation imposes a diurnal cycle on much life. Revolution, in conjunction with the axis of the earth tilted at 23.5°, produces seasonal climatic changes and variations in length of daylight.

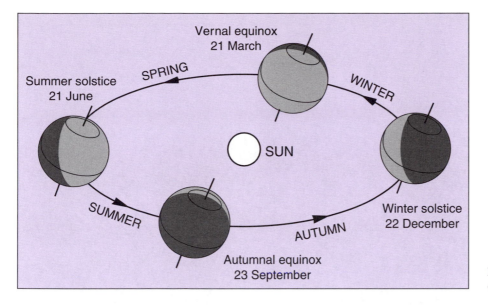

Figure 3.1 Revolution of the earth around the sun.

ROTATION

The end points of the axis of rotation are defined as the poles, north and south. All locations on the earth's surface, except the poles, move as the earth rotates and one circuit of the earth is a line of latitude. Because the earth is a sphere, the longest line of latitude lies midway between the poles, the equator. Lines from one pole to the other are lines of longitude. Unlike lines of latitude, these are not parallel; rather, they converge at the poles and are farthest apart at the equator. Lines of latitude are measured in degrees with the equator defined as 0° and the poles therefore as 90°N and 90°S; lines of longitude are similarly measured in degrees, in this case from an arbitrary 0° line known as the prime meridian that runs through the Greenwich Observatory in London, England. The geographic grid that is commonly used to describe locations is based on latitude and longitude.

REVOLUTION

Revolution of the earth around the sun is the second important movement. Figure 3.1 illustrates the four seasonal positions of the earth relative to the sun and introduces the terms 'equinox' and 'solstice'. The summer **solstice** occurs on 21 or 22 June in the Northern hemisphere and on 21 or 22 December in the Southern hemisphere. This is the time when the hemisphere reaches maximum inclination towards the sun so that at latitude 23.5°, the sun is directly overhead. The winter solstice occurs on 21 or 22 December in the Northern hemisphere and on 21 or 22 June in the Southern hemisphere. This is the time when the hemisphere reaches maximum inclination away from the sun. The two **equinoxes**, vernal and autumnal (spring and fall) are the halfway times between the solstices, and on both occasions the axial inclination is exactly at 90° with respect to the sun.

Much human activity is related to these global physical factors. Diurnal, seasonal, and annual cycles are basic to much social and economic organization. Durations of the day and year are defined by earth movements, while the month is defined by lunar movement around the earth. The hour is an arbitrary human division. Another arbitrary definition is the use of the equinoxes and solstices as the beginning of seasons.

PHYSICAL PROCESSES

THE EARTH'S CRUST

The crust or outer layer of the earth is composed of rocks, aggregates of minerals, that have been created and changed for at least 3,000 million years. It is usual to classify the many rock types as sedimentary — formed at low temperatures on the earth's surface; as igneous — solidified from a molten state; or as metamorphic — rocks that have changed under high temperature and pressure. But why are these rocks distributed as they are?

One of the most important features of the earth's surface is that it is part land (29 per cent) and part water (71 per cent). The present distribution of these two is the result of a long ongoing process of movement by the **lithosphere** — the earth's crust and the uppermost mantle below the crust. This lithosphere is divided into six major and several minor plates that move relative to one another, movements that are caused by other movements deep inside the earth in the liquid mantle. Our continents quite literally float in this liquid mantle and have some four-fifths of their mass below the surface. Suggestions that the land surfaces, the continents, move appeared as long ago as the seventeenth century when Francis Bacon (1561–1626), an English philosopher, observed that the Atlantic coasts of South America and Africa fitted together like a jigsaw. Early scientific evidence was gathered by Wegener (1880–1930), a German meteorologist, but it was not until the 1960s that this idea of moving continents,

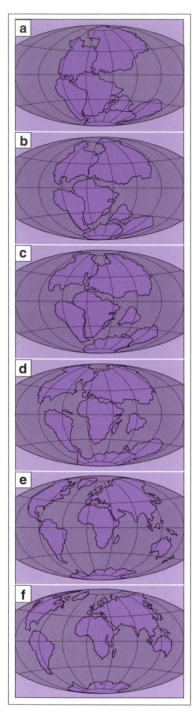

3.1 *(above)* A mountain landscape, Banff, Alberta, in the Canadian Rockies. (Canadian Pacific)

Figure 3.2 *(left)* In (a), beginning 222 million years ago, there is one land mass, a supercontinent called Pangea. By 180 million years ago, (b) Pangea is breaking up; there is a movement north of Laurasia while the southern land mass, Gondwanaland, is breaking away. In (c) the process continues, with the beginnings of the north and south Atlantic Oceans 135 million years ago. In (d) Madagascar has separated from Africa 65 million years ago; India is moving towards Asia and Australia remains linked to Antarctica. The present distribution of land and sea is shown in (e), while (f) shows a likely scenario 50 million years in the future. Australia continues to move north; there is an eastward movement of East Africa, a separation of California (west of the San Andreas fault) from North America, and a smaller Mediterranean, as Africa moves north.

known as **continental drift**, achieved scientific credibility. Figure 3.2 illustrates the positions of the continents since outward movements began about 225 million years ago (mya). At the present time, the Atlantic Ocean is widening about 2 cm (3/4 in) each year; Australia is moving northward, as is Africa, while parts of Africa and western North America may separate from their respective continents.

A second consequence of moving continents is the creation of mountain chains, such as the Himalayas. When continents collide, as India and Asia did perhaps 150 mya,

mountain chains are thrust upward. The Appalachians are much older, resulting from a collision of North America and Africa perhaps 350 mya.

Earthquakes are a third consequence of crustal movement when parts of the crust move over other parts as a result of movements in the liquid mantle. Once stress exceeds rock strength, a sudden movement — an earthquake — occurs. Such movements typically take place along fault lines, such as the San Andreas fault, that are lines of crustal weakness. Volcanoes also occur along lines of crustal weakness, often plate boundaries. A volcanic eruption involves liquid from the mantle, magma, being forced upwards and onto the earth's surface (Box 3.1).

As already noted, much of the earth's crust is covered by water — approximately two-thirds. Beneath the water is a physical landscape that is ever changing, while the water itself is also involved in a series of movements. Waves are caused by wind; tides result from the gravitational pull of the moon (and to a lesser extent the sun). The most significant movements, however, are ocean currents (Figure 3.3). Warm water moves away from the equator and cold water moves towards the equator.

Not all of the earth's water is in oceans; some is in the form of ice, a leftover from the most recent glacial period. Most of this ice is in two continental ice sheets, the Antarctic and Greenland sheets that cover about 10 per cent of the land surface and reach thicknesses of several kilometres. As this ice gradually melts, the sea level rises. Some 20,000 years ago (ya), the sea level was 100 m (328 ft) lower than today and the current rise is about 1 cm (0.39 in) a year.

The physical processes and circumstances identified so far are global and/or long-term, but they are important to us. Understanding past migration of life-forms, for example, requires an appreciation of land movement and climatic change. One example of the possible long-term effect of a physical process — exacerbated by human activity — is the melting of all ice that would result in a sea level rise of 60 m (197 ft). Such a rise would flood many of the largest cities in the world.

MAKING THE PHYSICAL LANDSCAPE

The physical landscape on which human activities take place and that humans continually modify is a product of a range of processes. Crustal movement and the effects already noted are followed by weathering and gradational processes that combine to produce specific *land-forms*. Weathering is the reduction in size of surface rock, while gradation is the movement of surface material under the influence of gravity, and typically involves water, ice, or wind. In addition to creating land-forms, these processes combine with the activities of living organisms to develop soil. Plant and animal life in turn are closely related in a circular fashion to the landscapes created.

Weathering occurs when rock is mechanically broken and/or chemically altered. Temperature change is an important cause of mechanical weathering, while water is the principal cause of chemical weathering. Once rocks are weakened in this way, they become especially susceptible to gradational processes. Water, ice, and wind can erode, move, and deposit material. Many of our distinctive physical landscapes are clear evidence of these processes at work.

Water is probably the most important cause of gradation. A continuous transfer of water from sea to air to land to sea is taking place. River erosion creates V-shaped valleys and waterfalls, while river deposition creates flood plains and deltas. Sea water is also affecting land surfaces as coastlines undergo constant change; cliffs are evidence of erosion, while beaches result from deposition. Other parts of the earth's surface show the results of ice and snow; during the

Box 3.1: Earthquakes and volcanic eruptions

If we were able to peel away the thin shell of air, water, and solid ground that is the outer layer of the earth, we would find that our planet is a furnace. Many of our major cities are a mere 35 km (22 miles) above this furnace and in the deep ocean trenches, the solid crust above the furnace is as little as 5 km (3 miles) thick. Earthquakes and volcanic eruptions, not surprisingly, are regular occurrences. There are about 200–300 earthquakes each year in Canada alone, while several major volcanoes have recently made headlines (Mount Pinatubo in the Philippines and Mount Unzen in Japan, both in 1991). Perhaps the most dramatic such event in recent times was the 1963 birth of an island, Surtsey, near Iceland in the north Atlantic. Surtsey was the result of a massive eruption of molten rock through a split in the crust. Over the course of a few months, 300 million m³ (almost 400 million cu yd) of lava, twice as much ash, and large amounts of carbon dioxide and steam literally poured out of the sea.

Needless to say, contemporary scientists are continually attempting to improve our ability to predict earthquakes and volcanic eruptions. We know the likely locations of 95 per cent of earthquakes and most volcanoes occur at the margins of the tectonic plates when plates move against each other, but specific predictions are more elusive. After two major earthquakes in central Asia in the 1950s (20,000 died in Ashkhabad and 12,000 in Tajikistan), scientists began to focus on prediction. Today, some earthquakes are correctly predicted while others, tragically, are not. The 1978 Oaxaca earthquake in Mexico was predicted, while a 1976 earthquake in Tangshan, China, was not and 250,000 died.

Predicting volcanic eruptions is even more difficult. Although we know where they are located, we do not know when an eruption will occur. One of the best-known eruptions was that of Vesuvius in AD 79 when the town of Pompeii was covered by 3 m (10 ft) of hot cinders, pumice stone, and burning rock. Since then, there have been many other eruptions of Vesuvius, most recently in 1906 and 1944. The 1980 eruption of Mount St Helen in Washington state was successfully predicted (Arkell 1991).

3.2 *(above)* Davidson Glacier, Alaska, USA. (Canadian Pacific)

3.3 *(right)* The Grand Canyon of the Colorado River, Arizona, USA. The world's best-known example of a gorge or canyon caused by fluvial erosion. It has a mean depth of 1.6 km (1 mile) and is 32 km (20 miles) long. (Ewing Galloway)

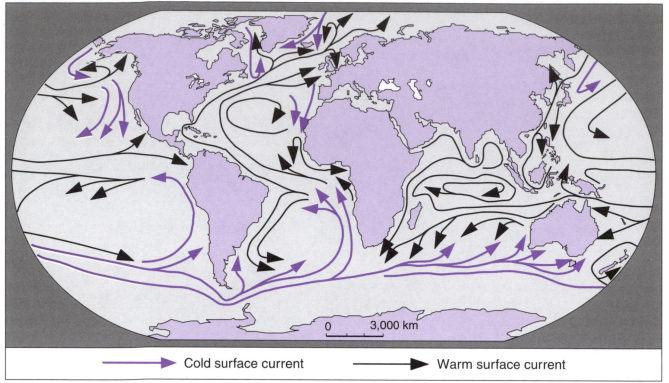

Cold surface current　　　**Warm surface current**

Figure 3.3 *(top)* Major ocean currents: The prevailing winds and the rotation of the earth combine to cause gyres (large whirlpools). These gyres move warm water away from the equator and cold water towards it. There are five major gyres and thirty major currents.

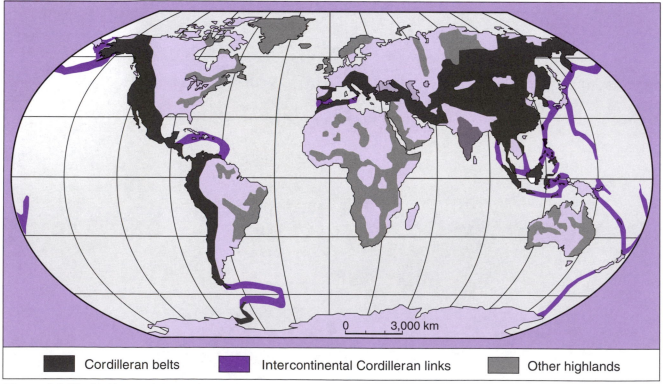

Cordilleran belts　　　**Intercontinental Cordilleran links**　　　**Other highlands**

Figure 3.4 The backbones of the continents, these global Cordilleran belts also extend through ocean areas.

Source: Adapted from R.C. Scott, *Physical Geography.* (New York: West, 1989):295.

most recent glacial period, ice covered an area approximately three times greater than that covered today, including northern Europe and Canada. Areas formerly covered by ice show such distinctive erosional features as U-shaped valleys and such evidence of deposition as drumlins (low hills). Wind erosion results from wind-blown materials colliding with stable materials (sandblasting) or more often by the removal of loose particles. Sand dunes are one of the best-known results of wind deposition of material.

Each of these processes is affecting parts of the earth's surface at all times (Box 3.2). Physical landscapes, then, are subject to continuous modification, almost a fine tuning. In principle, the eventual outcome, without crustal movement or climatic change, would be an almost featureless physical landscape.

Our physical environment is of course much more than the distribution of land-forms. Soils, vegetation, and climate are additional components of a physical environment. Figures 3.4, 3.5, 3.6, and 3.7 provide a generalized view of global land-forms, soils, vegetation, and climate respectively.

Figure 3.4 maps the major *cordilleran belts*, the backbones of the continents. The slope and ruggedness of local land-forms in particular are major considerations in human decisions about the location of economic activities, such as agriculture, industry, and transport.

Soil types are distributed in a systematic fashion (Figure 3.5). There are close general relationships to both climate and natural vegetation, two of the most important controls on soil. Soils are important for their agricultural potential and capacity to support buildings. Much of the European movement to the New World was closely tied to the availability of soils suitable for commercial agriculture, for example, the late nineteenth-century expansions into the grassland soils of Canada, Argentina, and Australia.

Figure 3.6 maps the global distribution of *natural vegetation*. The four principal types are grasslands, forests, desert shrubs, and tundra. Much of the history of agriculture is intimately tied to their distribution. In circumstances of limited agricultural technology, for example, forested areas can be a significant barrier.

Climate is the characteristic weather of an area over an extended period of time and can be usefully summarized by reference to mean monthly and annual statistics of temperature and precipitation. These means are a response to

---

Box 3.2: The geomorphology of Canada

Canada is the product of three primary geologic developments. First, the formation of the Canadian or Laurentian Shield, which was formed about 2,000 mya as a result of the collision and welding together of seven ancient microcontinents. Much of the shield consists of Archaean rocks 2,500 my old (mostly granite or granite gneiss). Second, there are three regions where mountains developed around the shield margins as a result of sediment accumulation in long, narrow basins: these are the Cordilleran mountain system in western Canada, which has a long and complex history related to the presence of an active continental margin (including volcanic and earthquake activity); the Appalachian-Acadian system in eastern Canada, which is older than the Cordilleran and which has been substantially lowered by erosion; and the Innuitian system, across the High Arctic from Alaska to northern Greenland. Third, the deposition of sediments in shallow seas in the intervening areas, often called epeiric seas. This third geologic development involves accumulation of chemical and organic sediments about 500 mya.

These three structural developments, the tectonic history, help us understand the various physiographic regions of Canada; the shield, the three mountain areas, and the several lowlands and plains that resulted from deposition in epeiric seas (such as the prairies and the St Lawrence lowland). But the landscapes of each of these regions also developed through time as weathering and gradational processes, closely related to climate, affected the physical surface. Most important for

3.4 The landscape of the Canadian Shield, an undulating, lake-dotted area with bare rock, thin soils, and muskeg. (Ontario Ministry of Natural Resources)

much of Canada is the impact of glacial periods; almost all of Canada was under ice at some time during the last 1.5 my (only 1 per cent is under ice today), and Canada was the site of the largest ice sheet in the Northern hemisphere during the most recent glacial period that ended about 10,000 ya. As a result many of Canada's land-forms are primarily glacial in origin (Trenhaile 1990).

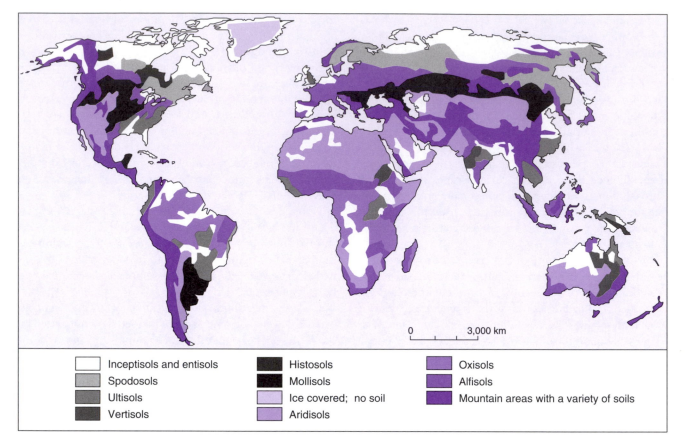

Inceptisols and entisols	Histosols	Oxisols
Spodosols	Mollisols	Alfisols
Ultisols	Ice covered; no soil	Mountain areas with a variety of soils
Vertisols	Aridisols	

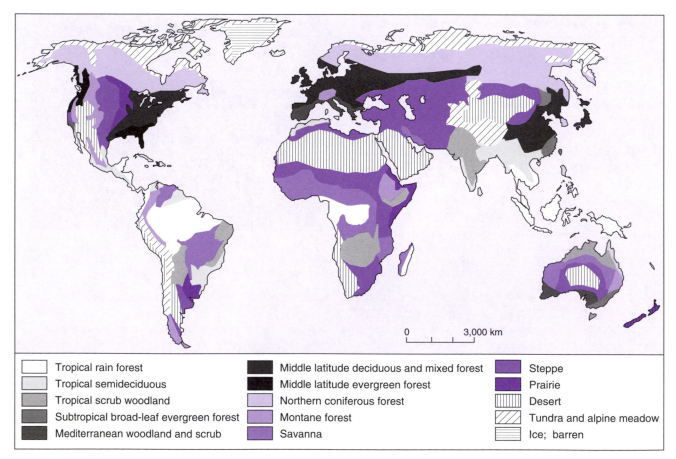

Tropical rain forest	Middle latitude deciduous and mixed forest	Steppe
Tropical semideciduous	Middle latitude evergreen forest	Prairie
Tropical scrub woodland	Northern coniferous forest	Desert
Subtropical broad-leaf evergreen forest	Montane forest	Tundra and alpine meadow
Mediterranean woodland and scrub	Savanna	Ice; barren

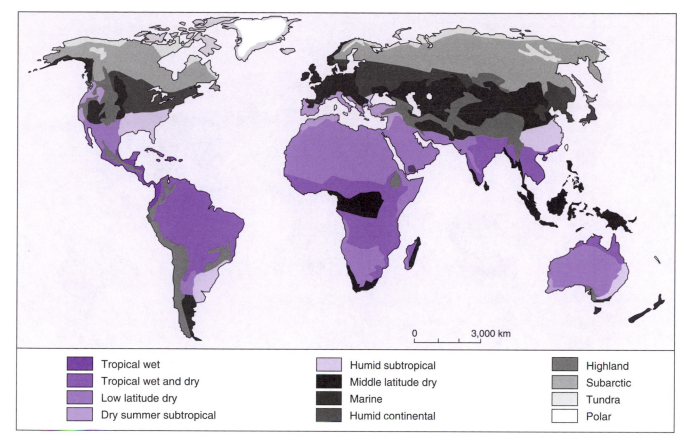

⬛ Tropical wet	⬛ Humid subtropical	⬛ Highland
⬛ Tropical wet and dry	⬛ Middle latitude dry	⬛ Subarctic
⬛ Low latitude dry	⬛ Marine	⬛ Tundra
⬛ Dry summer subtropical	⬛ Humid continental	☐ Polar

Figure 3.5 *(top opposite page)* Global distribution of soil types. Oxisols are typical of tropical rain forests. Ultisols, in the humid tropics, are low in fertility. Alfisols, dispersed throughout low- and middle-latitude forest/grassland transition areas, are of moderate to high fertility. Spodosols are associated with the northern forests of North America and Eurasia; low fertility is typical. Mollisols, associated with middle-latitude grasslands, are often very fertile. Aridisols, located in areas of low precipitation, lack water. Vertisols, limited to areas of alternate wet and dry seasons, may be very fertile. Histosols are most common in northern areas of North America and Eurasia. Inceptisols and entisols are immature soils found in environments that are poorly suited to soil development.
Source: Adapted from R.C. Scott, *Physical Geography* (New York: West, 1989):276–7.

Figure 3.6 *(bottom opposite page)* Global distribution of natural vegetation. Differences in forest types are related to climate. As they move away from the equator, forests become less dense and have fewer species. Deciduous trees dominate the middle latitudes, evergreens the higher latitudes. Savanna, steppe, and prairie are composed chiefly of grasses. Desert vegetation is closely linked with an arid climate, while tundra is the most cold-resistant type of vegetation.
Source: Adapted from R.C. Scott, *Physical Geography* (New York: West, 1989):232–3.

Figure 3.7 *(above)* Global distribution of climate. These types of climate, distinguished on the basis of moisture and temperature, are derived from the Köppen classification. Tropical wet, tropical wet and dry, and low-latitude dry climates may be grouped as low-latitude climates; dry summer subtropic, humid subtropical, mid-latitude dry, marine, humid, and continental climates as middle latitude climates; and subarctic, tundra, and polar climates as high-latitude climates.
Source: Adapted from R.C. Scott, *Physical Geography* (New York: West, 1989):158–9.

latitude. In Figure 3.7, it is clear that climatic distributions are generally linked to latitude. Three low-latitude, two subtropical, three mid-latitude, and three high-latitude climates are mapped along with a highland climate. Low-latitude climates are characteristically hot; the distinguishing factor of each of the three is precipitation. The two subtropical climates are less hot, and again the distinguishing feature is precipitation. The three low-latitude climates lack any significant seasonality. The subtropical and mid-latitude climates vary according to the degree of continentality and, of course, latitude. All five of these climates have seasonal variations. Proximity to oceans tends to increase precipitation

and reduce temperature variations. The three high-latitude climates are cold most of the year and have low precipitation. Finally, highland climates are a diverse group with both temperature and precipitation affected by altitude in addition to latitude. These climates display many local variations.

Climate is clearly a key consideration in many human decisions. But *weather* is also important. Weather can have a major impact on agricultural yields. Atypical or violent weather can cause floods or droughts. For example, hurricanes — vast tropical storms — can cause major damage to populated land areas.

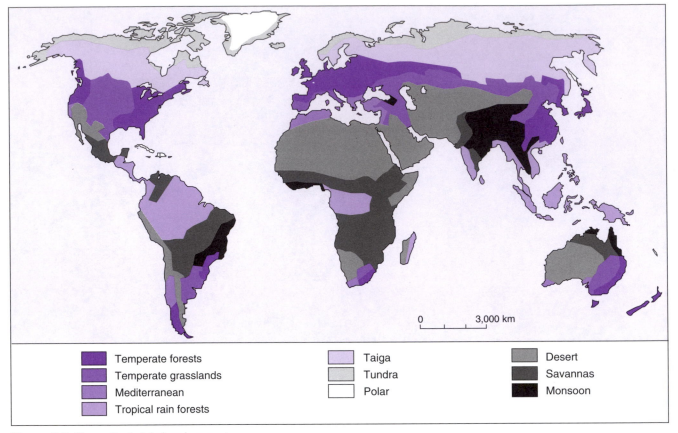

Figure 3.8 Generalized global environments.

GLOBAL ENVIRONMENTS

The preceding section has highlighted a wide variety of physical processes and environments. Their relationship to human activities is often intimate, but these physical factors do not in any sense determine our human activities. We need to be aware of physical factors at all levels, but we must not assume any causal relationships. As the remainder of this book demonstrates, we humans make the decision, albeit with many physical factors in mind! In this section, a brief summary of the major global environments (Figure 3.8) is provided to integrate much of the earlier material in a more applied human context.

The notion of global environmental regions is a most useful one and builds on the material presented in earlier sections of this chapter. Indeed, the regions defined are closely correlated with climatic regions in particular. Humans have favoured certain environments and have made changes to all areas they have settled in.

TROPICAL RAIN FORESTS

These areas are located close to the equator and experience high rainfall — typically exceeding 2,000 mm (78 in) per annum, and high temperatures — 24° C to 30° C (75° F to 86° F). They lack seasonal variation. Broad-leaf trees dominate and the vegetation is dense and highly varied. Despite this, once the vegetation is removed, the soil is characteristically shallow and easily eroded. Tropical rain forests have not become major agricultural regions. There are a number of economically valuable forest products, such as mahogany and rubber, and a number of starchy food plants, such as manioc, yams, and bananas.

MONSOON AREAS

Monsoon areas are characterized by marked seasonality, especially an extended dry season and heavy rainfall. Rainfall may be as high as 2,000 cm (780 in) in some areas. The typical natural vegetation is deciduous forest, but all monsoon areas have been dramatically changed as a result of human activity. Agriculture has been practised for a long time with rice as the basic crop in many of the areas. Population densities are usually high.

SAVANNAS

Savannas are located to the north and south of the tropical rain forests. The distinctive climate involves a very wet season and a very dry season. The natural vegetation is a combination of trees and grassland. Africa has the most extensive savanna areas, which are rich in animal species. To date, most savannas are relatively undeveloped economically, but they offer great promise for increases in animal

3.5 *(left)* Workers in rice paddies, south China.

3.6 *(below)* Herding cattle, Brazil. (World Bank)

3.7 Farming the Canadian prairies, near Regina, Saskatchewan, Canada. (Canadian National Railways)

and crop yields. Important grain crops include sorghum in Africa and wheat in Asia.

DESERTS

A desert area is one that has low and infrequent rainfall, perhaps as low as 100 mm (4 in) per annum and perhaps no rain for ten years. Areas are typically defined as arid if they receive less than 255 mm (10 in) of rain per annum and as semiarid if they receive less than 380 mm (15 in). There are hot, cold, sandy, and rocky deserts. In all deserts, water is a crucial consideration in human activities. Irrigation and nomadic pastoralism are two human responses to desert environments.

MEDITERRANEAN AREAS

There are only a few limited Mediterranean environments. They are characterized by long, hot, dry summers and warm, moist winters. The typical location is on the western side of a continent in a temperate latitude. Evergreen forests are the usual natural vegetation, although human activity means that scrub land prevails today. Important crops, especially in the Mediterranean proper, are wheat, olives, and vines.

TEMPERATE FORESTS

Much of Europe and eastern North America has a temperate climate with a distinct seasonal cycle and a natural vegetation comprised primarily of mixed deciduous trees and fertile soils. The seasonal cycle gives this area its distinctive character. The many animals and plants have adapted to the variations in heat, moisture, food, and light. Annual loss and growth of leaves is an especially striking adaptation that affects all other forest life forms. Particularly since the beginnings of agriculture, these areas have been attractive to and significantly altered by humans and are now dominated by agriculture, industry, and urbanization. These changes began in Europe about 7,000 ya, in Asia about 6,000 ya, and in North America about 1,000 ya. Indeed, the removal of forest cover has been so dramatic that today there is a general recognition of the need to preserve and even expand the remaining forests.

TEMPERATE GRASSLANDS

North American prairies, Russian steppes, South African veld and Argentina pampas are four of the large temperate grasslands. Located in continental interiors, they have insufficient rainfall for forest cover and tend to have great annual temperature variations. Such grasslands were once occupied by a particular large migrating herbivore such as the bison (North America), antelopes, horses, and asses (Eurasia), and pampas deer (South America). After remaining almost devoid of humans until the late seventeenth century, these grasslands became the granaries of the world in the twentieth century. Their soils are typically suited to grains; wheat became dominant in most areas.

BOREAL FORESTS

Boreal forests, often called taiga, are located only in the Northern hemisphere — in North America and Eurasia — and have long, cold winters, brief, warm summers, and limited precipitation. The natural vegetation is evergreen forest. Agriculture is very limited and the major economic importance of taiga is for forest products and minerals.

TUNDRA

Located north of the boreal forest, this area experiences long, cold winters; temperatures average about –5° C (23° F) and there are few frost-free days. Perennially frozen ground — permafrost — is characteristic and proves a severe challenge to human activities. There is a paucity of plants and animals adapted to this environment because the land has been exposed for only about 8,000 years since the last retreat of ice-caps. There has been little time even for soil to form.

POLAR AREAS

Polar conditions prevail in both hemispheres and are best represented by the Greenland and Antarctic ice sheets. Antarctica is a continental land mass under an ice sheet that reaches thicknesses of 4,000 m (about 13,123 ft). These areas lack vegetation and soil and have proven hostile to human settlement and activity. Indigenous inhabitants of this area lived in close relationship to the physical environment, subsisting on hunting and fishing products. In recent years, non-indigenous inhabitants have greatly increased in number particularly because of the discovery of oil.

These brief descriptions of ten global environments are not attempts to describe all areas, nor do these areas have uniform characteristics. Regions merge into one another; the tropical rain forest merges into savanna and then into desert, for example. Rarely are there clear, sharp divisions. Many areas are difficult to classify as a result. What has been emphasized is that the earth comprises many environments with which humans have to cope. Some have been greatly changed, while others have been little affected. All environments need to be considered in any attempt to comprehend our human use of the earth.

LIFE ON EARTH

Air-breathing life cannot exist without oxygen, which was not a part of the original atmosphere. Life, however, in the form of primeval bacteria and algae, evolved without oxygen; this life emitted oxygen as a waste, consuming carbon dioxide and nitrogen, which were in the original atmosphere. Not only did this process provide oxygen in the atmosphere, it also formed the ozone layer, which filters out harmful ultraviolet radiation from the sun. These first life-forms evolved in the seas. As indicated in Table 3.1, there is evidence of life-forms as early as 3,500 mya, of an ozone layer 2,500 mya, and of a breathable oxygenated atmosphere 1,700 mya. Such early life-forms were not affected by the absence of an ozone layer as they lived below the surface of the water. Oxygen-breathing life, initially single- and

subsequently multicelled, appeared following the creation of a suitable atmosphere. Soft-bodied animals, comparable to jellyfish, evolved 650 mya and shelled animals about 70 million years later.

About 400 mya, fishes diversified considerably and gradually moved into water areas subject to drying up. Amphibious species were evident shortly thereafter, moving onto land already occupied by plants and insects, especially close to water. Much of the earth at this time comprised swamp environments, which were well-suited to plant growth and provided food for amphibians. One group of amphibians evolved into reptiles, animals with waterproof skins capable of remaining out of the water indefinitely. These reptiles could move away from water and thus colonized large areas, soon dominating much of the earth for about 135 million years. During this period, mammals and birds appeared and when the reptiles vanished 65 mya, mammals and birds remained. The cause of the extinction of reptiles remains uncertain (Lewin 1982).

With the absence of reptilian competition, mammals and birds flourished, expanding and diversifying, filling the ecological gaps left by the dinosaurs. One new evolutionary line that began at this time was the primates; today, humans are one of about 200 species of living primates. Apes, a new

TABLE 3.1 *Basic chronology of life on earth*

Time	Event
Physical origins	
10–20 billion years ago?	Creation of universe
5,000 mya	Solar system forms
4,500 mya	Earth forms
Origins of life	
3,500 mya	Oldest life-forms
3,200 mya	Oxygen-creating bacteria
2,500 mya	Ozone shield forms
2,200 mya	Oxygen in atmosphere
1,700 mya	Atmosphere breathable
Origins and evolution of life-forms	
1,400 mya	Oxygen-using animals
800 mya	Multicellular life
650 mya	Soft-bodied animals
580 mya	Shelled animals
500 mya	First fishes
430 mya	First land plants
360 mya	First insects
350 mya	First amphibians
280 mya	First reptiles
225 mya	Breakup of Pangea
220 mya	First mammals
200 mya	Reptiles dominant for next 135 my
190 mya	First birds
65 mya	Dinosaurs extinct; birds and mammals diversify; first primates
25 mya	Grasslands become widespread; first apes
10 mya	Maximum diversity of mammals
6 mya	Emergence of earliest human ancestors

group of primates, evolved about 25 mya. Some 10 mya, mammals reached their greatest diversity. Available evidence suggests that our human evolutionary line separated from that of the apes about 6 mya (Feder and Park 1993:104).

Three factors need to be highlighted in this account of the emergence and spread of life on earth. First, the earth was a changing environment during this time. Climatic change, especially a series of glacial periods, was typical. Second, the land masses have moved. The most recent movements began with the initial breakup of Pangea (see Figure 3.3) some 225 mya. The spread of life has been greatly affected by both of these factors as land routes around the world have changed. Third, the relationship between the physical earth and the emergence and evolution of life is not fully understood (Box 3.3).

HUMAN ORIGINS

Although our knowledge continues to increase, there is still much that we do not know about human origins and early history (Table 3.2). Current evidence suggests that the large apes evolved as one primate line in Africa about 25 mya and then split into a number of relatively distinct evolutionary lines. One of these gave rise to gorillas, which in turn split into two further lines, chimpanzees and humans. We can suggest with reasonable confidence, using a combination of fossil, geological, and genetic evidence, that this split occurred; that is, human ancestors diverged from the ape line about 6 mya. The most compelling evidence is that humans and chimpanzees differ in only about 1 per cent of their genes, a fact that means that these two species could not have been evolving separately for more than about 6 million years (Feder and Park 1993:104).

OUR EARLIEST ANCESTORS: *AUSTRALOPITHECUS*

What was it that characterized these earliest human ancestors from other primates? Unlike all earlier primates, they were bipedal — they walked upright. This is a critical adaptation that permitted efficient movement over great distances while carrying objects at the same time. Not surprisingly, there is fossil evidence of their presence throughout eastern and southern Africa. These first **hominids** are known as *Australopithecus*. Most authorities currently agree that the first definite fossil evidence of the common ancestor of all hominids is that of *Australopithecus afarensis*, a species discovered in east Africa and dated to more than 3.5 mya. This species had a small brain of about 440 ml (modern human brains average 1,450 ml).

Box 3.3: Gaia

Gaia is the Greek name for the goddess of the earth. Today, the term is used to describe a self-regulating system with all components of the system, or ecosphere, in a stable balance. This remarkable concept was first introduced in a book by James Lovelock (1979). In brief, Gaia is seen as a self-regulating entity that keeps the environment relatively constant and comfortable for life; thus, all life and earth have evolved together as one. Specifically, the Gaia hypothesis supposes 'that the atmosphere, the ocean, the climate, and the crust of the Earth are regulated at a state comfortable for life because of the behaviour of living organisms' (Lovelock 1988:19). According to Gaia, the earth is 'alive'. This revolutionary idea contradicts the standard view that sees life adapting to earth conditions as both life and the earth evolved in their separate ways.

The Gaia concept obliges us to assume a global perspective. It is the planet that matters, not only one species of life (such as humans). The Gaia global perspective thus differs somewhat from that adopted by environmentalists. A focus on Gaia is a concern for the planet, not for people. If we humans continue to change the earth, then we may precipitate our replacement. If, on the other hand, we act as part of a living organism, then perhaps we will remain on earth for a long time.

The Gaia concept has helped to explain some crucial scientific issues. For example, the surface temperature of the earth has remained relatively constant since life first appeared, even though the heat from the sun has increased by about 25 per cent. How has this occurred? The level of CO_2 in the atmosphere has decreased over this time period, thus reducing the natural greenhouse effect and allowing surface temperatures to remain relatively constant. Life, specifically photosynthesis, is the cause of the long-term CO_2 decline. Without life, CO_2 would probably increase once again. (These comments help place our present greenhouse crisis, discussed in the next chapter, in perspective — by adding CO_2 and other greenhouse gases to the atmosphere, we are countering a 3.5-billion-year trend of CO_2 reduction.) A second example that supports the Gaia concept concerns the salinity of sea water. This has barely changed in the long term, which is puzzling given that salt is continually being added to the sea via rivers. How is salt being removed from the seas? The Gaia response is that various life-forms in the seas promote the segregation of sea water in lagoons, allowing the sun to evaporate the water and remove the salt.

Both of these examples illustrate the meaning of Gaia as a self-regulating entity, keeping the environment fit for life. The existence of Gaia is not a proven fact, indeed many scholars consider it too general to be tested (Kirchner 1989). Nevertheless, the concept is both scientifically stimulating (in a period of twenty years, Gaia has assumed legitimate scientific interest) and valuable for the global perspective it involves.

TOWARDS MODERN HUMANS: *HOMO HABILIS*

The next critical evolutionary event in the human lineage occurred about 3 mya when this hominid line split into two types, one of which became extinct and one of which evolved into modern humans. The first representative of the line leading to modern humans appears to be the species *Homo habilis*. Both *Australopithecus* and *Homo habilis* were probably restricted to Africa, notably the east and south African savanna areas, eating small animals and plants. But *Homo habilis* used their increased brain size (about 680 ml) to begin a new and important cultural behaviour — they made stone tools. Tool making is a major technological advance. Brain size distinguished *Homo habilis* from other previous hominids and from other hominid lines living at the same time. Some evidence suggests that *Homo habilis*, unlike *Australopithecus*, possessed the capacity for speech.

Why did *Australopithecus* evolve perhaps as long ago as 6 million years? Why did *Australopithecus afarensis*, our common ancestor, split into two types about 3 mya? Why did one of these become extinct about 1 mya? Although we are uncertain of the answers, there is evidence to suggest that global climatic changes are a causal factor. A decline in global temperatures corresponds with each of these evolutionary events. Such declines prompted drying trends in Africa and hence decreases in tropical forest and related increases in open woodland and dry savanna. Evolutionary changes can be seen as varying adaptations to these changing physical environments.

OUT OF AFRICA: *HOMO ERECTUS*

Beginning about 1.8 mya, *Homo erectus* appeared. This new species was distinguished from *Homo habilis* primarily because of the increased brain size of about 1,000 ml, a brain size at about the lower limit of modern humans. *Homo erectus* first appeared in Africa and subsequently spread over much of Africa and into the warm temperate areas of Europe and Asia. Currently this hominid species is thought to be the first to move out of Africa. Some recent evidence suggests that *Homo erectus* may have reached Asia shortly after the first appearance in Africa. Reasons for this movement are not known, but the explanation probably relates to climatic change, population increases, and the associated search for food. Again, the basis of subsistence was hunting and gathering.

After the appearance of *Homo erectus*, human biological evolution stabilized for some time, although cultural adaptation, including the use of clothing and shelters, was considerable. Different cultural adaptations were devised for different environments. By about 400,000 ya, hunting strategies had progressed to include the use of fire and possibly planned hunting of larger game. An innovation at this time was the introduction of relatively permanent pair bonding — monogamy. Human language also increased substantially to encourage the gradual development of human culture. All of these adaptations also occurred against a background of changing physical environments as the geological period beginning about 1.5 mya, the **Pleistocene**, was one of considerable fluctuations in global temperatures. Several extremely cold periods, when large areas of the earth were covered in ice, were punctuated by periods with average temperatures comparable to those of today — an alternation of glacial and interglacial phases. These fluctuations continued until as recently as 10,000 ya and may indeed be continuing, as it is possible that we are in an interglacial period at present.

MODERN HUMANS: *HOMO SAPIENS SAPIENS*

Some 400,000 ya, a new hominid evolved from *Homo erectus*. This species is similar to modern humans, but because there are physical differences between them and modern humans, they are often called *archaic Homo sapiens*. Their mean brain size of 1,220 ml was about 85 per cent of that of modern humans. The best-known subset of these archaics was the Neanderthals who first appeared about 130,000 ya and

TABLE 3.2 *Basic chronology of human evolution*

Time	Event
6 mya	Earliest date for evolution of first hominids, the bipedal primate, *Australopithecus* in Africa
3.5 mya	First fossil evidence of *Australopithecus afarensis* in east Africa
3 mya	Probable appearance of *Homo habilis* in east Africa; evidence of first tools
1.8 mya	First appearance of *Homo erectus* in east Africa
1.5 mya	Evidence of *Homo erectus* in Europe; evidence of chipped stone instruments in Africa; beginning of Pleistocene
700,000 ya	First evidence of chipped stone instruments in Europe
400,000 ya	*Homo erectus* uses fire; develops strategies for hunting large game; first appearance of *archaic Homo sapiens*
100,000 ya	Appearance of *Homo sapiens sapiens* in Africa only according to favoured replacement hypothesis
80,000 ya	Beginnings of most recent glacial period
40,000 ya	Arrival of humans in Australia
25,000–15,000 ya	Arrival of humans in America
12,000 ya	First permanent settlements
10,000 ya	End of most recent glacial period; beginning of Holocene period

Notes: mya = million years ago
ya = years ago

lasted until about 30,000 ya. In the early twentieth century, following the first fossil discoveries, Neanderthals were regarded as subhuman in intellect. By the middle of this century, it was not uncommon for them to be regarded as virtually modern in appearance and behaviour. The current interpretation favours a compromise interpretation — neither brutish apes nor modern humans.

We have now reached a critical point — the emergence of modern humans, often called *Homo sapiens sapiens*. Where and how did this occur? There are two rather different explanations. The *multiregional* hypothesis proposes that modern humans evolved independently from archaics at more or less the same time in a number of different geographic areas. However, the currently available palaeontological, genetic, and archaeological evidence all favour a second hypothesis. This *replacement* hypothesis proposes a single evolution in a limited geographic area and subsequent spread to replace archaics elsewhere. Specifically, modern humans, *Homo sapiens sapiens*, evolved in a single place, most probably Africa, south of the Sahara. This evolution was from a group of archaics before 100,000 ya. These humans spread throughout the world, replacing existing archaic groups because of some adaptive advantage. From this time onwards, human evolution is cultural, not biological.

Modern humans settled much of Africa, Asia, and Europe, and reached Australia 40,000 ya and America perhaps 25,000 ya, but certainly more than 15,000 ya. These movements of humans were of course related to the continuing climatic changes associated with the glacial and interglacial periods. Modern humans began to evolve and spread across the earth during a period of harsh climate. Evidence suggests that favoured locations were at the boundaries of geographic regions where the first permanent settlements appeared about 12,000 ya. The most recent glacial period ended 10,000 ya, which marks the end of the Pleistocene and the beginning of the **Holocene** period.

The fact that early humans effectively settled all areas of the earth is most significant. It suggests that humans were able to adjust to new environments via their culture. Migration to previously unsettled areas enlarged the overall resource base and allowed slow but steady increases in numbers of humans. We can only estimate the number of people in these early stages. *Australopithecus* was restricted to Africa and population estimates range from 70,000 to 1 million. *Homo erectus* covered much more of the globe and possibly attained a population of 1.7 million. A figure of 4 million seems reasonable for the human population about 12,000 ya.

Box 3.4: Species and races

There is only one species of humans in the world, namely modern *Homo sapiens*. That all humans are members of the same species is biologically confirmed by the fact that any human male is able to breed with any human female. All species, including humans, display variation among their members. Indeed, many plant and animal species exhibit sufficient variation that they can be conveniently divided into *subspecies* or races, geographically defined aggregates of local population. Such races of a species arise if groups become isolated for the necessary length of time in different environments. They then exhibit some genetic differences from other races as a consequence of the selective breeding that results from the isolation and adaptations to different environments.

To what extent is it possible to delimit distinct races, geographically defined aggregates of local populations, in the human species? To answer this question, we identify three stages. First, we know that modern *Homo sapiens* evolved at least 100,000 years ago, probably only in Africa. Second, members of this species then moved across the surface of the earth, replacing archaic *Homo sapiens* and gradually dividing into a number of groups; as a consequence of these movements and groupings, some selective breeding and adaptation to different environments occurred, causing some genetic differences to become apparent. Third, because of population increases and further migrations, these groups are no longer isolated.

It was during the second stage that the groups of humans often known as Negroid, Mongoloid, Caucasoid, and Australoid appeared. But are these groups races? *The answer is no, because none of these human groups was ever isolated long enough or completely enough to allow separate independent genetic changes to occur.* Use of the word 'race' is therefore mistaken because genetically speaking, humans cannot be divided into clear-cut stable subspecies.

Because race is a biological concept referring to a genetically distinct subgroup of a species, there are today no such things as races within our human species. There are genetic differences between members of the human species as a consequence of the earlier period of isolation of groups in different environments. Biologically, however, these differences are not sufficient to merit designation as races.

We are a biologically variable species and our variability results from the same processes that produce variability in all living things, namely, natural selection in different physical environments. These biological facts are of basic importance. They immediately inform us, for example, that it is incorrect to equate a cultural group with a racial group. Thus, there is a Jewish religion, but not a Jewish race. Similarly, there are Aryan languages, but not an Aryan race. There are no subspecies, races, within our human species.

UNDERSTANDING HUMAN EVOLUTION: THREE MYTHS

A *first myth* concerns the idea of human evolution, or evolution generally, as some sort of ladder of progress. As the above account suggests, evolution is not a history of steady progress to a finished product; it is not correctly seen as some sort of ladder. Rather, it is helpful to employ a metaphor popularized by the distinguished evolutionary biologist, Steven Jay Gould. For Gould, evolution is best seen as a bush with endlessly branching twigs, each twig representing a different species. Using this metaphor we can see humans as 'a fragile little twig of recent origin' (Gould 1987:19).

A *second myth* concerns the supposed existence of several different human **races** (Gould 1985; Montague 1964). As modern humans, *Homo sapiens sapiens*, moved from Africa to different physical environments and as they splintered into distinct spatial groupings, there were adaptations in culture and certain body features or **phenotypes**. Thus, relatively distinct phenotypes emerged with different skin colours, head shapes, blood groups, and so forth. Certain characteristics became dominant in particular areas, either because they were environmentally appropriate or because they resulted from chance genetic developments. The resultant variations in human physical characteristics are quite minor and do not in any respect represent different types of humans (Box 3.4).

A *third myth* concerns the idea that races not only exist but can be classified according to their level of intelligence. We are fortunate today in that we have clear scientific evidence of the unity of the human race. Such knowledge was not at hand until relatively recently. The characteristic response to physical variations in humans, particularly different skin colours, has been to regard these as evidence of different types of humans with different levels of intelligence. Box 3.5 summarizes the history of the erroneous idea that the human population can be divided into distinct and unequal groups, usually called races. The reality is that all humans are members of the same species and the differences within the species are differences of secondary characteristics, such as skin colour (Kennedy 1976).

CONCLUSION

The physical environments that are now occupied by humans are both diverse and changing. Diversity is easily identified, while change is more localized in the short term and global in the long term. In this chapter we have identified some major global characteristics of the earth, some key physical processes that affect physical landscapes, and some characteristic landscapes. We have also commented on the types of life-forms that appeared and we presented an account of our present understanding of the evolution of humans.

The physical geographic content is much more than mere background or stage setting for the chapters to follow. Human activities are closely related to physical environments. This is not an environmental determinist assertion. Rather, it is an acknowledgement of a central and pervading relationship. Humans make the decisions as individuals and groups, but they make these decisions with a multitude

---

Box 3.5: A history of racism

In Box 3.4 we learned of the unity of the human species and that there are no distinct subspecies, races, within that species. Notwithstanding these biological facts, the concept of distinct racial groupings has long been popular in lay and scientific circles. Ideas about the existence of races and the relative abilities of the supposed races have been central to many cultures.

Racism is the belief that human progress is irrevocably linked to the existence of distinct races. Such ideas of racial variations in ability appear to have been common in most cultures. They flowered especially in Europe beginning in the Middle Ages because of: (a) the rise of Protestantism which involved notions of good versus bad peoples; (b) European overseas movements during which different groups were encountered; and (c) polygenetic theories of human origins, which became popular at the expense of traditional Christian Augustinian thought that insisted (correctly, as we know from Box 3.4) upon the unity of the human species.

By the eighteenth century, racist explanations for the increasingly evident variety of global cultures were normal. Such philosophers as Voltaire, David Hume, and Lord Kames all saw clear differences in ability between what were commonly seen as racial groups. Undoubtedly the key racist thinker and writer was Joseph Arthur de Gobineau whose *Essay on the Inequality of Human Races* was published in 1853–5. Gobineau ranked races as follows: Whites, Asians, Negroes. Within the Whites, the Germanic peoples were seen as the most able. A second influential racist was Houston Stewart Chamberlain. But racist logic was apparent also among many of the greatest scientists. Darwin wrote of a future when the gap between human and ape would increase because such intermediaries as the chimpanzee and Hottentot would be exterminated (see Gould 1981:36). As we have seen in Box 3.4, the racial categories employed in all such discussions have no biological meaning.

One explanation for the popularity of racist thought is the fact that most cultures classify other cultures relative to themselves. The consequences of racist thinking are varied and considerable. Belief in the inferiority of a specific group has led to mass exterminations, slavery, restrictive immigration policies, and, most generally, unjust treatment.

of physical matters in mind. Successful adaptation — that is, coping — is adapting to the physical environment as well as to relevant human variables.

In a recent book dealing with geography and social theory, the following statement appears:

> The journey along Mulholland Drive, atop the Hollywood Hills, provides one of the world's great urban vistas. To the south lies the Los Angeles basin, a glittering carpet. To the north, the San Fernando Valley (still part of the City of Los Angeles) unfolds in an equivalent mass of freeways, office towers, and residential subdivisions. There is probably no other place in North America where such an overpowering expression of the human impact on landscape can be witnessed. And yet, the physical landscape cannot be denied. Even in this region of almost 12 million people, the landscape still contains and molds the city (Dear and Wolch 1989:3).

The 12 million people in the Los Angeles region today is about three times the number on the entire earth 12,000 ya. The story of human biological and early cultural evolution before this date is a complex and sometimes uncertain one. The account provided in this chapter emphasizes the known facts of the story and, where the facts are uncertain, this is acknowledged and current consensus thought is reflected.

SUMMARY

The earth

A habitable environment, the earth is a sphere rotating on its axis; one rotation defines a day. The earth orbits the sun and one revolution is a year. Revolution, combined with an axial tilt, produces seasons and variation in length of daylight hours. Diurnal, seasonal, and annual cycles are basic to much human activity. Locations on the earth can be identified by reference to an imposed geographic grid, lines of latitude and longitude.

The earth's crust

The present distribution of land and water is the consequence of a long and ongoing process of movement by the lithosphere, known as continental drift. These movements are also responsible for major mountain chains, such as the Himalayas and Appalachians, and for earthquakes.

Physical landscape makers

The details of a physical landscape result from ongoing weathering and gradational processes. Once rock is weathered, either mechanically broken or chemically altered, water, ice, or wind can erode, move, and deposit the weathered material. Water, ice, and wind each produce distinctive landforms. Our physical environments thus contain land-forms, but are also characterized by particular soils, vegetation, and climates. Each of these three is distributed in a relatively systematic fashion, related to latitude and to each other.

Global environments

The general correspondence between soils, vegetation, and climate permits a classification of the earth's surface into typical environments. Ten such environments are noted and the typical physical content identified. Human activities are often closely related to but are not caused by these physical environments. Although global environments can be distinguished, there are rarely clear divisions between different environments and any one environmental type can include a wide variety of physical conditions.

Life on earth

Life on earth evolved 3,500 mya and has undergone a series of additions, deletions, and gradual changes since that time. The first life-forms were under water and their presence led to the development of an oxygenated atmosphere that could support air-breathing life and of an ozone layer that acts as a protective shield from the sun's ultraviolet rays. During the long history of life on earth, numerous physical changes have taken place, especially the movement of continents and climatic change.

Human origins

The earliest humans, the first bipedal primates, were *Australopithecus* (6 mya), who emerged in east Africa. Subsequent species had increasingly larger brains. *Homo habilis* (3 mya) also evolved in east Africa; these early humans used tools. *Homo erectus* (beginning 1.8 mya) moved out of Africa to Europe and Asia. Archaic *Homo sapiens* emerged 400,000 ya. Modern humans, *Homo sapiens sapiens*, appeared in Africa no later than 100,000 ya. These modern humans reached Australia 40,000 ya and North America possibly 25,000 ya; they settled in environmentally suitable locations.

The unity of the human race

Humans are members of one species — *Homo sapiens sapiens*. Early spatial separations of groups of humans facilitated the development of physical variations. These are of minimal relevance to our understanding of both people and place. The often-used concept of races is of negligible value.

WRITINGS TO PERUSE

FEDER, K.L., AND M.A. PARK. 1993. *Human Antiquity: An Introduction to Physical Anthropology and Archaeology*, 2nd ed. Toronto: Mayfield.

> An introductory physical anthropology textbook that contains a detailed discussion of the origin of humans; a balanced presentation.

MCKNIGHT, T.L. 1993. *Physical Geography: A Landscape Appreciation*, 2nd ed. Toronto: Prentice-Hall.

> This is a detailed and especially well-illustrated physical geography textbook; physical-human relations are frequently discussed.

PASSINGHAM, R.E. 1982. *The Human Primate*. Oxford: Freeman.

An excellent book that discusses humans as an animal species.

SCOTT, R.C. 1989. *Physical Geography*. New York: West.

A physical geography textbook with an especially clear account of our planetary setting.

STRAHLER, A.N., and A.H. STRAHLER. 1989. *Elements of Physical Geography*, 4th ed. New York: Wiley.

A physical geography textbook that is particularly sensitive to the relevance of human and land interactions.

TRENHAILE, A.S. 1990. *The Geomorphology of Canada: An Introduction*. Toronto: Oxford University Press.

A textbook presenting a systematic explanation of the land-forms of Canada.

4

The earth: A fragile home

IN THIS CHAPTER we begin to discuss the human use of and changes to the earth related to improving our well-being. For most of the time that human ancestors and humans have been using the earth, such changes have been slight for two reasons — limited technologies and a small population. The overview of human evolution in Chapter 3 noted such technological advances as the use of stone tools by *Homo habilis* some time after 3 million years ago and the use of fire as a hunting strategy by *Homo erectus* before 400,000 years ago. Such technologies when combined with limited numbers of people (probably as few as 4 million as recently as 12,000 years ago), resulted in few and temporary local-scale environmental changes. Today, the situation is vastly different with a substantial array of agricultural and industrial technologies and a population of 5.5 billion (1993). Our interactions with and impacts on the environment are now numerous, relatively permanent, and often on a global scale.

A GLOBAL PERSPECTIVE

Our concern is with the impact of human groups upon land. But what is the value of a global perspective on this question? *The short answer is that everything is related to everything else; one cannot simply change one aspect of nature without directly or indirectly affecting other aspects.* In principle, then, human activity in any one area has the potential to affect all other areas; the evidence is now overwhelming. One way to approach these issues is to consider the question of how a civilization survives. Smil (1987:1) answered, 'It survives by harnessing enough energy and providing enough food without imperilling the provision of irreplaceable environmental services. Everything else is secondary.'

This chapter considers how harnessing energy and producing food have changed our environment and whether or not the changed environment threatens human survival. For many human geographers, the most valuable tools for analysing these global issues are those relating to systems and ecology.

SYSTEMS, ECOLOGY, AND ECOSYSTEMS

In principle, the concept of a **system** is simple and widely applicable. In practice, the concept appears to have prompted few insights into the workings of humans and land. Despite this major limitation, it is useful to identify what we mean by systems and consider their potential value.

Systems can be usefully defined as sets of interrelated parts. The attraction of the systems concept is its ability to describe a wide range of phenomena and offer a simplified description of what is usually a complex reality. Descriptions of systems typically rely upon distinctions between the parts of the system and the relationships between the parts. Open systems are related to elements outside of the system, necessitating the study of inputs and outputs of energy and matter. Closed systems lacking inputs and outputs of matter are less common. Finally, relationships between the parts of a system, or between a system and some external elements, are often described as feedbacks. Positive feedback reinforces some change that is occurring; negative feedback counters some change.

The term **ecology** is derived from two Greek words: *oikos* refers to 'place to live', while *logos* refers to 'the study of'. Thus ecology is the study of organisms in their homes. The concept of an ecosystem integrating systems and ecology was formally conceived by an English botanist, Tansley, in 1935.

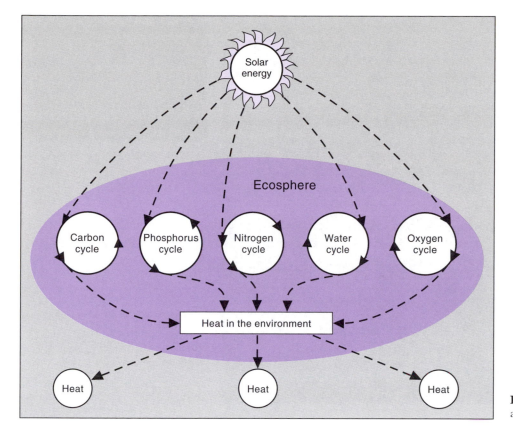

Figure 4.1 Chemical cycling and energy flows.

Ecosystems can be identified on a wide variety of scales. The global ecosystem (often called an ecosphere or biosphere) is the home of all life on earth. It is a thin (about 14-km/8.5-mile) shell of air, water, and soil. Ecosystems can also be identified within the larger ecosphere as any self-sustaining collection of living organisms and their environment. Thus, the ecosystem concept is concerned with distinct groupings of things and the relationships between these things.

On the global scale, the three basic parts of the ecosystem are the atmosphere, hydrosphere, and lithosphere — air, water, and land. This global ecosystem has one key input — the sun as a source of energy warming the earth, providing energy for photosynthesis, and powering the water cycle to provide fresh water. In order to be self-sustaining, our global ecosystem depends on the cycling of matter and the flow of energy. Figure 4.1 outlines the cycling of critical chemicals and the one-way flow of energy. Matter must be cycled; the law of conservation of matter states that matter can neither be created nor destroyed, only changed in form. Energy flows through the system because of the second law of thermodynamics — energy quality cannot be recycled. We call the circular pathways of chemicals biogeochemical cycles; 'bio' refers to life, 'geo' refers to earth. Our global ecosystem, indeed all ecosystems, are dynamic. As we are about to discuss, one of the most important — and least understood — causes of ecosystem change is the human population.

HUMANS AS SIMPLIFIERS OF ECOSYSTEMS

Any human change to an ecosystem is typically a simplification, and a simplified ecosystem is usually vulnerable (Box 4.1). Ecologists agree that our global ecosystem must comprise a balance of human-modified systems and non-human-modified systems. For example, a cultivated lowland ecosystem is dependent on nearby upland forests that release water and minerals to the cultivated area. Removal of the forests affects the adjacent area. This small example confirms the interrelatedness of local ecosystems and emphasizes that reality is complex, that — as the basic principle of ecology states — all things are related. One of our great needs today is to live in our ecosystems in a cooperative and not dominating manner. Today, human use of the earth is one of domination, not cooperation. Nothing illustrates this better than our use of energy and our high valuation of technology.

ENERGY AND TECHNOLOGY

Humans have basic physiological needs for food and drink and a host of culturally based wants that have no upper limit. Needs and wants are satisfied largely as a result of humans using their energy to utilize other forms of energy. **Energy** is the capacity to do work. The more successfully we can utilize other energy, the more easily we can achieve our wants and the more we can crucially affect the environment. The

4.1 *(left)* Nuclear power station, Pickering, Ontario, Canada. (Ontario Hydro Corporate Archives 94141)

4.2 *(above)* Dead fish — one consequence of acid rain. (Ontario Ministry of the Environment and Energy)

Box 4.1: Lessons from Easter Island

One of the most remote inhabited places on earth, Easter Island in the Pacific Ocean is 3,747 km (2,328 miles) off the coast of South America and 2,250 km (1,400 miles) southeast of Pitcairn Island, the nearest inhabitable land. When Europeans first visited Easter Island on Easter Sunday, 1722, they encountered a population of about 3,000 living in a state of warfare, minimal food available from limited resources, and a treeless landscape (Ponting 1991:1–7). They also encountered clear evidence of a once-flourishing earlier culture — between 800 and 1,000 huge stone statues, or *moai*. Because it was clear to visiting Europeans that these statues could not have been carved, transported, and erected by the local population, it was long assumed that Easter Island must have been visited by a more culturally and technologically advanced group.

There is a simpler but more disturbing answer to this 'mystery' of Easter Island. Radiocarbon dating of evidence suggests that Easter Island was settled by the seventh century, and possibly as early as the fifth century, with the first settlers likely being Polynesians and not South Americans. They arrived at an island with few species of plants and animals and limited opportunities for fishing, but with considerable areas of woodland. Their diet comprised sweet potatoes (an easy crop to cultivate) and chicken. The considerable amount of free time available allowed the Easter Islanders to engage in elaborate rituals and construct the *moai*. Agricultural activities required the removal of trees, but most of the deforestation occurred because of the need to move statues. It is likely that statue construction involved competition between different groups. By about 1550, the population peaked at about 7,000. Deforestation was probably complete by 1600, which probably prompted soil erosion, a reduction of crop yields, and a shortage of building materials for both homes and boats. It was these impoverished Easter Islanders, not their remarkable ancestors, that Europeans encountered.

It appears that the Easter Island culture and economy began to fall into disarray because of the total deforestation. What is the significance of this lesson? Bahn and Flenley (1992:212–13) suggest:

We consider that Easter Island was a microcosm which provides a model for the whole planet. Like the Earth, Easter Island was an isolated system. The people there believed that they were the only survivors on Earth, all other land having sunk beneath the sea. They carried out for us the experiment of permitting unrestricted population growth, profligate use of resources, destruction of the environment and boundless confidence in their religion to take care of the future. The result was an ecological disaster leading to a population crash.

means by which we become aware of other energy sources and acquire the ability to use those sources is through the development of new technology.

Technology is our ability to convert energy into useful forms. An early major technological change was human use of fire to convert inedible plants for human use. The domestication of plants is an example of our acquiring control over a natural converter, for plants convert solar energy into organic material via photosynthesis. Similarly, our domestication of animals involves control over another natural converter; animals change one form of chemical energy (inedible plants, usually) into another form usable by humans (such as animal protein). In this sense, the domestication of plants and animals, sometimes known as the first **agricultural revolution**, involved the use of new energy sources as a result of technological change.

The period after the agricultural revolution through to the eighteenth century saw many less dramatic technological changes that permitted increasing human control of energy sources: thus new plants and animals were domesticated and new tools and techniques were invented. In addition, three new energy converters were invented to utilize the energy in water and wind: the water-mill, the windmill, and sailing craft. But it was the **Industrial Revolution** of the eighteenth century that allowed us to begin large-scale use of new energy sources via inanimate converters. The key development was the adoption of the steam engine in the late eighteenth century, while the new energy sources were coal in the second half of the eighteenth century, oil and electricity in the second half of the nineteenth century, and nuclear power in the middle of the twentieth century.

This brief account raises two key points. First, we are using energy sources many of which, such as coal and oil, are irreplaceable. Second, we are using these sources in a way that is often harmful to our ecosystem. Deteriorating environments resulting from **pollution** are now commonplace in industrial societies. According to Smil (1989:10), 'Environmental pollution, previously a matter of regional impact, started to affect more extensive areas around major cities and conurbations and downwind from concentrations of power plants as well as the waters of large lakes, long stretches of streams and coastlines, and many estuaries and bays.' Several twentieth-century developments have contributed to this situation, among them thermal generation of electricity, mass use of automobiles, use of plastics, and use of nitrogenous fertilizers and pesticides.

Energy use is spatially variable. Canadians lead the world in the per capita use of energy, followed by Americans, while Europeans and Japanese use less; in general, these and other countries in the **more developed world** use greater amounts of energy per capita than do countries in the **less developed world**. Energy sources are also spatially variable. Globally, oil accounts for 36 per cent of energy used, coal for 25 per cent, gas for 17 per cent, electricity for 5 per cent, nuclear for 4 per cent, and other sources for 13 per cent. In many parts of the less developed world, instead of oil, coal, and gas, the primary energy sources are the traditional **biomass** sources of wood, crop waste, and animal dung. And energy use is increasing. In 1989, it was estimated that in one year humans use an amount of fossil fuel that took some 1 million years to produce. It is clear that we cannot continue indefinitely in this fashion because we are draining the supply of non-renewable energy sources. We must realize and act upon the realization that our human survival depends on a continuing access to appropriate energy sources.

NATURAL RESOURCES ARE CULTURAL APPRAISALS

Humans continually evaluate physical environments. As human culture (especially technology) changes, so do those evaluations. A resource is thus a human perception interpreted in technological, political, economic, and social terms. But groups do not necessarily agree on what is and is not a resource. An area of wetland in prairie Canada can be seen by some group members as a valuable scientific or recreational landscape and by others as potential farm or building land. Different interest groups evaluate according to different criteria and hence may hold radically different views.

Traditionally, geographers divide resources into two types. **Stock resources** include all minerals and land and are essentially fixed as they take long time periods, by human standards, to be created. **Renewable resources** are those that are continually created, such as air and water. Such a clear distinction is generally useful, but does blur some key issues. There are, of course, a whole series of resources that lie somewhere between the two extremes of stock and renewable and whose continued availability is dependent on the way we manage those resources. Obvious examples include animal populations for a hunting economy and fish populations for much of the contemporary world. In both cases, there may be a perceived need for conservation, but also some compelling reasons to continue to deplete the resources. In Canada, the east coast fishery has been especially susceptible to these issues.

ENVIRONMENTAL ETHICS
WESTERN ENVIRONMENTAL CONCERN BEFORE 1900

Use or abuse? While it is not always easy to distinguish use and abuse — they are, after all, relative terms — there is considerable evidence to suggest that currently we are causing damage, perhaps irreparable damage, to our environment. Concerns about the consequences of human activities were raised by the Greeks with Plato noting the detrimental effects of agricultural activities on soil. Despite such early observations, this general question received relatively little attention in the western world until the eighteenth century. Before that time, geographers' favourite questions concerned, first, the earth as a home for humans made by God and, second, the land as a cause of human activity (environmental determinism). Significantly, the Europeans' general failure to appreciate the possible dangers of some human activities does not appear to be paralleled elsewhere; in many cultures, the protection of natural resources has long been promoted.

The eighteenth-century origins of western environmental concern were related to the overseas movement of Europeans, particularly movement to tropical areas. The image of pristine tropical environments was associated with European ideas of a utopia and it was quickly realized that European activity in colonial areas was proving to be environmentally destructive. Scientists, typically employed by commercial groups such as the East India Company, responded by cataloguing the newly recognized floras and faunas. In Mauritius, the French introduced a number of conservationist measures, particularly concerning the location and amount of forest that could be removed, and these were imitated by the British in several West Indian islands (Grove 1992:44–5).

The general question of human impact on the land first received scholarly attention from Buffon (1707–88) in discussions of the contrasts between settled and unsettled areas and of the human domestication of plants and animals. Buffon adopted the perspective that humans inhabit the earth in order to transform the earth. Malthus (1760–1834) identified the terms of the current debate by discussing the relationship between available resources and numbers of people. On a more specific level Humboldt, during his travels in South America, explicitly identified such human impacts as lowered water levels in lakes, and explained these in terms of agricultural activities involving deforestation.

Probably the earliest systematic work on human impacts was that of Marsh (1801–82), an American geographer and congressman. This work, *Man and Nature, or Physical Geography as Modified by Human Action* (1864 with revised editions in 1874 and 1885), aimed:

> to indicate the character and, approximately, the extent of the changes produced by human action in the physical conditions of the globe we inhabit; to point out the dangers of imprudence and the necessity of caution in all operations which, on a large scale, interfere with the spontaneous arrangements of the organic and of the inorganic worlds; to suggest the possibility and the importance of the restoration of disturbed harmonies and the material improvement of wasted and exhausted regions; and, incidentally, to illustrate the doctrine that man is, in both kind and degree, a power of a higher order than any of the other forms of animated life, which, like him are nourished at the table of bounteous nature (Marsh [1864]1965:iii).

This powerful statement resonates of the 1960s rather than the 1860s. The late nineteenth-century western world, heavily involved in colonial expansion, was concerned with environmental change only when that change had negative

impacts on their economic interests. 'If a single lesson can be drawn from the early history of conservation, it is that states will act to prevent environmental degradation only when their economic interests are shown to be directly threatened. Philosophical ideas, science, indigenous knowledge and people and species are, unfortunately, not enough to precipitate such decisions' (Grove 1992:47).

THE CURRENT DEBATE: ORIGINS

Environmental concern is one of the great developments of the second half of the twentieth century and for many students, this is a topic about which they possess much general knowledge garnered from various sources. Environmental concerns are for the first time at the forefront of public consciousness. In 1989, *Time* magazine declared earth to be 'Planet of the Year'; many throughout the world celebrate Earth Day each 22 April, and many people make conscious efforts to conserve and recycle.

The first signs of this real shift in our appreciation of human impacts on ecosystems came in the 1960s with the publication of Rachel Carson's *Silent Spring* (1962), although this was preceded by a seminal academic work, totalling 1,194 pages and entitled *Man's Role in Changing the Face of the Earth* by Thomas et al., published in 1956. The 1960s also witnessed increasing pressure from advocates of wilderness preservation and new scientific evidence about worsening air pollution. The problem of **acid rain** was recognized at this time and solutions for it continue to be sought. Two popular suggested explanations for the environmental 'crisis' concerned, first, the Judaeo-Christian belief that humans were on earth to subjugate nature and, second, ideological claims about the failings of capitalism. The first of these is lacking because it ignores the complexity of Christian attitudes. The second explanation is one aspect of an approach that stresses the links between different parts of the world: 'Clearly there are problems, many of them — and all of them intertwined in the operations of a capitalist world economy, which is hell-bent on annihilating space and place. Those problems are severe now at a global scale, and life-destroying in some places' (Johnston and Taylor 1986:9). Our increased knowledge of environmental impacts beyond the capitalist world economy, as in eastern Europe and the former Soviet Union prior to the major political changes that began in 1989, leads us to question such assertions, although the basic idea of interrelatedness is sound.

THE CURRENT DEBATE: POLITICAL OVERTONES

Environmental issues entered the political arena in the early 1970s with the creation in the United States of the Environmental Protection Agency, while the first major international meeting, the United Nations Conference on Human Environment, was held in Stockholm in 1972. By 1980, the western world was increasingly aware of environmental and related food supply problems in the less developed world with food shortages in India and droughts in the Sahel region of Africa. A series of disasters and discoveries during the 1980s ensured that the environment was always in the news. These included the 1984 leak of methyl isocyanate from a pesticide plant in Bhopal, India, that killed perhaps as many as 10,000 and disabled up to 20,000; the 1986 nuclear disaster in Chernobyl, Ukraine, that killed up to 7,000 and will cause many more to die of radiation poisoning or related cancer; the 1985 discovery of a seasonal ozone hole over Antarctica; the realization of both the rapidity and consequences of tropical rain forest removal, especially in Brazil, and, more generally, an increasing concern for numerous local environmental problems.

By the late 1980s, the environment was on the national agenda of many countries as well as on the international political agenda. At the *national* level, green political parties first appeared in West Germany in 1979 and were present in most countries in the more developed world by 1990. Further, many countries have some form of green plan; an encyclopedic survey of the Canadian environment is available as a part of Canada's Green Plan (Supply and Services Canada 1991).

International agreement is the best way to solve those environmental problems that transcend international boundaries and there have been many calls for the creation of international institutions and policies, most notably by the Brundtland Commission that reported in 1987 (World Commission on Environment and Development 1987). Not surprisingly, although many governments agree with the need for international policies, most are unwilling to sacrifice their sovereignty; to reach agreements, countries need to resolve conflicting goals and priorities. Major international developments include the Montreal protocol of 1987 aimed at the reduction and eventual elimination of chloro-fluorocarbons (CFCs) that are one cause of global warming and the United Nations Conference on Environment and Development (the Rio Earth Summit) of June 1992, attended by an estimated 30,000 people and some 170 national leaders.

THE CURRENT DEBATE: THREE CONTENTIOUS ISSUES

Before addressing the impacts humans are having, three issues are identified. The first concerns relationships between the environment and the economy. Market forces are unlikely to solve environmental problems as they rarely result in what is best for the environment even at the national level. A detailed analysis of economic decision makers in Canada showed that only 6 per cent gave significant consideration to

---

Box 4.2: The tragedy of the commons

Imagine that you and a group of friends are dining at a fine restaurant with an unspoken agreement to divide the check evenly. What do you order? Do you choose the modest chicken entrée or the pricey lamb chops? The house wine or the Cabernet Sauvignon 1983? If you are extravagant you could enjoy a superlative dinner at a bargain price. But if everyone in the party reasons as you do, the group will end up with a hefty bill to pay. And why should others settle for pasta primavera when someone is having grilled pheasant at their expense? (Glance and Huberman 1994:76).

This lighthearted situation accurately depicts the clash between individual and collective attitudes that, in the context of human use of the environment, was so forcefully put forward by Hardin (1968) as follows:

- a group of graziers use an area of common land
- they continually add to their herds so long as the marginal return from the additional animal is positive, even though the common resource is being depleted and the average return per animal is falling
- indeed, individual graziers are obliged to add to their herds because the average return per family is increasing
- clearly, efficient use of the common resource requires restricted herd sizes

- tragically, individuals will not reduce herd sizes on the common land unless all other group members similarly reduce their herd numbers
- hence the metaphor, 'the tragedy of the commons'

Both our lighthearted and serious examples prompt the same question. How do we ensure that individuals behave for the common good rather than for personal gain? With reference to the environment, three solutions are proposed (Johnston 1992). First, resources may be privatized, with the private owners implementing strategies for environmental preservation not available to group owners. Second, the group owners may be able to devise an agreement about the use of the common resource that they are able to implement themselves. According to some recent work in social theory, the success of local and regional recycling programs results from such group cooperation (Glance and Huberman 1994:80). Third, the common resource may be subject to some external control. However difficult it may be to implement, it is the third proposed solution that appears most necessary for ensuring the reduction of deleterious human impacts on the environment.

Individually rational behaviour does not result in a collectively prudent outcome when individuals have access to common resources.

the environment and thus suggested that a major future challenge is the integration of economic and environmental concerns (Gale 1992).

Second, environmental problems are increasingly affecting relationships between countries not only because of the international implications of many human impacts but also because of the attempts by environmentalists in one country to impose their standards on another country. Payment is one solution; the 1987 Montreal accord included a fund to reward those countries most likely to suffer economically as a result of the accord. International disapproval is another possible solution; Britain eventually agreed to cease dumping sewage sludge in the North Sea because of the political costs of the dumping. Finally, trade policies are one of the few weapons available to a national government to convince another national government to amend its environmental behaviour.

The third issue, closely related to the second, concerns the behaviour of individuals as group members (Box 4.2). The ecophilosopher Arne Naess argued that humans need to develop a new world view that we are all connected, that sees a need to work with and not against nature, and that sees a central goal of human activity as the preservation of ecosystems. 'Deep ecology' is a term sometimes used to describe this viewpoint. Table 4.1 provides a comparison of deep and shallow views of ecology. Both represent improvements over some traditional attitudes towards land in which humans are seen as the source of all value, land exists for human use, and energy and other resources are unlimited. Deep ecology is now often called 'sustainable development', a term introduced by the 1987 *Brundlandt Report*.

Before reviewing the important concept of sustainable development and evaluating what actions need to be taken to safeguard the environment for future generations, it is appropriate to summarize what we do and do not know about human impacts on earth.

HUMAN IMPACTS

Human history is one of impacts on land. We must have an impact on land to survive, but only recently have we realized that some of our impacts actually threaten our continued survival:

1. Small, often insignificant, changes to the environment can have major impacts if they are repeated sufficiently often. Arable and pastoral activities can lead over time, to major environmental problems.
2. Technological changes, largely through demands for energy, continually change our environment.
3. Technological changes, through the lifestyles that they promote, again continually change our environment.
4. Increasing numbers of people are a threat to the environment.

IMPACTS ON ECOSYSTEMS IN GENERAL

Hunter gatherers affect ecosystems in particular areas and on a short-term basis. Impacts, both spatially and temporally, are limited because of few people, low levels of technology, and some deliberate conservation strategies. It seems probable that most ecosystems recover from hunter-gatherer activities. Natural energy flows are little altered except in cases of regular use of fire or animal overkill.

Agriculturalists, both cultivators and pastoralists, affect far more extensive areas over longer time periods. Cultivators change normal energy flows, often permanently, and often consciously direct new flows. Pastoralists have less effect on energy flows, as domesticated animals often merely replace previously wild populations.

TABLE 4.1 *Shallow and deep ecology compared*

Shallow ecology (Spaceship earth)	Deep ecology (Sustainable earth)
Views humans as separate from nature	Views humans as part of nature
Emphasizes the right of humans to live (anthropocentrism)	Emphasizes the idea that every life-form has in principle a right to live; recognizes that we have to kill to eat, but that we have no right to destroy other living things without sufficient reason based on ecological understanding
Concerned with human feelings (anthropocentrism)	Concerned with the feelings of all living things; deep ecologists feel sad when another human or a cat or dog feels sad and grieve when trees and landscapes are destroyed
Concerned with the wise management of resources for human use (anthropocentrism)	Concerned about resources for all living species
Concerned with stabilizing the population, especially in less developed countries	Concerned not only with stabilizing the human population worldwide, but also with reducing the size of the human population to a sustainable minimum without revolution or dictatorship
Either accepts by default or positively endorses the ideology of continued economic growth	Replaces this ideology with that of ecological sustainability and preservation of biological and cultural diversity
Bases decisions on cost-benefit analysis	Bases decisions on ethical intuitions about how the natural world really works
Bases decisions on short-term planning and goals	Bases decisions on long-range planning and goals and on ecological intuition when all facts are not available
Tries to work within existing political, social, economic, and ethical systems	Questions these systems and looks for better systems based on the way the natural world works

Source: G.T. Miller Jr, *Living in the Environment*, 4th ed. (Belmont, CA: Wadsworth, 1982):356.

4.3 Landscape affected by fire and logging, Whitefish, Montana, USA. (K.S. Swan, US Forest Service)

Industrialists have affected virtually the entire surface of the earth. Energy flows are substantially altered by the addition of fossil fuels and nuclear-based power. Today, oil has sometimes been considered almost a world currency because of political circumstances and its limited supply sources. Ecosystem stability or instability is now very much a product of human decision making rather than natural processes. This is the first century in which we have 'advanced' sufficiently to cause global instabilities, primarily because of our depletion of the ozone layer and increasing concentration of atmospheric carbon dioxide.

IMPACTS ON VEGETATION

When we consider human impacts on vegetation, animals, land and soil, water, and climate, it is appropriate to discuss vegetation first, as our modification of plant cover results in changing soils, climates, geomorphic processes, and water: 'Indeed, the nature of whole landscapes has been transformed by man-induced vegetation change' (Goudie 1981:25). Figure 4.2 summarizes some of the consequences of vegetation change.

Fire

For perhaps at least 1.5 million years, humans used fire deliberately to modify the environment. Initially, vegetation removal could result in increased animal numbers and increased mobility of human hunters. Fire offered security and a social setting at night, and encouraged movement to colder areas. For later agriculturalists, fire was a key method of clearing land for agriculture and improving grazing areas; fire continues to serve these and similar functions today. Indeed, deforestation by fire or other means has been prompted largely by the need to clear land for agricultural activities, both pastoral and arable. In Europe, large-scale deforestation occurred from about the tenth century onwards and, in temperate areas of European overseas expansion, occurred largely in the nineteenth century. Box 4.3 describes the European attitude to forest in one part of the New World. Deliberate burnings and the much more common natural fires — it has been estimated that lightning strikes some 100,000 times each day (Tuan 1971:12) — have drastically modified vegetation cover.

Fire has played a principal role in creating some of the vegetation systems discussed in the previous chapter — savannas, mid-latitude grasslands, and Mediterranean shrub lands are prime examples — and has probably affected all vegetation systems except tropical rain forests. Areas significantly affected by fire typically possess considerable species variety.

Plant domestication

Domestication is a process whereby a plant is so modified that the modified plant fulfils a specific human aim; once domesticated, the plant is permanently different from the original. This process is ongoing and is an important part of agricultural research today. Associated with plant domestication has been the labelling of other plants (that are not domesticated) as weeds — and their removal. Once again, such human activity contributes to ecosystem simplification. Early domesticates included wheat, barley, oats (southwest Asia), sorghum, millet (west Africa), rice (southeast Asia), yams (tropical areas), potato (Andes), and manioc and sweet potato (lowland South America). Other domesticates include such pulses as peas and beans and trees/shrubs such

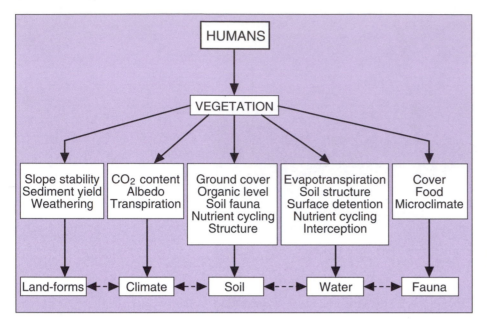

Figure 4.2 Some consequences of human-induced vegetation change.
Source: Adapted from A. Goudie, *The Human Impact: Man's Role in Environmental Change* (Oxford: Blackwell, 1981):25.

as peach and grape. Domestication involves the introduction and perpetuation of human-induced selection at the expense of natural selection. The choice of domesticates results from a combination of physical and cultural (preference) variables. Pastoral activities resulting from animal domestication can increase species diversity, but are characteristically detrimental to vegetation cover, causing overall soil deterioration and erosion.

Data collected by remote sensing are a versatile and effective tool for monitoring forestry operations. In Canada, Landsat and other imagery allow forest managers to collect information on forest inventory, depletion, and regeneration in areas as small as 2 ha (5 acres) and with boundary accuracy within 25 m (82 ft). Such data are invaluable in the accurate mapping of, for example, clear-cut areas.

Tropical rain forest removal

Human removal of vegetation — particularly tropical rain forest depletion and desertification — continues to be of major concern. Without human activity, forests would cover most of the land surface. Large-scale deforestation accompanied the rise of Chinese, Mediterranean, and western European civilizations and the nineteenth-century expansion of settlement in North America and Russia. Today, deforestation is concentrated in the tropical areas of the world. Viewed in this historical perspective, the current removal of rain forest is not excessive, but there are important differences between temperate and tropical deforestation. Tropical forests typically grow on much poorer soils that are unable to sustain the permanent agriculture now practised in temperate areas. Also, tropical rain forests now

---

Box 4.3: The threatening forest

Europe, once heavily forested, was by the eighteenth century a largely agricultural region with few remnants of the once-dominant forest. For eighteenth-century and later Europeans, cleared land represented progress and the triumph of technology.

When Europeans moved to temperate overseas areas, they confronted a very different environment. Eastern North America in particular was a densely wooded region. Aboriginal populations typically cleared only small areas and then often moved elsewhere, allowing the forest to regrow. Among the nineteenth-century British in Ontario, the prevailing attitude to the forest was one of antagonism: 'Consequently, settlers stripped the trees from their land as quickly as possible, shrinking only from burning them as they stood. They attacked the forest

with a savagery greater than that justified by the need to clear the land for cultivation, for the forest smothered, threatened and oppressed them' (Kelly 1974:67).

Forests were seen as indicative of a lack of progress. They were removed not simply because cleared land was needed for agriculture, but also because for the majority of settlers, the forest was oppressive and threatening, a symbolic triumph of nature over humans.

Deforestation in Ontario proceeded apace as humans triumphed over land. By the 1860s, settlers became aware of the disadvantages of deforestation, such as lack of shelter belts, fuel, and building materials. However, by then the damage was done.

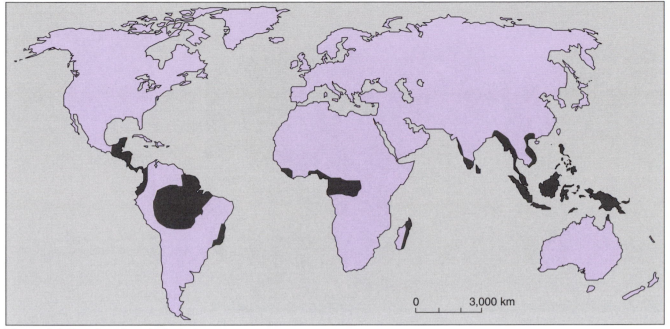

Figure 4.3 Location of tropical rain forests.

play a major role in the health of our global ecosystem — a fact that has only been significantly acknowledged in recent years. Figure 4.3 shows the distribution of tropical rain forests for the late 1980s. These rain forests cover about 8.6 million km^2 (3,320,720 square miles).

The current rate of depletion and the loss to date are both open to dispute. The rate of depletion may be as high as 100,000 km^2 (38,613 square miles) per year; a rate that if continued will result in a complete elimination of rain forest in the first half of the next century. Fortunately, remote sensing using Landsat and other satellite data permits some objective assessment of rates of clearance. These data showed that annual clearance rates in Brazil and elsewhere were several times greater than the early 1980s estimates made by the United Nations Food and Agricultural Organization (Repetto 1990). On the other hand, the World Bank's 1989 estimate of a 12 per cent loss of Brazilian rain forest was shown to be about twice the actual loss. Clearly, there are still uncertainties about the details of rain forest removal. Why are the rain forests being removed, and what are the ecological impacts?

Rain forests are located primarily in less developed areas of the world, such as Bolivia, Brazil, Colombia, Venezuela, Gabon, Zaire, Indonesia, and Malaysia, but the more developed areas of the world are a leading cause of deforestation as they have an enormous appetite for tropical timber and the inexpensive beef produced in areas cleared of tropical timber. In addition, poor people in the less developed world use cleared land for some subsistence farming. Such farming is often only possible for a few years because of cultivation techniques that rapidly deplete the soil of key nutrients. And cattle ranching becomes rapidly less profitable as the rain

forest soil supports grazing for only a few short years. Nevertheless, as already noted, from the perspective of the countries experiencing rain forest clearance, such activity can be seen as a means of reducing population pressure elsewhere and as generally equivalent to the resource and settlement frontier of, say, North America during the past 200 years and of Europe during the past 1,000 years.

There are two principal ecological consequences of rain forest removal. First, it is the major cause of species extinction because the rain forests are the home of at least 50 per cent of all species, which may total some 30 million. It is not easy to appreciate what such a statistic means. One way to interpret it is to acknowledge that:

> The human species came into being at the time of greatest biological diversity in the history of the earth. Today as human populations expand and alter the natural environment, they are reducing biological diversity to its lowest level since the end of the Mesozoic era, 65 million years ago. The ultimate consequences of this biological collision are beyond calculation and certain to be harmful. That, in essence, is the biodiversity crisis (Wilson 1989:108).

Another interpretation involves recognizing that many tropical forest species are very important or potentially important as foods, medicines, and sources of fibres and petroleum substitutes.

The second principal ecological consequence of rain forest removal involves global warming. Carbon is stored in trees and when burning occurs, the carbon is transferred to the atmosphere as carbon dioxide. In addition, soil is a source of carbon dioxide, methane, and nitrous oxide, all of

4.4 *(left)* Amazon rain forest, Brazil. (E.G. Munroe)

4.5 *(opposite page)* Badly eroded land in the Tennessee Valley, USA. (Tennessee Valley Authority)

which are released into the atmosphere as a result of forest removal and farming. Each of these gases contributes to what we now call the greenhouse effect — a topic that we will consider shortly.

Desertification

Desertification is land deterioration caused by climatic change and/or by human activities in semiarid and arid areas: 'It is the process of change in these ecosystems that can be measured by reduced productivity of desirable plants, alterations in the biomass and the diversity of the micro and macro fauna and flora, accelerated soil deterioration, and increased hazards for human occupancy' (Dregne 1977:324). The significance of desertification is not simply the clearing of vegetation but rather the consequences of clearing that include soil erosion by wind and water and possible alterations of the water cycle.

Deserts are a natural phenomenon whereas **desertification** is the expansion of desert areas as a result of either or both physical and human activity. The human causes are complex, but typically involve removal of vegetation as a result of overgrazing, fuel needs, intensive cultivation, and waterlogging and salinization of irrigated lands. These causes characteristically occur because of population pressure and/or poor land management. Fortunately, the technology to combat desertification is available, although only occasionally is there the will to use it (Box 4.4). Unfortunately, a major international effort to combat desertification, the 1977 United Nations Plan of Action, has been a failure. Two reasons for this failure were, first, that technical solutions were applied to areas where the key causes were economic, social, and political and these underlying causes were not addressed; second, local populations were not involved in the search for solutions. A potentially important recent development is discussion concerning an international convention to combat desertification as first proposed at the 1992 Earth Summit in Rio de Janeiro.

Estimates of the spatial extent of desertification in 1977 and 1992, both by the United Nations Environment Program, arrived at a figure of 3.5 million ha (8,648,000 acres) affected. Remote-sensing imagery and local area surveys have not confirmed this figure and there is therefore considerable confusion concerning the actual spatial extent. Most publicized of the areas experiencing desertification is the Sahel zone of West Africa, an area that was first brought to world attention following the 1968–73 drought. Here, desertification is caused by population pressure, inappropriate human activity, human conflict, and periods of drought. Elsewhere, desertification similarly has multiple causes and no simple solution. Many of the human causes of desertification may be related to the pressures placed on local people by the introduction of capitalist imperatives into traditional farming systems.

As is the case with the tropical rain forests, areas subject to desertification are the home of poor people who are not necessarily able to adopt appropriate remedies and lack the necessary political influence. A proper solution requires that the dry land ecosystems be treated as a whole — land management is needed. Further, population pressures need to be reduced; land needs to be equitably distributed, and greater security of land tenure is required.

IMPACTS ON ANIMALS

Animal domestication serves many purposes, providing foods such as meat and milk (cows, pigs, sheep, goats), as well as draught animals (horses, donkeys, camels), and pets (dogs,

Box 4.4: Defeating desertification

The Kenyan Green Belt Movement had its origins in a 1974 Nairobi tree-planting scheme that focused on the value of working on a community level and especially with women (Agnew 1990). The objectives of the movement are many and varied, as befits any effort to solve so difficult a problem as desertification.

The central objective is to reclaim land lost to desert and to guarantee future fertility. Specific objectives include conserving water, increasing agricultural yields, limiting soil erosion, and increasing wood supplies. The ideas of working on the local level and involving women are crucial. Reclaiming land is important to the local people and their direct involvement makes the exercise much more meaningful to them, while women (the principal wood gatherers in Kenya and in most of the Sahel) are increasingly aware of future needs. Planting crops around trees improves yields — an important and direct consequence in a subsistence economy.

Tree planting to combat desertification is far from being a panacea, but it is proving to be a positive development. Other parts of the world are adopting the strategies of the Kenyan movement, particularly its community-based focus. The parallels with the Grameen Bank (Box 6.5) are intriguing.

cats). Once domesticated, animals have often been moved from place to place, both deliberately and accidentally. Some deliberate introductions, such as that of the European rabbit to Australia, have had drastic ecological consequences. Whalers and sealers were probably the first to introduce rabbits into the Australian region, but the key arrival was in 1859 when a few pairs were introduced into southeast Australia to provide 'sport' for sheep station owners. Following this introduction, rabbits spread rapidly across the non-tropical parts of the continent, prompting a series of 'unrelenting, devastating' (Powell 1976:117) rabbit plagues. Rabbits consume vegetation needed by sheep and remain a problem today despite the introduction of the disease, myxomatosis, and of rabbit-proof fences (Figure 4.4). The European rabbit in Australia has probably caused more damage than any other introduced animal anywhere in the world, but rabbits are only one of many unwanted guests (Box 4.5).

As human numbers and levels of technology have increased, so also has the number of animal extinctions. It is possible that humans have caused species extinction since perhaps 200,000 BC, but the evidence is uncertain. It is certain, however, that hunting populations have caused extinctions. After Europeans arrived in New Zealand, all moa species became extinct, and there is much evidence to suggest that animal extinctions in North America coincided with human arrival. The 1859 publication of *The Origin of Species* by Darwin helped place the extinction of plant and animal species in context and was followed by protectionist legislation in several of the British colonial areas; an early example is the 1860 law in Tasmania to protect indigenous birds.

Today, our role is clearer. When natural animal habitats are removed, as in the case of tropical rain forest, extinction follows. This is one component of what we can regard as a major threat to biodiversity.

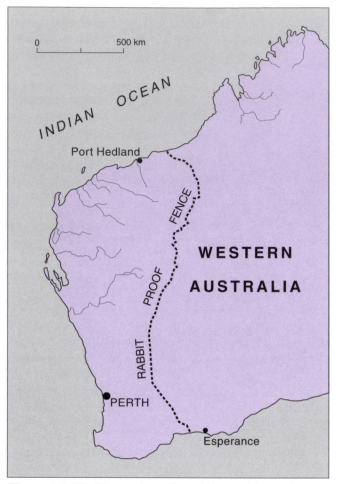

Figure 4.4 The Western Australian rabbit-proof fence. This 1,139-mile (1,833-km) long fence was built in the 1890s in an attempt to confine rabbits to the central desert, but it was built too late; rabbits had moved west of the fence before it was completed.

Box 4.5: Unwanted guests

Beginning in 1788, the European colonization of Australia ended a long period of isolation and introduced numerous animals and plants. Many of these introductions, both deliberate and accidental, were able to proliferate because they were placed in a new and different ecosystem. Inevitably, they have affected the native animals that evolved for millions of years in isolation from other areas and other animals.

Rats and mice that travelled in the holds of ships multiplied rapidly after being accidentally introduced. Sheep and cattle were brought over to satisfy British economic needs, while animals brought over to satisfy the 'sporting' needs of British settlers included deer, fox, and, of course, the rabbit. The interior deserts prompted the introduction of pack animals such as burros, asses, and camels, and soon there were more camels in Australia than in all of Arabia. Many varieties of introduced livestock that have reverted to a feral (wild after previously being domesticated) existence include pigs, horses, camels, and water-buffalo.

Feral water-buffalo do considerable damage to forest ecosystems by destroying trees and eroding soil. More generally, all feral species require a supply of that most precious resource in the interior, water, thus reducing the supply for native species. Many of the native animals of Australia have disappeared because of the presence of the introduced species. Of the smaller marsupials, seventeen have become extinct and another twenty-nine are considered endangered.

A recent introduction that is causing considerable damage is the cane toad, first brought to Queensland in the 1930s to combat the cane grub that was damaging the sugar cane crop. Unfortunately, cane toads do not only eat cane grubs — they eat almost anything. They are also capable of much more rapid reproduction than native toad species, with females producing up to 40,000 eggs a year; they compete effectively with the native species, and they have no natural predators. Cane toads have moved far out of the sugar cane area and there are no signs of an end to their movement.

4.6 Aerial view of pipe open-pit mine. (INCO)

IMPACTS ON LAND AND SOIL

We live on the land and use soil extensively; hence, humans are a geomorphic agent changing land and affecting that thin and vulnerable resource, soil.

Most human activities actually create land-forms. Excavation of resources such as common rocks (limestone, chalk, sand, gravel), clays (stoneware clay, china clay), minerals (dolomite, quartz, asbestos, alum), precious metals (gold, silver), and fossil fuels (oil, coal, peat) can have major impacts: changing ecosystems, lowering land surface, flooding, building waste heaps, creating toxic wastes, and leaving scenic scars. In addition to excavation and related dumping, other human activities make us geomorphic agents: river channels are often modified and sand dunes are affected; other effects include coastal erosion and coastal deposition. There is a multitude of causes and consequences of human impact on land.

Degradation and loss of arable land are occurring throughout the world as a result of population increases, industrialization, and improper agricultural practices. Smil (1993:67) reported that the average annual loss of farmland in China between 1957 and 1980 was a mammoth 1 million ha (2,471,000 acres).

By its very nature, soil is especially susceptible to abuse, and humans use and abuse soils extensively. Agricultural activities are the major human activities affecting soil, while major negative consequences include chemical changes involving salinization, laterization, soil erosion. Humans increase soil salinity largely through irrigation with increased

4.7 Dust Bowl landscape in the 1930s. (Agriculture Canada, Lethbridge, Alberta)

salinity negatively affecting plant growth. Humans increase the laterite content of soil (laterite is an iron- or aluminum-rich duricrust naturally present in tropical soils) by removing vegetation. Laterite is essentially hostile to agriculture.

Soil erosion is associated with deforestation and agriculture. Forests protect soil from run-off and roots bind soil. Probably the best-known example of human-induced soil erosion is the 'Dust Bowl' in the North American mid-latitude grasslands. The various causes of this 1930s phenomenon included a series of low rainfall years, overgrazing, and inappropriate cultivation procedures associated with wheat farming. Combined, these causes created 'black blizzards' that in turn led to out-migration.

IMPACTS ON WATER

Water is an essential ingredient of all life. We know this and yet we choose to ignore it. Rather than carefully safeguarding water quantity and quality, we cause shortages and continually contaminate it. These two issues of scarcity and contamination dominate our consideration of human impact on water.

The global water cycle

How much water is available? Figure 4.5 outlines the global water cycle, identifying three principal paths — precipitation, evaporation, and vapour transport. Total annual global precipitation is estimated at 496,000 km³ (118,990 cubic miles), most of which (385,000 km³/92,361 cubic miles) falls over oceans and cannot be easily used. Water returns to the atmosphere via evaporation from the oceans (425,000 km³/101,957 cubic miles) and from inland waters and land along with transpiration from plants (71,000 km³/17,032 cubic miles combined). In addition, some of the precipitation that falls on land is transported to oceans via surface run-off or groundwater flow (41,000 km³/9,836 cubic

miles), and some water evaporated from the oceans is transported by atmospheric currents and subsequently falls as precipitation over land (again, some 41,000 km³/9,836 cubic miles). In principle, 41,000 km³ (9,836 cubic miles) are available each year globally, but this figure is reduced by 27,000 km³ (6,477 cubic miles) lost as flood run-off to the oceans and by another 5,000 km³ (1,199 cubic miles) flowing into the oceans in unpopulated areas. Perhaps 9,000 km³ (2,159 cubic miles) are readily available for human use.

This total of 9,000 km³ (2,159 cubic miles) is possibly enough water for 20 billion people. However, some states have a plentiful amount and others have an inadequate amount. Where water is abundant, it is treated as though it were virtually free; where it is scarce, it is a precious resource. Thus the average citizen of the United States annually consumes seventy times more water than does the average citizen of Ghana.

Using and polluting water

Agriculture makes the main use of water, consuming 73 per cent of global supplies, often highly inefficiently. Industry consumes perhaps 10 per cent of global supplies, while the third principal use is basic human needs. As each of these needs increases in any given area, whether through agricultural and industrial expansion or population increases, water quantity may be a problem. Shortages are common in many areas on either a continuous or periodic basis. Bahrain, for example, has virtually no fresh water and relies on the desalinization of sea water. Groundwater depletion is common in the United States, China, and India, while water levels have fallen in both Lake Baikal and the Aral Sea.

The second major issue, in addition to that of water quantity, concerns water quality. As water passes through the cycle described in Figure 4.5, it is polluted in two ways. Organic waste (from humans, animals, and plants) is

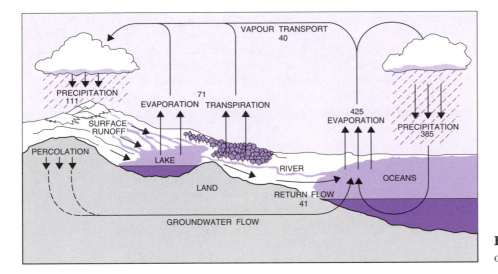

Figure 4.5 The global water cycle.

biodegradable, but can still cause major problems in the form of oxygen depletion in rivers and lakes and in the form of water contamination, causing such diseases as typhoid and cholera. Much industrial waste is not easily degraded (paper, glass, and concrete are exceptions) and such wastes are a major cause of deteriorating water quality. These pollutants enter water via pipes from industrial plants, diffuse sources (run-off water containing pesticides and fertilizers), and the atmosphere (acid rain).

Both inland waters and oceans are suffering the consequences of pollution (Box 4.6). Notorious examples include the Aral Sea and the toxic chemical waste in waters at Love Canal in the United States (refer to Fig 2.19). Both Lake Erie and the Rhine, especially in the Netherlands, are suffering from more general urban industrial pollution; such surface water pollution is reversible, as evidenced by the positive results of the Rhine Action Plan. Acid rain is a more difficult issue because the pollution source can be in one location and the actual pollution at another distant location. Acid rain is one general consequence of urban and industrial activity that releases large quantities of sulphur and nitrogen oxides into the atmosphere. The effects of acid rain are not entirely clear, but there is no doubt about its negative impacts on some aquatic ecosystems. One of the most severe problems involved in reducing water pollution is the need for international cooperation. The Rhine Action

---

Box 4.6: Changing and polluting water bodies in Asia, Europe, and North America

The Aral Sea on the border between the central Asian states of Kazakhstan and Uzbekistan (both formerly a part of the Soviet Union) was the fourth largest inland water body in the world until the 1960s. Today, remote-sensing imagery shows that the coastline has retreated about 80 km (50 miles) because of a reduced inflow, resulting in a 40 per cent reduction in size. The explanation is that, since the 1950s, much of the inflowing water has been used to irrigate cotton. Further, because cotton requires large quantities of defoliants, pesticides, and fertilizers, the Aral Sea has become dangerously polluted. In the local area, there are high rates of infectious diseases, cancers, miscarriages, and fetal abnormalities.

The Rhine River receives pollutants from heavily industrialized areas in addition to domestic sewage and shipping discharges. Considerable treatment is necessary before this water can be used for drinking or crop irrigation. The Netherlands, which requires Rhine water for vegetable crops, receives water polluted by six countries over 1,300 km (807 miles). Polluted water also changes local ecosystems, destroying much animal and plant life.

When Europeans first reached Lake Erie, it was clear and supported a large fish population. Today, Erie is often considered the most seriously polluted large water body in the world, a consequence of having about 13 million people in its watershed. Domestic sewage, household detergents, agricultural wastes, and industrial wastes have gradually combined to pollute Lake Erie. Clear water is now green, algae thrive, and fish have perished. The biogeochemical cycle of Lake Erie has been disturbed. Untreated sewage begins a cycle of oxygen loss, water layering according to temperature, and the death of cold-water fish species such as lake trout, herring, blue pike, and whitefish. Even when sewage is treated, oxygen loss continues. Algae growth is enhanced by the addition of phosphates and nitrates. Ultimately a lake so afflicted will become a marsh full of weeds.

Saving the Aral Sea, the Rhine River, and Lake Erie is not easy. Ways must be found to balance legitimate economic interests and valid environmental concerns. For both the Rhine River and Lake Erie, international cooperation is of prime importance; the Rhine Action Plan has already achieved some successes.

Plan involves four countries, but attempts to combat acid rain, especially in Europe, require much greater cooperation.

Nowhere is the need for international agreement more evident than in combating ocean pollution. Many states exploit oceans, but no state is prepared to assume responsibility for the effects of human activities. More than 50 per cent of the world's population live close to the sea and most of the world's ocean fish harvest is gathered from coastal waters. Water quality in the oceans, especially coastal zones, is seriously endangered and many ocean ecosystems have been damaged. Remote sensing enables objective assessments of water pollution; images of the Mediterranean Sea show a marked contrast between the northern shore, heavily polluted by water from major European rivers and coastal towns, and the southern shore. Remote sensing permits effective monitoring of oil spills, such as the Exxon Valdez incident off the Alaskan coast in 1989. Once again, we do not know enough about the consequences of our activities, but we do know that restoring ocean water quality is likely to be much more difficult than is the case with surface inland water.

IMPACTS ON CLIMATE

Nowhere is the concept of interrelations better demonstrated than in a consideration of human impacts on climate. Our impacts on other components of the ecosystem affect climate and many environmentalists feel that our most damaging impact is on global climate. Unfortunately, as with our discussions of other human impacts, we are confronted with two general areas of uncertainty relating to the role played by humans (as opposed to physical factors) and to the extent of any human-induced change.

Physical causes of climatic change

Climatic changes not prompted by humans occur on various time scales and thus there are a multitude of causes. It appears that some changes result from external variables, such as variation in the input of solar radiation, but also that climate experiences ongoing oscillations regardless of such factors. It has recently been determined, for example, that long-term changes are related to the changing distribution of land and sea with the current distribution conducive to the emergence of glacial periods in North America and Eurasia. The most recent such glacial period terminated some 12,000 years ago. Other climatic changes occur at time scales of a few hundred to a few thousand years and these changes are not well understood. During our current interglacial period (see Chapter 3), there has been a cool phase 10,000–11,000 years ago, a distinctly warm phase 6,000 years ago, and another warm period about 1,000 years ago. Historical evidence shows wine cultivation abandoned in fourteenth-century England and a cold Europe in the sixteenth and seventeenth centuries.

It is important to consider human impacts in the above perspective. We know that we are capable of changing climate, but we are less sure of precisely how this will affect future climates because we are uncertain about future changes caused by non-human variables. We know that we are moving towards another glacial period, but we do not know what shorter-term changes may naturally occur prior to the onset of a new ice age.

The natural greenhouse effect

Temperatures on the earth's surface result from a balance between incoming solar radiation and loss of energy from earth to space. If the earth had no atmosphere, then the average surface temperature would be about –19° C, (–3° F) but the presence of an atmosphere results in an actual average surface temperature of 16° C (61° F). The atmosphere causes an increase of surface temperatures because it prevents about one-half of the outgoing radiation from actually reaching space; some is absorbed and some bounces back to earth. This natural 'greenhouse' effect is not related to human activity but is the result of the presence in the atmosphere of water vapour, carbon dioxide (CO_2), ozone

---

Box 4.7: Chloro-fluorocarbons and the atmosphere

Perhaps nothing better exemplifies the devastating effect that our human needs, satisfied by technological advances, can have on the environment. Chloro-fluorocarbons (CFCs) were not even synthesized until the late 1920s; at that time they were a remarkable advance. They are ideal as coolants because they vaporize at low temperatures and also serve well as insulators. Most importantly, they are easy and therefore inexpensive to produce, hence their widespread use since the Second World War as coolants in refrigerators, as propellant gases in spray cans, and as an ingredient in a wide range of plastic foams.

Realization of the impact that CFCs have on the atmosphere, especially on ozone, prompted twenty-four countries to gather in Montreal in 1987 and agree to limit CFC production by reducing production by 35 per cent by 1999. Most environmental experts argue that this reduction is inadequate. Subsequently, a 1990 London agreement set the goal of eliminating CFC production by 2000. Because CFCs have a long atmospheric lifetime, peak levels will be reached about a decade after production ceases. There is also a need to recycle CFCs.

What is really needed is, of course, a substitute for CFCs. So far, possible substitutes are more expensive to produce than CFCs and world economies, whether capitalist or socialist, are not keen to sacrifice short-term economic benefits for long-term planetary stability.

(O_3), and other gases. These greenhouse gases are but a fraction of the atmosphere as nitrogen and oxygen combine to make up 99.9 per cent of the atmosphere (excluding the widely varying amounts of water vapour), but their impact is nevertheless considerable. Today we are increasing the greenhouse effect by adding CO_2 and some other gases that perform a similar function, such as sulphur dioxide (SO_2), nitrous oxide (N_2O), methane (CH_4), and a variety of chloro-fluorocarbons (CFCs). How are we doing this?

Human-induced global warming

In the most general sense, our increasing the natural greenhouse effect is a result of our increasing numbers and more advanced technology. More specifically, it is the result of burning fossil fuels, increased fertilizer use, increased animal husbandry, and deforestation. Until recently, much carbon was stored in the earth in the form of coal, oil, and natural gas. Burning these resources releases CO_2, water vapour, SO_2, and other gases that are then added to the atmosphere. Estimates suggest that the concentration of CO_2 in the atmosphere has increased from 260 ppm (parts per million) 200 years ago to 350 ppm today and might be 400 to 550 ppm in 2030. Burning wood also adds CO_2 to the atmosphere. Further, soil contains large quantities of organic carbon in the form of humus, and agricultural activity speeds up the process by which this carbon adds CO_2 to the atmosphere. Agricultural activity also adds CH_4 and N_2O to the atmosphere. CH_4 is increasing particularly as a result of paddy rice cultivation.

Each of the greenhouse gases noted so far (CO_2, N_2O, SO_2, CH_4) is naturally present in the atmosphere and our impact is that of increasing natural concentrations. But we are also adding new greenhouse gases, the most important being two of the CFCs — $CFCL_3$ and CF_2CL_2. These gases are added primarily by our use of aerosol sprays, refrigerants, and foams (Box 4.7).

The direct result of our increasing the greenhouse effect is a change in average temperatures, with an increase of between 1.5° C (2.7° F) to 4.5° C (8° F) suggested for the year 2030. During the last 120 years, the average temperature has increased 0.5° C (1° F), but we are unclear as to whether this is a result of human activity. The best way to detect global temperature change is to use satellites to measure ocean temperatures; ocean temperatures are not subject to the same diurnal and seasonal fluctuations as are air temperatures.

If a general warming does indeed occur, then it will be the polar regions that are most affected. Other than this observation, we can anticipate little about the spatial variations of any global warming. Again, it is worth emphasizing that our inability to predict detailed consequences is partly related to natural spatial and temporal climatic change. Suggested consequences of human-induced change include a poleward retreat of cold areas with resultant expansion of forests and agriculture, changing distribution of arid areas, and, of course, a rising sea level as a result of ice-cap melting (Box 4.8 and Figure 4.6).

Damaging the ozone layer

A second atmospheric concern is that of the ozone layer. Ozone (O_3) is a form of oxygen that occurs naturally in the cool upper atmosphere. It serves as a protective sunscreen for the earth by absorbing ultraviolet solar radiation that is potentially harmful, causing skin cancer, cataracts, and a weakening of the human immune system. Ultraviolet radiation is also damaging to vegetation. Ozone depletion was first recognized in 1985 by scientists with the British Antarctic Survey. An ozone hole over Antarctica appears to be about one-half the size of Canada. The principal culprits are the CFCs already noted and some greenhouse gases ($CFCL_3$ and CF_2CL_2). As these rise into the atmosphere, chemical reactions occur and ozone is destroyed.

Urban climates

Any discussion of human impacts on climate must also consider urban climates. Not surprisingly, any urban development affects local climate. Built-up areas store heat during the day and release it at night, and also generate artificial heat, creating an 'urban heat island'. Built-up areas also affect cloud formation and precipitation, but these impacts are more difficult to determine and measure. Perhaps the most obvious impact of all involves the creation of smog (smoke-fog). Chemical smogs are common in areas such as Los Angeles and result largely from automobile emissions.

EARTH'S VITAL SIGNS

As the preceding account makes clear, there is no doubt that our current impacts on ecosystems, from the global to the local, are greater than ever before and are increasing as a general result of the growth of population and technology. Further, despite much contradictory evidence, there is little doubt that our global ills are serious and that they require international cooperation if they are to be appropriately addressed (see Box 4.2); as already noted, the first such international agreement was the 1987 Montreal accord limiting the production of CFCs. There is much debate, however, concerning the present condition of the environment and the probable future scenario. As Smil (1993:35) stated, 'Confident diagnoses of the state of our environment remain elusive.'

APOCALYPSE NOW, DEFERRED, OR NEVER?

There are many different opinions about the impact of human activities on the environment. At one extreme are catastrophists who view the current situation and future prospects in totally negative terms, a view recently articulated by Kaplan (1994), while at the other extreme are cornucopians who feel that current problems are greatly exaggerated and that human ingenuity and technology will overcome these problems (Simon and Kahn 1984). A balanced view of such an emotional topic is not easy, but there is no doubt that the environment is sustaining considerable damage and that both individual and collective action are needed to rectify this situation. On the other hand, as Smil (1993:5) convincingly argues, since the 1960s the environment has become the subject of many ill-informed commentaries and predictions: 'Once the interest in environmental degradation began, the western media, so diligent in search of catastrophic happenings, and scientists whose gratification is so often achieved by feeding on fashionable topics, kept the attention alive with an influx of new bad news.' As our discussion of human impacts has demonstrated, it is not an easy task to separate the many very real problems from the numerous exaggerated claims.

RESPONDING TO UNCERTAINTY

In view of the many uncertainties, Smil (1993:36) asserted that:

> The task is not to find a middle ground: the dispute has become too ideological, and the extreme positions are too unforgiving to offer a meaningful compromise. The practical challenge is twofold. First, to identify and to separate the fundamental long-term risks to the integrity of the biosphere from less important, readily manageable concerns. The second task is to separate effective solutions to such problems from unrealistic paeans to the power of human inventiveness.

The first challenge is continually being addressed by geographers and other environmental scientists. Given our current

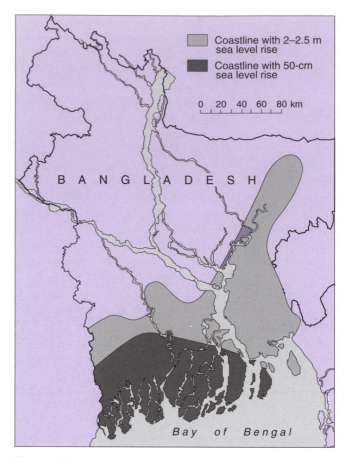

Figure 4.6 Impact of sea-level change on Bangladesh.
Source: G.A. McKay and H. Hengeveld, 'The Changing Atmosphere', in *Planet Under Stress*, edited by C. Mungall and D.J. McLaren (Toronto: Oxford University Press, 1990):66.

Box 4.8: Rising sea levels

Global warming is predicted to affect sea levels. A rise of as much as 1.5 m (almost 5 ft) has been suggested for 2050. If this occurs, the consequences are potentially catastrophic for many currently populated areas. Up to 15 per cent of Egypt's arable land would be at risk, and many coastal cities such as New York and London would be below sea level. The consequences for two of the most densely populated areas in the world — coastal Bangladesh and the Netherlands — would be disastrous unless we are able to adapt.

The Netherlands has successfully adapted to a situation where much of the current area is already almost 4 m (13 ft) below sea level. In principle, the Netherlands model — construction of levees and dikes — can be followed. Venice is already pursuing a similar tactic, constructing a flexible seawall to protect the city against Adriatic storms. But what are the costs of such endeavours? One estimate places the cost around US $300 billion to protect major areas only, not general coastal margins.

There is a second alternative to adaptation — moving away from the threatened areas. For many people in the less devel-

oped world, this is barely an option; for many in the more developed world, it is culturally and economically unthinkable at the present time.

But our prediction of sea level rise is at best uncertain. In 1985, one major United States scientific committee predicted a sea level rise of 1 m (3 ft) by 2100, and subsequently amended the prediction in 1989 to a rise of .3 m (1 ft). Even the higher of these two is less than the prediction referred to earlier. Why such uncertainty? The principal unknown concerns the effects of warming on the Antarctic ice sheet. Rather than increased ice melt, it is possible that warmer weather could increase snowfall, which would in turn help build up the ice sheet.

The problem is a very serious one. We need to adapt soon to an unknown situation. Dikes and population movements may or may not be needed. If they are needed, we do not know the extent of the need. Global uncertainty becomes local political uncertainty.

4.8 Yosemite National Park, California, USA. (Yosemite Concession Services Corporation)

understanding of global environmental problems, there are three possible general responses to the second challenge. First, we can attempt to develop new technologies to counter our deleterious impacts. For example, some scientists have suggested that dust might be deliberately spread in the upper atmosphere to reflect sunlight — replacing the depleting ozone layer. Overall, it seems inappropriate to rely too much on such endeavours. Despite massive evidence that we are changing climate on a variety of scales — local to global — we still know too little to deliberately change climate in ways we desire. The results of attempts at rain making, hurricane modification, and fog dispersal have been mixed. Second, we might acknowledge that environmental impacts are inevitable and emphasize our own need to adapt to such changes. In the case of possible climatic change, adaptation might involve new water-supply systems and coastal defences.

The third, probably most popular and most logical response, involves the conservation of resources and the prevention of harmful impacts. **Conservation** refers generally to any form of environmental protection. Prevention involves limiting the increase of greenhouse gases, reducing the use of certain materials, reusing and **recycling**. The reason for recycling may be economic or environmental.

Prevention is central to many current moves to protect environments. It can be argued that economic systems, including capitalism, do not properly reward the efficient use of resources and that the physical environment has been seen as irrelevant in the final economic accounting. Only recently have we begun to understand that price and value are not equal. But how do we assign a price to the value of a wilderness landscape, for example? One answer, although imprecise, is to assert that certain ecosystems or landscapes are sufficiently distinct as to merit protection or preservation. The largest protected ecosystem today is probably that of Antarctica. Examples of other protected landscapes include many national and other parks and wilderness regions with the first such preserved areas being Yosemite (1864) and Yellowstone (1872) in the United States. Canada has national and provincial parks.

What seems to be needed — to improve the health of the earthly patient of which we are all a part — is education about and understanding of the need for reducing harmful human impacts. Solutions are needed at all spatial and social scales.

SUSTAINABLE DEVELOPMENT

To conclude this chapter, we return to the key concept of sustainable development. Clearly, we are transforming the earth in ways that we are not intending and, equally clearly, we need to manage the earth along appropriate pathways — specifically along the pathway of sustainable development. Management requires us to understand what kind of earth we want, to reach consensus, and to assign appropriate values on matters of economic change and conservation. Such aims will be difficult to achieve, if only because people in different areas live in very different circumstances and have very different value systems. But regardless of such variations, relations between humanity and the land need to be such that sustainable development is achieved — that environment and economics be central concerns.

Simply put, **sustainable development** is economic development without harming the environment. A more detailed definition is 'Sustainability is the nascent doctrine that economic growth and development must take place, and be maintained over time, within the limits set by ecology in the broadest sense — by the interrelations of human beings and their works, the biosphere and the physical and chemical laws that govern it' (Ruckelshaus 1989:167). This definition is very similar to that of deep ecology, introduced earlier and summarized in Table 4.1. It can also be explained using earlier systems concepts. The earth can be regarded as a closed system in the sense that although energy enters and leaves the system, matter only circulates within the system. This type of system can reach a dynamic equilibrium state — one that involves optimal energy flow and matter cycling so that the system does not collapse. Thus, we can equate a dynamic equilibrium with a sustainable development situation.

How do we move towards a sustainable world, a world where changes are in accord with sound ecological principles? Any such move clearly requires a significant and deliberate change in attitude. There are three essential components to the new attitude.

1. We need to recognize that humans are a part of nature. To destroy nature is to destroy ourselves.
2. We need to account for environmental costs in all our economic activities.
3. We need to understand that all humans deserve to achieve acceptable living standards. A world with poor people cannot be a peaceful world.

There are, of course, vast differences between acknowledging the need for a new attitude, acquiring that attitude globally, and, finally, achieving those attitudes in practice.

Many strides have been made in the right direction. Clean air acts, environmental impact assessments, and departments of the environment are now standard in many countries. It would be an exaggeration to assert that solutions are in sight, or even that they are always being sought (Box 4.9). However, environmental issues are recognized as relevant at all levels, from individuals to governments. This is not equally so for all countries and it is the clear responsibility of the more developed world to demonstrate environmental concern by example:

> ... in creating the consciousness of advanced sustainability, we shall have to redefine our concepts of political and economic feasibility. These concepts are, after all, simply human constructs; they were different in the past, and they will surely change in the future. But the earth is real, and we are obliged by the fact of our utter dependence on it to listen more closely than we have to its messages (Ruckelshaus 1989:174).

SUMMARY

Use or abuse?
Today, there is considerable evidence that we are damaging our home and an increasing realization that our home is a fragile one.

Scale
The value of a global perspective in any discussion of human use of the earth is clear — everything is related to everything else.

The increasing human impact
Two variables — sheer numbers of people and increasing technology and energy use — help us understand why we currently exert major impacts on the earth.

Ecosystems
Combining systems logic and ecological principles results in the ecosystem concept. An ecosystem is any self-sustaining collection of living organisms and their environment. We have a global ecosystem — ecosphere or biosphere — which is the home of all life on earth, and a whole series of ecosystems

---

Box 4.9: Religion and environmental concern

Religious borders in Europe, particularly between Protestantism in the north and Catholicism in the south, are reflected in the success of recent environmental movements. Green parties are important today in the Protestant north — Scandinavia, Germany, and the Netherlands — while they are less evident in the Catholic south — France, Spain, and Italy.

In Chapter 8 we briefly consider the impact of religion and tradition on attitudes towards land. According to some, Protestantism has a much greater concern with individual freedom and hence an ultimate responsibility for human actions. Catholicism turns to confession, penance, and a belief in purgatory and thus there is less of a sense of responsibility for actions. With reference to the environment, then, Protestants may see themselves as having authority over the environment and, at the same time, being responsible for its welfare. Catholics, on the other hand, may have a respect for the environment that is less conducive to correcting deleterious impacts on land.

that combine to comprise this global version. The ecosphere includes air, water, land, and all life; it is a closed system as energy flows through while matter is cycled within.

Ecosystem simplification

When humans change an ecosystem, the characteristic consequence is a simplification. We cause changes largely because of our numbers, our technology, and energy use. Today, we number in excess of 5.3 billion and our technology — that is, our ability to convert energy into forms useful to us — is ever increasing. Examples of technological change include the deliberate use of fire, animal and plant domestication, and the use of fossil fuels associated with the Industrial Revolution.

As human culture (especially technology) changes, so does the evaluation of resources. Further, different interest groups evaluate resources according to different criteria and hence have differing views. Stock resources are finite in quantity, while renewable resources are relatively unlimited in quantity.

Our simplification of ecosystems is now sufficiently evident that a new environmental ethic is needed — one involving cooperation with, not domination of, nature.

Human impacts

Living on the earth results in changes to the planet. Some changes are inevitable and necessary to human survival. Other changes result from inappropriate actions that reflect human greed and misunderstandings. A consideration of changes through time confirms our increasing ability to make an impact on all levels of ecosystems. It is convenient — but necessarily misleading — to consider impacts on specific parts of ecosystems such as vegetation, animals, land and soil, water, and climate. In each case we have caused many and varied changes.

Vegetation change is especially related to our use of fire, grazing, and deforestation. Two principal issues today involve tropical rain forest depletion and desertification; in both cases, human activity is at least partially explained by the problems experienced by poor people in less developed countries. Impacts on animal life include those associated with domestication and species extinction. Extensive changes to land surfaces result from a host of human activities such as resource excavation and the creation of urban industrial complexes. Soil is especially susceptible to abuses involving salinization, laterization, and erosion. Water, an essential ingredient for life, is typically used so that shortages result and contamination occurs. Even the oceans are being polluted, a problem that requires international cooperation if solutions are to be found.

The ecosystem concept of interrelatedness is especially useful in a consideration of climate. We are having a significant effect on the atmosphere by increasing the quantity of greenhouse gases and by damaging the ozone layer; such impacts need to be considered in the light of our understanding of non-human-caused climatic change. We will probably cause a general warming of the earth's surface in the order of 1.5° C (1.5° F) to 4.5° C (8° F) by about 2030.

The health of the patient

Our global, human-induced ills are serious. Probably the best solution is that of prevention — an argument advanced by many environmentalist groups. It can be argued that economic systems, both capitalist and communist, do not have a reward system in place for proper use of the environment.

Sustainable development

Sustainable development serves our present needs, but does not compromise the ability of later generations to meet their needs.

WRITINGS TO PERUSE

BROWN, L.R., et al., eds. 1994. *State of the World*. New York: W.W. Norton.
   An annual publication by the Worldwatch Institute on progress being made or not made towards a sustainable society.

CIPOLLA, C.M. 1974. *The Economic History of World Population*, 6th ed. Harmondsworth: Penguin.
   An excellent survey with a strong focus on the importance of energy sources as a means of understanding economic change.

FRIDAY, L.E., and R.A. LASKEY, eds. 1988. *The Fragile Environment*. Cambridge: Cambridge University Press.
   A series of essays by specialists in a variety of disciplines addressing many of our basic themes, but always with an ecological focus.

GOODIN, R.E. 1992. *Green Political Theory*. Oxford: Polity.
   A novel and stimulating book arguing that environmental policy needs to be guided by an appreciation of the value that humans sense for themselves in relation to nature.

JOHNSON, C. 1991. *The Green Dictionary: Key Words, Ideas and Relationships for the Future*. London: McDonald Optima.
   A useful source for definitions, and sometimes discussions, of a wide variety of terms relevant to human impacts on the environment.

JOHNSON, D.L., ed. 1977. 'The Human Face of Desertification'. *Economic Geography* 53:317–432.
   A theme issue that contains nineteen articles many of which offer specific examples of desertification in a wide variety of spatial, cultural, and technological contexts.

MANNION, A.M. 1991. *Global Environmental Change*. New York: Wiley.
   A substantial book that reviews environmental change over the last 3 million years.

MUNGALL, C., and D.J. MCLAREN, eds. 1990. *Planet under Stress*. Toronto: Oxford University Press.

This volume is produced for the Royal Society of Canada and is an excellent up-to-date account of a wide range of issues. It includes a full-colour pictorial introduction.

PHILLIPS, D.E., A. WILD, and D.S. JENKINSON. 1990. 'The Soil's Contribution to Global Warming'. *Geographical Magazine* 62, 4:36–8.

One example of a brief popular piece that succinctly summarizes the important links between soils and the greenhouse effect.

PORTER, G., and J.W. BROWN. 1991. *Global Environmental Politics*. Boulder: Westview.

A good discussion of the links between environmental issues and other aspects of public policy; includes evaluations of how well national governments and international organizations are responding to environmental challenges.

SIMMONS, I.G. 1989. *Changing the Face of the Earth*. Oxford: Blackwell.

A detailed text that gives coherence to a vast array of factual information with an ecological viewpoint by focusing on energy flows through ecosystems.

SMIL, V. 1990. 'Planetary Warming: Realities and Responses'. *Population and Development Review* 16:1–29.

A clear and objective account of global warming that concludes that warming may lead to desirable changes in economic and population policies.

5

The human population: History and concepts

AT THE TIME OF WRITING, the world population is estimated at 5.6 billion. By the time you read these words, that figure will be much higher; the United Nations estimates 7 billion by the year 2010, 8.4 billion by 2025, and a stable population of about 10 billion by 2100. More significant than this numerical increase, however, is the fact that those parts of the world currently least capable of supporting increased numbers are the areas experiencing the greatest increases. Hence, an ever-present theme in this and the following chapter is that of increasing numbers of people and of the subsequent uncertainties that this increase implies for human well-being. Some human geographers and other scholars view our increasing numbers with great concern because of suggested links between population numbers and such problems as famine, disease, and, as discussed in the previous chapter, environmental deterioration. Others are more optimistic about our ability to cope with increasing numbers, arguing that past increases in population have been accompanied by improvements in human well-being and technological change.

The present distribution of humans on earth, with most of the population living on a small part of the land area, reflects both physical and human circumstances. There are, for example, very few people in cold and hot desert areas because the environments are so inhospitable, while there are relatively large numbers of people in temperate areas. The areas that are experiencing rapid increases are typically in the less developed world, whereas there are relatively stable populations in the more developed world; this fundamental difference is a direct result of a multitude of political, social, and other human considerations.

There are three basic goals to be achieved in our discussion of the human population in this and the following chapter: to understand why global and regional population numbers change through time; to understand why people are located where they are; and to debate the consequences of our increasing numbers. Much of this discussion relies on measures and procedures developed in **demography**, which is concerned with the size and make-up of populations (according to such variables as age and sex), with the processes that influence the composition of populations (notably fertility and mortality), and with the links between populations and the larger human environments of which they are a part.

FERTILITY AND MORTALITY

All changes in world population size can be understood by reference to the two processes of fertility and mortality. Thus,

$$P_1 = P_0 + B - D$$

where

P_1 = population at time 1

P_0 = population at time 0 (before time 1)

B = number of births between times 0 and 1

D = number of deaths between times 0 and 1

Fertility and mortality rates vary significantly according to time and location with both rates affected by many different variables.

If the population size of some subdivision of the world is considered, then a third process, migration, is relevant.

Thus,

$$P_1 = P_0 + B - D + I - E$$

where

I = number of immigrants to area between times 0 and 1

E = number of emigrants from area between times 0 and 1

The effects of migration are discussed in the next chapter.

FERTILITY MEASURES

The simplest and most common measure of **fertility** is the *crude birth rate* (CBR), the total number of live births in a given period (usually one year) for every 1,000 people already living. Thus:

$$CBR = \frac{\text{number of live births in one year}}{\text{mid-year total population}} \times 1,000$$

The CBR for the world in 1993 is 26. Historically, measures of CBR have typically ranged from a minimum of 15 (the theoretical minimum is of course 0) to a maximum of 55, which is a rough estimate of the biological maximum. Clearly, the CBR is a very useful statistic, but it may be somewhat misleading because births are related to the total population and not to that subset of the population that is able to conceive (it is called crude for this reason). There are other more sophisticated measures that more accurately reflect underlying fertility patterns by reference to the biological concept of **fecundity**. Two such measures are noted; both relate births as directly as possible to that segment of the population responsible for them and thus exclude those unable to conceive, namely males, and those unlikely to conceive, namely very young girls and postmenopausal women.

First, the *general fertility rate* (GFR) refers to the actual number of live births per 1,000 women in the fecund age range, those years in which a woman has the ability to conceive, typically defined as ages 15 to 49 (sometimes 15 to 44). It is calculated as follows:

$$GFR = \frac{\text{number of live births in a one-year period}}{\text{mid-year number of females aged 15–49 years}} \times 1,000$$

Second, and more usefully, the *total fertility rate* (TFR) is the average number of children a woman would have, assuming she has children at the prevailing age-specific rates as she passes through the fecund years. This is an age-specific measure of fertility that is useful because child-bearing during the fecund years varies considerably with age. It is calculated as follows:

$$TFR = 5 \sum_{A=1}^{7} \frac{\text{number of births to women in age group A in a given period}}{\text{mid-year number of females in age group A}}$$

where, A refers to the seven five-year age groups of 15–19, 20–4, 25–9, 30–4, 35–9, 40–4, and 45–9. The 5 preceding the summation sign is necessary because each age group

covers five years. The world TFR for 1993 is 3.3. This means that the average woman has 3.3 children during her fecund years.

There are a variety of other measures of fertility, with different measures used in different circumstances and with different aims in mind. During our discussions, we will limit ourselves to two measures, the CBR and the TFR. The CBR statistic reflects what has actually happened in a given time period — x number of children per 1,000 members of the population were born. The TFR, on the other hand, reflects an assumption that as a woman reaches a particular age, she will have the same number of children as do the women of that age today — it is a predictive measure.

The impact of a particular CBR or TFR on population totals is related to mortality. Generally, a TFR of between 2.1 and 2.5 is considered a replacement level; that is, it maintains a stable population. The lower figure, 2.1, applies to areas with relatively low levels of mortality and the higher figure, 2.5, to areas with relatively high levels of mortality.

FACTORS AFFECTING FERTILITY

Fertility, the reproductive behaviour of a population, is affected by biological, economic, and cultural factors. It is not difficult to identify and comment on many of these factors; it is difficult, however, to assess their specific effects.

Biological factors

Age and related fecundity is the key biological factor. Fecundity begins at about age 15 for females, reaches a maximum in the late 20s, and terminates in the late 40s. The pattern for males is less clear with fecundity commencing at about age 15, peaking at about age 20, and then declining, but without a clear termination age. Some females and males are sterile — incapable of reproduction; data are not readily available, but a figure of 10 per cent of all married couples in a developed country having one sterile partner is probably not unrealistic.

Reproductive behaviour is also affected by nutritional well-being; populations in ill health are very likely to have impaired fertility. Thus, periods of famine reduce population growth by lessening fertility (as well as by increasing mortality). Box 5.1 reviews the situation in contemporary tropical Africa, a region of generally high birth rates but containing particular areas of low fertility.

A third biological factor affecting fecundity is the level of fatness. Nomadic pastoral societies are typically low on body fat as they have a low starch diet, and are characterized by low fertility.

Economic factors

Another group of variables affecting fertility are principally economic. Indeed, until recently, fertility changes in developed countries were essentially one part of the process of economic development. With increasing industrialization, fertility declines. This economic argument suggests that traditional societies are strongly pronatalist as the family is a total production and consumption unit, whereas modern societies emphasize small families and individual independence. Once economically valued for their contribution, children are now an expense. The economic argument is that the decision to have children is essentially a cost-benefit decision. In the traditional, often extended-family setting, children are valuable as productive agents and sources of security in old age — hence large families. Neither of these factors is important in modern societies.

This economic argument implies that any reductions in fertility are essentially caused by economic changes, an argument that is central to the demographic transition, one of the six explanations of population growth through time to be discussed later in this chapter.

Cultural factors

A host of complex and interrelated cultural factors affect fertility and there are now reasons to suggest that current reductions in fertility are occurring for primarily cultural as opposed to economic reasons; this suggestion is central to the fertility transition, another of the six explanations of population growth through time to be discussed later in this chapter.

Most cultural groups recognize *marriage* as an appropriate setting for reproduction. There are several measures of **nuptiality**, the simplest of which is the nuptiality rate:

$$\text{nuptiality rate} = \frac{\text{number of marriages in one year}}{\text{mid-year total population}} \times 1{,}000$$

The age at which females marry is important as this may reduce the effective fecund years. A female marrying at age 25, for example, has 'lost' ten fecund years. Perhaps the most obvious instance of such loss was Ireland until recently. In the 1940s, the average age at marriage for Irish women was 28 and for Irish men 33; in addition, high percentages of population lost all their fecund years as a result of not marrying — as high as 32 per cent of women and 34 per cent of men. This situation of late marriage or non-marriage is usually explained by reference to social organization and economic aspirations. Interestingly, since about 1971, the Irish people have tended to marry earlier and more universally. Other cultures actively encourage early marriage. Some Latin American countries, such as Panama, have legal marriage ages as low as 12 for females and 14 for males. Overall, delayed marriage and celibacy are uncommon throughout Asia, Africa, and Latin America. Some of the recent success in reducing fertility in China can be explained by the government requirement that marriage be delayed until age 25 for females and 28 for males.

Fertility is closely related to marriage, but is also affected by *contraceptive use*. Attempts to reduce fertility within marriage have a long history. The civilizations of Egypt, Greece, and Rome all utilized contraceptive techniques. Today, contraceptive use is common and is highest in developed countries. The less developed countries typically have lower rates of contraceptive use, but most evidence shows that the rates are increasing. Data for 1993 contraceptive use are shown in Table 5.1; two statistics are included for each area: the

---

Box 5.1: Fertility in tropical Africa

Fertility is typically high throughout tropical Africa with several countries having a total fertility rate (TFR) between 6 and 8. But the national statistics mask some remarkable variations within countries. For some areas, TFRS are as low as 2 to 5. Two geographers working at Syracuse University identified a belt of low rates from southwestern Sudan, through the Central African Republic, and into Zaire. Other low-fertility areas include parts of Cameroon and Gabon, the Lake Victoria area, isolated locations in the savanna zone of west Africa, parts of Namibia and Botswana inhabited by San peoples, parts of Ethiopia inhabited by nomadic pastoralists, and parts of the east African coast (Doenges and Newman 1989). In each of these cases, there appears to be a close correlation between a low TFR and a particular ethnic group.

There are a variety of explanations for the presence of areas of abnormally, relatively speaking, low rates of fertility. Four reasons are noted. First, there are cultural variations in the length of time a baby is breast-fed. Breast-feeding limits ovulation and extends the period of infertility following birth from about two months to eighteen months. The second cause of localized areas of low fertility is the impact of diseases such as gonorrhoea and syphilis, which reduce the likelihood of sexual intercourse and cause sterility. Poor nutrition is the third cause: thus fertility falls following famine and is consistently low in areas experiencing chronic undernutrition. The fourth cause relates to aspects of marriage. Almost all women marry and the age at first marriage is consistently in the mid-teens. But there are notable exceptions that clearly cause reduced fertility; for example, among the Rendille of northern Kenya, cultural practices result in one-third of the women not marrying until their mid-thirties. Among nomadic pastoralists, prolonged spousal separation may reduce fertility.

Doenges and Newman (1989:111) concluded their study of tropical African fertility by noting: 'If in the future impaired fertility ceases to be a significant concern for African populations, deliberate birth control will have a better chance of becoming an accepted social norm.' In this respect, impaired fertility acts very much like high infant mortality: both are important limitations on social well-being that require resolution before the state of regulated fertility can be reached.

5.1 Urban street scene, Peshawar, Pakistan. (United Nations)

percentage of married women using contraception, and the percentage of married women using modern contraceptive methods such as the pill, IUD, and sterilization. It is noteworthy that for many countries, and for many of the larger regions of the world, data are not available. If we focus on individual countries, we find percentages as low as 3 and 1 (Ivory Coast) and as high as 81 and 78 (United Kingdom). Contraception use is closely related to government attitudes and religion. Most of the world's population today live in countries that actively encourage limits to fertility, whereas as recently as 1960 only India and Pakistan had active programs to reduce fertility.

Another important cultural factor affecting fertility is *abortion*. Abortions actually outnumber deaths today. Further, for every two births in the world, at least one pregnancy is deliberately terminated. Abortion is an even more complex moral question than contraception (Box 5.2). It is a long-standing practice designed to terminate an unwanted pregnancy and yet typically subject to widespread condemnation. Opposition to abortion is essentially religious and moral. The reality, however, is that as many as 60 million abortions are performed each year, only about half of which are legal.

In many countries, governments influence abortion — pragmatism is often the motive. China has a liberal abortion law in response to the need for population control; Sweden has a liberal abortion law on the grounds that abortion is a human right. Other countries, dominated by Catholicism or the Muslim faith, either deny access to abortion or limit access to instances where a woman's life is threatened. In some other countries, such as Canada, the abortion question is one of the most hotly debated biological, ethical, social, and political issues.

VARIATIONS IN FERTILITY

Spatial variations in fertility in the 1990s correspond closely to spatial variations in level of economic development, a fact that offers general support for the suggested importance of the economic factors noted earlier. It appears that, especially since the onset of the Industrial Revolution in the mid-eighteenth century, modernization and economic development

TABLE 5.1: *Contraceptive use by regions, 1993*

% Married women using contraception		
	Total	Modern
World	57	49
More developed world	69	54
Less developed world	54	49
Less developed world (excluding China)	42	35

Source: Population Reference Bureau, *1993 World Population Data Sheet* (Washington, DC: Population Reference Bureau, 1993).

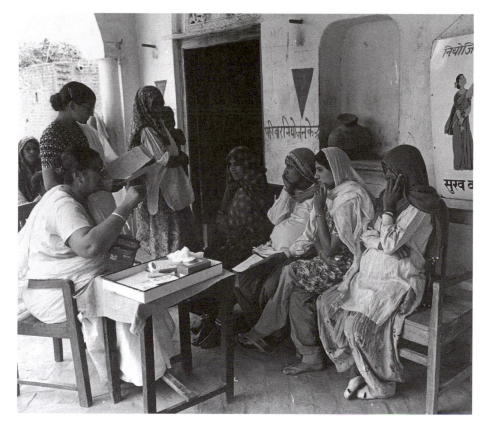

5.2 A family planning fieldworker provides information about contraceptives to village women. (United Nations)

Box 5.2: Primitive abortion

It appears that abortion has long been practised to limit population numbers or, more specifically, family numbers. A passage from one of the first (1925) realist Canadian novels shocked not only the central character of the novel but also the Canadian reader of the time:

'So, from many things that were said and from some that I saw I inferred that mother expected shortly to have another child and that that greatly worried her; but even more did it worry my father. He began to speak still more curtly to my mother; and he treated her as if she were at fault and had committed a crime. He prayed even more than before, both more frequently and longer. Gradually my mother began to get into a panic about her condition.

'So, one day, taking me along, she went over to see that woman in the potato-patch once more.

'A number of things were said back and forth which I remember with great distinctness but which have nothing to do with my story.

'At last Mrs Campbell laughed out loud. Of course, she said, it's plain to be seen by now. It's a curse. But I can tell you I wouldn't be caught that way. Not I! I'm wise.

'But what can you do? my mother exclaimed. He comes and begs and says that's what God made them male and female for. And if you want to hold your man ...

'I was only ten years old. But I tell you, I knew exactly what they were talking about. And right then I vowed I should never marry. I was furious at the woman and afraid of her.

'You're innocent all right, she said at last contemptuously. I don't mean it that way, child. But when I'm just about as far gone as you are now, then I go and lift heavy things; or I take the plow and walk behind it for a day. In less than a week's time the child comes; and it's dead. In a day or two I go to work again. Just try it. It won't hurt you. Lots of women around here do the same.

'So, when we came home, my mother took some heavy logs, dragged them to the saw-buck, and sawed them. I begged her not to do it; but even I could see that she was desperate.

'Next day she was very sick. I was sent to the house of the German preacher in the village. And when I was allowed to come back, my mother was at work again on the land. She looked the picture of death; but she was cheerful. My father prayed more than ever.

Source: F.P. Grove, *Settlers of the Marsh* (Toronto: McClelland and Stewart [1925] 1966):108–9.

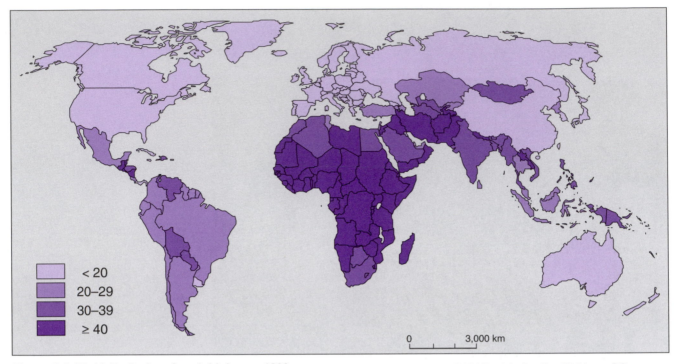

Figure 5.1 World distribution of crude birth rates, 1993. Source: Population Reference Bureau, *1993 World Population Data Sheet*, (Washington, DC: Population Reference Bureau, 1993).

have prompted lower levels of fertility. Thus for 1993, at the broad regional level, the more developed world has a CBR of 14 and a TFR of 1.8, while the less developed world has a CBR of 29 and a TFR of 3.7. More specifically, in the more developed world, Spain is among the group of countries with low fertility as it has a CBR of 10 and a TFR of 1.3, while in the less developed world, Malawi has a CBR of 53 and a TFR of 7.7. Figure 5.1 maps CBR by country. In the general sense, variations of this type reflect some economic considerations.

Current evidence, however, indicates that the high measures of fertility in the less developed world are decreasing rapidly for cultural rather than economic reasons (Box 5.3). Thus, as recently as 1990 the less developed world had a CBR of 31 and a TFR of 4 compared to the 1993 figures of 29 and 3.7 noted earlier. There is also evidence of declining fertility in several countries in the more developed world that may be prompted by cultural factors (Box 5.4).

Fertility also varies significantly within any given country. There is, for example, usually a clear distinction between urban areas with relatively low fertility and rural areas with relatively high fertility. This distinction applies to countries regardless of level of development. Similarly, fertility within a country is higher for those with low incomes and for those with limited education.

MORTALITY MEASURES

As is the case with fertility, mortality may be measured in a variety of ways. The simplest, equivalent to the CBR, is the crude death rate (CDR), the total number of deaths in a given period (usually one year) for every 1,000 people living. Thus:

$$\text{CDR} = \frac{\text{number of deaths in one year}}{\text{mid-year total population}} \times 1{,}000$$

The CDR for the world in 1993 is 9. Measures of CDR have typically ranged from a minimum of 5 to a maximum of 50.

The CDR does not take into account the fact that the probability of dying is closely related to age (it is called crude for this reason). Usually, death rates are highest for the very young and the very old, producing a characteristic J-shaped curve (Figure 5.2). There are other mortality measures that consider the age structure of the population. The most useful of these for our purposes is the infant mortality rate (IMR): this is the number of deaths of infants under 1 year old per 1,000 live births in a given year. Thus:

$$\text{IMR} = \frac{\text{number of infant deaths under 1 year old}}{\text{number of births in that year}} \times 1{,}000$$

The world IMR for 1993 is 70, although figures for individual countries range from as low as 5.5 for Iceland to as high as 151 for Mozambique. The IMR is sensitive to cultural and economic conditions as it declines with improved medical and health services and better nutrition; reductions in the IMR usually precede overall mortality decline.

Although not correctly identified as a mortality measure, another useful statistic that reflects mortality is life expectancy (LE), the average number of years to be lived from birth.

World life expectancy in 1993 is 65, although several European countries have expectancies in the high 70s, while several African countries are in the low 40s.

FACTORS AFFECTING MORTALITY

Humans are mortal. It is possible in principle for the human population to attain a CBR of 0 for an extended period; the same cannot be said for the CDR. While fertility is affected by many biological, cultural, and economic variables, mortality is somewhat easier to explain, despite the fact that the World Health Organization has recognized some 850 specific causes of death! Historically, causes of death also include war and famine. In general terms, mortality varies inversely with socio-economic status. The CDR and LE figures already noted bear witness to this statement. High LE

statistics are related to high-quality living and working conditions, a high level of nutrition, good sanitation, and availability of medical services.

VARIATIONS IN MORTALITY

The world pattern of CDR shows much less variation than does that for CBR (Figure 5.3). This is a reflection of the general availability of at least minimal health care facilities throughout the world. The central African region is the one remaining area of high mortality. Figure 5.4 maps LE by country. In this case, major variations are evident: as already noted, LE figures are more sensitive to food availability and health facilities than are death rates. The map of LE is a fairly good approximation of the health status of populations. Low LE statistics are found in tropical countries in Africa

Box 5.3: Declining fertility in the less developed world

The less developed world is currently experiencing a decline in fertility that began in the 1970s and appears to have little to do with economic factors and everything to do with cultural factors (Robey, Rutstein, and Morris 1993). Birth rates are falling rapidly without any prompting from economic growth and related improved living conditions.

The magnitude of this declining fertility is indicated by Thailand, which had a 1975 TFR of 4.6 and a 1993 TFR of 2.4; similar reductions apply to many other countries in Asia and Latin America, including Indonesia, Turkey, Colombia, and Morocco. The fact that several sub-Saharan African countries are also experiencing reduced fertility is especially significant as this area has always appeared to be resistant to any such trend; decline is evident in Kenya, Botswana, Zimbabwe, and Nigeria.

The explanation for these declines appears to be related to the availability and acceptance of new contraceptive technologies, the success of family-planning programs and the educational power of the mass media. Traditional methods of contraception, such as periodic abstinence and withdrawal, are being replaced by modern methods. Female sterilization is the favoured method of contraception in Asia and Latin America, but is less popular in Africa and the Middle East.

The fact that this decline in fertility is occurring and has been in many countries in the less developed world for about twenty years was not well understood until recently because data on such matters are not readily available. The data now being used are derived largely from forty-four surveys of more than 300,000 women over the past eight years; the data are substantial and well documented. These fertility declines are in addition to the well-documented case of China where government policies, often compulsory, have resulted in a below replacement fertility level of 1.9 only twenty years after initial realization of a problem.

One detailed analysis of Botswana (VanderPost 1992) showed that the fertility decline is proceeding in a geographically uneven manner; as is so often the case, national statistics disguise local details. Indeed, in some areas of Botswana, fertility is actually increasing, whereas in other areas it is declining. The difference is essentially an urban-rural difference: in 1988, the TFR in urban areas was 4.1 and in rural areas 5.4.

It appears that some economic change may be prompting the Botswana fertility decline. On the basis of detailed data collection at a variety of scales, VanderPost (1992) argued that the fertility transition in Botswana is a two-stage process. First, fertility tends to increase somewhat because of cultural changes involving the abandonment of traditional practices of marriage and child-bearing. These changes are linked to improvements in local infrastructures such as the construction of schools, clinics, water supply systems, and roads. Such changes influence fertility behaviour because of the related introduction of a different value system. Second, fertility declines as modern contraceptive techniques become accepted as one critical component of the new value system.

The importance of these declines in the less developed world is difficult to exaggerate, both practically and in terms of our understanding of fertility change. Because family planning is now practised through much of the less developed world (typically it appears without any substantial prior improvement in economic circumstances), many earlier accounts of anticipated population growth are being revised. The conventional wisdom in the 1970s was well expressed by Demeny (1974:105): countries in the less developed world 'will continue their rapid growth for the rest of the century. Control will eventually come through development or catastrophe.' It now appears that rapid growth is slowing down not for either of these two reasons but because of cultural changes concerning willingness to employ modern contraceptive methods in order to have smaller families.

and in south and southeast Asia. The reverse is particularly evident in Europe and North America.

Mortality measures also vary markedly within countries. In countries such as Canada, the United States, and Australia, the Aboriginal populations have higher levels of CDR, IMR, and a lower LE than the population as a whole. These reflect what we might call the social inequality of death. Two examples suffice. In Australia in 1980, the IMR for Aboriginals was 32.7 compared to 10.2 for non-Aboriginals. In the United States, the national IMR in 1993 is 8.6, but Washington, DC, has a figure of 21.1 while South Dakota and Alabama have a figure of 13.3. Black babies in the United States are almost twice as likely to die in their first year as are White babies, a state of affairs that has remained essentially unchanged during the last fifty years.

Box 5.5 provides two examples of LE and IMR data that clearly reflect environmental and health problems at least partly related to inefficient political systems.

NATURAL INCREASE

The rate of natural increase (RNI) is determined by subtracting the CDR from the CBR and is therefore a measure of the

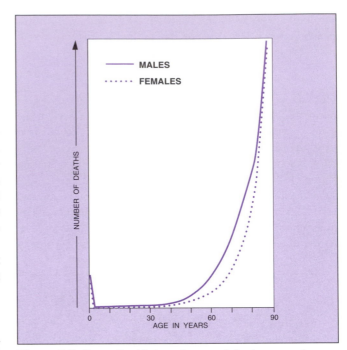

Figure 5.2 Generalized J-shaped curve of death rates and age.

Box 5.4: Declining fertility in the more developed world

The lowest fertility in the world today is in several southern European countries where the TFR is as low as 1.3 (Italy and Spain) and 1.4 (Greece, Portugal, and Germany).

The extraordinarily low figures for southern Europe reflect a desire for postponing starting a family and a desire for smaller families. For many Europeans, the ideal family size is now one child; a 1979 survey in France showed only 3 per cent favouring a one-child family compared to 19 per cent in 1989. Such preferences may be prompted by uncertain economic circumstances and the growth of employment opportunities for women (Hall 1993; King 1993).

The figure for Germany reflects a remarkable and unprecedented trend in the former East Germany, which appears to have come as close to a temporary suspension of child-bearing as any large population in the human experience. Eastern Germans have virtually stopped having children. The explanation is not an increase in abortions, which have also fallen abruptly. One possible explanation is the trauma associated with the transition from communism to capitalism and, specifically, concern about employment opportunities for future generations in a region with high levels of unemployment. It may be appropriate, then, to interpret the low fertility rate in Germany as one temporary outcome, actually a form of demographic disorder, of the transition from the 'old' to the 'new' political order. This suggestion appears to have some merit as fertility has also declined between 1989 and 1993 by 20 per cent in Poland, 25 per cent in Bulgaria, 30 per cent in Romania and Estonia, and 35 per cent in Russia (*The Economist* 1993:54).

Whether or not the low fertility evident in these countries continues remains to be seen. In some other European countries, rates have increased slightly between 1990 and 1993; for example, Norway (from 1.8 to 1.9), Sweden (from 2 to 2.1), Denmark (from 1.6 to 1.7), Finland (from 1.7 to 1.8), Austria (from 1.4 to 1.5), and the Netherlands (from 1.5 to 1.6). Remember that even these increased rates typically imply net losses in population as the replacement level in such countries is a TFR of about 2.1.

Although Canada does not belong to the group of lowest fertility countries, both the 1993 CBR and TFR statistics are declining or stable in recent years. Many European countries have slightly lower CBR and TFR data than Canada, and the United States (with a CBR of 16 and a TFR of 2) has slightly higher measures of fertility. Table 5.2 summarizes the current demographic situation in Canada.

TABLE 5.2: *Population data, Canada, 1993*

Total population	28.1 million
CBR	15
CDR	7
RNI	0.8%
Doubling time	87 years
IMR	6.8
TFR	1.8

Source: Population Reference Bureau, *1993 World Population Data Sheet* (Washington, DC: Population Reference Bureau, 1993).

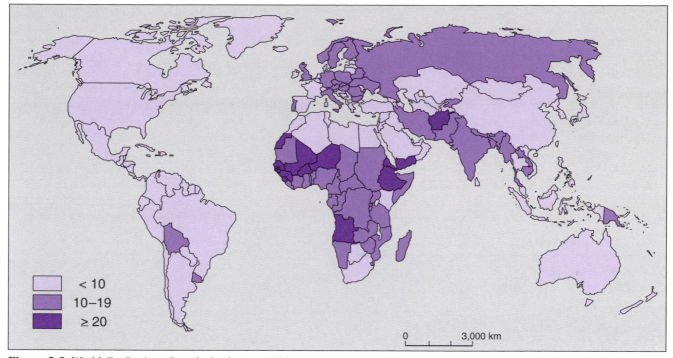

Figure 5.3 World distribution of crude death rates, 1993. Source: Population Reference Bureau, *1993 World Population Data Sheet*, (Washington, DC: Population Reference Bureau, 1993).

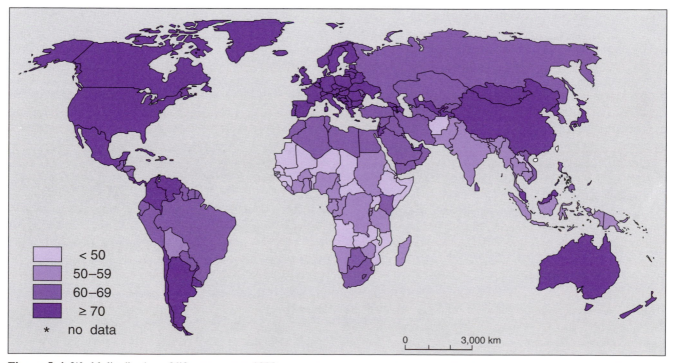

Figure 5.4 World distribution of life expectancy, 1993. Source: Population Reference Bureau, *1993 World Population Data Sheet*, (Washington, DC: Population Reference Bureau, 1993).

rate (usually annual) of population growth. In 1993, the world CDR was 9, the CBR was 26, and hence the RNI was 16 per 1,000 (not 17 because of arithmetic rounding); this figure is typically expressed as a percentage of total population and in 1993 was therefore 1.6 per cent. This percentage has been relatively constant in recent years. The RNI takes into account only mortality and fertility, not migration, and is therefore not usually a correct figure for the population growth of a given area smaller than the earth. Given this dependence on CDR and CBR, it is not surprising that the highest RNI figures are for those countries in the less developed world. Tables 5.3 and 5.4 include RNI data for two groups of countries, while Figure 5.5 maps RNI data on an individual country basis.

Combined, these three sets of information provide a useful commentary on the present world population. Low RNI statistics are essentially restricted to Europe, and especially to some former communist countries and several bordering on the Mediterranean. High RNI statistics are concentrated in the western Asian Islamic world and tropical Africa (a number of countries that have temporarily inflated RNI statistics because of refugee in-movement are excluded from Table 5.4).

Although, as suggested in boxes 5.3 and 5.4, there are fertility declines in parts of both the less and more developed worlds, the number of females of reproductive age continues to rise. The total world population continues to grow rapidly.

Consideration of RNI introduces the useful concept of doubling time — that is, the number of years needed to double the size of a population, assuming a constant RNI. For example, at the current 1.6 per cent growth rate, the world population will double in approximately forty-two years. Relatively minor variations in RNI can significantly affect doubling time. An RNI of 0 (where CDR and CBR are equal) results in a stable population with an infinite doubling time; on the other hand, an RNI of 4.1 per cent produces a doubling time of a mere seventeen years.

TABLE 5.3: Countries with highest rates of natural increase (3.5 and above), 1993

Country	Natural increase (percentage)	Population (millions)
Syria	3.8	13.5
Iraq	3.7	19.2
Kenya	3.7	27.7
Togo	3.6	4.1
Jordan	3.6	3.8
Iran	3.5	62.8
Ivory Coast	3.5	13.4
Yemen	3.5	9.8
Oman	3.5	1.6
Comoros	3.5	0.5

Source: Population Reference Bureau, *1993 World Population Data Sheet* (Washington, DC: Population Reference Bureau, 1993).

TABLE 5.4: Countries with lowest rates of natural increase (0.1 and below), 1993

Country	Natural increase (percentage)	Population (millions)
Hungary	-0.2	10.3
Bulgaria	-0.2	9.0
Latvia	-0.2	2.6
Germany	-0.1	81.1
Ukraine	-0.1	51.9
Italy	0.0	57.8
Estonia	0.0	1.6
Russia	0.1	149.0
Spain	0.1	39.1
Romania	0.1	23.2
Greece	0.1	10.5
Czech Republic	0.1	10.3
Portugal	0.1	9.8
Denmark	0.1	5.2
Croatia	0.1	4.4

Source: Population Reference Bureau, *1993 World Population Data Sheet* (Washington, DC: Population Reference Bureau, 1993).

Box 5.5: Problems in central Asia and Russia

Measures of infant mortality and life expectancy are especially useful indicators of quality of life. Both statistics reflect the quality of the diet available, the character of the health care system, and the availability of public sanitation and disease control. Two examples that indicate serious problems are noted, both in the former USSR.

In one area on the shores of the Aral Sea in central Asia, the IMR is estimated to be as high as 111 (see Box 4.6). This high IMR rate reflects a number of basic health problems — a contaminated water supply, an abominable sewage system, and unsanitary hospitals. These problems are exacerbated in the summer as a consequence of increased temperatures and related disease problems. These basic health problems, in turn, reflect a host of social problems resulting from years of neglect by political leaders.

A remarkable example of an unexpected decline in LE is that of Russia. In 1987, the LE was 67 for Russian males and 74 for Russian females, but since then, there have been significant decreases. The 1993 estimates released by the Russian State Committee on Statistics show that LE has fallen to 59 for males and to 73 for females. These figures are below those of all major industrialized countries; comparable Canadian data for 1993 are 74 for males and 81 for females. The explanation appears to be a combination of issues. The transition from communism to capitalism may have created a new class of wealthy entrepreneurs, but it has also resulted in economic problems for many other Russians. Diets for many are worsening because of high prices for meats, fruits, and vegetables; the industrial policies of the previous communist regime, which are a cause of the high IMR noted earlier, are undoubtedly affecting the health of many Russians with polluted air and polluted drinking water; there is an epidemic of alcohol abuse, and the high rate of abortions may have left many females unable to have children.

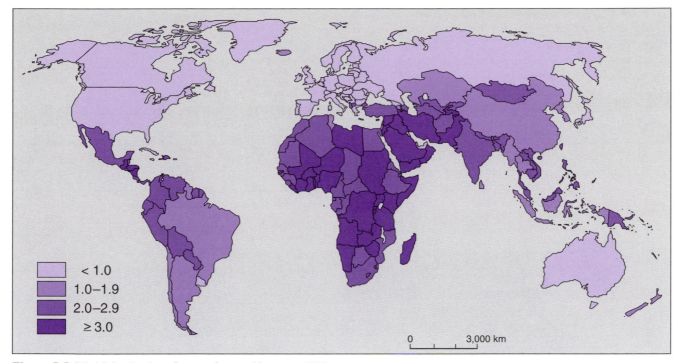

Figure 5.5 World distribution of rates of natural increase, 1993. Source: Population Reference Bureau, *1993 World Population Data Sheet*, (Washington, DC: Population Reference Bureau, 1993).

GOVERNMENT POLICIES

In our discussions so far, we have paid only incidental attention to governments' impact on population. Most governments attempt to influence various aspects of population such as total numbers and spatial distribution. Governments can also directly or indirectly attempt to control deaths, births, and migrations.

All death control policies have identical objectives, namely, to reduce mortality. Such policies include work safety measures and medical care and are largely for economic and humanitarian reasons. Despite the near universality of such policies, many governments actively raise mortality levels at specific times, such as in times of war. Further, many governments do not ensure that all members of their population have equal access to comparable health care. This applies to countries throughout the world regardless of the level of economic development.

Unlike death control policies, those related to birth control have varying objectives. Many governments choose not to establish any formal policies, either because of indifference to the issue or because they recognize that there is little clear evidence to support a particular policy.

Other governments are actively *pronatalist*. Such policies are typical of countries dominated by a Catholic or Islamic theology (e.g., France, Iraq), in countries where the politically dominant ethnic group is in danger of being numerically overtaken by an ethnic minority (e.g., Israel), and in countries where more people are perceived as necessary for economic and strategic reasons (see Box 5.6).

Two countries in Asia are actively encouraging fertility increases at the present time (Dwyer 1987). Both Singapore and Malaysia succeeded in lowering their TFR by following an antinatalist line until new government policies were developed in 1984 to increase fertility. In Singapore, this reversal was related to the perceived economic difficulties as a result of reducing the TFR to 1.6 by 1984; there was a perceived need to provide a larger market for domestic industrial production and a larger workforce. The Malaysian case is similar in that the motivation was economic, although the TFR remained above replacement level in 1984 at 3.9. Both Singapore and Malaysia belong in the Asian group of newly industrialized countries, which will be discussed in Chapter 12. The extent to which such pronatalist policies are successful is debatable; the 1993 TFR for Singapore is only 1.7 (a slight increase since 1984), and for Malaysia 3.6 (a decrease since 1984).

In addition to such direct attempts to increase fertility, many countries indirectly support increased fertility through baby bonuses, establishment of day care centres, and generally widespread high-quality housing; Canada is an example.

The best-known policies relating to births, however, are those that are *antinatalist*. Since about 1960, many of the less developed countries have initiated policies designed to reduce fertility. The rationale behind such policies is that overpopulation is seen as a legitimate danger and that, in some areas, the **carrying capacity** has been neared, attained, or exceeded. Regardless of whether or not this is reality, there is a common assumption that certain areas are too densely populated and that the best solution is reduced fertility. As noted earlier, birth rates are indeed falling in many of the less developed countries as a result of changes in attitudes, government programs, and such general social advances as improvements in literacy and rural development.

India was the first country to actively intervene to reduce fertility. Beginning in 1952, government programs encouraged contraceptive use and sterilization and offered financial incentives. Fertility rates have declined, but with a 1993 RNI of 2.1 per cent, the various programs have not achieved the desired result. Since 1952, there have been changes in

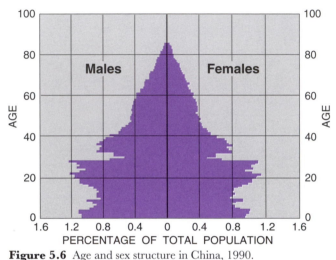

Figure 5.6 Age and sex structure in China, 1990.
Source: J. Jowett, 'China's Population: 1,133,709,738 and Still Counting', *Geography* 78 (1993):405.

the types of program; sometimes coercion has been favoured, at other times education. India also continues to experience significant interregional differences in cultural and economic variations. India, with the second largest population in the world, is thus continuing to grow rapidly, with a doubling time of thirty-four years.

China, on the other hand, has made great strides towards reducing fertility. The current RNI is 1.2 per cent and the doubling time is sixty years. Much of this success can be attributed to the programs, mostly introduced in the late 1970s, of China's communist government. Families are restricted to having one child, and marriage is prohibited until the age of 27 for men and 25 for women. Contraception, abortion, and sterilization are free. There are financial incentives for those with only one child and penalties for those with more than one. It is interesting, if a little perplexing, to note that communist China has a strict antinatalist policy, while the conventional Marxist argument holds that such human problems as apparent resource limits and apparent overpopulation reflect inappropriate modes of production and social organization. A classic Marxist solution to population issues focuses not on reducing fertility but on correcting our economic and social organizations. All of this does not really mean that Marxists are pronatalist. It does mean that social change is the first priority and that the issue of fertility is secondary. It does not preclude the possibility that an antinatalist policy may be a step in the right direction. The current demographic situation for China is summarized in Box 5.7.

Governments may also attempt to influence migration, especially between countries. Most countries formally restrict

---

Box 5.6: Fertility in Romania, 1966–89

In the October 1986 issue of the German magazine, *Der Spiegel*, the president of Romania was quoted as saying, 'The fetus is the socialist property of the whole society. Giving birth is a patriotic duty, determining the fate of our country. Those who refuse to have children are deserters, escaping the laws of national continuity.'

Because of alarm over a birth rate of 16 in 1966, the communist government in Romania ended all legal access to abortion and set a goal of increasing the national population to 24–5 million by 1980 (a 30 per cent increase). Contraceptives were banned and all Romanian employed women under 45 had to take a monthly gynaecological examination. Unmarried people over 15 and married couples without children or a valid medical reason were assessed an additional 30 per cent income tax. The total fertility rate increased from 1.9 in 1966 to 3.6 in 1968, but since that time, the rate steadily dropped and in 1990 stood at 2.3. There is reason to believe that many women had illegal abortions despite the establishment of a special unit within the State Security Police whose job was to

combat abortion. Interestingly, all of the strategies to increase the fertility rate were negative. There was no use of positive inducements such as maternity leave, family income supplements, and so forth.

The principal motivations for desiring to increase fertility were probably national security (more people equal more strength) and economy (to alleviate labour shortages).

Romania prior to 1989 is one example of a particular fertility policy enforced in a distinctive manner. In general, countries vary substantially in their approach to fertility. Among previously communist countries, the former USSR encouraged higher fertility, but at the same time probably had the highest national legal abortion rate in the world (more than one abortion for every birth). In general, it seems reasonable to say that Romania had the most regulated fertility policy in the world.

Most countries have direct or indirect influences on fertility. What type of policies are evident in Canada? Does the Canadian government encourage or discourage births? What role do churches play?

immigration and until about 1989, some countries, especially communist countries in eastern Europe, limited emigration.

AGE STRUCTURE

Fertility and mortality vary significantly with age. Inevitably, then, the growth of a population is affected by what we might call the age composition of the population. Such age compositions are dynamic. As we have seen in Box 5.7 it is usual to represent age and sex compositions with a **population pyramid**. Three general categories can be suggested (Figure 5.7). First, if a population is rapidly expanding, a high proportion of the total population will be in the younger age groups. Because fertility is high, each age group is larger than the group that precedes it. Second, if a population is relatively stable, then each age group, barring the older groups that are losing numbers, is similarly sized. Third, if a population is declining, then the younger groups will be small relative to the older groups.

The three generalizations in Figure 5.7 are useful, but specific age and sex composition pyramids offer more precise applications. Usually, these population pyramids distinguish males and females and divide them into five-year categories. Each bar in the pyramid indicates the percentage that a particular group, such as 30- to 34-year-old females, is of the total population. Figure 5.8 presents the situation for Brazil for the period 1950 to 1987 in a series of three pyramids.

A distinctive feature of this century is the *aging* of the population. In 1900, less than 1 per cent of the world population was over age 65; today, the total is more than 6 per cent, and the estimate for 2050 is 20 per cent (Olshansky,

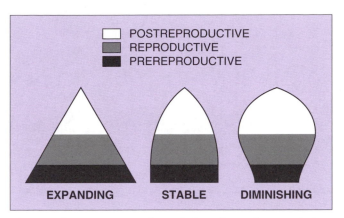

Figure 5.7 Age structure of populations. Expanding populations have a high percentage in the prereproductive age group. Stable populations have relatively equal prereproductive and reproductive age groups. Diminishing populations have a low percentage in the prereproductive age group.

Box 5.7: Population in China

Data from the 1990 census provide a relatively sound factual description of the Chinese demographic situation (Jowett 1993). China continues to enjoy a declining CBR that is low by Asian standards (18), a CDR that is among the lowest in the world (7), an RNI of 1.2, and an LE that is close to the average for the more developed world (70).

A useful way to depict the recent history of Chinese population growth is with a population pyramid (Figure 5.6). The indentations in the pyramid show that there have been several changes in fertility and mortality during the past forty years. The 1960s were a period of generally high fertility and low mortality, now reflected by the bulge for the ten-year cohort, 18–27; these were people born in the high-fertility period, 1963–72. In 1964, more than 40 per cent of the population was under age 15. The high fertility of 1972 was followed by dramatic decreases beginning in 1973 that have resulted in a narrowing of the pyramid and the 1990 percentage under 15 was reduced to 28.

The broadening of the base of the pyramid reflects the fact that fertility is highest for women in their late 20s and, even if the number of births to married women of this age remains constant, the CBR will remain high. As Jowett (1993:405) noted, 'China is currently faced with the "echo" effect from demographic developments that occurred 20–25 years ago. Thus the large-small-large-small cohorts of population aged eight, six, three and zero are a condensed form of the fluctuations from the previous generation aged 32, 29, 27 and 23.'

The 1990 census also provides details of spatial variations in the demographics of the Chinese population that are masked by the national data. For example, the western and southwestern interior includes ethnic minorities exempt from the one-child policy and the resultant high fertility means the populations are relatively youthful. From 1964 to 1990, the majority Han group (92 per cent of the population) increased by 60 per cent while the ethnic minorities increased by 128 per cent. These contrasting figures are explained by the differing levels of fertility and by the fact that some Han have chosen to be reclassified as a member of one of the minority groups probably because of the resultant exemption from the one-child policy.

The census also indicates a strong negative correlation between fertility on the one hand and literacy, income, industrialization, and urbanization on the other. For example, fertility varies substantially between urban areas with a TFR of 1.7, comparable to many European countries, and rural areas with a TFR of 2.7. Overall, the information provided by the 1990 census reflects the various government policies and priorities that typically favour urban rather than rural areas, the coastal rather than the interior areas, men rather than women, and the majority Han rather than the various ethnic minorities.

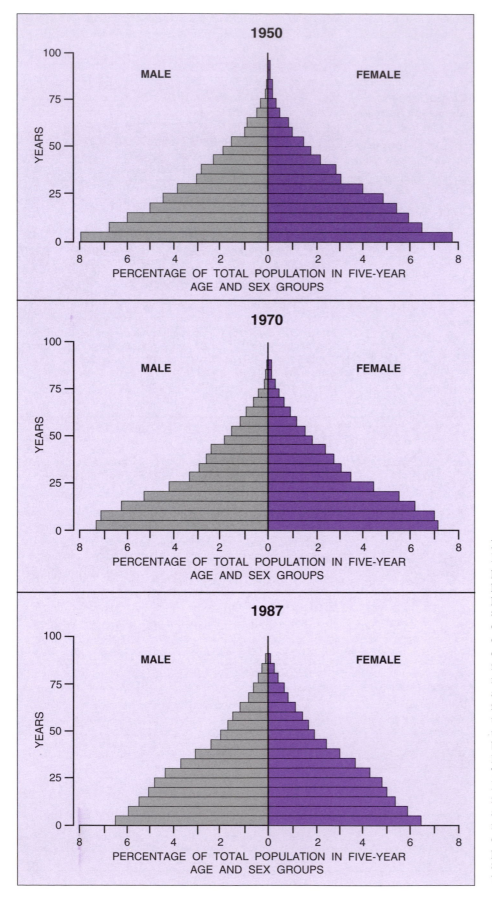

Figure 5.8 Age-sex structure in Brazil: 1950, 1970, 1987. Age-sex pyramids show the changes in Brazil's population over a thirty-seven-year period. Wide bases and gently sloping sides indicate high birth and high death rates. In the first pyramid much of the population was under 20, ensuring increased birth rates in the future. Later pyramids show a slight reduction in the number of births, and a decrease in the death rate: the first stage in the demographic transition. Throughout this period, the population increases rapidly because of the difference between high birth rates and low death rates. The 1987 pyramid indicates that Brazil's population growth has not yet reached its maximum acceleration — a fact that does not bode well for a nation already experiencing difficulties supporting its current population.

Source: United Nations Demographic Yearbooks, various years.

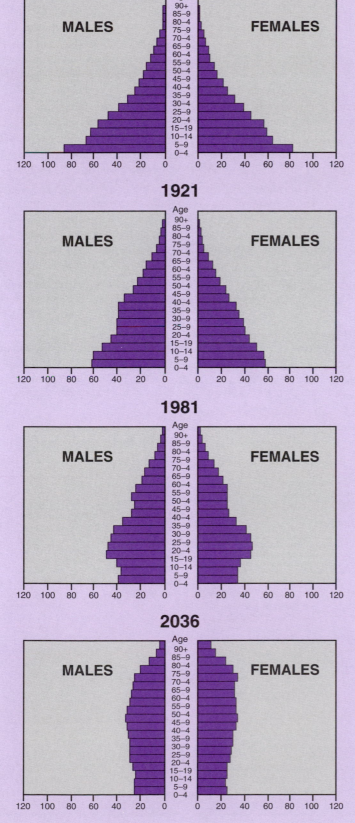

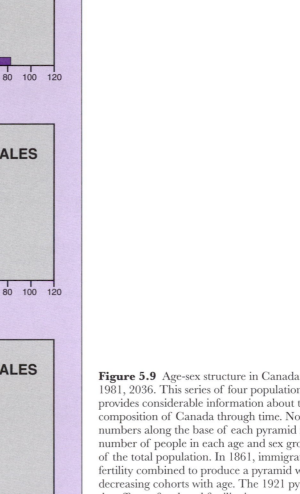

Figure 5.9 Age-sex structure in Canada: 1861, 1921, 1981, 2036. This series of four population pyramids provides considerable information about the population composition of Canada through time. Note that the numbers along the base of each pyramid indicate the number of people in each age and sex group per 1,000 of the total population. In 1861, immigration and high fertility combined to produce a pyramid with steadily decreasing cohorts with age. The 1921 pyramid shows the effects of reduced fertility in a narrowing of the base, but the effects of immigration are still apparent. In 1981, the Canadian population is characterized by extremely low birth and death rates; the age-sex structure exhibits a 'beehive' shape. The pyramid bulge between the ages of 15 to 35 represents the postwar baby boom that accelerated population growth. The birth rate has long since returned to its normal low; without the baby boom, the pyramid would have had almost vertical sides. A continuing low birth rate and aging population are reflected in the proposed pyramid for 2036; this age-sex structure may well be characteristic of what will occur in most of the more developed world.

Note: As figures 5.6, 5.8, and 5.9 demonstrate, there are varying styles for the presentation of population pyramids.

Source: Statistics Canada, *Report on the Demographic Situation in Canada, 1992*, Catalogue 91–209E Annual (Ottawa: Statistics Canada, 1992).

Carnes, and Cassel 1993). This aging occurs because of declines in infant, child, and maternal death rates that initially make a population younger. Through time, this youthful population, when combined with declining fertility, leads to population aging as fewer births narrow the base of the population pyramid; this results in a relative increase in the older population. Aging also results from continual improvements in health care and lifestyle. Figure 5.9 shows the expected impact of aging on the Canadian population.

This population aging will result in quite different patterns of disease and disability with the increasing importance of degenerative diseases associated with aging, such as cancer and heart problems. Aging may require a radical revision of economic and social institutions such as social security and health care programs. At the present, social structures have not evolved to accommodate the rapidly changing age structures.

HISTORY OF POPULATION GROWTH

As with all animal populations, early human populations were subject to constraints, growing on occasion, declining on others. The principal constraint was climate and the related availability of food. Unlike other animals, however, humans gradually increased their freedom from such constraints as a consequence of the development of culture. Cultural adaptation has enabled humans to increase in numbers and, so far, to avert extinction. For most of our time on earth, our numbers have increased very slowly.

Examples of cultural change that permitted increased numbers prior to 12,000 years ago are the use of speech, which encouraged cooperative searches for food; the introduction of monogamy, which encouraged the survival of children; and the use of fire and clothing, both of which encouraged expansion into cooler areas. Nevertheless, the cumulative effect of these advances was not great; 12,000 years ago, the human population totalled perhaps 4 million. The relatively rapid growth since that time has not been regular. Rather, there have been a number of relatively brief growth periods with longer intervals of slow growth. Each growth period can be explained by a major cultural advance.

The first such advance, about 12,000 years ago, can be labelled the agricultural revolution. This 'revolution' took place over thousands of years and involved the domestication of animals and plants; successful diffusion of agriculture resulted in earlier economic activities, hunting and gathering becoming marginal. The 'revolution' also took place in a number of different centres, occurring first in the Tigris and Euphrates valleys of present-day Iraq. Some 9,000 years ago (7000 BC) the first region of high-population density appeared. It stretched from Greece to Iran and included Egypt. By 4000 BC, agriculture and related population centres were evident on the Mediterranean coast, in several European locations, in Mexico, Peru, China, and

India. At the beginning of the Christian era, additional centres rose throughout Europe and Japan with the total world population estimated at 250 million, a dramatic increase in the 10,000 years since the beginnings of agriculture.

Prior to the cultural innovation of agriculture, population numbers changed little, increasing as cultural advances permitted and decreasing in difficult physical environments. Birth and death rates were high, about 35 to 55 per 1,000, and life expectancy was short, about 35 years. An agricultural way of life maintained these high birth rates while death rates fluctuated, reflecting generally inadequate control over the environment (Table 5.5). This state of affairs prevailed until the seventeenth century. The 250 million at the beginning of the Christian era slowly increased to 500 million by 1650; such an increase indicates generally improved control of environment, but without any major cultural advances.

From 1650 onwards, however, population increased rapidly as a response to the second major cultural advance, the Industrial Revolution. Originating in England, the Industrial Revolution initiated a rapid diffusion and growth of technologies, such that industry became the dominant productive sector. The agricultural revolution had involved more effective use of solar energy in plant growth. This second revolution involved the large-scale exploitation of new sources of energy — coal, oil, and electricity. The impact on death rates was a rapid reduction with a delayed but equally significant drop in birth rates. World population thus increased as follows: 680 million in 1700, 954 million in 1800, and 1.6 billion in 1900.

The growth initially spurred by industrialization has largely ceased in the developed world, but continues in the less developed world. Thus, world totals continue to inflate

TABLE 5.5: *Major epidemics, 1500–1700*

Date	Event
1517	Smallpox in Mexico, death of one-third of population, followed by other diseases in Mexico and other American countries
1522–34	Series of epidemics in western Europe resulted in significant decrease of population
1563	Plague in Europe
1580–98	Epidemics in France
1582–3	Epidemics and famines in southern Sudan
1588	Epidemics and famines throughout China
1618–48	German population declined from 15 million to 10 million because of epidemics and war
1628–31	Disastrous plague epidemics in Germany, France, and northern Italy
1641–4	Epidemics, famines, and conflicts reduced Chinese population by 13 million
1648–57	Plague in Spain; 1 million died
1650–85	Plague and conflict in Britain; 1 million died
1651–3	Plague and conflict in France; 1 million died
1690–4	Food shortages and epidemics throughout Europe

with the 1.6 billion of 1900 increasing to 2.5 billion in 1950, 4 billion in 1975, and 5.5 billion in 1993. The dramatic nature of these recent increases relative to earlier growth is evident in Figure 5.10.

This brief summary explains the basis of population growth. Numbers have increased in response to cultural, specifically technological, advances. The dramatic increases since about 1650 reflect the fact that we reduced the death rate before reducing the birth rate. In what is now the more developed world, populations thus grew rapidly after about 1650; elsewhere numbers increased rapidly in this century, especially since the 1940s.

We now tackle two questions. First, is it possible to make sound projections of future population numbers? Second, how do we begin to make sense of past changes?

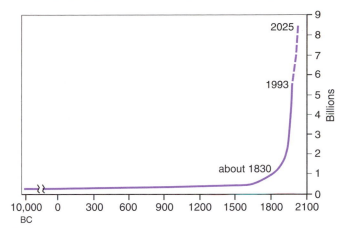

Figure 5.10 World population growth.

POPULATION PROJECTIONS

Predicting population growth is a hazardous task. In the early 1920s, Pearl predicted a stable population of 2.6 billion persons by 2100. A second example is the 1945 prediction by an eminent American demographer, Frank Rotestein, of 3 billion by 2000. These are just two of many examples. But despite some unimpressive precedents, there is good reason to suggest that we are now in a position to make better forecasts. The principal reason is that both fertility and mortality rates now lie within narrower ranges than was previously the case. In the more developed world, fertility and mortality rates are now relatively stable and in the less developed world, mortality rates can decrease further while fertility rates are presumed to be declining.

Current United Nations 'medium' projections suggest a world population of 8.4 billion for 2025. This is assuming that the mortality transition will be complete at that time such that the CDR is approximately equal throughout the world. A look further ahead might see the current fertility transition that is taking place in the less developed world complete by the year 2100 with a relatively stable world population above 10 billion.

This discussion of population projections does not take into account the possibility that there may be **limits to growth**, a phrase that became popular after being used as the title of a report published by the Club of Rome (Meadows et al. 1972). (We raised these questions briefly in the previous chapter under our discussion of 'Earth's vital signs'.) Many environmentalists and ecologists argue cogently that there are definite limits to growth of population and economies. The earth is finite and many resources are not renewable. The classic work in this genre is, of course, that of Malthus, but the 1972 Club of Rome report broadened the argument to include natural resources and environmental impacts. Basing their arguments on trends in the early 1970s, the authors of the report predicted the likelihood of world population exceeding world carrying capacity with a resultant collapse of population and economy. An earlier

well-publicized work, *The Population Bomb* (Ehrlich 1968), similarly anticipated widespread famine, raging pandemics, and possibly nuclear war by about 2000 as the world squabbled over scarce resources.

There is an alternative view to the limits-to-growth argument, a view known as the **cornucopian thesis** that is typically advocated by economists. According to this second view, technology will continue to advance sufficiently so that new resources are created as old resources are depleted: 'there are limits to growth only if science and technology cease to advance, but there is no reason why such advances should cease. So long as technological development continues, the earth is not really finite, for technology creates resources' (Ridker and Cecelski 1979:3–4).

Indeed, some feel that people themselves are the ultimate resource rather than the source of pollution envisaged by the limits-to-growth thesis. Japan, for example, has few resources other than people and, with a 1993 RNI of 0.3 per cent, that resource is increasing only slowly. According to some commentators, this will lead to a slow-down in economic growth.

We cannot reconcile this issue today. If we ask ourselves if the world is close to being overpopulated, then a range of answers is forthcoming. The limits-to-growth thesis and the cornucopian thesis could not be more different. At this time it is important to recognize the range of views and seek evidence to help us find the right answer.

EXPLAINING POPULATION GROWTH

Making sound predictions of future population numbers is not easy, nor is it easy to explain past population growth. In this section, six different proposed explanations are described and evaluated.

S-SHAPED CURVE

All species, including humans, have a great capacity to reproduce, which is regulated by many constraints such as space, food supplies, disease, and social strife. Our first explanation chooses to simplify our problem enormously by proposing a simple and natural statement of population growth. This is known as the S-shaped curve (see Figure 3.8).

This curve is produced under carefully controlled experimental conditions. The growth process begins slowly, then increases rapidly (exponentially), and finally levels out at some ceiling. It seems probable that such growth curves never actually occur in nature. It may not be unusual for growth to be first slow, then rapid, but it does appear to be unusual for any population (plant or animal) to remain steady at some ceiling level. A more characteristic final stage would involve a series of oscillations. Various scientists nevertheless predicted that world human population growth would correspond to the S-shaped curve. We have already referred to the prediction made by Pearl in the 1920s of a stable population of 2.6 billion by 2100. Clearly, using the curve as a predictive tool is hazardous. Simply put, use of this curve does not consider the variety of relevant cultural and economic factors that affect human populations. As we have already ascertained in our discussion of fertility, stable populations result not from some natural law but from human choice to knowingly and willingly control birth rates.

MALTHUSIAN THEORY

Our second explanation, unlike the S-shaped curve, is rooted in real world facts. Malthusian theory was first presented in 1798 and remains relevant today. Thomas Robert Malthus (1766–1834) was born into a changing England, one of industrialization and rapid population growth. He was a humane individual who was greatly concerned about the welfare of the growing numbers of poor people. He opposed the prevailing economic thought, mercantilism, which was explicitly pronatalist — more births meaning more wealth, on the logic that a large labour force was needed in England to enhance national income and was needed overseas to increase English strength in colonies.

Malthus expressed his views in a book titled *An Essay on the Principle of Population*, first published in 1798 with a final, seventh edition in 1872. Basic Malthusian logic is straightforward. It can be usefully presented in terms of two axioms and an hypothesis.

Axiom 1: Food is necessary for human existence; further, food production increases at an arithmetic rate, i.e., 1, 2, 3, 4, 5 ...

Axiom 2: Passion between the sexes is necessary and will continue; further, population increases at a geometric rate, i.e., 1, 2, 4, 8, 16 ...

Given these two axioms, regarded by Malthus as reasonable assumptions, the following hypothesis is deduced.

Hypothesis: Population growth will always create stress on the means of subsistence.

This dramatic hypothesis, Malthus argued, was the inevitable response to the differential growth processes of food and population. This scenario applied to all plants and animals. Malthus noted that only humans, as rational beings (that is, with culture), could actually anticipate the hypothesis and thus avoid stress. Plants and other animals increase their populations until such increases are limited by lack of space and/or food. Human populations can be contained within the limits set by subsistence, Malthus further argued, only by two types of checks. *Preventive checks* are conscious reductions in fertility by such means as moral restraint and delayed marriages. *Positive checks* include war, pestilence, and famine. Malthus believed that the preventive checks would only operate under the threat of positive checks. Thus, he concluded, humans were essentially incapable in practice of controlling reproduction except under extreme circumstances. The human future was to be one of famine, vice, and misery.

Malthus's central concern was imbalance between population and food. His arguments are intriguing, if not especially prophetic. In considering his theory, we must first note that there is no particular justification for the specific rates of increase proposed in the two axioms. Indeed, Malthus himself acknowledged in later editions of his work that the rates of population and food increase could not be definitely specified. Malthus was not able to anticipate that food supplies could be increased not only by increasing the supply of land (which he correctly saw as finite) but also by improving fertilizers, crop strains, and so forth. Malthus also failed to realize that contraception was to become normal and accepted. Malthus was aware of contraception, of course, but he saw it as an immoral preventive check (along with homosexuality), unlike the moral preventive checks of delayed marriage and moral restraint of which he approved.

Events have proven Malthus incorrect in his predictions and his theory lost favour in the mid-nineteenth century as birth-rate reductions and emigration eased population pressures in Europe. Today, what might be called neo-Malthusian theory is often argued as relevant in the less developed world. China is the prime example of a country that is in this situation. Because its political philosophy actively encourages state intervention into matters of everyday life, it has been able to implement a rigorous population policy.

MARXIST THEORY

One of the earliest and most powerful critics of Malthus was Marx, who objected to the rigorous axioms and hypothesis used by Malthus and saw a need to relate population growth to the prevailing mode of production in a given society. For Marx, Malthus represented a bourgeois viewpoint whose primary aim was to maintain existing social inequalities. The difference is fundamental: Malthus saw population growth as the primary cause of poverty; Marx saw the capitalist system as the primary cause — the problem was not a

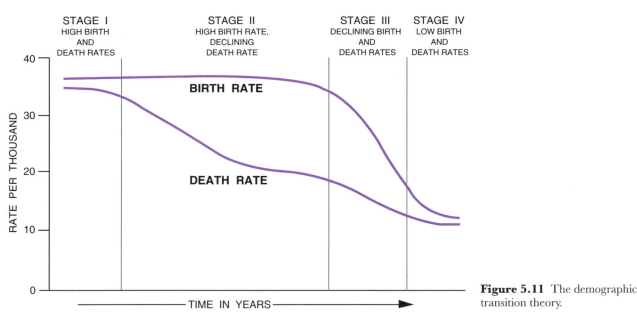

Figure 5.11 The demographic transition theory.

population problem but a resource distribution problem resulting from capitalism.

Another crucial distinction between these two writers concerns the concepts of overpopulation and surplus population. Malthus regarded overpopulation as the result when food and other necessities of human life were in such short supply that life-threatening circumstances arose. Marx regarded surplus population as an inevitable consequence of capitalism, literally an unemployed group or a reserve workforce. Both concepts are valuable to us as geographers. Overpopulation may prevail in some areas of the contemporary world, while surplus population is indeed an inevitable concomitant of capitalism.

BOSERUP THEORY

A fourth explanation is available in the writings of Ester Boserup on historical changes in agriculture in subsistence societies (Boserup 1965). According to this argument, subsistence farmers select farming systems that permit them to maximize their leisure time and they will only change these systems if population increases and it becomes necessary to increase the food supply accordingly.

She argued that in subsistence societies, population growth requires that farming be intensified so that the supply of food is increased to feed the additional population. Thus population growth has positive effects, improving human welfare. However, although population increase prompts an increase in gross food output, there is actually a decline in food output per capita. Further, there is also a series of negative changes because most of the agricultural areas with high and increasing populations are already areas of poverty and often limited agricultural technology.

The contrast with Malthus is especially clear. Malthus saw population as dependent on food supply; Boserup

reversed this relationship by proposing population as the independent variable. Evidence suggests that this theory is applicable in a subsistence context, but not in a developed world context because of higher technology and declining agricultural populations.

DEMOGRAPHIC TRANSITION

So far in our discussion of proposed explanations of population growth, we have considered one natural law, the S-shaped curve, and three influential writers: Malthus, Marx, and Boserup. Each has something to offer; each is at least partially flawed. Clearly, population growth is not easily explained. Our fifth explanation is in a somewhat different category.

The demographic transition is a model or descriptive generalization. It is based on known facts rather than on specific axioms or general assumptions. As such, it has a major advantage: it is clearly more realistic, albeit simplified, than most of the previous explanations. It also has an inevitable major disadvantage: because it is descriptive, it does not offer a bold provocative perspective or hypothesis. The demographic transition simply describes changing levels of fertility and mortality, and hence of natural increase, over time. The description is based on the experience of the contemporary more developed world.

Figure 5.11 presents the demographic transition in conventional graphical form. The first stage is characterized by a high CBR and a high CDR, with the two rates being approximately equal. The CDR fluctuates in response to war and disease. This stage involves a low-income agricultural economy. Population growth is limited. In the second stage, there is a dramatic reduction in CDR as a result of the onset of industrialization and related medical and health advances. This reduction is not accompanied by a parallel reduction in

CBR and hence the RNI is high; population growth is rapid. The principal feature of the third stage is a declining CBR, the result of voluntary decisions to reduce family size facilitated by advances in contraceptive techniques. The evidence suggests that voluntary birth control is related to increased standards of living. The RNI falls during this third stage as the CBR approaches the already low CDR. In the fourth (and final?) stage, the CBR and CDR are once again, as in the first stage, approximately equal and a low RNI reappears. The transition involves a change from wastefully high birth and death rates to a much more socially acceptable state of affairs.

Growth rates are similar in the first and fourth stages, but the manner in which the rates are generated are very different. The transition also involves two central stages during which birth and death rates are unequal and a high rate of increase prevails. All three population pyramids in Figure 5.7 exhibit the characteristics of a country at the onset of the transition. The four population pyramids in Figure 5.9 show a country that has proceeded through the transition, with the added complications of the baby boom and an aging population.

This demographic transition correctly describes in a simplified form the experience of the more developed world. Overall, these areas achieved the fourth stage during the first half of this century, but what about the less developed world? Throughout much of Asia, Africa, and Latin America, the situation still parallels that of the second stage. Can we assume that passage through the third stage and entry to the stable fourth stage are merely a matter of time and will be prompted by economic development? Expressed another way, is the demographic transition a valuable predictive tool? The answer appears to be no.

The current situation in the less developed world is not truly parallel to that prevailing in nineteenth-century Europe; populations are increasing by the billions, not by the millions, and prospects for industrialization in many countries are questionable. Further, countries that are currently in the second stage have experienced reductions in CDR because of the diffusion of technology from the more developed areas. Clearly, it is one thing to successfully introduce methods for reducing death rates and quite a different thing to successfully introduce methods for reducing birth rates. Lower death rates are welcomed by all societies; lower birth rates are not. Reducing death rates is essentially a technological achievement; reducing birth rates does involve technological issues, but is fundamentally a cultural achievement.

But, as we have already discussed, there is compelling evidence to suggest that the fertility decline evident in the third and fourth stages of the demographic transition is indeed occurring in the less developed world, but for very different reasons (see Box 5.3).

THE FERTILITY TRANSITION

Since about 1970, fertility has declined throughout much of the less developed world and is continuing to decline. To help understand this change, it is appropriate to refer not to the experience of the more developed world, which is summarized in the demographic transition, but to an explanation that is based on the current circumstances of the areas being affected. Although the current fertility declines accord with the description provided by the demographic transition, it seems clear that they do not accord with the logic behind that explanation with respect to either timing or circumstances. Fertility rates in the less developed world are falling more rapidly than previously in the more developed world. The fertility transition is an explanation that appears to accord with currently available data for the less developed world (Robey, Rutstein, and Morris 1993).

Since about 1970, the most powerful influence on fertility in the less developed world has been the extent to which modern contraceptive methods are employed. If family planning increases, fertility drops. In the less developed world (not including China), about 38 per cent of married females now practise family planning and about 80 per cent of these employ modern as opposed to traditional methods of contraception. Although the better-educated females are those most likely to limit family size, education is not a prerequisite for using contraception as evidenced by a decreasing correlation between education and use of contraceptives. Information about family planning is widespread because of the influence of the mass media.

According to the fertility transition argument, the appeal of large families has fallen because of the obvious problems associated with rapid and substantial population increases; these problems include pressure on agricultural land and the poor quality of urban life. The evidence also suggests that females are increasing in status and that they favour marrying later, having smaller families, and spacing births.

The principal reason why fertility declines are occurring so rapidly in the less developed world is probably because once the cultural impetus is in place, effective contraception is available; such was not the case when the more developed world was first culturally predisposed to smaller families and abstinence, withdrawal, and abortion were the only techniques available. The fertility declines occurring today may be best described as a 'reproductive revolution' because they are rapid and substantial.

Economic development does help create a climate conducive to reductions in fertility, but the key point of the fertility transition argument is that these reductions are caused by a new cultural attitude, the willingness to employ modern contraceptive methods, and by the ready availability of these methods. Development is not the best contraceptive, 'contraceptives are the best contraceptive' (Robey, Rutstein, and Morris 1993:65).

THE NEED FOR BETTER EXPLANATIONS

How do the ideas in the preceding sections aid our understanding of population growth? Each explanation is struggling with a complex issue and two of them, those by Malthus and Marx, must be evaluated in the light of the circumstances prevailing at the time of their initiation. Human geographers are far from satisfied with the explanatory and predictive power of these theories and, as we saw in Chapter 2, the various and varied philosophies of human geographers each place a different emphasis and meaning on explanation.

EVALUATING THE AVAILABLE EXPLANATIONS

An *empiricist* philosophy focuses on facts, not theory. Remember that much traditional regional and cultural geography was at least implicitly empiricist. Much the same can be said for population geography; its focus is on facts to the general exclusion of interpretation, understanding, or explanation. Both the demographic and fertility transitions are empiricist as they are descriptions based on available data, although both incorporate suggested explanations for the trends shown by the data. It is now reasonably clear that the demographic transition, which is explained by changing economic circumstances, provides a good account of the post-1750 experience of what is now the more developed world, but is not an appropriate predictive tool for the present less developed world. The fertility transition that is currently occurring in the less developed world is explained by changing cultural preferences.

A *positivist* philosophy focuses on theory construction, related hypothesis and law formulation, and statistical analyses. The S-shaped curve, Malthusian theory, and the Boserup thesis all have positivistic overtones in that they explicitly relate cause and effect. In the S-shaped curve, population size is related to time and the availability of space and food. In Malthusian theory, population growth is similarly explained, while Boserup sees population as cause and food supplies as effect. All three contributions are useful, but clearly far from being of universal value. The S-shaped curve is of limited usefulness; Malthusian theory has clear limitations, while the Boserup argument is basically limited to subsistence societies. A problem with the use of a positivist philosophy is that it requires data to test hypotheses statistically, data that are rarely available at a variety of spatial and temporal scales.

A *humanistic* philosophy has been little employed in analyses of population growth. There are some obvious applications, given the humanistic emphasis on the experiences of individuals. What, for example, was the reaction of Romanian women to the strict pronatalist policies of that country prior to 1989? Or how do Chinese peasant couples respond to the one-child policy? The understanding of the fertility transition seems to be an area where a humanist approach might be applicable.

A *Marxist* philosophy applied to population issues sees population growth as an outcome of the particular productive mode of society. Fertility behaviour, for example, is seen as caused by society and not as the outcome of free decision making. It seems that the value of Marxist theory is essentially restricted to the period of early industrial capitalism.

ALTERNATIVE EXPLANATIONS?

It is worth noting that our discussions have rarely chosen to subdivide human populations on the basis of age, ethnicity, class, religion, gender, or other relevant variables. The rationale for this decision to treat humans as a cohesive whole relates to our favoured scale of analysis, typically a world or regional scale. Material focusing on population subgroups will be introduced in later chapters. Nevertheless, we must acknowledge that the human experience is not unitary, being closely tied to a number of variables. One important variable that merits discussion at this time is that of gender.

Feminist geography has developed as an area of interest in human geography quite recently. In common with an approach such as Marxism, it is concerned with criticizing and changing established procedures. Specifically, feminist geography is involved in the question of gender inequality in our human experience. The aim is not only to recognize that there are different gender geographies but, much more fundamentally, to develop gender-specific theory.

How might such a feminist geography add usefully to our discussions of population? Answers are many and varied. We will briefly consider gender differences in the more developed world as an example. The principal change currently occurring involves a changing role from wife and mother to wife, mother, and worker outside the home. In the more developed world, the typical family structure remains the conjugal variety, where a person's family of procreation (being a parent) is independent of a person's family of orientation (being a child). The family has responsibility for reproduction, and the changing role of women in society is clearly linked to fertility. The more developed countries all have low fertility rates and increased participation of women in the labour force. Thus, women and men both have links with a world outside the family, a fact that ought, in principle, to be fostering an increasing similarity of the human experience. Whether or not this is actually taking place is debatable.

A brief review of philosophies and related attempts at explaining changes in fertility and mortality cannot do full justice to their existing and potential value. It does appear, however, that our current explanations, while useful, are far from adequate. There is a need for improved theory. What does such a theory need to accommodate?

Regardless of philosophical affiliation, new explanations need to explain reality at a variety of spatial, social, and temporal scales. As an example, let us consider the reality of the demographic transition. How are we to explain changing death and birth rates? An appropriate theory needs to incorporate economic circumstances (such as industrialization) and cultural issues (such as desirable family size). But we also need to consider individual behaviour of males and females and the motivations of their behaviour. We further need to recognize that the transition occurred differently in different regions; in France, the decline in fertility was evident by the 1830s, while in England, it was not evident until the 1890s. Thus there are significant spatial variations in the timing and details of what we previously presented as a uniform transition. Detailed analysis of such variations might facilitate improved theory construction by identifying specific causal variables. A full theory to explain the transition is still wanting.

SUMMARY

How many?
A population of 5.6 billion in 1993, possibly 6.3 billion in 2000.

Fertility
There are various measures of fertility; these include the crude birth rate, the general fertility rate, and the total fertility rate. In 1993, the world birth rate was 26 (26 live births per 1,000 members of the population), while the total fertility rate was 3.3 (the average woman has 3.3 children). A total fertility rate of about 2.1 to 2.5 is sufficient to maintain a stable population. Fertility is affected by many variables: age and related fecundity, nutrition, level of industrialization, age of marriage, celibacy, governmental policies, contraceptive use, and abortion. Spatial variations in fertility are closely related to level of development; a total fertility rate of 1.8 is evident in the more developed world and 3.7 in the less developed world. At present, there is evidence of declining fertility in much of the less developed world because of widespread acceptance of family planning, and in much of the more developed world possibly because of uncertain economic prospects.

Mortality
Mortality measures include the crude death rate, the infant mortality rate, and life expectancy. In 1993, the death rate was 9 and life expectancy was 65 years. Major causes of death are old age, disease, famine, and war. The world pattern of death rates shows much less variation than that of birth rates; central Africa is the last region of high mortality. Infant mortality and life expectancy data are good approximations of health. Many countries exhibit significant internal variations in mortality measures.

Natural increase
In 1993, the rate of natural increase was 1.6 per cent. This rate will result in a doubling of world population in about forty-two years. Natural increase is affected by the age composition of a population.

Government intervention
Most governments intervene directly and/or indirectly to influence growth. Death control policies are normal and designed to reduce mortality. Birth control policies can be *laissez-faire*, pronatalist, or antinatalist. A pronatalist approach may be related to a dominant religion or economic and strategic motives. Antinatalist policies, of varying degrees of success, are common in many of the less developed countries. The two most populous countries, China and India, have employed different approaches with China being highly successful and India less successful.

Predicting growth
Past forecasts have often erred seriously. A current projection has a population in excess of 8.4 billion by 2025 and a relatively stable population of about 10 billion by 2100. There is much uncertainty concerning the relative merits of two arguments. The limits-to-growth thesis sees definite limits to population and economic growth because the earth is finite. The cornucopian thesis sees technology as continually creating new resources and hence accommodating increased numbers of people.

World population growth
The world population increases in response to cultural change. An agricultural revolution began about 10,000 BC and prompted increases from about 4 million to 250 million by the beginning of the Christian era. A second major change, the Industrial Revolution, included an increased population from 500 million in 1650 to 1.6 billion in 1900.

Explaining growth
The S-shaped curve is a useful biological analogy, but is clearly not applicable based on current evidence. Malthus, Marx, and Boserup have each contributed usefully to our understanding of population growth. Malthus saw population as limited by food supplies; Marx saw population as a response to a particular social and economic structure; Boserup saw population increases as prompting food supply increases. Each of these three is of value, but none is of general applicability. Malthus is the most discussed theorist and a contemporary version of his theory is known as neo-Malthusianism. One very useful descriptive model is the demographic transition that is a summary of birth and death rates over the long period of human history in what is now the developed world. Debate centres on the predictive power of the model and the fertility transition appears to be a more appropriate description for current changes in the less developed world. There is a clear need for more and better theory — whether it be positivist, humanist, or Marxist.

WRITINGS TO PERUSE

CIPOLLA, C.M. 1974. *The Economic History of World Population*, 6th ed. Harmondsworth: Penguin.

A stimulating and provocative overview of world population growth emphasizing our use of energy sources; dated but still interesting.

DAY, L.H. 1992. *The Future of Low Birth-Rate Populations*. London: Routledge.

An assessment of the demographic situation, policy alternatives, and likely future changes in an aging and stable or declining population.

LUTZ, W., S. SCHERBON, and A. VOLKOV, eds. 1994. *Demographic Trends and Patterns in the Soviet Union Before 1991*. London: Routledge.

A compendium of demographic research on the former USSR; detailed reviews of fertility, mortality, age, marriage, and the family. Contrasts developed west and north with the less developed central Asian region.

NEWMAN, J.L., and G.E. MATZKE. 1984. *Population: Patterns, Dynamics and Prospects*. Englewood Cliffs: Prentice-Hall.

This population geography textbook can be used by students with minimal previous study of the topic.

OMRAN, A.R. 1992. *Family Planning in the Legacy of Islam*. London: Routledge.

An examination of the Islamic view of marriage, family formation, and child rearing.

PACIONE, M., ed. 1986. *Population Geography: Progress and Prospect*. London: Croom Helm.

An advanced book with good chapters on, among other topics, theory, fertility, and mortality.

PETERS, G.L., and R.P. LARKIN. 1983. *Population Geography: Problems, Concepts and Prospects*, 2nd ed. Dubuque: Kendall Hunt.

An excellent textbook in population geography covering all of the topics in this chapter in detail, including clear accounts of the various measures of fertility and mortality.

Population Reference Bureau. 1993. *1993 World Population Data Sheet*. Washington, DC: Population Reference Bureau.

This invaluable factual document is published annually and is a comprehensive source of basic demographic data.

Scientific American. 1974. *The Human Population*. San Francisco: Freeman.

A collection of articles all originally published in *Scientific American*. The data are somewhat dated, but most articles remain of value, especially those dealing with the genetics of human populations.

WOODS, R. 1979. *Population Analysis in Geography*. London: Longman.

Very useful and straightforward text on population geography.

————. 1981. *Theoretical Population Geography*. London: Longman.

A stimulating and critical overview of conceptual issues.

6

The human population: An unequal world

OUR DISCUSSION in the previous chapter emphasized that the current rate of population growth is a significant and possibly temporary deviation from the rates that prevailed during most of human history, that this high growth rate is from the highest base number in human history (5.6 billion in 1993), and that there are significant spatial variations in growth rates. In this chapter we examine three general questions that are related to these population circumstances.

First, where do we live and why? Second, why have so many people, both past and present, moved from one location to another and why are there so many people today who are effectively without a home country? Third, why is the world demographically divided between a stable population in the more developed world and a rapidly increasing population in the less developed world and what are the consequences of this division? To help answer these questions, we attempt to explain why the world is divided into more and less developed areas. This is a concern that includes discussion of the quantity and quality of food supplies in different parts of the world, the spatial distribution of debt, the regional selectivity of disasters, the measures employed to distinguish levels of development, and the political and economic relationships between countries.

DISTRIBUTION AND DENSITY

Determining where people are located and in what numbers is one of our central concerns. We need to determine population distributions and densities. Achieving these basic factual goals is not quite as straightforward as you might anticipate. There are three related problems. First, the required data are not available for all countries and regions within countries and even if they were available, they may not be reliable. Second, we must use data collected for predetermined divisions that are not necessarily those in which we are interested. Third, distribution and density statistics are related to spatial scale and often portray an inaccurate picture of reality.

MEASURING DENSITY

Distribution refers to the spatial arrangement of geographic facts, while **density** refers to the frequency of occurrence of geographic facts within a specified area. Conventionally, population density is arithmetic density, the total number of people in a unit area. Frequently, maps of population combine characteristics, distribution, and density. Figure 6.1 is a good example; this map depicts population density and distribution for the world. A good (although simplified) picture is provided.

Use of a single measure of population density for a country is often quite misleading. Canada is a prime example of a country with some areas of very high density and some of very low density, so that one statistic for the entire country is less than helpful. We referred to this type of problem in Chapter 2 when considering the question of spatial scale. For Canada, the 1993 population density is 3.05 people per square kilometre, but this single statistic is merely an average that results from large areas with virtually no population combined with relatively small areas of much higher density. Viewed from this perspective, the single density figure for a country such as Canada is rather meaningless. Indeed, density is most appropriately used as a measure at the local scale

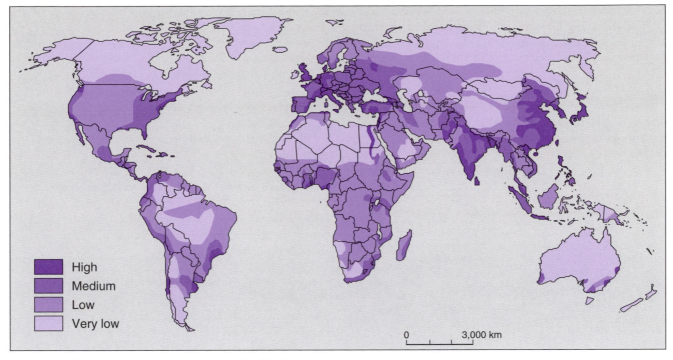

Figure 6.1 World population distribution and density, 1993.

where there are least likely to be significant spatial variations, as we have noted.

Density is sometimes calculated in a number of more refined versions by relating population to cultivated or cultivable land or to some other economic measure. **Physiological density,** for example, is the relation between population and that portion of the land area deemed suitable for agriculture.

Although a simple density measure employed at the national or multinational scale may be quite misleading, there is value in considering such information to acquire a broad outline of the pattern of world population. Table 6.1 provides 1993 population data using basic continental divisions

and highlights Asia with 59 per cent of the world population and the relatively small region of Europe with 11 per cent (1983 data are included for comparative purposes).

Table 6.2 identifies the ten countries with the largest total populations. China ranks first, followed by India; these two countries have much larger populations than any other country. Three other Asian countries — Indonesia, Bangladesh, and Pakistan — are included in this top ten. Intriguingly, had this table been constructed using the 1990 estimates, then the population of Nigeria would have appeared to be much larger — 119 million and not 95 million. To understand this dramatic 'reduction' between 1990 and 1993, consult Box 6.1.

Table 6.3 shows the number of people per square kilometre in each of these ten countries; perhaps surprisingly, the

TABLE 6.1: *Population by major world regions, 1983 and 1993*

Region	1983 Population (millions)	1983 % of world population	1993 Population (millions)	1993 % of world population
Asia	2,730	58	3,257	59
Africa	513	11	677	12
Europe	489	10	613	11
Latin America	390	8	460	8
North America	259	6	287	5
Former USSR	272	6	285	5
Oceania	24	0.5	28	0.5
World	4,677	100	5,607	100

Source: Population Reference Bureau, *1983* and *1993 World Population Data Sheets* (Washington, DC: Population Reference Bureau, 1983 and 1993).

TABLE 6.2: *The ten most populous countries, 1993*

Country	Population (millions)
China	1,179
India	897
United States	258
Indonesia	187
Brazil	152
Russia	149
Japan	125
Pakistan	122
Bangladesh	114
Nigeria	95

Source: Population Reference Bureau, *1993 World Population Data Sheet* (Washington, DC: Population Reference Bureau, 1993).

countries with the largest populations are not those with the highest densities. Of the ten most populous countries, Bangladesh is by far the most densely populated, followed by Japan and India.

MAPPING WORLD POPULATION

Although the world map of population distribution and density (Figure 6.1) is of great value, it is a static picture of a dynamic situation, providing no indication as to how the distribution developed or how it might change in the future. It suggests three areas of population concentration: eastern

TABLE 6.3: *Population densities of the ten most populous countries, 1993*

Country	Population density per square kilometre
Bangladesh	875.0
Japan	331.0
India	301.8
Pakistan	158.8
China	126.4
Nigeria	104.4
Indonesia	102.7
United States	28.2
Brazil	18.0
Russia	8.7

Source: Calculated from data in *1993 World Population Data Sheet* (Washington, DC: Population Reference Bureau, 1993).

Asia, the Indian subcontinent, and Europe. Both of the Asian areas are long-established population centres, the location of early civilizations and of early participation in the agricultural revolution. Today, both include areas of high rural population density, especially in the coastal, low-land, river valley locations, and large urban centres. In Europe, densities are high, but lower than in the high-density areas of Asia and are related primarily to urbanization. In all three areas, population densities show clear relations to land productivity.

There are other scattered areas of high density such as northeastern North America, in the vicinity of large cities in Latin America, the Nile Valley, and parts of west Africa. But perhaps the most compelling impression conveyed by Figure 6.1 is that a large percentage of the earth's surface is only sparsely populated.

EXPLAINING THE MAP OF WORLD POPULATION

Physical variables

There is a simple but potentially misleading correlation between Figure 6.1 and basic physical geography. Once again, remember that physical geography does not cause human geography; what we are acknowledging is that there are close relationships between the two resulting from human recognition of the relative attractiveness, especially productivity, of areas. We have favoured certain physical environments and not others. Even a cursory comparison of Figure 3.8 (our map of global environments) with Figure 6.1

Box 6.1: The 1991 Nigerian census

Censuses are the principal sources of information about the demographic characteristics of a country. Enumerations of populations have a long history, but the first modern censuses were conducted in some European countries in the eighteenth century and relatively reliable data are available for most European countries from at least the early nineteenth century. Canada conducted a first comprehensive census in 1851. Almost all countries have conducted censuses during the second half of this century. In order to be really useful, censuses need to be regular (in many countries, it is now usual to conduct a census at least once every ten years) and also comprehensive (that is, to include all members of the population).

Nigeria conducted an official census in 1991, the first since 1963 (Porter 1992). The long delay between counts is explained by reference to both the cost involved, at least Can. $150,000, but more importantly to two related political issues. First, the spatial distribution of government spending on services and amenities has been linked to population numbers. Indeed, the 1963 census contained many errors because of inflated returns from some regions and a 1973 census was then declared invalid because the results, both total and

regional, were clearly false. Second, since independence from Britain in 1960, population numbers have been used as an argument in the ongoing debate over which ethnic group exercises political control; the principal conflict is between a primarily Muslim north and a primarily Christian south.

During the early 1990s, the Population Reference Bureau based in the United States estimated the total population of Nigeria at about 119 million, an estimate based on the recorded 1963 total and assumptions about subsequent natural increase rates. The 1991 census thus provided a major surprise when it reported a total population of only 88.5 million. Based on this census information, the 1992 estimate for Nigeria was 90 million and the 1993 estimate, as detailed in Table 6.2, is of a total population of 95 million. Although the possibility of significant errors in the 1991 census cannot be discounted, there are sound reasons to be reasonably confident as to the quality of the data.

The lesson to be learned from this example is an important one; that is, all data have to carefully assessed for their reliability and any uses made of the data need to acknowledge possible limitations.

suggests relationships. Three of the global environments — monsoon, Mediterranean, and temperate forest areas — are associated with high population densities while three other environments — desert, tundra, and polar areas — are associated with very low densities.

At the global scale, four physical variables are relevant: temperature, availability of water, relief, and soil quality. High-density areas typically have temperatures permitting an agricultural growing season of a minimum five to six months. Water is essential either as precipitation or in the form of irrigation water. Optimum temperature levels and precipitation amounts are not easy to specify as they vary with technology, but it is clear that extremely high temperature and precipitation together, as in tropical rain forests, are not associated with dense population.

Cultural variables

A second set of factors, namely aspects of human cultural organization, is related to global population densities. The high-density areas apparent in Figure 6.1 experienced relatively early state organization — China, India, southern Europe, Egypt, Mexico, and Peru were all centres of early civilizations. Establishing state organization facilitates population concentration; similarly, the collapse of such organization operates against continued concentrations, as in the southern Iraq area. Some areas of high density today are not explicable by reference to matters of early state development. In western Europe and northeastern North America, high densities were initiated at a much later date in conjunction with the Industrial Revolution and associated urbanization.

The distribution and density of world population today is the dynamic outcome of a long historical process. Is it possible that the pattern is stabilizing today? We know that certain areas are continuing to experience high rates of natural increase and are therefore likely to experience density increases. We also know that population continues to move from one location to another; indeed, an area such as northeastern North America has grown largely as a consequence of in-movement.

MIGRATION

Humans have always moved from one location to another. Early preagricultural movements involved expansion to all major areas of the world with the exception of Antarctica. This mobility expanded the total resource base, facilitated overall population increases, and stimulated cultural change by requiring ongoing adaptations to new environmental circumstances. Migrations have continued to be a characteristic of human occupation of the earth.

Migration is a version of human mobility that involves a spatial movement of residence. Thus we do not regard the journey to work, shopping trips, or the movements of some

pastoralists and agriculturalists as migration. This definition and the associated exclusions are not precise, but are adequate for our purposes. Our focus is on those movements that involve a change of residence. We are particularly concerned with the distance moved, the time spent in the new location, the political boundaries crossed, the geographic character of the two areas involved, the causes of the migration, the numbers involved, and the cultural and economic characteristics of those moving.

WHY PEOPLE MIGRATE

To ask why people migrate is to seek a single simple answer to explain the multitude of migrations that have taken place in human history. Of course, there is no such answer, but we are able to produce a valuable generalization.

Push-pull logic

People move from one location to another because they consider the old location as less favourable in some crucial aspect than the new location. It is not possible to specify how much more favourable the new location needs to be, for this is clearly a matter of individual judgement.

In mid-nineteenth-century Ireland, for example, there was virtually unanimous agreement that an overseas location, such as the United States, Canada, or Australia, was preferable to Ireland and hence migration was the typical decision. In this situation, it was perhaps those who chose not to move who were making a conscious decision. In other cases, migration involves only a small percentage of a population, suggesting that the perceived difference between old and new locations is not so substantial. If we rank each area (however defined) of the world on an attractiveness scale from, say, 1 to 10, with 1 being the least attractive and 10 being the most attractive, who will move where in this simple scenario? The answer is that those living in low-ranked areas will wish to move to high-ranked areas; thus there will be large-scale movements from areas ranked low to areas ranked high. This is a simple way of saying that migration decisions can be conceptualized as involving a push and a pull. Being located in an unattractive area is a push; being aware of an attractive area is a pull.

Typical push-pull factors can only be identified in a relative context. Three factors are noted: economic, political, and environmental.

A simple *economic* argument is that migration is a consequence of differences in wages with people moving from low- to high-wage areas. Relatively low wages are a push; relatively high wages are a pull. Another economic argument involves the relative availability of agricultural land. Throughout the old European world from perhaps the sixteenth to the twentieth century, land was in short supply relative to the situation in overseas temperate areas. A third and perhaps the most fundamental economic argument involves situations where life itself is threatened because of inadequate food supplies; in a relative sense, any alternative area is more attractive.

In addition to economic aspects, areas can also be seen as unequal *politically*. The post-1945 desire of many eastern Europeans to migrate reflected an assessment of the relative merits of communist versus democratic political orders. In some extreme cases, people feel obliged to seek refuge in a country other than their own. In recent years, Afghanistan, Ethiopia, the former Yugoslavia, and Rwanda have proved to be political environments sufficiently unattractive as to prompt massive refugee movement. This topic is so important to us as human geographers that it is dealt with separately in the following section.

Relative attractiveness can also be measured *environmentally*. Migration may be induced by flooding and desertification, for example, and the historical experience suggests that most migration has been to the temperate climatic areas.

Laws of migration

Areal inequality is the logic behind the push-pull concept. However simplistic the concept may be, it is a valuable form of general explanation, applicable in a wide variety of different circumstances. Clearly, it is not especially profound: to assert that people move from A to B because B is in some way preferable to A is not a very insightful assertion. Fortunately, we are able to produce other more specific generalizations.

A classic attempt to develop laws of migration was that of Ravenstein in the late nineteenth century. Ravenstein (1876, 1885, 1889) based his generalizations on analyses of population movements in Britain. The laws he proposed are still some of the best developed migration concepts. They are not laws in a formal positivistic sense, but rather generalizations with varying degrees of applicability. There are eleven such generalizations (Box 6.2).

The mobility transition

A third attempt to explain why people migrate, introduced by Zelinsky (1971), is labelled the mobility transition. This term is derived from the demographic transition concept discussed in Chapter 5 and proposes five phases of temporal changes in migration:

Box 6.2: The Ravenstein laws

E.G. Ravenstein wrote three articles on migration over 100 years ago and these have proven to be highly influential in much subsequent research on migration issues. These articles identified eleven laws, more correctly described as generalizations, and were based on information contained in the British censuses of 1871 and 1881. The following list first states the generalization and then provides some brief comments.

1. The majority of migrants travel only a short distance: the tyranny of space. The general concept of distance friction, noted in Chapter 2, is one of the most fundamental of all geographic concepts.
2. Migration proceeds step by step. Thus, a migrant from Europe to Canada may first move to a port city such as Montreal and then to rural Quebec.
3. Migrants going long distances generally go to one of the great centres of commerce or industry. This reflects the fact that large centres are usually the best-known long-distance locations.
4. Each current of migration produces a compensating countercurrent. Any such countercurrent is usually relatively small.
5. The natives of towns are less migratory than those of rural areas. This law reflects the fact that rural to urban migration is common.
6. Females are more migratory than males within their country of birth, but males more frequently venture beyond. Females often move within a country in order to marry. International migration is usually of young males.

7. Most migrants are adults — families rarely migrate out of their country of birth. Overseas family migration is atypical as overseas migration usually involves males only.
8. Large towns grow more by migration than by natural increase. (Remember that Ravenstein was writing at a time of dramatic industrial and urban growth.)
9. Migration increases in volume as industries and commerce develop and transport improves. This reflects the attractiveness of urban centres and the reduction of distance friction.
10. The major direction of migration is from agricultural areas to centres of industry and commerce. This movement continues to be important in the late twentieth century as evidenced by the ongoing rural depopulation in the Canadian prairies.
11. The major causes of migration are economic (again, a law that reflects Ravenstein's time). This text provides several examples of socially and politically motivated migrations.

These eleven statements have been modified, not disproven, by subsequent research. Clearly, they are somewhat time specific, but their validity is spatially general (Grigg 1977). Probably the principal limitations are the exclusion of reference to various forms of forced (as opposed to voluntary) migrations and the failure to account for the current exodus from large cities. What is your impression of these laws? Do they appear to make sense to you, both intuitively and with reference to the larger text content? Or are they so general as to be valueless?

1. the premodern traditional society
2. the early transitional society
3. the late transitional society
4. the advanced society
5. a future superadvanced society

Each of the five has particular migration characteristics. In Phase One there is minimal residential migration and only limited human mobility. Zelinsky sees this phase as temporally parallel to the first stage of the demographic transition (high birth rates, high and fluctuating death rates). In Phase Two numerically significant migration commences in the form of rural to urban and overseas movements. This phase is temporally parallel to the second stage of the demographic transition (continuing high birth rates, rapidly falling death rates). During Phase Two, the mass movements associated with industrialization and European overseas expansion occur. In Phase Three rural to urban mobility declines somewhat, but remains numerically significant, while overseas migration is drastically reduced. This phase is temporally parallel to the third stage of the demographic transition (declining birth rates, low death rates). In Phase Four residential mobility continues apace, rural to urban movement lessens but continues, urban to urban movement is significant, and international migration involves both unskilled and skilled workers moving from the less developed to the more developed world. This phase is temporally parallel to the fourth stage of the demographic transition (low birth rates, low death rates). Phase Five is Zelinsky's attempt to predict future trends. Most migration is between urban centres. In our version of the demographic transition, there is no equivalent stage.

The mobility transition is a very useful categorization of temporal changes in mobility. Essentially, Zelinsky is proposing two causal factors: first, the demographic context and, second, technology. As with most model formulations, the approach is a good means of generalizing about migration and migration trends. It does not offer specific explanations of migrations and is primarily valuable as a classification scheme. The links with the demographic transition model are particularly intriguing.

But such an approach is not without its critics. The mobility transition can be regarded as one example of **developmentalism**. It can be seen as a 'geography of ladders' where 'the world is viewed as a series of hearth areas out of which modernisation diffuses, so that the Third World's future can be explicitly read from the First World's past and present in idealised maps and graphs of developmental social change' (Taylor 1989:310). An alternative view sees the mobility history of a particular country as a reflection of global as well as country processes. We recognized the same problem, somewhat less formally, in our discussion of the demographic transition and we will pursue this argument in a more general and fuller context later in this chapter.

A behavioural explanation

In addition to the earlier attempts to explain why people migrate, we may also approach the question from a more humanistic perspective. None of the three methods outlined place primary attention on the people themselves. Push-pull logic, the eleven laws of Ravenstein, and the mobility transition model all have positivistic overtones; they seem to imply that each individual human responds in an identical fashion to various external factors. Of course, any human geographer who uses these approaches knows this is not so, but is willing to sacrifice some reality to gain some simplicity. Our fourth and final attempt to explain why people migrate shifts attention to the people themselves and away from the forces presumed to be affecting their decisions.

For the sake of convenience, we will label this approach as the behavioural view of migration as it centres on the behaviour of individuals rather than on aggregate, usually large group, behaviour. Interestingly, behavioural approaches usually employ a version of push-pull logic, but typically do so at the level of the individual.

Place utility is the extent to which an individual is satisfied with particular locations. Typically, a person's place utility for his or her current residence is founded on relative certainty, whereas that same person's place utility for most other locations is less grounded in fact. Place utility is thus an individually focused version of push-pull logic. This concept was first introduced by Wolpert (1965) who argued that it was necessary to research an individual's **spatial preferences**.

All too clearly, such preferences are based on perceptions and not objective facts. All people have mental images of different places and these 'mental images' contribute to migration decisions. We use the term 'mental maps', which was first introduced in Chapter 2, when we talk about these images and we typically determine these maps by using questionnaires — asking individuals for their perceptions of different areas. Pioneering work has been done by Gould and White (1986) and group perceptions (spatial preferences) have been generated. Figure 6.2 provides examples of the mapped data. In principle, we can use these maps to predict migration behaviour.

THE SELECTIVITY OF MIGRATION

We have asked why people migrate from one area to another, and a number of useful generalizations have been forthcoming. As the discussion of place utility emphasized, areas are neither unattractive or attractive in an objective sense. Rather, individuals perceive their home area and other areas according to their own judgements. What is unattractive to one resident may not be to another — hence migration is typically selective. Migrations are, in the final analysis, decisions made by individuals, and migration is a selective process; we are not all equally affected by the general factors prompting migration. Who moves and who stays?

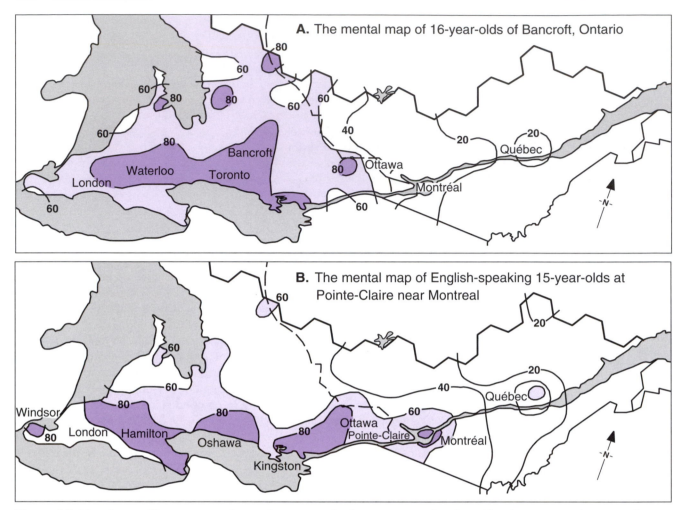

Figure 6.2 Mental maps. These two mental maps demonstrate that in some cases, cultural considerations are more important than distance. Map A presents the mental map of 16-year-olds in English-speaking Bancroft, Ontario. The larger urban areas are seen most favourably, but the most notable feature is the sharp drop into Quebec. Map B presents the mental map for English-speaking 15-year-olds at Pointe-Claire near Montreal in the predominantly French-speaking province of Quebec. Rather than favouring their local area, these 15-year-olds favour the English-speaking areas of Ontario.

Source: Adapted from P. Gould and R. White, *Mental Maps*, 2nd ed. (Boston: Allen and Unwin, 1986):77–9.

Migration selectivity is evident on the basis of age (most migrants are older adolescents or young adults); on the basis of marital status (most historical and current less developed world migrations involve single adults); on the basis of gender (males are typically more migratory, but there are many exceptions to this generalization); on the basis of occupation (higher-skilled workers are most likely to move); and on the basis of education (migrants have higher levels of education than non-migrants). In the most general sense, there is a useful relationship between **life cycle** and the likelihood of an individual migration.

In addition, there is often a substantial difference between what people would *like* to do and what they are *able* to do. People who want to migrate may be unable to leave their home for political reasons, and people who want to move to a specific new area may not be able to do so because of the immigration policies of that area. In a similar fashion, potential migrants need to consider the economic and personal costs of movements; some may be unable to pay for the move, while others may not move because of health, age, or family circumstances.

These ideas can be usefully conceptualized as shown in Figure 6.3. Origins and destinations both have good (+), bad (–), and neutral (0) attributes. Movement away from the origin may be affected by political issues, and movement to the destination may be affected by immigration policies. Between the two are personal and cost obstacles (Lee 1966).

TYPES OF MIGRATION

One of the most useful attempts to classify migration is that by the sociologist, Petersen (1958), who noted four classes of migration. First, there is primitive migration associated with preindustrial peoples and caused by some ecological necessity. Second, there is forced migration when people have little

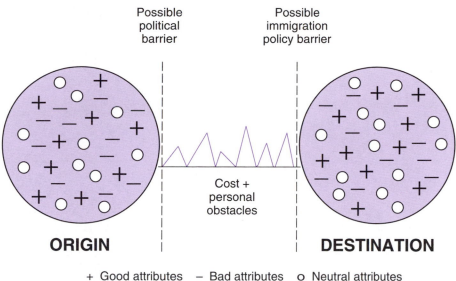

Possible political barrier

Possible immigration policy barrier

Cost + personal obstacles

ORIGIN

DESTINATION

+ Good attributes   − Bad attributes   o Neutral attributes

Figure 6.3 The push-pull concept and relevant obstacles.

or no alternative but to move, usually as a result of political circumstances. Third, there is free migration when people decide to move or stay on the basis of place utility. Fourth, illegal migration can occur either when a country prohibits out-movement or when people enter a country without official approval.

Primitive migration

This is really a specific instance of adaptation to environment; adaptation is achieved by rejecting one environment in favour of another. Preindustrial societies tended to make such adaptation decisions in a group, not an individual, context. Thus, hunting and gathering groups might migrate on a regular basis as the resources of an area are depleted or as game move. Similarly, some agricultural groups might move as soil loses fertility. Another instance of primitive migration occurs when populations increase in size so that additional land is needed. In preindustrial societies, primitive migration was normal behaviour as groups searched for appropriate environments; this normal human behaviour was responsible for our human occupation of most earth environments. As we noted in Box 4.1, by the fifth century AD, even Easter Island, perhaps the most isolated location in the world, was settled.

Forced migration

Forced migration has a long history. **Slavery** was an indispensable institution in such early civilizations as those in Greece and Rome. It appears that the population of Rome at its peak was generated primarily by the forced migration of slaves. Europeans found slavery necessary to fulfil their objectives when settling such areas as the Caribbean and the warm coasts of North and South America — areas not conducive to large-scale European migration. Perhaps as many as 11 million slaves were moved out of Africa between 1451 and 1870.

A second example of forced migration is that involving late-nineteenth-century labour movements from China, Java, and India to the new European-controlled plantations of Malaysia, Sumatra, Burma, Sri Lanka, and Fiji. These labour movements were supposed to involve contracts, but quite often force was used. A third and quite differently motivated example was the post-1938 movement of Jewish populations in areas controlled by Nazi Germany. In each of these instances, movement was quite literally forced on people.

A variant of forced migration involves those cases where the migrant has some voice, however small, in the decision making. This *impelled* migration would include the many cases where people choose to flee oppressive political regimes, war zones, and areas of famine. The dividing line between forced and impelled is not clear, nor is the line between impelled and free. Most contemporary refugee movements would typically qualify as impelled.

Free migration

Free migration means a person clearly has the option of either staying or moving. Historically such migrations were from densely settled countries to less densely settled ones. European overseas expansion to temperate areas such as the United States, Canada, Australia, New Zealand, South Africa, and Argentina involved free migration. These free movements occurred primarily in the nineteenth century; for the preceding 300 years, migration was a relative trickle. As we have seen in our discussion of the mobility transition, such migration was related to the demographic and technological changes in Europe after about 1650. Large-scale European overseas movement occurred primarily from 1800 to 1914, a period during which about 70 million people migrated overseas.

Thus, Europe was able to relieve the population pressures related to the second stage of the demographic transition at

6.1 Eastern European immigrants to North America on board ship, *c.*1910–14. (National Archives of Canada C68842)

least partly by emigration. In 1800, people of European origin totalled 210 million; by 1900, the total was 560 million — a 166 per cent increase. In 1900, one of every three people in the world was of European origin. Most European countries participated in these migrations. Irish, English, Scottish, Germans, Italians, Scandinavians, Austro-Hungarians, Poles, and Russians all moved overseas in large numbers. Many Russians also moved east to the Caucasus and Siberia. This historically brief period of movement had massive impacts on areas of both origin and destination, involved a major redistribution of peoples, and significant contacts between previously separate groups. Indeed, many of the political issues that face countries today have their roots in these circumstances, especially those problems involving minority group territorial claims.

Contemporary free migration continues, but in rather different directions. Popular destinations today are all of the more developed countries. This situation confirms our earlier observations that individuals' migration decisions largely stem from assessments of relative place utility. Much potential free migration, however, does not take place because of government policies. Most developed countries have implemented restrictive immigration policies. In the early twentieth century, countries such as the United States, Canada, Australia, New Zealand, and South Africa practised explicitly racist policies; today, this is not usually the case (Box 6.3).

Free migration also continues to occur within rather than between countries. In the United States, the south and west are attractive destinations for a combination of environmental and job opportunity reasons; a similar situation prevails in England, with the southeast being the most attractive area, although the late 1980s saw evidence of increased out-migration from London (Champion 1992). In many countries people continue to move from rural to urban areas and from central to suburban zones within urban areas. These movements often lead to significant changes in the structure and composition of local populations as they are selective according to stage in life cycle, gender, and ethnic background.

6.2 Vietnamese boat people rescued in the South China Sea, 1985. (UNHCR)

Indeed, there is a clear need to reconceptualize migration research to accord with some of the new social formations in developed countries. Most migration continues to be primarily economic in motivation, being related to employment, income potential, and the housing market, and it is now a key component of local population change (Green and Owen 1991).

Illegal migration

Illegal migration is a relatively new type of migration. Two versions are possible. Some countries, typically communist, impose restrictions on emigration. When these restrictions are lifted, as in East Germany in the second half of 1989, previously thwarted immigrants are able to emigrate legally. Some countries experience great difficulties in preventing illegal migration. For example, India has difficulty with Bangladeshi migrants and the United States with Mexican migrants.

REFUGEES

As we have seen, migration has a long history, many different forms, a multitude of causes, and operates at different spatial scales. Viewed as a whole, migrations are attempts by populations to settle the surface of the earth in a rational fashion. Once politically unconstrained, contemporary migrations between countries are now related to a host of restrictions; interestingly, a different political partitioning of space would likely result in a different population map. Migrations within countries are not usually affected by political matters and thus we can interpret most intranational movements as a good reflection of relative place utilities.

REFUGEE MOVEMENTS, 1945–90: A GROWING PROBLEM

The first major refugee movements to take place after the Second World War were readjustments to changing political circumstances. Immediately after 1945, about 15 million Germans relocated. The partition of India in 1947, which

---

Box 6.3: Restrictive immigration

Discriminatory immigration policies were typically introduced in the nineteenth and early twentieth century in those areas experiencing large-scale immigration. Overall, the basic intent of such policies was to limit immigration from Asia once it was clear that without policies, such immigration would probably be considerable.

In Australia, restrictions limited immigration of Chinese labourers and goldfield workers. The first legislative action, in Victoria in 1855, was followed by similar bills in other states. By 1901, the in-movement of all coloured peoples was limited as a consequence of requiring literacy in a European language. New Zealand introduced a restrictive immigration policy in 1881 and replaced this with a literacy requirement in 1899. In

Canada, rather similar tactics were pursued in British Columbia where Chinese immigrants were at first welcomed as cheap labour; discriminatory policies began in 1885 with the introduction of a head tax until, in 1923, Chinese immigration was virtually prohibited by an act that was not repealed until 1947. British Columbia employed similar tactics to limit the numbers of incoming Japanese and East Indians.

The reasons for implementation of such policies were various. Racist attitudes are the most obvious reason, with the British stereotype typically regarded as most desirable. In addition, there was often concern that abundant cheap labour would cause incomes to fall and that competition would harm retail business.

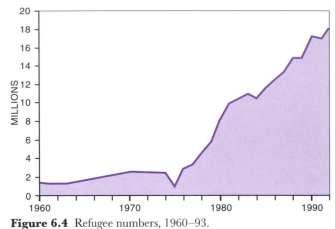

Figure 6.4 Refugee numbers, 1960–93.

Source: United Nations High Commissioner for Refugees, *The State of the World's Refugees: The Challenge of Protection* (London: Penguin, 1993).

included the creation of the Muslim state of Pakistan, caused the movement of about 16 million people — 8 million Muslims fled to Pakistan and 8 million Hindus and Sikhs fled to India. A rather different type of movement took place prior to the construction of the Berlin Wall in 1961 with about 3.5 million moving from then-communist East Germany to democratic West Germany. Even after the wall was in place, about 300,000 succeeded in fleeing west.

From about 1960 to the mid-1970s, the annual total number of refugees in the world was relatively low at 2–3 million (Figure 6.4). However, in 1973 the end of the Vietnam War resulted in one of the first major transcontinental movements; some 2 million refugees fled Vietnam, which necessitated a major international relief effort. Because of its involvement in the war, the United States received about half of these refugees. The problem of refugees fleeing Vietnam was finally resolved only in the early 1990s. More generally, the 1980s saw a huge increase in numbers of refugees for various political, economic, and environmental reasons, an increase that continues in the mid-1990s.

TABLE 6.4: *Refugees and asylum seekers by region of asylum, end of 1991*

Region	Number
Africa	5,340,800
East Asia	688,500
Middle East/south Asia	9,820,950
Latin America	119,600
Europe/Canada	677,700
Total	16,647,550

Sources: United States Department of State, *World Refugee Report* (Washington, DC: Bureau for Refugee Affairs, June 1992), Department of State Publication 9998, 178.

United States Committee for Refugees, *World Refugee Survey: 1992* (Washington DC: US Committee for Refugees, 1992):32–3.

TABLE 6.5: *Countries with more than 250,000 refugees and asylum seekers in need of protection and assistance, end of 1991*

Country	Number	Total
Pakistan from	Afghanistan 3,591,000	
	Others 3,000	
		3,594,000
Iran from	Afghanistan 3,000,000	
	Iraq 150,000	
		3,150,000
Malawi from	Mozambique 950,000	
		950,000
Sudan from	Ethiopia 690,000	
	Chad 20,000	
	Zaire 4,500	
	Uganda 2,700	
		717,200
Guinea from	Liberia 397,000	
	Sierra Leone 169,000	
		566,000
Ethiopia from	Somalia 519,000	
	Sudan 15,000	
		534,000
Thailand from	Cambodia 370,000	
	Burma 70,000	
	Laos 59,000	
	Vietnam 13,700	
		512,700
Zaire from	Angola 310,000	
	Sudan 104,000	
	Burundi 45,000	
	Rwanda 12,000	
	Uganda 10,000	
	Others 1,300	
		482,300
India from	Sri Lanka 210,000	
	Tibet 100,000	
	Bangladesh 65,000	
	Bhutan 15,000	
	Afghanistan 9,800	
	Burma 2,000	
	Others 800	
		402,600
Germany from	Yugoslavia 74,854	
	Romania 40,504	
	Turkey 23,877	
	Bulgaria 12,056	
	Iran 8,643	
	Nigeria 8,358	
	Vietnam 8,133	
	Afghanistan 7,337	
	USSR 5,690	
	Sri Lanka 5,623	
	Others 61,025	
		256,100
Tanzania from	Burundi 131,000	
	Mozambique 72,000	
	Rwanda 22,300	
	Zaire 16,000	
	South Africa 9,600	
	Others 200	
		251,100

Sources: United States Department of State, *World Refugee Report* (Washington, DC: Bureau for Refugee Affairs, June 1992), Department of State Publication 9998, 179–201.

United States Committee for Refugees, *World Refugee Survey: 1992* (Washington, DC: US Committee for Refugees, 1992):32–3.

THE PROBLEM TODAY: NUMBERS AND CAUSES

It is generally agreed that refugees are people forced to migrate, usually for political reasons. There is, however, no one accepted definition as to who is a refugee; there are also considerable logistical difficulties in counting refugees, and finally governments often have vested interests in providing incorrect information on numbers. Accordingly, estimates of the total number vary.

The most reliable source of information on refugee numbers today is the United Nations High Commissioner for Refugees (UNHCR). UNHCR's data demonstrate all too clearly that the number of refugees is increasing rapidly; the estimated number for 1993 is about 18 million and there are about 10,000 new refugees each day. Clearly, the growing problem since 1945 that began to increase dramatically about 1980 is now a tragedy of epic proportions.

But even the figure of 18 million underestimates the seriousness of the problem because it refers only to those refugees who have crossed an international boundary. The estimated number of refugees displaced within their own countries because of, for example, civil wars, persecution, or drought, is an additional 24 million. Further, in addition to the 18 million displaced across an international border and the 24 million displaced within a country, there are other people in refugee-like circumstances, such as undocumented aliens or asylum seekers. In Bangladesh 258,000 Pakistanis have been stranded since partition in 1947, and in Turkey there are 800,000 Iranians who are not recognized as refugees.

Although there are refugee problems in parts of the developed world, the problem is most intense in those countries that are least able to cope. Several of the poorest countries in the world today have the additional problems of both funding wars and caring for refugees. The war in Afghanistan in the 1980s prompted about 6 million refugees, including 4 million people moving to neighbouring Pakistan, while the problems in Somalia and neighbouring countries continue (Box 6.4).

Table 6.4 provides the 1991 (end of year) estimates of refugee numbers by region of asylum, and Table 6.5 provides data on those countries that have received in excess of 250,000 refugees and asylum seekers in need of protection. Table 6.4 identifies two principal problem regions: (1) the Middle East and (2) south Asia and Africa. Refugee numbers are relatively small elsewhere in the world. Table 6.5 provides more detailed data. Pakistan and Iran have huge numbers of refugees, almost entirely from Afghanistan, while several African countries also have large numbers of refugees within their borders. In Asia the numerically largest refugee problems are in Thailand and India. Table 6.6 is a retabulation of the world refugee data that are partially used to construct Table 6.5; Table 6.6 more clearly indicates the source areas of refugees with Afghanistan as the principal source, followed by Palestinian refugees and those from Mozambique.

Not included in Table 6.5, but included in the general regional number for the Middle East/south Asia in Table 6.4 and also in Table 6.6, are those countries with a large Palestinian population: Israel has 528,700 Palestinians in the Gaza Strip and 430,100 in the West Bank, Jordan has 960,200, Lebanon has 314,200, and Syria has 293,900. The Palestinians represent a special problem as they remain without a territory of their own.

In the more developed world, the collapse of communism in eastern European countries in the late 1980s and the breakup of the former USSR in 1991 encouraged mass refugee movements elsewhere in Europe. The inclusion of Germany in Table 6.5 as a country with more than 250,000 refugees, many from elsewhere in Europe, reflects these changing post-1989 political circumstances. Notwithstanding this example, the global refugee problem is essentially a problem in the less developed world.

SOLUTIONS?

Different refugees have different reasons for moving and there are no simple ways to resolve the many different cases. The UNHCR has traditionally proposed three solutions: voluntary repatriation, local settlement, and resettlement. More recently, the focus has turned to attacking the underlying causes of refugee problems. All attempts at solutions are problematic.

Voluntary repatriation tends to be favoured by the UNHCR, but is actually impossible for most refugees as the circumstances that prompted movement have rarely changed.

Local settlement is difficult in areas that are poor and lack resources. Whether people are refugees for political or some other reason, nearby areas are unlikely to offer a solution as the immediate problems are similar. Refugees are often in

TABLE 6.6: *Principal sources of the world's refugees and asylum seekers, end of 1991*

Country	Number
Afghanistan	6,600,800
Palestinians	2,525,000
Mozambique	1,483,500
Ethiopia/Eritrea	752,400
Somalia	717,600
Liberia	661,700
Angola	443,200
Cambodia	392,700
Iraq	217,500
Sri Lanka	210,000
Burundi	208,500
Rwanda	203,900
Sudan	202,500
Sierra Leone	181,000
Western Sahara	165,000
Vietnam	122,650
Yugoslavia	120,000
China (Tibet)	114,000
Burma	112,000

Source: United States Committee for Refugees, *World Refugee Survey: 1992* (Washington, DC: US Committee for Refugees, 1992):34.

6.3 Rwandan refugees fleeing ethnic violence, the Rwanda–Tanzania border, 1994 (UNHCR)

Box 6.4: Refugees in the Horn of Africa

One area of the world experiencing continuing refugee problems is the Horn of Africa (Figure 6.5). This area comprises three countries — Somalia, Ethiopia, and Djibouti. Today, along with neighbouring Sudan, the Horn is a land of refugees. There are an estimated 700,000 Ethiopian refugees in the Sudan, 60,000 in Somalia, and 1,400 in Djibouti. In addition, there are 485,000 Somali and 7,000 Sudanese refugees in Ethiopia, a further 29,200 refugees in Sudan from countries outside of the Horn, and increasing numbers of refugees in Kenya.

The cause is a tragic mix of human and physical geographic reasons. In 1962, Ethiopia annexed Eritrea and since then, a bitter war has continued. Ethiopia sees the conflict as a secessionist war, while Eritrea sees the conflict as a fight for self-determination. The two positions are irreconcilable. For several years in the 1980s, Eritrea and the neighbouring Ethiopian province of Tigray experienced severe droughts; the Ethiopian government responded by refusing international aid access to these areas.

A second political factor relates to Somali irredentism. Although it is one of Africa's poorest countries, Somalia places economic and social development second to the political ideal of integrating all Somali people into one nation. Since independence in 1960, this has involved conflict with Ethiopia (in the Ogaden) and Kenya especially. A full-scale war between Somalia and Ethiopia in 1977–8 coincided with drought.

Both of these political conflicts are closely related to European colonial policies and subsequent American and Russian (former USSR) involvement. They are not simply a result of regional matters.

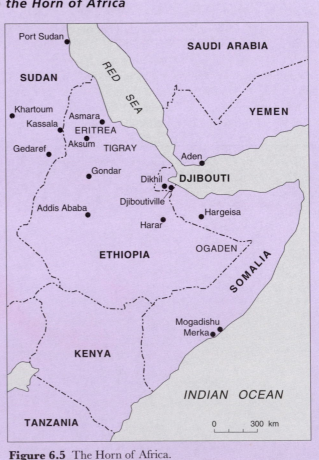

Figure 6.5 The Horn of Africa.

need of food, water, and shelter, usually in an area that is already poor and possibly suffering from some environmental problem, such as the frequent droughts in parts of eastern and southern Africa.

Resettlement in some other country is an option for only a few. No country is legally obliged to accept refugees for resettlement and only about twenty countries do so on a regular basis. Table 6.7 details the numbers of UNHCR-assisted cases and their countries of resettlement for 1990 to 1992. Inevitably, there is a large gap between the number of places sought for refugees by UNHCR and the places available. More generally, for the year 1990, the number of persons resettled or granted asylum in Germany (which had about 200,000 refugees at the time) was 11,597; the comparable figure for the United States was 115,155, and 30,560 for Canada.

Attacking the root causes of refugee problems is a mammoth and unenviable challenge. The root causes are often a complex mixture of political, economic, and environmental issues for which no simple solutions are available. Further, as the data presented in this section indicate, the biggest refugee problems are in parts of the less developed world that already face huge challenges.

THE LESS DEVELOPED WORLD

WHAT IS THE LESS DEVELOPED WORLD?

Reference to a Third World was first made in the early 1950s to suggest that former colonial territories could follow a third and different route to both the capitalist (First World) and socialist (Second World) economies. By 1960, the term was being used to refer to a group of African, Asian, and Latin American countries that were all meeting the sad standard set in 1969 by Prime Minister Lee Kuan Yew of Singapore: 'poor, strife-ridden, chaotic'. As is the case with any such general grouping, the Third World was characterized by tremendous variations rather than similarities. Despite

this major limitation of any such general grouping of countries, there is some value in the classification and other terms have been used in much the same way: some prefer to distinguish between rich and poor, between north and south, or between developed and developing (one of the problems with the term 'developing' is that it is implicitly developmentalist). This text uses the terms 'more developed' and 'less developed' as these are employed in a formal classification by the United Nations and hence are the basis for much of the data provided by various international organizations (both terms were first introduced in Chapter 4 and are included in the glossary).

The less developed world contains a great variety of places and peoples, but in the worst case scenario may be characterized as experiencing relatively high levels of mortality and fertility, relatively low levels of literacy and industrialization, and is often beset by political problems stemming from ethnic conflict or other rivalries. The basic demographic data are often uncertain. Censuses, which are the principal data source, may be incomplete and incorrect (refer again to Box 6.1) as the poorest countries have limited capital to conduct censuses and low literacy levels also affect the quality of the data collected. Although this and the two previous chapters have frequently found it useful to distinguish between the more and less developed worlds, further insights can be provided by examples; boxes 6.5, 6.6, and 6.7 provide capsule commentaries on three case-studies, one each from Africa, Asia, and Latin America.

It is all too clear that any discussion of the less developed world quickly moves into areas other than population. Although the less developed world is often identified by population characteristics, the problems go far beyond these. The most compelling and immediate of these issues is subsistence.

POPULATION AND FOOD

Undernutrition and malnutrition

There is a world food problem despite the fact that if the world's food production was equally divided among the world population, nobody would lack food. Today, the world food problem is largely associated with the less developed world, although hunger and malnutrition were evident in Europe until the nineteenth century.

A diet may be deficient in quantity and/or quality. Insufficient quantity results in **undernutrition**. A sufficient quantity varies according to age, sex, weight, average daily activity, and climate. It would be misleading to suggest a particular quantity (in calories, for example) needed to avoid undernutrition.

A diet that is deficient in quality results in **malnutrition**. Food is needed not only for energy but also for health. An appropriate diet includes protein to facilitate growth and replace body tissue, and includes various vitamins to combat disease.

TABLE 6.7: *UNHCR-assisted cases by country of resettlement, 1990–2*

Country	1990	1991	1992
Australia	7,275	3,303	3,481
Canada	10,340	6,294	4,486
Denmark	525	394	415
Finland	634	390	542
Netherlands	546	612	529
Norway	800	1,038	1,082
New Zealand	837	587	338
Sweden	1,082	1,117	2,459
Switzerland	254	269	353
United States	24,462	15,720	19,463
Others	5,274	2,537	3,500

Source: R. Colville, 'Resettlement: Still Vital After All These Years', *Refugees* 94 (1993):7.

Examples of ill health that can be caused by undernutrition or malnutrition include kwashiorkor (too few calories), poor sight (inadequate vitamin A), poor bone formation (inadequate vitamin D), and beriberi (inadequate vitamin B_1). Most such diseases are on the decline as their specific causes have been identified. Unfortunately, the most extreme consequence of undernutrition and malnutrition, death, has not been eliminated.

The extent of the problem

How much food is lacking? How many people suffer from undernutrition or malnutrition? These are not easy questions to answer. We all know that various parts of the world, especially in Africa, seem to experience a series of food-related tragedies, but it is not so well known that world food supplies continue to increase. Hunger and famine continue in a world where increasing food supplies exceed increasing populations. In the period 1961 to 1980, food production in the less developed world increased 3.1 per cent per year, while population increased 2.4 per cent. Clearly, world improvements do not translate into improvements at the regional level. Further, many of the less developed countries have sufficient food for their populations, but do not succeed in allocating food according to individual need.

The World Bank estimates that over 630 million poor people live on the very edge of existence, and that over 1 billion people receive only 90 per cent of their daily nutritional requirements, which is insufficient to support normal activity and work (Kates and Chen 1993:6). In 1993, the highest levels of undernutrition were in Somalia, Mozambique, Sierra Leone, Bangladesh, and Bolivia.

Food aid

Food aid has not proven to be a solution to famine. First, it tends to be directed to urban areas, although the greatest problems are usually in the rural areas; indeed, much of the food aid goes to governments to sell. Second, food aid tends to lower food prices, thus reducing the incentive to grow crops and increasing dependence. Third, in practical terms, food aid often fails to be effectively distributed because of inadequate transport. Finally, the typically undemocratic governments in receiving countries control food supplies and may feed armies before others.

Population and food: The future

In Chapter 4, during a discussion of the likely future state of the environment, and in Chapter 5, during a discussion of future human population growth, we recognized that there are widely divergent opinions. Much the same is true in a discussion of the future of the world food problem. There are those who see Malthusian overtones and anticipate catastrophe, and those who argue that increasing food supplies sufficient to feed an increasing population is not a problem.

The problem is a complex one. It is true that the number of undernourished people in the world is increasing. It is also true that there is enough food produced today to feed all 5.6 billion people in the world. There seem to be good reasons to believe that the world food problem is not caused

---

Box 6.5: The less developed world: Ethiopia

As was discussed in Box 6.4, Ethiopia is one of the countries in the Horn of Africa that is currently suffering through a tragic combination of human and environmental problems. In this box we describe population circumstances in Ethiopia on the basis of the 1984 census — its first national census (see Kloos and Adugna 1989).

Ethiopia was first settled by Hamitic peoples of north African origin, but an in-movement of Semitic peoples from southern Arabia occurred in the first millennium BC, and a Semitic empire was founded at Axsum. This empire became Christian in the fourth century AD. The rise of Islam displaced the empire southwards and established Islam as the dominant religion of the larger area; the empire was overthrown in the twelfth century. Ethiopia escaped European colonial rule, but its borders were determined by Europeans occupying the surrounding areas. The Eritrea region was colonized by Italy from the late nineteenth century until 1945. Attempts by the central government in Addis Ababa to gain control over the Eritreans and over the Somali group in the southeast generated much conflict. In 1974–5, a revolution displaced the long-serving ruler, Haile Selassie, and established a socialist state.

Ethiopia is slightly larger than Ontario — 1.2 million km² (463,356 square miles). Much of the country is tropical highlands, typically densely populated because of good soils and an absence of many diseases. The 1993 population is estimated at 56.7 million; the CBR is 47, the CDR is 20, and the RNI is 2.8. There is no clear national population policy, nor family-planning program. Population growth is highly uneven with the central, western, and southern areas growing at the expense of the northern and eastern areas. Low urbanization rates are characteristic; urban growth has actually declined since 1975 because of socialist land reform policies and the low quality of life in urban areas.

The majority of Ethiopians have little or no formal education, especially the predominantly rural population. The key social and economic unit is the family, with women as subordinates. Women are subservient first to their fathers, then to their husbands, then, if widowed, to adult sons. Health care varies substantially between urban and rural areas.

6.4 Food aid being provided to Vietnamese refugees in a refugee camp in Thailand, 1988. (UNHCR)

by inadequate supplies today and that supplies will be adequate in the future (Bongaarts 1994; Smil 1987). Rather, the problem, in common with refugee and other similar problems, is associated with specific areas and a multitude of complex interrelated causes, physical, political, economic, and cultural.

Explaining the world food problem

Three sets of ideas are now outlined, each of which contributes to an explanation of this most fundamental of problems facing the world today.

Overpopulation is often suggested as the cause of the food problem, but there is really no evidence to suggest that there

---

Box 6.6: The less developed world: Sri Lanka

Local considerations are important components of any situation evident in the various countries of the less developed world. Sri Lanka (Ceylon until 1972), for instance, is located in the Indian Ocean south of India. It is mostly low-lying, has a tropical climate and a limited resource base. Minerals are limited, as are sources of power. The dominant economic activity is agriculture with about 36 per cent of the country under cultivation.

Sri Lanka has a 1993 population of 17.8 million with a CBR of 20, a CDR of 6, and an RNI of 1.4. These figures are good by less developed world standards, but the problems of Sri Lanka are greatly exacerbated by ethnic conflict. The Sinhalese (74 per cent of the population) moved from north India and conquered the island in the sixth century BC; Tamils (18 per cent of the population) arrived in the eleventh century and Arabs (7 per cent of the population) in the twelfth and thirteenth centuries. Portugal ruled the island from 1505 to 1655, followed by the Dutch until 1796, and the British until independence was achieved in 1948.

Sri Lanka has experienced a rapidly increasing population, is experiencing regular ethnic conflict, and continues to experience problems of colonial dependency. Combined, these three factors contribute to the less developed status. Thus, although the country has effectively passed through the demographic transition (along with only a few other low-income countries such as China), it has a per capita gross national product of US $420 (1988). This figure of US $420 compares unfavourably to the less developed world average of US $710.

Ethnic conflict centres around two problems. First, the Sri Lanka Tamils seek recognition as an indigenous people to protect themselves from persecution by the Sinhalese majority. Tamils do not have citizenship status or voting rights. Today guerrilla activity is commonplace.

Second, problems of colonial dependency developed especially under British rule in the nineteenth century with the development of plantation agriculture benefiting the British rather than the local population. By 1945, tea plantations covered about 17 per cent of the cultivated area.

Despite passing through the demographic transition, Sri Lanka is a very poor country that still suffers from the colonial era and severe ethnic conflict. According to one recent commentator, 'There appears to be no solution to the problem in sight' (Simpson 1989:4). Contemporary Sri Lanka is a far cry from the tropical Indian Ocean island called 'Serendip' by the first Arab visitors and 'Paradise' by many later Europeans.

are too many people in the world. Overpopulation is a relative term, and areas that are densely populated are not necessarily overpopulated. Table 6.8 confirms that dense populations (in Hong Kong, the Netherlands, and South Korea) are not necessarily cases of overpopulation. Less densely settled countries such as Ethiopia and Mexico may very well be relatively overpopulated. A compelling example of this general issue is China; a country that experienced regular famines when it had a population of 0.5 billion is now essentially free of famine with a population of over 1 billion.

Another suggested cause of the world food problem concerns the inadequate distribution of available supplies. Most countries have the transportation infrastructure to guarantee the intranational movement of food, but other factors prevent satisfactory distribution.

A third suggested cause of food problems is some specific physical or human circumstance. A 1984 drought in central and eastern Kenya is estimated to have caused food shortages for 80 per cent of the population. Other causes in specific cases include flooding (Bangladesh) and wars (Ethiopia). Again, however, these are not the root causes. One school of thought in geography sees the root cause as the larger world political system.

This provocative assertion suggests a fourth possible explanation of the world food problem that is more general in character and that is more correctly viewed as an explanation of the presence of global inequalities in general. For this reason, this fourth explanation, known as world systems theory, is discussed later in this chapter.

THE WORLD DEBT PROBLEM

It is estimated that in 1993, the less developed world owed a total of US $1.5 trillion, an almost unimaginable figure (World Bank 1993). This money is owed to international lending agencies and commercial banks in the more developed world. The 1970s was a decade of high lending to less developed countries by the commercial banking sector, development agencies, and governments in the more developed world. These loans had long payback terms and were meant to set up and support economic and social programs. There was a general expectation that less developed economies would boom and there would be good returns on these investments.

Unfortunately for all concerned, the situation soured in the 1980s as debt steadily increased during a recession while the more developed world adopted monetarist economic policies when world trade declined and interest rates increased. Many countries are now so poor and owe so much that they need to borrow more money just to maintain existing interest payments, a situation that first arose in 1982 in Mexico (Sowden 1993).

The significance of debt to a country depends on the country's economy. The United States is the biggest net debtor in the world, but the effect on the economy is lessened because it is a high-income country with a large volume of exports. Other countries, such as South Korea, borrowed heavily to finance industrialization and have been able to repay because of their industrial and export success. In many less developed countries, however, the cost of servicing the

TABLE 6.8: *Population densities, selected countries, 1993*

Country	Population density per square kilometre
Hong Kong	5,893.1
South Korea	451.7
Netherlands	448.0
Ethiopia	51.5
Mexico	47.2

Source: Calculated from data in the *1993 World Population Data Sheet* (Washington, DC: Population Reference Bureau, 1993).

Box 6.7: The less developed world: Haiti

The 'nightmare republic' of author Graham Greene, Haiti is a third example of a politically troubled country in the less developed world (Barberis 1994). The basic demographic data for 1993 show the highest CBR (43), the highest CDR (15), and the highest RNI (2.8) in the Western hemisphere. The use of modern contraceptive techniques is low, although there are good reasons to believe that the demand is far from being met. There is a total population of 6.5 million with a high population density of 236 per square kilometre.

Haiti has a long history of political turmoil since independence from France in 1804 after a twelve-year rebellion. After being ruled for thirty years by the Duvalier family, the first free election was held in 1990, but the victor was quickly overthrown (in 1992) and the country was once again ruled by a military despot until democracy was re-established in 1994.

There is a long history of conflict between the poor (95 per cent of the population) and the rich (5 per cent) and also extreme economic disparity between the capital city of Port-au-Prince and the rural areas. A 1990 estimate suggested that the wealthiest 1 per cent of the population hold 44 per cent of the wealth. The rural population practise subsistence agriculture on often poor soils, leading to erosion and declining soil fertility. Access to clean drinking water is a problem and hunger is widespread; there are high rates of infant mortality, tuberculosis, and HIV.

foreign debt accounts for up to 30 per cent of all income from exports. Debt is therefore a crushing burden on many impoverished countries and some, such as Colombia, have actually exported food to help repay debt, although their own populations are malnourished. Not surprisingly, there is a concentration of severely indebted low-income countries in Africa. Box 6.8 offers an example of a country where an innovative program to help poor individuals, rather than the country itself, is being implemented.

THE SELECTIVITY OF DISASTERS

Human geographers have long used the term 'natural disasters' to refer to such physical phenomena as floods, earthquakes, and volcanic eruptions and their human consequences. In recent years, however, it has been recognized that this is an inappropriate label. Certainly, the event itself is natural in the sense that it is a part of the larger physical environment, but not all such natural events become disasters. To understand the extent to which a natural event becomes a human disaster, we need to understand the larger cultural, political, and economic framework. In short, some parts of the world and some people are more vulnerable to natural events than are other places and other people. This is simply analogous to acknowledging that, on average, the rich live longer than the poor. Just as rich people are able to afford a good diet and medical care, so some parts of the world are able to protect themselves against natural events.

The United Nations has designated the 1990s as a decade to focus on reducing the human disasters that so often accompany natural events, especially in the less developed world. This is in response to the facts that between the 1960s and the 1980s, the number of major disasters has increased fivefold, there has been a considerable increase in loss of life, and there are great regional disparities between the more and less developed worlds. Table 6.9 provides data for three regions on the average loss of life associated with disasters; there is clear evidence the less developed world is especially vulnerable and that this vulnerability is increasing through time.

As with food supplies and national debts, again the less developed world suffers the most. Adverse cultural, political, and economic conditions combine to place increasing numbers of people at great risk in the event of any environmental extreme. 'The problems of staggering population/urban growth and crippling overseas debt that face many Third World countries have often become manifested in poor construction standards, poor planning and infrastructure, inadequate medical facilities and poor education — all of which exert a direct influence on vulnerability to natural hazards' (Deg 1992:201). Overall, there is a close relationship between vulnerability and poverty. Box 6.9 provides details on a poor and densely populated area of the world that is especially susceptible to devastating floods on a regular basis.

TABLE 6.9: *Natural events and human disasters, regional data, 1947–67 and 1969–89*

Region	Average loss of life per disaster
1947–67	
North America	38
Western Europe	230
Less developed world	984
1969–89	
North America	19
Western Europe	99
Less developed world	2,066

Source: M. Deg, 'Natural Disasters: Recent Trends and Future Prospects', *Geography* 77 (1992):201.

Box 6.8: Grameen Bank, Bangladesh

A remarkable transformation is occurring in parts of Bangladesh, a transformation brought about by a program begun by Grameen Bank. It is a program that focuses exclusively on improving the status of landless and destitute people — especially women.

Bangladesh is one of the poorest countries in the world. It has a large population of 113.9 million, an exceptional population density of 875 per square kilometre, a CBR of 37, a CDR of 13, and an RNI of 2.4. The per capita gross national product is one of the lowest in the world at US $170 (1988). The Grameen Bank program is operating within this context, and is doing so with the most disadvantaged of the Bangladeshi population in mind.

The Grameen Bank was established in 1976 and by 1988, it had 571 branches covering 17 per cent of the villages in Bangladesh; the aim is to serve the entire country by 1995. Its program is based on the assumption that there is a crucial need to enhance the social and economic status of women. Loans are issued to those most in need and the bank has a recovery rate of 98 per cent. The bank requires that loan applicants first form groups of five prospective borrowers and then meet regularly with bank officials. Two of the five then receive loans and the others become eligible once the first loans are repaid. The focus is on the collective responsibility of the group. As the 98 per cent recovery rate shows, this innovative approach works.

This successful technique is being replicated elsewhere in the world, hopefully with as much success as is currently being experienced in Bangladesh (see Mahmud 1989).

MEASURING AND EXPLAINING GLOBAL INEQUALITIES

PROBLEMS OF DEFINING AND MEASURING

There is often debate regarding the appropriateness, for purposes of measuring economic and social development, of such macroeconomic indicators as **gross national product** (GNP) per capita. Some feel that GNP expressed in monetary terms provides reliable comparative data for the relative economic performance of countries while also serving as a reliable surrogate measure for social development in the areas of health, education, and overall quality of life. Others feel that using GNP as a measure of development is unsound because it does not consider the spatial distribution of economic benefits and fails to acknowledge the real measures of development relating to issues such as those discussed in the previous two sections, for example, population displacement, inadequate food supplies, the impacts of debt, and vulnerability to environmental extremes. It can be argued that for the less developed countries, GNP is a measure of how the minority wealthy population are progressing, completely failing to reflect the state of the poor majority. Why such divergent opinions? Simply because there is little agreement as to what is meant by development.

A principal problem is that definitions of development often reflect ethnocentrism; typically, an unconscious acceptance of some particular point of view is then generalized as universal. The traditional and still most popular point of view equates development with economic growth and modernization and thus regards GNP as a valuable means of measuring development.

Measuring development using GNP

The *World Development Report* (World Bank 1993), an annual publication that first appeared in 1978, measures development based on economic criteria. Countries are divided into three categories according to GNP per capita — low income, middle income, and high income. As noted, such measures are valuable, but certainly not fully adequate. Indeed, 'GNP per capita does not, by itself, constitute or measure welfare or success in development. It does not distinguish between the aims and ultimate uses of a given product, nor does it

TABLE 6.10: *Extremes of human development, 1992*

Top ten			Bottom ten		
Country	Rank	HDI value	Country	Rank	HDI value
Canada	1	0.982	Somalia	151	0.088
Japan	2	0.981	Guinea-Bissau	152	0.088
Norway	3	0.978	Djibouti	153	0.084
Switzerland	4	0.977	Gambia	154	0.083
Sweden	5	0.976	Mali	155	0.081
United States	6	0.976	Niger	156	0.078
Australia	7	0.971	Burkina Faso	157	0.074
France	8	0.969	Afghanistan	158	0.065
Netherlands	9	0.968	Sierra Leone	159	0.062
United Kingdom	10	0.962	Guinea	160	0.052

Source: United Nations Development Programme, *Human Development Report, 1992* (Toronto: Oxford University Press, 1992):19–20.

Box 6.9: Flooding in Bangladesh

Bangladesh is a low-lying country (mostly only 5–6 m/16–20 ft above sea level) that lies at the confluence of three large rivers, the Ganges, the Brahmaputra, and the Meghna. Floods are normal in this region; indeed, they are an essential part of everyday economic life as they spread fertile soils over large areas. Tragically, however, floods during the monsoon season often have catastrophic consequences, especially if they coincide with tidal waves caused by cyclones in the Bay of Bengal (see Figure 4.6). Floods are related to three causes:

1. Deforestation in the inner catchment areas that results in more run-off.
2. Dike and dam construction in the upstream areas that reduce the storage capacity of the basin.
3. Coincidental high rainfall in the catchment areas of all three rivers.

Regardless of specific cause, floods are difficult to control because of their magnitude and because the rivers often change channels.

There are various disastrous consequences of flooding. The most extreme is death; in a 1988 flood, over 2,000 died. Impelled migration is another consequence; again in 1988, over 45 million people were uprooted and forced to flee the flood waters. But, as already noted, human disasters occur on a regular basis. In a normal year, over 18 per cent of Bangladesh is flooded and even these normal floods result in shifting of river courses and erosion of banks that, in turn, result in population displacement. In an already poor country, such displacement further aggravates problems such as landlessness and food availability. Most displaced persons move as short a distance as possible for cultural and family reasons. During these impelled migrations, the poorest suffer the most, women-headed households are especially vulnerable. Some migrants opt to move to towns, hoping to become more economically prosperous. Typically, however, such rural-to-urban migrants become disadvantaged inhabitants of squatter settlements, clustering together and are generally unwelcome in the established urban setting.

say whether it merely offsets some natural or other obstacle, or harms or contributes to welfare' (World Bank 1993: 306–7). It is also possible to see the error of developmentalism in the use of such data with the implicit suggestion that countries can increase their GNP level.

Use of GNP per capita data clearly identifies Africa as the least developed region within the less developed world. Using 1990 data, there are twelve African countries with values less than US $240 (Mozambique is the lowest with US $80), while the average figure for sub-Saharan Africa is only US $330. Several Asian countries (notably Bangladesh, Bhutan, and Nepal) and Latin American countries (notably Guyana and Haiti) have a GNP per capita under US $350. By comparison, the figure for Japan is US $25,890 and for the United States is US $21,790 (World Bank 1992:2–5).

Measuring human development

An annual *Human Development Report* that first appeared in 1990 aims to complement GNP measures of development. There are three distinctive characteristics to this newer publication (United Nations Development Programme 1992). First, it uses a wide variety of data to construct a human development index (HDI) based on three goals of development: LE, education, and income. Second, there is an explicit concern with how development affects the majority poor populations of the less developed world; it is recognized that there is a need to enlarge the range of individual choices. Third, the concept of development employed focuses on the satisfaction of basic needs, gender inequality, and environmental issues. The HDI does not measure absolute levels of human development but ranks countries in relation to one another.

Table 6.10 provides data on those countries with the ten highest HDI values and on those with the ten lowest HDI values in 1992 (out of the 160 countries for which data are available). The HDI has a maximum value of 1.000 and a minimum value of 0.000. Again, as with GNP per capita data, African countries are the least developed.

RELATIONS WITH THE MORE DEVELOPED WORLD: WORLD SYSTEMS THEORY

Perhaps the single most important factor that explains the plight of the countries in the less developed world is their relationship with more developed countries. Most of the less developed countries have a colonial history; even those countries such as China, Thailand, Liberia, Saudi Arabia, Iran, and Afghanistan that have not been colonies of European countries have been affected by Europe's world dominance between about 1400 and 1945. Why is this consideration significant?

First, on the world scale, **colonialism** has resulted in **dependence**: many former colonies became dependent on the more developed countries. Aid that is meant to promote development can be seen as actively encouraging increased dependence. Second, the internal distinctiveness of former colonies has largely been relegated to secondary status,

dominated by European structures. In the broadest sense, the less developed countries lack power, including the power to control and direct their own affairs.

The notion of world systems, an exciting contribution to current human geography, centres on a description of the dynamic capitalist world economy from 1500 onwards. It examines the roles that individuals can play in a state and the roles that specific states can play in the larger set of state interrelationships. The world systems framework is the work of Wallerstein (1979, for example) and can be briefly summarized as follows. The capitalist world economy emerged gradually from feudalism in the sixteenth century, consolidated up to 1750, expanded to cover the world in the form of industrial capitalism by 1900, and is now, since 1917, experiencing a crisis and may change to something closer to socialism. Interestingly, events in Europe from 1989 onwards appear to make the prospect of an enlarged socialism less likely, but this in no way detracts from the general Wallerstein argument.

The contemporary result of this historical process is a division of the world into three principal zones: core, semi-periphery, and periphery. The *core* states benefit from the current situation as they receive the surplus produced elsewhere. Principal core states are Britain, France, the Netherlands, the United States, Germany, and Japan. World business and financial matters are centred in these countries. The *semi-periphery* comprises states that are at least partially dependent on the core: Argentina, Australia, Brazil, and South Africa are examples. The *periphery* includes those states that are dependent on the core and are effectively colonies. All those countries that we regard as less developed belong in this group.

Although this world system is dynamic, it is extremely difficult for a state to move out of peripheral status, for the other states have vested interests in maintaining its dependency. In principle, it is possible to devise a threefold structure for non-capitalist countries, although the mechanism of dependency in this case is not economic but political status.

A world systems explanation of the world food problem

People can either grow or purchase food. According to a world systems-inspired perspective, it is becoming increasingly difficult for many in the less developed world to pursue either of these apparent options. In the less developed world, the percentage of the population involved in agriculture is declining (from 81 per cent in 1950 to 63 per cent in 1985), but at the same time, those remaining in agriculture are not benefiting from technological advances.

The typical scenario in the less developed world is one in which a very few commercial agriculturalists are technologically advanced while the vast majority are incapable of competing. Indeed, this vast majority have lost control over their own production because of larger global causes. As an example, in Kenya, farmers are actively encouraged to grow export crops such as tea and coffee at the expense of basic food production. The result is that a staple crop, such as maize, is not being produced in a sufficient amount. Other examples of how peasant farmers are losing their freedom involve the increasing control of credit by large corporations and the policy of cheap food for urban populations at the expense of the peasant farmer. The essential argument here — and it is an argument, not a fact — is that the capitalist mode of production is affecting peasant production in the less developed world in such a way as to limit the production of staple foods, thus causing a food problem.

A second and related argument involves the increasing inability of people in the less developed world to purchase food. Food is a commodity throughout the world. Hence, the factors that determine production — everywhere — are those related to the making of a profit. As already noted, the world has the ability to feed all of us, but too many of us are too poor to purchase food. Food is produced only for those who can afford to buy. In 1972, a year when famine was widespread in the Sahel region of Africa, farmers in the United States were actually paid to take land out of production in order to increase world grain prices.

If we combine these arguments, we can suggest that the cause of the world food problem is the peripheral areas' dependency on the core area. Any attempts other than clearly humanitarian actions to correct the problem are doomed to failure as the more developed world is not likely to initiate changes that might create problems for itself. If this argument is followed to the logical conclusion, the world food crisis can only get worse regardless of technological change because the cause of the problem lies in global political and economic patterns.

The questions remain. Feast and famine live side by side in our global village. There are rich and poor countries and in the poor countries especially, there are rich and poor people. Increasingly, agriculture is a business and as such, not all of those involved compete effectively. Food problems are not typically caused by overpopulation, and population densities are a poor indicator of pressure on resources. We will not solve food problems by merely decreasing human fertility.

What is needed is a concerted and cooperative international effort to improve the quality of peasant farming and to reorient it to the production of staple foods. There is little indication that this will be achieved in the foreseeable future. Undernutrition and malnutrition are not disappearing. Famines, typically prompted by events such as droughts and wars, will continue to occur. The less developed world is not about to develop whatever criterion of development is employed.

SUMMARY

Distribution and density

Distribution refers to the spatial arrangement of a phenomenon, while density refers to the frequency of occurrence of a phenomenon within a specified area. Population maps frequently combine both characteristics. Asia has 58 per cent of the world population; China is the most populous country, followed by India. A third area of population concentration is Europe. The Asiatic regions include very high rural population densities. Distribution and density of population are related to land productivity and cultural organization.

Causes of human mobility

Human populations have always been mobile. Migrations from one location to another are often explained in terms of the relative attractiveness of areas. This idea can be expressed in simple push-pull terms, by reference to laws, or by using the concept of place utility. Migration can also be related to the demographic transition such that a mobility transition is identified.

Types of migration

Primitive migration refers to the gradual movement of humans over the surface of the earth; it is normal behaviour involving ongoing adaptation to new environments. Forced migration occurs when people have little or no choice but to move; slaves and refugees are examples of forced migrants. Refugees are prompted to move for human and environmental reasons. Major problem areas today are Ethiopia and neighbouring countries, Israel and neighbouring countries, Iraq, and various Asiatic countries. Free migration is the result of decision making, of an evaluation of alternative available locations. Illegal migration is increasingly common in a world where countries have immigration quotas and are selective as to who is admitted.

The less developed world

These countries can be identified using solely economic criteria such as GNP or by a number of human development criteria; all have similar relationships with the more developed world. Many have been colonies and most are in a dependent situation today; massive debt loads are normal and they are especially vulnerable to environmental extremes such as droughts, floods, and earthquakes.

The world food problem

There is a world food problem. It is not new, but although evidence is contradictory, solutions are not being achieved. Undernutrition is caused by insufficient food; malnutrition is caused by low-quality food lacking the necessary protein and vitamins. One possible cause of the world food problem is overpopulation where the carrying capacity of the land has been exceeded; another is the inadequate distribution of food; another is a natural or human disaster such as drought or war. One argument is that the world food problem is a consequence of the increasing inability of many to either grow or purchase food as a result of the imposition of a capitalist mode of production on the less developed world.

World systems theory

Wallerstein's world systems approach posits a world divided into three zones: core, semiperiphery, and periphery. The world is organized such that core countries benefit and peripheral countries suffer. The less developed world comprises only peripheral countries. It can be argued that the world food problem can only get worse regardless of technological change because the cause of the problem lies in global political and economic patterns.

WRITINGS TO PERUSE

BLACK, R., and V. ROBINSON, eds. 1993. *Geography and Refugees: Patterns and Processes of Change*. New York: Belhaven Press.
A collection of detailed case-studies including examples from both the developed and less developed worlds.

BRADLEY, P.N. 1986. 'Food Production and Distribution — and Hunger'. In *A World in Crisis: Geographical Perspectives*, edited by R.J. Johnston and P.J. Taylor. New York: Blackwell, 89–106.
A readable and provocative article explaining the world food problem in terms of a world economy focus.

CHAMPION, A., and A. FIELDING, eds. 1992. *Migration Processes and Patterns, Volume 1*. New York: Belhaven.
A series of detailed analyses of contemporary migration trends; strong emphasis on links to cultural and economic change; focus on British examples.

DEMENY, P. 1989. 'World Population Trends'. *Current History* 88:17ff.
A factually based discussion of trends and projections.

GRIGG, D.B. 1977. 'Ravenstein and the "Laws of Migration"'. *Journal of Historical Geography* 3:41–54.
An excellent summary and discussion of the topic, which helps our understanding of the contemporary relevance of both Ravenstein specifically and the value of laws generally.

———. 1985. *The World Food Problem*. Oxford: Blackwell.
A comprehensive survey emphasizing history and regional issues.

HENDRY, P. 1988. 'Food and Population Beyond Five Billion'. *Population Bulletin* 43.
A valuable overview of world population situation with emphasis on the less developed world.

LEWIS, G.J. 1982. *Human Migration*. New York: St Martin's Press.
A useful book covering the many facets of the topic.

LOESCHER, G. 1993. *Beyond Charity: International Cooperation and the Global Refugee Crisis*. Toronto: Oxford University Press.
A powerful book that argues for a reform and strengthening of organizations such as UNHCR and for a concerted strategy by the industrial nations to tackle refugee problems effectively.

NEWLAND, K. 1994. 'Refugees: The Rising Flood'. *World Watch* 7, no. 3:10–20.
One example of a brief, readable account that covers basic data; information on political, environmental, and economic causes of specific problems; discussions of ethnic tensions, human rights violations, preventive measures, and possible responses.

POOLEY, C.G., and I.D. WHYTE. 1991. *Migrants, Emigrants and Immigrants: A Social History of Migration*. New York: Routledge.
Includes a series of case-studies with emphasis on Britain and areas of British settlement overseas.

ROGGE, J.R. 1985. *Too Many, Too Long: Sudan's Twenty-Year Refugee Dilemma*. Totowa, N.J.: Rowman and Allenfeld.
A detailed empirical study of Sudanese refugees.

STILLWELL, J., P. REES, and P. BODEN, eds. 1992. *Migration Process and Patterns, Volume 2: Population Redistribution in the United Kingdom*. New York: Belhaven Press.
A series of empirical analyses of the migration component of population change between and within the various regions of the United Kingdom.

United Nations High Commissioner for Refugees. 1993. *Refugees: Resettlement*. No. 94, December.
One example of the regular magazine publication, *Refugees*. Each issue has a topical focus and includes a series of short, informative articles.

———. 1993. *The State of the World's Refugees: The Challenge of Protection*. London: Penguin.
The first of a new series of reports; provides detailed accounts and statistical data. Argues for the need to prevent the conditions that cause refugee problems.

ZELINSKY, W. 1971. 'The Hypothesis of the Mobility Transition'. *Geographical Review* 61:219–49.
Links mobility to the demographic transition and technological change.

7

Cultures: The evolution and regionalization of landscape

THE CONTENTS of this and the following chapter belong to three related subdisciplines of human geography, namely historical geography, cultural geography, and social geography. These three are not neatly separated; indeed, all three share concerns with the behaviour of humans as individuals and as group members and with the landscapes that humans create.

In this chapter our principal concern is with what we call the geographic expression of culture on landscape. This includes discussion of cultural evolution in general, specific cultural groups, the regions and landscapes associated with specific cultures, language, and religion. Overall, the content of this chapter is strongly empiricist and reflects a longstanding tradition in human geography.

In Chapter 8 our principal concern is with what we call the spatial constitution, the symbolic expression and the social significance of culture. This includes discussion of relevant social theory, the links between landscape and power, popular and folk culture, gender, ethnicity, and class. Overall, the content of Chapter 8 reflects Marxist and humanist emphases and comprises methods and material that are relatively novel in human geography.

A WORLD DIVIDED BY CULTURE?

An introductory geography textbook by the eminent American geographer, Preston James, was first published in 1964 under the inspired title, *One World Divided*. We have already recognized some physical divisions on the basis of such variables as climate, relief, soil, and vegetation and some human divisions based on demographic differences. Our concern now is with human divisions that derive from differences in culture. Human variables in general create barriers far more difficult to cross than any physical barrier. The divided world we noted in our accounts of the human population is but the tip of the iceberg. James (1964:2) wrote that humans have 'brought into being mountains of hate, rivers of inflexible tradition, oceans of ignorance.'

Today, we live in a world of tremendous cultural diversity, but also one of increasing interaction between cultures. In many areas, traditional groups are struggling to maintain established ways of life; in other areas, the often uncontrolled passions of language, religion, and ethnicity emerge as cultural and sometimes political conflicts. Ironically, then, it is our greatest human achievement — our culture — that has been responsible for the erection of these barriers between peoples.

INTRODUCING CULTURE

Humans differ from all other forms of life in that they have developed biologically and culturally. Other forms of life are limited to biological adaptation and so are highly specialized and typically restricted to particular physical environments. Indeed, environmental change has often resulted in species extinction. So far, humans have avoided such a fate primarily because of our culture, our ability to analyse and change the physical environments that we encounter. Unlike other animals, humans can form ideas out of experiences and then act on the basis of these ideas. Humans can not only change physical environments, they can knowingly change them in directions suggested by experience. This is what we mean by the term 'culture'.

Why is it, then, that this remarkable human ability has had the unfortunate tendency to produce the consequences noted by James? Each cultural change, each new idea prompted by experience, involves the addition of new knowledge and new responsibilities that at least initially are placed in a cultural framework that is inappropriate and that cannot easily accommodate them. Additions to our cultural heritage are obliged to fit into existing patterns of attitudes and behaviour. Our contemporary world remains one that values traditions and the past, a situation that is not always conducive to the efficient use of developments in culture. Once in place, differences in attitude and behaviour tend to be self-perpetuating and can only be removed by the creation of new sets of values that are able to reduce cultural variations between groups of humans. Creating new sets of values is far from an easy task as it involves human engineering, literally changing ourselves in accord with these new values.

We can summarize this important introductory argument as follows:

1. Our world is divided, most importantly as a consequence of spatial variations in culture.
2. By culture, we mean the human ability to develop ideas from experiences and subsequently act on the basis of those ideas.
3. Unfortunately, once cultural attitudes and behaviours are in place, there is an inevitable tendency for all new ideas (developments in culture) to be evaluated with reference to a possibly inappropriate set of existing ideas (culture).
4. One task humans may choose to tackle today is creating new sets of values, engineering ourselves.

HUMANS IN GROUPS

The terms 'culture' and 'society' are two of the most awkward employed by human geographers — indeed, two of the most awkward in social science. Both terms are primarily identified with disciplines other than human geography and are subject to a variety of interpretations. Culture is particularly associated with anthropology and society with sociology. In the most general sense, both terms refer to a similar human scale of analysis. Thus, psychologists study individual human behaviour while anthropologists and sociologists study group human behaviour. The distinction between the two group-oriented disciplines is not an easy one to articulate. Anthropology and sociology evolved as separate disciplines in the nineteenth century with the former centring on non-western and rural issues and the latter centring on western and urban issues. As recently as 1958, eminent representatives of the two disciplines found it necessary to clarify what they meant by the terms 'culture' and 'society' (Box 7.1).

HUMAN GEOGRAPHY AND CULTURE

As early as 1889, Vidal argued for geography as the analysis of human and land relations with emphasis on way of life, or *genre de vie* (see Box 1.7). This possibilism was explicitly opposed to environmental determinism and similarly opposed to any version of human determinism. Both determinisms were unacceptable to Vidal who saw physical and human landscapes as one, a tradition pioneered by earlier geographers, especially Humboldt. A principal outcome of human and land relations is distinctive areas of human occupance, regions or *pays*. In addition to being uncomfortable with environmental determinism, Vidal actively engaged in a lively debate with Emile Durkheim, the father of modern sociology, concerning the relevance of social determinism. Durkheim favoured the society concept as a causal factor, while Vidal saw it as a relevant concept in the study of human and land relationships, but not as a cause.

A view comparable to that of Vidal did not develop in North America until the 1920s when Sauer put forward a series of ideas collectively known as the landscape school

---

Box 7.1: The concepts of culture and society

The emerging disciplines of anthropology and sociology focused on the concepts of culture and society respectively, but did not define or use these concepts in any significantly different way. The classic early definition of culture originated by Tylor — 'that complex whole which includes knowledge, belief, art, morals, law, customs and any other capabilities and habits acquired by man as a member of society' (see Friedl and Pfeiffer 1977:288) — and developed by Boas was little different from the idea of society presented by the great early sociologists, Comte, Spencer, Durkheim, and Weber. Thus, the core concepts of late nineteenth- and early twentieth-century anthropology and sociology were not clearly distinguished; rather, the distinction between the disciplines was operational — anthropologists applied the concepts largely to non-literate groups, and sociologists applied the concepts to literate ones.

In 1958, a distinguished representative of each of the two disciplines collaborated to clarify the terms, proposing that culture be used in a narrower sense than previously. Thus culture is the 'transmitted and created content and patterns of values, ideas and other symbolic-meaningful systems as factors in the shaping of human behavior and the artifacts produced through behavior.' Society, or social system, on the other hand is the 'specifically relational system of interaction among individuals and collectivities' (Kroeber and Parsons 1958:583).

Although the distinction proposed by Kroeber and Parsons has been highly influential, members today of both disciplines make various use of both terms. Human geographers have traditionally had their closest links with anthropology, and hence with the culture concept, but are today turning to sociology and the society concept for new inspirations.

(see Box 1.9). These ideas were derivative of possibilism and centred on the creation of cultural landscapes from earlier physical landscapes. Sauer particularly emphasized that the object of geographic study was the landscape, not culture. There have been suggestions that Sauer was influenced by a school of thought in American anthropology espoused by Alfred Kroeber and known as the **superorganic** (Box 7.2). The evidence on this count is contradictory. There is little doubt, however, regarding the impact of the landscape school on North American human geography — it has survived largely intact from the 1920s to the present.

The geographic use of the term **culture** is now becoming apparent. We have already introduced culture as referring to our unique human ability to knowingly change physical environments in directions suggested by experience thus resulting in the creation of distinct ways of life. But what does culture comprise? Culture can be usefully divided into two categories, non-material and material, that involve three components (Huxley 1966; Zelinsky 1973:72–4). First, *non-material* culture comprises both key attitudinal elements or values, such as language and religion, known as **mentifacts,** and also the norms involved in group formation, such as rules about family structure, known as **sociofacts**. *Material* culture comprises all the elements related to people's livelihood, known as **artefacts**, and thus includes the human landscape.

Traditionally, then, following Vidal and Sauer (but especially the latter), human geographers have seen the landscapes created by humans through their cultures as a prime object of study. In order to study human landscapes, both mentifacts and sociofacts facilitate our understanding of landscape. Anthropologists study human cultures, whereas human geographers have traditionally paid attention to the landscape expressions of these cultures.

HUMAN GEOGRAPHY AND SOCIETY

Simply put, society is to sociology what culture is to anthropology — the central concept of the discipline. As with the term 'culture', there are several varied definitions depending on the theoretical perspective employed. The traditional view of **society** is as a cluster of institutionalized ways of doing things. Thus, while culture refers to the way of life of the members of a society, society refers to the system of interrelationships that connect individuals together as members of a culture (Giddens 1991:35). Thus, society refers to attitudes and behaviours that occur and recur. All too clearly, there is overlap with culture here; society includes non-material culture — languages, religions, and social structures are institutionalized ways of doing things.

Let us turn to the concept of society as used in human geography. North American geographers who adhere to Sauer's landscape school have employed the term 'culture' rather than 'society' in their analyses of the human landscape; British and European geographers have done the reverse, employing the term 'society' rather than 'culture' in their analyses of landscapes. Notwithstanding these different terminological preferences, we as human geographers need to learn that there is no clear and absolute distinction

---

Box 7.2: The superorganic concept of culture

Sauer regarded culture as 'the impress of the works of man upon the area' (Sauer 1925:30). His highly influential view was based on earlier French and German statements, but it also incorporated a rather different dimension. Unlike Vidal, who rejected Durkheim's overtures concerning the need to be socially deterministic, Sauer appears to have accepted the popular North American view of culture as cause — cultural determinism.

In North America a leading anthropologist, Alfred Kroeber, argued for culture as superorganic, meaning that it is at a higher level than the individual and that it constrains individual behaviour. Kroeber taught anthropology at Berkeley. Sauer, who taught geography there, sent most of his students to study with Kroeber. The suggestion that Sauer accepted Kroeber's view of culture is further evidenced by his various references to culture as the 'factor', the 'agent', and the 'shaping force' (Sauer 1925). But the evidence is far from unequivocal on this point for Sauer does not downplay the role of physical geography, and his own prolific writings are most assuredly not examples of the application of the superorganic concept.

The question of Kroeber's influence does not end there. There are many indications that Sauer's students often subscribed to this view and they were a highly influential group in mid-twentieth-century human geography. It is therefore reasonable to suggest that the human geographic focus on landscape, especially the visible landscape, is in direct response to an acceptance of culture as a causal variable.

The issues raised in this box are not straightforward. For some scholars, Kroeber is not a cultural determinist at all, while others see him as the archetypal cultural determinist; for some Sauer followed Kroeber, while for others Kroeber was but a minor player in Sauer's ideas; for some Sauer's legacy is an inflexible cultural determinism, while for others that legacy is rich and varied. It is useful to acknowledge such disagreements, but hardly useful for us to try to negotiate acceptable resolutions in this introductory context! What we can usefully conclude is that Sauer's landscape school paid much attention to landscape as a geographical expression of the impact of culture, and little attention to culture itself.

7.1 Petroglyphs in Petroglyphs Provincial Park, Ontario, Canada. (Ontario Ministry of Natural Resources)

between these two terms, either in social science generally or in human geography specifically. Not surprisingly, then, current research clearly indicates that human geographers now recognize the need to integrate North American cultural and European social geography. Relatively nominal differences in emphasis cannot disguise a basic unity. Human geographers are able to utilize the two related concepts of culture and society to analyse landscape.

THE IMPORTANCE OF HUMAN SCALE

As human geographers, we are particularly aware of the importance of selecting appropriate spatial scales of analysis when conducting research. Our concern here is with a second choice of scale, human scale.

As students of human populations, we may in principle choose to analyse at any scale from individuals to the world population, through the intermediate scales of nuclear families, extended families, friendship circles, voluntary associations, involuntary associations, institutions, and nations. Selecting the appropriate scale is a function of the particular type of study being conducted. In order to enhance our understanding of our divided world, our typical approach has been to focus on cultures. Thus, we study groups delimited on the basis of common operating rules (or what we might call common mentifacts or common sociofacts). Human geographers characteristically note that an appropriate social scale is one that allows delimitation of a group that is meaningful given the aims of the work. This not very profound conclusion is actually more important than it might appear.

Philosophical emphasis and human scale

The whole question of choice of human scale is a thorny issue in social science. A humanistic focus recognizes the need to study both individuals and groups; in human geography, then, we need to study humans, their actions, and intentions. A Marxist focus typically centres on groups because it is argued that any individual must be defined within an appropriate larger cultural context. Most traditional cultural geography has favoured a group scale because such a scale is in accord with the typical geographic interest in the world or regions of the world. Finally, most contemporary social theory favours the group scale, largely because our individual actions result from ideas and beliefs grounded in groups defined on the basis of interaction and communication.

THE EVOLUTION OF CULTURE

EARLY HUMAN GROUPS

Evidence regarding the earliest cultural attainments of humans and the earliest organization of human groups is limited. Evidence of language evolution dates from about 2.5 million years ago, as does evidence of stone tool making. Fire was possibly first used some 1.5 million years ago, prompting the identification of specific locations as home bases. Language appears to have expanded substantially about 400,000 years ago. Combined, these developments facilitated group formation. Such early human groups probably numbered ten to thirty (occasionally as high as 100) people linked by

7.2 Iroquois longhouse construction during the prehistoric era. (Royal Ontario Museum)

kinship. Groups of this size prevailed until the agricultural revolution.

The factors that prompted language, tool making, and group formation are unclear, but probably include information sharing, mating, child care, basic subsistence requirements, cooperative hunting, and minimizing conflict. Early human groups are typically labelled bands. Bands involved little labour specialization or variation in status. Age and gender were likely the two principal means by which individuals were differentiated. Separate political authorities were not evident; any authority was limited to family heads. Religion, similarly, did not have a separate existence, for any religious beliefs were simply a part of overall activity. Economic activities were limited to a basic search for food, a search that was largely related to the particular physical environment.

These early cultural developments resulted in a way of life that was not difficult and that did not involve long-term food shortages or major disease outbreaks. Nevertheless, it is a way of life that has been abandoned throughout the world with but a few exceptions, such as the Kazaks of central Asia or the Inuit of northern Canada. The cultural development that prompted this wholesale change is that of the agricultural revolution.

THE BEGINNINGS OF CIVILIZATION

Smil (1987:1) wrote:

> while the maintenance of a suitable environment with its myriad of irreplaceable services remains a critical precondition of human existence, much more is needed for the development of civilizations. The essence of these additional requirements is the appropriation of energies, first merely by better management of human labour, later by extensively harnessing various renewable transformations of solar radiation ..., finally by extracting fossil fuels and generating electricity.

The approach employed in Chapter 4, during our discussion of human impacts on environment, recognized the important role played by human use of energy sources. At this time our primary concern is with understanding specific reasons for the beginnings of civilization: where, how, and why did some human groups begin the necessary appropriation of energies?

Prior to the agricultural revolution, variations in culture were intimately related to the physical environment. The limited development of religious, political, and economic systems meant that little variation was evident in these arenas. This situation of limited variation, other than that tied to different physical environments, changed as the agricultural revolution appeared, which gave rise to what we might now term 'civilization'. Whatever its genesis, the agricultural revolution is considered the earliest stage of what we call civilization. By the term **civilization**, we mean a particular type of culture that includes a relatively sophisticated economy, political system, and social structure (Box 7.3). What specific changes caused the beginnings of civilization?

There are numerous potential responses to this question. Racist explanations such as that proposed by Gobineau were common in the nineteenth century, but are now discredited (see Box 3.5). Explanations that invoked an environmentalist determinist argument, such as that by Huntington (1924), are also no longer tenable as they are not supported by facts.

The nineteenth-century anthropologist, Morgan, proposed that all cultures evolve through similar stages; in this schema, culture is seen as evolving through successive levels of material achievement (Table 7.1). This is also an inadequate explanation, in this case because it has nothing to say about what caused changes in material achievement. Three other suggested explanations, all of which presuppose an increasing population, merit more detailed comment.

The hydraulic hypothesis

According to the German historian, Wittfogel (1957), an expanding population would require more agricultural land. In a semiarid area, this expansion could be achieved by irrigation. Any construction of canals and aqueducts could only be achieved if people work cooperatively and if there is some directing authority. According to Childe (1951), irrigation resulted in significant increased productivity. Both of these scholars see irrigation as requiring new cultural organization in the form of central control.

A coercive theory

It is possible that the necessary social changes, particularly central control, preceded agricultural development. Some scholars see such control arising as groups and their territories increased in size. The second suggested explanation proposes that the development of agriculture and civilization may have resulted from increases in population pressure in an area where it was not possible for the population to expand within the area available to them. Rather than intensifying agriculture, as is suggested in the hydraulic

---

Box 7.3: Defining civilization

There is no one clear, correct definition of civilization. The word itself derives from the Latin *civis*, meaning citizen. Related words such as 'political' and 'police' come from the Greek *polis*, meaning 'city'. It is clear that we generally associate civilization with an urban way of life.

Some anthropologists, such as Kroeber, view civilization as more internally complex than other forms of human organization. Others, such as Morgan, see the type rather than the complexity of organization as decisive. A third approach is to define civilization by identifying a set of characteristics. Thus, Childe (1951) identified ten characteristics, subsequently organized as follows:

Primary	Secondary
1. city settlement	6. monumental public works
2. labour specialization	7. long-distance trade
3. concentration of surpluses	8. standardized monumental artwork
4. class structure	9. writing
5. state organization	10. arithmetic, geometry, astronomy

The primary characteristics are aspects of human organization, while the secondary characteristics are features of material culture.

The first civilizations in the world evolved in areas where agriculture also evolved. For example, in Mesopotamia, the area between the Tigris and Euphrates rivers, which today is part of southern Iraq, irrigated agriculture; water management practices and land reclamation were being practised probably as early as 5,000 years ago. This 'cradle of civilization', as it is often known, demonstrated an early form of complex social organization with the introduction of regulations on water use. The city of Babylon became a great administrative, cultural, and economic centre.

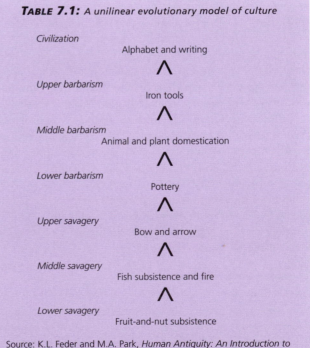

TABLE 7.1: *A unilinear evolutionary model of culture*

Civilization

Alphabet and writing

∧

Upper barbarism

Iron tools

∧

Middle barbarism

Animal and plant domestication

∧

Lower barbarism

Pottery

∧

Upper savagery

Bow and arrow

∧

Middle savagery

Fish subsistence and fire

∧

Lower savagery

Fruit-and-nut subsistence

Source: K.L. Feder and M.A. Park, *Human Antiquity: An Introduction to Physical Anthropology and Archaeology* (Toronto: Mayfield, 1993):397.

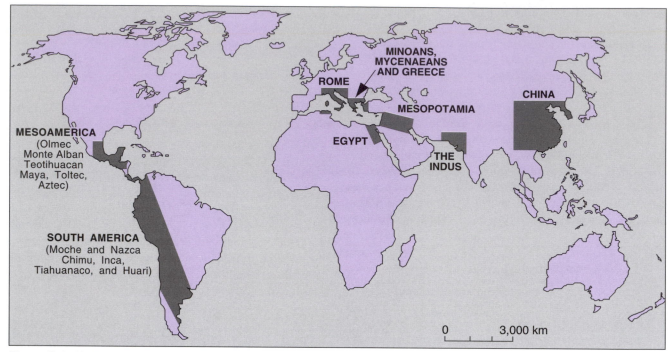

Figure 7.1 Civilizations in the ancient world.

hypothesis, such population increases are seen as prompting expansion by means of force. This process of expansion gradually leads to the creation of empires. According to Carneiro (1970:734), 'only a coercive theory can account for the rise of the state'.

Marxism

A Marxist view sees the development of class differences as a logical consequence of the increases in wealth for those individuals possessing domesticated animals. Once such class differences develop, class conflict follows as it is necessary for those with wealth to defend themselves from those without wealth. Also, exchange begins between those with and those without animals and, as population increases, a class of specialists become involved in the exchange process. By this means, people not directly involved in production begin to control economic life, adding a second dimension to the emerging class conflict. Such class conflict in turn generates central control as a means for the more dominant class to maintain and even enhance its dominance. Civilization follows from these beginnings in the form of a culture with rulers and the ruled.

EARLY CIVILIZATIONS

Civilization, then, is a major cultural advance involving a whole array of changes. Although each early civilization had its own distinctive character, they shared several characteristics. Agriculture and urbanization are the two most obvious developments that have impact on land, but civilization also involves changes in non-material culture: the rise of powerful élites and the disappearance of equalitarian groups; the

growth of bureaucratic systems and private land ownership. Religion was often the basis of early power, although all early civilizations became increasingly secular. Early civilizations that incorporated these changes are mapped in Figure 7.1, while a general time scale is depicted in Table 7.2.

As Table 7.2 makes clear, the emergence of a particular civilization, of a set of cultural advances, is not necessarily permanent; indeed, civilizations seem to have been especially vulnerable. Collapses are usually seen as the result of natural disasters, such as earthquakes or flooding, or of conflict with other groups or internal conflict. Specific collapses have

TABLE 7.2: *Basic chronology of early civilizations*

Egypt	3000–332 BC
Minoan	3000–1450 BC
Indus	2500–1500 BC
Mesopotamia	2350–700 BC
China	2000 BC–present day
Mycenaean	1580–1120 BC
Olmec	1500–400 BC
Greece	1100–150 BC
Rome	750 BC–AD 375
Monte Alban	200 BC–AD 800
Moche and Nazca	200 BC–AD 700
Teotihuacan	AD 100–700
Maya	AD 300–1440
Tiahuanaca	AD 600–1000
Toltec	AD 900–1150
Chimo and Inca	AD 1100–1535
Aztec	AD 1200–1521
Benin	AD 1250–1700

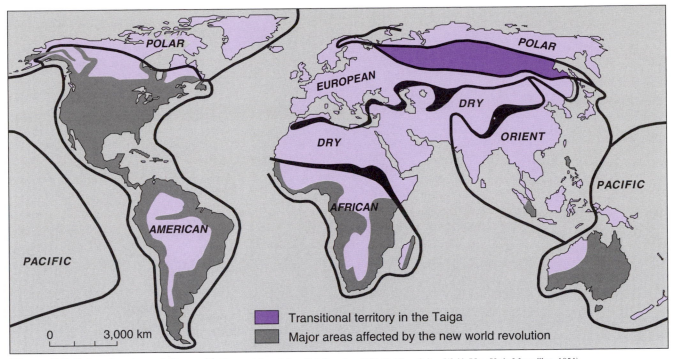

Figure 7.2 Cultural regions of the world. Source: Adapted from R.J. Russell and F.B. Kniffen, *Culture Worlds* (New York: Macmillan, 1951).

probably resulted from all three of the causes. Unlike earlier cultural groupings, civilization involves dense clusters of people and potential problems of organization and inequality.

CULTURAL REGIONS

The areas occupied by early civilizations are examples of cultural regions. In this section the focus is on the usefulness of the concept of **cultural regions** in understanding contemporary human geography.

As human geographers, our traditional concern is not with humans themselves or with their cultures but with human impacts on landscape. But, of course, these impacts vary considerably according to specific characteristics of human groups — their culture. Because different cultures have emerged in different areas and because the earth is a diverse physical environment, there is a wide variety of human landscapes. Human geographers have not simply described and explained these landscapes but have also delimited regions.

Delimiting any region involves at least four initial decisions. First, what criterion or criteria are to be employed? Second, because cultural regions are dynamic, what date is to be used? Third, what spatial scale is to be employed? Fourth, where are boundary lines or zones to be placed? These four interrelated issues are not easy to resolve.

WORLD REGIONS

On the world scale, there have been many notable attempts to delimit a meaningful set of cultural regions. One attempt by the historian Toynbee (1935–61) was a broad effort to

examine civilizations. He recognized a total of twenty-six civilizations, sixteen now dead, and ten surviving. Of the ten living civilizations, three are arrested — namely the Polynesian, Nomad, and Inuit — because they suffered the consequences of an overspecialized response to a difficult environment. The seven living civilizations are those of Western Christendom, Orthodox Christendom, the Russian offshoot of Orthodox Christendom, Islamic culture, Hindu culture, Chinese culture, and the Japanese offshoot of Chinese culture. At the time of writing, Toynbee saw most of Africa as primitive, meaning culturally uncommitted.

Toynbee's schema clearly has severe limitations as an attempt at world region delimitation, but it is instructive. The principal criterion for delimitation is religion in the sense that each civilization is defined on a religious basis, but the real criterion is the manner in which different cultures have responded to the environment — hence the acknowledgement of arrested and abortive civilizations.

A first attempt by geographers to delimit world regions was that of Russell and Kniffen (1951). Cultural groups were recognized, then related to areas such that cultural regions were delimited (Figure 7.2). Seven regions and a transitional area effectively cover the world, with each region regarded as the outcome of a long evolution of human and land relations. This valuable contribution has problems, regardless of how well justified the regions are, and it is important to emphasize that world regionalizations of this type are simply classifications striving to facilitate general world understanding.

Scale is the central problem. The larger the area to be divided, the more superficial or numerous the regions. World regionalizations are valuable and often necessary, but

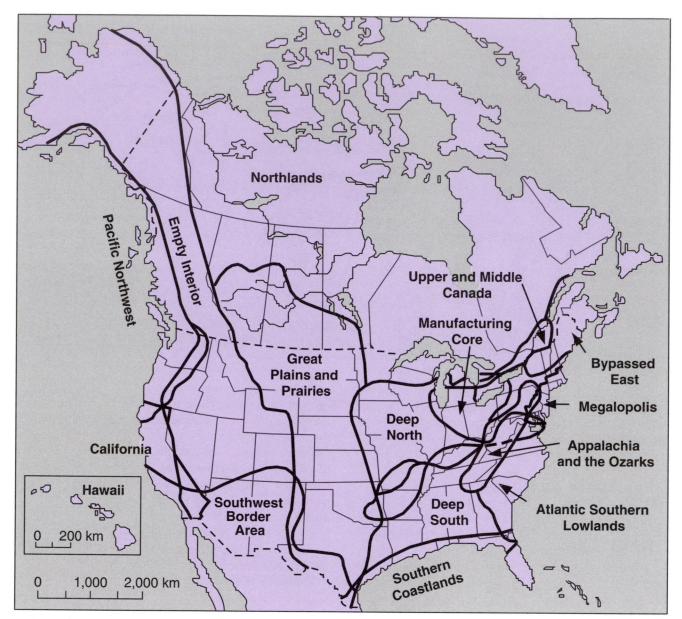

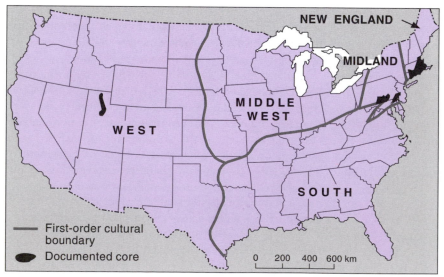

Figure 7.3 *(top)* Regions of North America.

Source: Adapted from S.S. Birdsall and J.W. Florin, *Regional Landscapes of the United States and Canada*, 2nd ed. (Toronto: Wiley, 1981):18.

Figure 7.4 *(bottom)* Cultural regions of the United States.

Source: Adapted from W. Zelinsky, *Cultural Geography of the United States*. (Englewood Cliffs: Prentice-Hall, 1973):118.

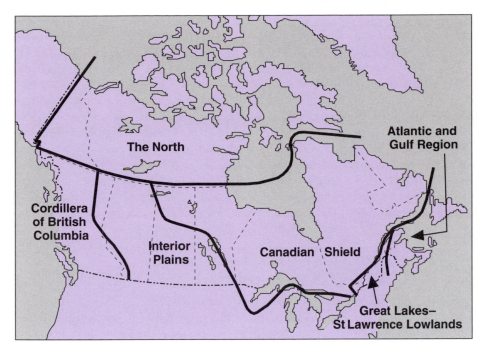

Figure 7.5 Regions of Canada.
Source: Adapted from J.L. Robinson, *Concepts and Themes in the Regional Geography of Canada*, rev. (Vancouver: Talon Books, 1989):17.

represent good examples of the use of the culture concept in geography. Like regionalizations based on physical variables (as detailed in Chapter 3), they are most appropriately seen as useful classifications rather than as insightful applications of geographic methods. The difficulties involved in delimiting Europe as a cultural region are outlined in Box 7.4.

Cultural regions and their associated distinctive landscapes are more effectively considered on a different scale. If, for example, we take North America as one example of a world region, we find that subdividing greatly enhances our appreciation of the culture and landscape relation.

NORTH AMERICAN REGIONS

North America is a relatively easy region to subdivide in that the development of contemporary culture regions is sufficiently recent to enable us to focus on origins. Figure 7.3 delimits sixteen regions in North America that are based on a variety of criteria, not simply culture. Each of these regions is likely to be generally recognizable and the authors of the map attempt to convey the 'feeling' of each region in their discussions. The regionalization is interesting because of the focus on regional themes and the recognition that the United States–Canada border is not a fundamental geographic division.

A regionalization of the United States
Figure 7.4 is a regionalization of the United States and is explicitly derived from the interesting concept of first effective settlement: 'Whenever an empty territory undergoes settlement, or an earlier population is dislodged by invaders, the specific characteristics of the first group able to effect a viable, self-perpetuating society are of crucial significance for the later social and cultural geography of the area, no matter how tiny the initial band of settlers may have been' (Zelinsky 1973:13).

This concept is used to demarcate five regions: the west, middle west, south, midland, and New England. Each of these five can be further subdivided. Only one of the five regions has a single major source of culture; the other regions have sources identified for various subregions. Thus, New England has England as the major source of culture; the midland is divided into two subregions, each of which has multiple sources; the south is divided into three subregions, each of which has multiple sources; the middle west is divided into three subregions, each of which has multiple sources; and the west is divided into nine subregions, each of which has multiple sources. In addition, a number of sub-subregions are noted, as are three regions of uncertain status (Texas, peninsular Florida, and Oklahoma). As a cultural regionalization, this example is especially useful because of the single clear variable employed. But the simplicity of the approach cannot disguise the complex cultural landscapes of the United States — hence the many subregions.

A regionalization of Canada
Our third North American example is a division of Canada into six regions (Figure 7.5). The regionalization process employs a variety of variables, but a basic cultural division is evident. The north is delimited according to political boundaries and can be easily subdivided into a northwest that is relatively forested and has a mixed European and Native population and an Arctic north that is treeless, mostly populated by Inuit people, and with limited resource potential by southern Canadian standards. The British Columbia region is characterized by major differences in

physical and human geography and typically has a resource-based economy; similar comments apply to the Atlantic and Gulf region, which also has added problems of low incomes and high unemployment rates. The interior plains are an agricultural region for cattle, wheat, and mixed farming, and its dispersed farmsteads and often regularly spaced towns show the results of government survey and planning before settlement. The core region of Canada is the Great Lakes–St Lawrence lowlands. Here are most of the people, the largest cities, major industry, and intensive agriculture, and here also is the cultural and economic heartland of Canada. The largest region is the Canadian Shield, an area that is sparsely settled with most settlements being resource towns.

THE MAKING OF CULTURAL LANDSCAPES

Our discussion of cultural regions has been descriptive rather than explanatory. To consider in more detail how regions of distinctive cultural landscape arise requires an

understanding of cultural variations over space and of the ability of culture to affect landscape.

Although there are certain basic similarities between all cultures — such as an agreed-upon emphasis on obtaining food and shelter, and on reproducing — how such goals are attained vary. As humans settled the earth, culture evolved initially in close association with the physical environment and gradually gained increased freedom from this constraint. As their ties to the physical environment gradually decreased, humans became more tied to their culture. One of the most important changes in human history is the change to a capitalist mode of production, which largely destroyed the ties to the physical environment, but created a new set of ties — to culture itself. Some human geographers see this change as an ongoing process that is effectively homogenizing world culture by minimizing regional variations — a view that is derived most clearly from Marx (see Box 2.5).

CULTURAL ADAPTATION

As humans settled the earth different cultures evolved in different locations. Each culture, despite many underlying

---

Box 7.4: Europe as a cultural region

Although generally seen as one major world region, Europe is all too clearly an area of diversity. Despite this diversity, Europe is typically seen as 'a *culture* that occupies a *culture area*' (Jordan 1988:6). Initially, Europe was erroneously viewed as a continent physically separated from Asia; by the sixteenth century, this error was corrected, but the image of a separate continent was so powerful that a new divide was needed. Thus it became generally accepted that the Caucasus, the Urals, and the Black Sea represented a meaningful boundary. The image of a separate Europe, an area that was culturally distinctive, was allowed to remain intact.

Using the political map of the mid-1980s, Europe was defined on the basis of twelve traits measured for each country (Figure 7.6):

- majority of population speak an Indo-European language
- majority of population have a Christian heritage
- majority of population are Caucasian
- more than 95 per cent of population are literate
- an IMR of less than 15
- an RNI of 5.0 per cent or less
- per capita income of US $7,500 or more
- 70 per cent or more of population are urban
- 15 per cent or less are employed in agriculture and forestry
- 100 km (62 miles) of railway plus highway for 100 km² (39 square miles)
- 200 or more kg (441 lbs) of fertilizer are applied annually per ha of cropland
- free elections are permitted

Figure 7.6 implies a regionalization; to proceed further would require a more explicit focus on variables such as language or religion. One particularly innovative delimitation, by Jordan (1988:402), combines three approaches: a north/south distinction, an east/west distinction, and a core/periphery distinction.

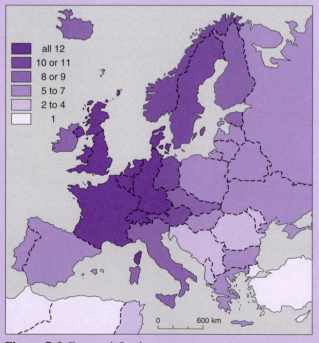

■	all 12
■	10 or 11
■	8 or 9
■	5 to 7
■	2 to 4
□	1

0   600 km

Figure 7.6 Europe defined.
Source: T.G. Jordan, *The European Culture Area: A Systematic Geography*, 2nd ed. (New York: Harper & Row, 1988):14.

similarities, varied in crucial ways in language, religion, political system, kinship ties, and economic organization. As we query how different cultures create different landscapes, we must also consider the means by which sound human/ physical environment relationships develop.

This question has long been central to human geography: witness environmental determinism, possibilism, and related viewpoints. Our answer centres on the concept of **cultural adaptation**. Humans continually adapt to the environment; they do so genetically, physiologically, and culturally. The cultural adaptations may be at an individual or group level and as human geographers, our concern is with both levels. Cultural adaptation involves, as the term suggests, changes in technology, organization, and ideology of the group. Such changes occur in response to both physical and human problems, permit solutions to these problems, improve the effectiveness of solutions, and provide adaptability or increase awareness of problems. Cultural adaptation thus refers to the process by which sound human and land relationships evolve.

We introduced earlier the idea of culture as the human ability to develop ideas from experiences and subsequently to act on the basis of those ideas. Cultural adaptation thus incorporates changes in attitudes as well as behaviour. An example of attitudinal change is our currently increasing awareness of the need for environmentally sensitive land use. An example of behavioural change is the implementation of such practices. Needless to say, behavioural changes do not always accompany or rapidly succeed attitudinal changes.

How do human geographers view the concept of cultural adaptation — cultural change responding to both environmental and cultural challenges? Four approaches are now identified.

Culturally habituated predisposition

In an analysis of the evolution of agricultural regions, it was suggested that such regions could be seen as the 'landscape expression of ... the totality of the beliefs of the farmers over a region regarding the most suitable use of land' (Spencer and Horvath 1963:81). According to this view, cultures are particular beliefs, psychological mind-sets, that result in a culturally habituated predisposition towards a specific activity and hence a specific cultural landscape.

Cultural preadaptation

An alternative view involves the concept of cultural preadaptation. According to this view, a cultural group moving into a new area may be preadapted for that new area. This situation arises if the source area of the culture is a diverse and varied environment such that any necessary adjustments to their culture either have been made prior to the move or are relatively easy to make following the move. Following this logic, Jordan and Kaups (1989) proposed that the American backwoods culture had significant northern European roots, specifically in Sweden and Finland.

Duplication, deviation, and fusion

A third view sees cultural landscapes as a result of duplication of, deviations from, and fusion of earlier cultures (Mitchell 1978). This view is a variant of the first effective settlement concept.

Core, domain, and sphere

Perhaps the best approach to this question is that involving a core, domain, and sphere model (Meinig 1965). According to this view, a cultural region or landscape can be divided into three areas: the core, the hearth area of the culture; the domain, the area where the culture is dominant; and the sphere, the outer fringe. Cultural identity decreases with increasing distance from the core. Figure 7.7 shows what most observers would agree is one of the easiest regions to delimit — the Mormon region of the United States. But Meinig has successfully extended these ideas, in modified form, to other regions such as Texas and the American southwest. There have been few attempts to apply these ideas outside North America.

Each of the four views articulated above is, however indirectly, very much in the Sauer landscape school tradition. There is a consistent concern with the landscapes created by cultural groups and particularly with the material manifestations of culture — the visible landscape. The two aspects of culture most significant for an understanding of our human world are language and religion. Not only are they important in themselves, they are also good bases for delimiting cultures, and hence regions, and they affect behaviour and landscape.

LANGUAGE

A CULTURAL VARIABLE

Language, probably the single most important human achievement, is of interest to human geographers for several reasons. First, language is a cultural variable, a learned behaviour. It initially evolved so that humans could communicate in groups, probably prompted by the need to organize hunting activities as a group (not an individual) activity. Gradually, as early humans moved across the earth, different languages arose in different areas. It is possible that early humans, concentrated in the area of origin, all spoke one early language, but such a situation changed as movement separated groups and offered new physical environmental experiences. Thus, from the beginnings of cultural evolution, we have language as a potential source of group unity (and therefore of total-population disunity), a topic of great interest to human geographers.

Second, language is much more than a source of group unity and possibly a useful means to delimit groups and hence regions. Language is the means by which a culture ensures continuity through time. Quite simply, the death of a language is the death of a culture. This knowledge prompts

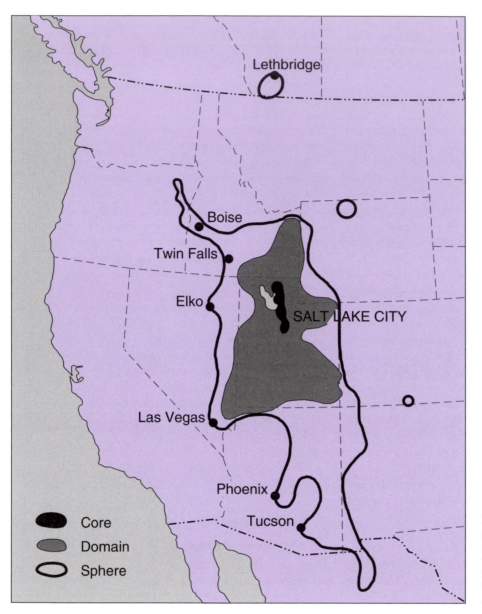

Figure 7.7 Core, domain, and sphere.

Source: Adapted from D.W. Meinig, 'The Mormon Culture Region: Strategies and Patterns in the Geography of the American West, 1847–1964', *Annals, Association of American Geographers* 55 (1965):214.

many attempts to ensure the continuity, or even revival, of language in such areas as Wales and Quebec. Language is often the fundamental and necessary building-block of a nation. It relates to virtually all activities, all aspects of our lives. As human geographers, we are interested in the rise and fall of languages, the plight of minority languages, and the relationships between language and nationalism.

Our third interest in language concerns relationships with the physical and human environments, relationships that function in both directions. Spatial variation of language is partly caused by physical and human environmental variations, while language itself is an effective moulder of the human environment. This third interest considers language's links with not only the visible landscape but also the symbolic landscape and group identity, thus returning us to our first interest.

CLASSIFICATION AND REGIONS

As is the case with many cultural variables, and indeed with culture itself, language began as one (or at least a few), splintered into many — perhaps 7,000, and now displays about 3,000 variations (not counting minor dialects). To make sense of this complex situation of change and variety, a classification has been devised using the concept of language family — a group of closely related languages that show evidence of a common origin.

Table 7.3 lists principal language families, subfamilies, and individual languages, while Figure 7.8 maps the families on a world basis. Note that Table 7.3 refers to the present day, whereas Figure 7.8 reflects language distribution prior to the period of European overseas expansion. Thus, Figure 7.8 maps the Americas and Australia not as Indo-European but as Amerind and Australian respectively.

TABLE 7.3: *An overview of language data, 1987*

Language family	Number of extant languages	Number of speakers	Location	Major languages
AUSTRIC	1,175	293,000,000	SE Asia, Oceania	
Miao-Yao	4	7,000,000	S. China, N. Vietnam, N. Laos, N. Thailand	Miao, Mien
Austroasiatic:	155	56,000,000	NE India, SE Asia	
Munda	17	6,000,000	NE India	Santali, Mundari
Mon-Khmer	138	50,000,000	NE India, SE Asia, Nicobar Is.	Vietnamese, Mon, Khmer
Daic	57	50,000,000	S. China, SE Asia	Thai, Lao, Li, Shan, Zhuang, Kam
Austronesian:	959	180,000,000	Oceania, S. Vietnam, Madagascar	
Western	533	179,000,000	Madagascar, Formosa, Philippines, S. Vietnam, Kampuchea	Malagasy, Javanese, Sundanese, Malay, Tagalog, Cebuano, Ilokano, Hiligaynon
Eastern (=Oceanic)	426	1,500,000	Melanesia, Micronesia, Polynesia	Fijian, Samoan, Tahitian, Hawaiian
INDO-PACIFIC	731	2,735,000	New Guinea, Timor, Alor, Pantar, Halmahera, New Britain, New Ireland, Bougainville, Solomons, Reef Islands, Santa Cruz, Andaman Island, Tasmania	Enga, Wantoat, Telefol, Iatmul, Asmat
AUSTRALIAN	170	30,000	Australia	Western Desert Language
NA-DENE	34	202,000	Alaska, W. Canada, Oregon, California, Arizona, New Mexico	Navajo, Apache
AMERIND	583	18,000,000	North, Central, and South America	
Kutenai	1	200 (in 1977)	Montana, Idaho	Kutenai
Algic:	16	91,000	Canada, US	
Ritwan	1	10 (in 1980)	N. California	Yurok
Algonquian	15	91,000	C. and E. Canada, C. and E. US	Ojibwa, Cree, Blackfoot, Cheyenne
Mosan:	27	9,500	NW US, SW Canada	
Chimakuan	1	10 (in 1977)	NW Washington	Quileute
Wakashan	6	2,700	SW Canada	Nootka, Kwakwala
Salish	20	6,800	SW Canada, NW US	Shuswap, Kalspel, Squamish
Keresan	2	7,000	New Mexico	Keres
Siouan-Yuchi	11	21,000	C. US	Dakota, Crow
Caddoan	4	1,000	C. US	Wichita, Pawnee
Iroquoian	7	15,000	E. US	Cherokee, Mohawk
Penutian	68	3,200,000	W. Canada, W. and SE US, S. Mexico	Chinook, Zuni, Muskogee, Quiche, Cakchiquel, Kekchi, Mam, Yucatec
Hokan	28	55,000	California, Arizona, Texas, Mexico, Colombia	Mohave, Yuma, Tlapenec, Tequistlatec
Tanoan	7	7,400	New Mexico, Oklahoma	Kiowa, Tewa
Uto-Aztecan	25	1,100,000	W. US, Mexico	Comanche, Hopi, Nahuatl
Oto-Manguean	17	1,700,000	S. Mexico	Otomi, Mixtec, Zapotec
Chibchan-Paezan	43	200,000	Florida, S. Mexico, Central America, W. South America	Tarascan, Yanomami, Guaymi, Cuna, Paez, Warao, Embera, Cayapa
Andean	18	8,500,000	W. South America	Quechua, Aymara, Mapudungu
Macro-Tucanoan	47	35,000	NW and E. South America	Ticuna, Tucano, Nambikuara, Puinave
Equatorial	145	3,000,000	South America, Caribbean	Guarani, Tupi, Goajiro, Arawak
Macro-Carib	47	50,000	N. South America	Galibi, Witoto
Macro-Panoan	49	50,000	W. South America	Toba, Tacana
Macro-Ge	21	10,000	E. South America	Bororo, Chavante

TABLE 7.3: *An overview of language data, 1987 cont.*

Language family	Number of extant languages	Number of speakers	Location	Major languages
KHOISAN	31	120,000	South Africa, Namibia, S. Angola, Botswana, N. Tanzania	Nama (=Hottentot)
NIGER-KORDOFANIAN	1,064	181,000,000	C. and S. Africa	Fula (=Fulani), Mandinka, Yoruba, Rwanda, Shona, Tswana, Xhosa, Zulu
NILO-SAHARAN	138	11,000,000	C. Africa	Kanuri, Luo, Nubian, Maasai, Songhai
AFRO-ASIATIC	241	175,000,000	N. Africa, Near East	
Berber	30	11,000,000	Algeria, Morocco, Tunisia, Libya, Mauritania, Senegal	Shilha, Kabyle, Riff, Tuareg, Tamazight
Chadic	123	30,000,000	Chad, Niger, Ghana, Nigeria, Cameroon, Central African Republic, Togo, Benin	Hausa
Omotic	34	1,000,000	W. Ethiopia, N. Kenya	Ometo
Cushitic	35	12,000,000	Somalia, Ethiopia, Sudan, Kenya, Tanzania	Somali, Oromo
Semitic	19	121,000,000	N. Africa, Near East	Arabic, Hebrew, Aramaic, Amharic, Tigrinya
CAUCASIAN	38	5,000,000	Caucasus (USSR)	Georgian
INDO-EUROPEAN	144	2 billion	Europe, SW Asia, India, Americas, Australia, South Africa, New Zealand	
Armenian	1	5,000,000	USSR	Armenian
Indo-Iranian	93	700,000,000	Iran, Afghanistan, Pakistan, India,	Romany, Farsi, Kurdish, Pashto, Punjabi, Gujarati, Hindi-Urdu, Marathi, Bengali
Albanian	1	4,000,000	Albania	Albanian
Greek	2	10,000,000	Greece	Greek
Italic	16	500,000,000	Rumania, Italy, France, Spain, Portugal, Central and South America	Rumanian, Italian, French, Prevençal, Catalan, Spanish, Portuguese
Celtic	4	2,500,000	Ireland, Wales, N. France	Irish, Welsh, Breton
Germanic	12	450,000,000	Germany, Holland, Scandinavia, Great Britain, North America, Australia, New Zealand, South Africa	German, Yiddish, Dutch, Afrikaans, English, Danish, Swedish, Norwegian, Icelandic
Balto-Slavic	15	290,000,000	USSR, Poland, Czechoslovakia, Yugoslavia, Bulgaria	Lithuanian, Latvian, Russian, Ukrainian, Byelorussian, Polish, Czech, Slovak, Serbo-Croatian, Bulgarian
URALIC-YUKAGHIR	24	22,000,000	Finland, Estonia, Hungary, USSR	Hungarian, Finnish, Saami, Estonian
ALTAIC	63	250,000,000	Asia	
Turkic	31	80,000,000	Turkey, USSR, Iran,	Turkish, Uzbek, Uighur, Azerbaijani, Turkmen, Tatar, Kazakh, Kirghiz, Chuvash, Bashkir
Mongolian	12	3,000,000	Mongolia, China, USSR	Khalkha
Tungus	16	80,000	E. USSR, China	Manchu, Evenki
Korean	1	55,000,000	Korea	Korean
Japanese-Ryukyuan	2	115,000,000	Japan	Japanese
Ainu	1	few	N. Japan, S. Sakhalin Is. (USSR)	Ainu
CHUKCHI-KAMCHATKAN	5	23,000	NE Siberia (USSR)	Chukchi

TABLE 7.3: *An overview of language data, 1987 cont.*

Language family	Number of extant languages	Number of speakers	Location	Major languages
ESKIMO-ALEUT	9	85,000	Alaska, N. Canada, Greenland, NE USSR	Eskimo, Aleut
ELAMO-DRAVIDIAN	28	145,000,000	S. and E. India, S. Pakistan	Telugu, Kannada, Tamil, Malayalam
SINO-TIBETAN	258	1 billion	China, Tibet, Nepal, India, Burma, Thailand, Laos	Mandarin, Wu, Yue, Tibetan, Burmese, Karen

Source: Adapted from M.C. Ruhlen, *A Guide to the World's Languages: Volume 1, Classification* (Stanford: Stanford University Press, 1987).

Numerically, the largest family is the Indo-European, followed by the Sino-Tibetan. Mandarin has more speakers than any other language, with English ranking as second. Interestingly the Indo-European family is spatially dispersed today, while the Sino-Tibetan is limited to one area. This distribution reflects the colonial expansions of many Indo-European-speaking countries compared to the relative immobility of Sino-Tibetan speakers. Thus the numerical importance of Mandarin is attributable to the large population of China, whereas the numerical importance of English is attributable to, among other things, English colonial activity.

The Indo-European language family

Rather than review each of the language families in detail, we choose to investigate one family, Indo-European, and one component language, English. This choice is not arbitrary. English is the 'language of the planet, the first truly global language' (McCrum, Cran, and MacNeil 1986:19).

The parent of all Indo-European languages, what we might call proto-Indo-European, probably evolved as a distinct means of communication in one of two cultures, the Kurgan culture of the Russian steppe region north of the Caspian Sea or a farming culture in the Danube valley. The likely period of origin is between 6000 BC and 4500 BC. Our knowledge of these origins is gleaned largely from analysing vocabulary and thus reconstructing culture. The evidence suggests a culture that was partially nomadic, that had domesticated various animals (including the horse), that used ploughs, and grew cereals. Contemporary Indo-European languages have similar words for snow but not for sea, thus placing the origin language in a cold region some distance from the sea.

As this cultural group dispersed, using the horse and the wheel, the language moved and evolved (Figure 7.9). Different environments and group separation resulted in many changes. The earliest Indo-European settlement in what is now England was by Gaelic-speaking people after 2500 BC. They were followed by a series of Indo-European invasions: the Romans in 55 BC, and the Angles, Saxons, and Jutes in the fifth century AD (Figure 7.10). The latter three groups introduced what was to become the English language, forcing Gaelic speakers to relocate in what is now the Celtic fringe. Surprisingly little mixing of Gaelic and the incoming language took place. Old English, as it is called, was a far from uniform language and was subsequently enriched by the addition of Latin via the introduction of Christianity in 597, by the invasions of Vikings between 750 and 1050, and the Norman (French) invasion of 1066. Each of these invasions prompted a collision of different languages, and English responded by diversifying. Today it is a language with an estimated 500,000 words (excluding an equal number of technical terms); by contrast, German has 185,000 words and French 100,000.

Beginning in the fifteenth century, this English language, along with other Indo-European languages, spread around

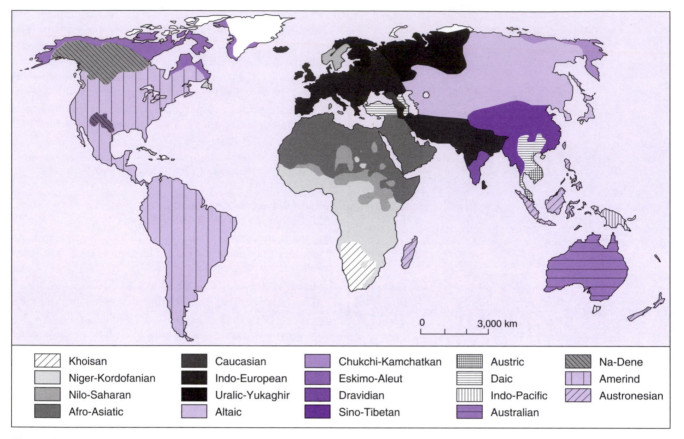

Figure 7.8 World distribution of language families.

Source: M.C. Ruhlen, *A Guide to the World's Languages: Volume 1, Classification* (Stanford: Stanford University Press, 1987).

the globe. Such language spread encouraged further varieties of English, including many American varieties. Not only is English the first language of many countries, it is also the second language of many others. It is estimated, for example, that there are perhaps 70 million speakers of English in India and that this is a vital unifying force in a country of many languages.

LANGUAGE AND IDENTITY

A common language facilitates communication; different languages create barriers. For many groups, language is the primary basis of identity, hence there are close links between language and nationalism, attempts to preserve minority languages, and even attempts to create a universal language. One consequence of this potent impact of language is undoubtedly to promote friction between groups, which divides our world.

Language and nationalism

The link between language and **nationalism** is very clear. In medieval Wales, there was one word for language and nation, and in present-day Ireland, survival of the Gaelic-speaking area is 'synonymous with retention of the distinctive Irish national character' (Kearns 1974:85). There are two reasons why language is often seen as the basis for delimiting a nation. First, a common language facilitates communication. Second, language is a most effective symbol or

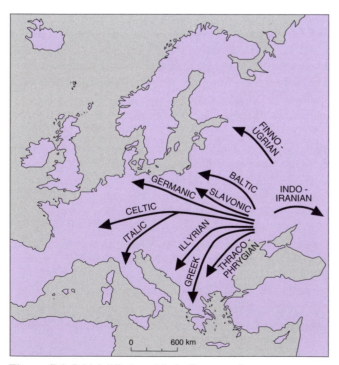

Figure 7.9 Initial diffusion of Indo-European languages.

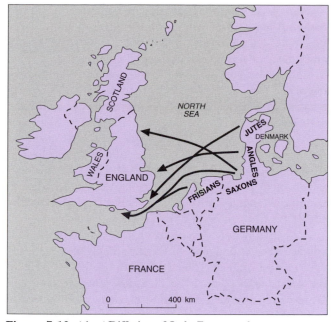

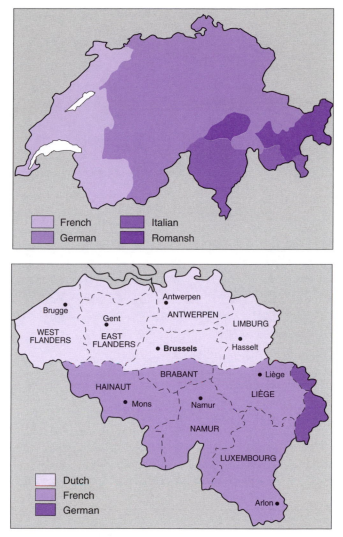

Figure 7.10 *(above)* Diffusion of Indo-European languages into England.

Figure 7.11 *(above right)* Four official languages in Switzerland.

Figure 7.12 *(right)* Dutch and French in Belgium.

emblem of 'groupness'. This symbolism is very powerful and is a sufficient reason for declaring a national identity even in those cases, such as Ireland, where the language does not serve a significant communication function.

Before the nineteenth century, state/language concordance was not typical. France and the United Kingdom were one-language states, but most of Europe was politically divided on the basis of factors other than language. From the beginning, the subsequent rise of nationalism was a process that strove to integrate language and state. Two principal consequences were the creations of the states of Italy in 1870 and Germany in 1871.

Multilingual states

Other areas in Europe did not succumb to these tendencies. Switzerland is a prime example of a viable political unit with 70 per cent German speakers, 19 per cent French, 10 per cent Italian, and 1 per cent Romansh, a clear (if not necessarily typical) indication that nations are not always comprised of a single language group (Figure 7.11). Explaining the stability of Switzerland requires an understanding of the pre-1500 evolution of the Swiss state, of the long-standing practice of delegating much governmental activity to local regions, and of its neutrality at times of major European conflicts.

More typical examples of **multilingual states** are Belgium and Canada, both of which appear weakened as a result of lacking a unifying language. In Belgium, there are Dutch speakers in the north and French speakers in the

south (Figure 7.12). Despite conscious efforts to follow the Swiss example, Belgium is characterized by disunity. Explaining Belgian instability involves recognizing that it was created as late as 1830 as an artificial state, favoured by other European powers but not necessarily by those who became Belgians. Further, Belgium has not succeeded in being detached from European conflicts. Today, Belgium is a country with an all too clear boundary between Dutch and French speakers. Although Belgium is a bilingual state, the two areas are regionally unilingual. The capital, Brussels, is primarily French-speaking, although located in the unilingual Dutch area, a fact that further aggravates a difficult situation. Thus, not only is Belgium not a bilingual country in practice, it does not have a bilingual transition zone.

Similar comments are in order for Canada (Figure 7.13). The language transition area between English and French Canada is not a continuous area but a series of pockets, especially centred on the cities of Montreal and Ottawa. This situation suggests an outcome comparable to that prevailing in Belgium (Box 7.5).

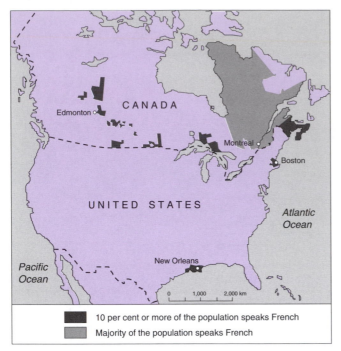

Figure 7.13 French and English in Canada.

Minority languages

Any language is most likely to survive when it serves in an official capacity. **Minority languages** without official status typically experience a slow but inexorable demise. Often minority-language speakers strive for much more than mere survival of their language, favouring instead to use the language as a justification for the creation of a separate state.

Examples of minority languages are numerous: Welsh and Irish in Britain, Spanish in the United States, French — despite official status — in Canada, Basque in Spain, Hausa and other languages in Nigeria, and Cantonese in China. This is a highly condensed list — most countries in the contemporary world have one official language plus one or more minority languages. The consequences are varied. In Britain, both Welsh and Irish continue to strive for survival and indeed independence (Box 7.6). In Spain, the Basque language, one of very few languages in Europe that do not belong to the Indo-European family, is the basis for a powerful and often violent Basque independence movement. In Nigeria, the official language is English, a result of colonial activity and of the fact that Nigeria has no indigenous language.

Undoubtedly, language is one of if not the most, important factor promoting unity or disunity. Stable states typically have a single language. There are, however, few countries in the world that can claim to be unilingual; Japan, Iceland, and Uruguay are examples. More than one language contributes to instability. Language is a justification for both internal conflict (Spain) and territorial expansion (Nazi Germany).

Communications between different language groups

Historically, some languages have assumed importance even when they are not the first language of populations. In India, following independence, English joined Hindi as an official language. In parts of east Africa, Swahili is an official language; Swahili is an example of a **lingua franca**, a language that developed to facilitate trade between different groups (Africans and Arab traders). In some other

---

Box 7.5: Linguistic territorialization in Belgium and Canada

Although the specific causes are quite different, the Belgian and Canadian language situations are basically similar. Belgium is clearly divided into two linguistic territories, while Canada appears to be approaching the same situation.

Since about the fifth century, the area that is now Belgium has been divided into a northern area of Flemish speakers (Flemish is the ancestor of modern Dutch) and a southern area where Walloon (French) is spoken. Thus Belgium lies on both sides of a major linguistic border, that between the Germanic and Romance subfamilies of the Indo-European family. To expect a viable state to emerge in this context as late as 1830 was quite ambitious, particularly as the French speakers explicitly wished to join France. Belgium is thus an artificial state created especially by the British and Germans who did not favour any French territorial expansion. Belgium thus comprises two distinct language areas with a minimal transition zone; it has attempted to follow the example of Switzerland, but without success.

Canada, settled by both the French and the English during the formative years, is also a bilingual state. With the formation of Canada in 1867, both French and English were adopted as official languages. The French formed a majority in Quebec and the English a majority in Ontario. Most of the remainder of Canada west of Ontario was settled either by English speakers or by others who settled into an English-speaking area. Since 1867, immigration of French speakers has been very limited. Interaction between the French and the English prompted the rise of a distinct transition zone in western New Brunswick, southern Quebec, and eastern Ontario. Recent research, however, suggests that the zone is disintegrating into a series of pockets such that the linguistic territorialization evident in Belgium since 1830 is becoming a reality in Canada (Cartwright 1988). This conclusion is reached following analysis of local migration and interaction between the two groups.

The political situations in Belgium and Canada are fraught with problems as a consequence of spatially delimited language differences.

colonial areas, **pidgin** languages appeared as simplified ways of communicating between different language groups. If a pidgin becomes relatively elaborate and serves as a mother tongue, it is known as a **creole**; these are relatively common in the Caribbean region.

There have been a number of attempts to promote use of a single universal language, among them the introduction of new languages. In 1887, the most popular such language — Esperanto — was introduced, but failed to make a significant impact, at least partially because such creations lack traditional roots. In one sense, human geographers reject the idea of a universal language because languages are tied to cultures and environment. But human geographers also recognize that in principle, a universal language could promote communication and understanding between groups, thus minimizing division and friction. It is unlikely, however, that any of the artificial languages will ever succeed in playing such a role. Today, the best hope for such a language rests with English.

LANGUAGE IN LANDSCAPE

Place names

Our discussion so far has emphasized the centrality of language to culture and group identity, and its effects on our partitioning of the earth. But language plays another key role — language is *in* landscape. Place names or toponyms are the clearest such expression. We name places for at least two reasons: first, in order to understand and give meaning to landscape. A landscape without names would be like a group of people without names; there would be little basis for distinguishing one location or person from another. Second, naming places probably serves an important psychological need — to name is to know and control, to remove uncertainty about the landscape. For these two reasons, humans impose names on all landscapes that they occupy and on many that they do not occupy (the moon is a prime example).

Place names, then, are a feature of our human-made landscapes, often visible in the form of road signs and an integral component of maps, our models of the landscape. Many place names combine two parts, generic and specific. Newfoundland, for example, has *Newfound* as the specific component, *land* as the generic component — the type of location being identified.

There are many ways of classifying place names. In a detailed analysis of Finnish settlement in Minnesota, Kaups (1966) found a classification based on the mechanisms of naming useful. Possessive names indicate an association, possibly ownership, by an individual (personal) or a group (ethnic). Of ninety-two Finnish place names, fifty-two were

---

Box 7.6: The Celtic languages

The Celts were one of the major early groups to diffuse from the Indo-European core area. They spread across much of Europe, but gradually as other more organized groups came into contact with them, they found it necessary to retreat into some of the more inhospitable and isolated areas of western Europe. Today, the remains of the once-large Celtic group are in a series of four pockets: western Wales, western Scotland, western Ireland, and northwest France. Each of these four areas still has Celtic speakers: about 500,000 speak Welsh in Wales, about 80,000 speak Gaelic in Scotland, about 70,000 speak Erse (Irish Gaelic) in Ireland, and about 675,000 speak Breton in northwest France.

The retreat of Celtic speakers followed a 500-year period of expansion that began about 500 BC from a southern German core. In Britain, the Celts were pushed to the western limits by the Anglo-Saxon in-movement. Over time, various Celtic languages, such as Cornish (southwest England) and Manx (Isle of Man in the Irish Sea), disappeared, leaving the four pockets already noted.

The remaining Celtic languages are all in precarious positions, basically because the languages are not associated with political units. In Ireland, the Irish language area, called the Gaeltacht, covered about 33 per cent of the country in 1850 and had perhaps 1.5 million speakers; today it covers about 6 per cent and has, as noted, about 70,000 speakers. The first conscious attempts to save the Gaeltacht came in 1956: 'To a great extent, survival of the Gaeltacht has become synonymous with retention of the distinctive national character' (Kearns 1974:85). The basic assumption is that language is a key requisite for any group that aspires to retain a traditional culture. Consequently, the Irish government has, since 1956, actively encouraged retention of the language and cautious development of the rural Gaeltacht region. So far, the various government attempts to foster social and economic development have been quite successful and have not resulted, as some thought they might, in loss of the Irish language.

Welsh, a minority language identified with a rural way of life in an increasingly Anglicized environment, is in a similar position. Again, it is evident that the language is intimately tied to traditional Welsh culture. Welsh Wales is restricted to the extreme northwest and southwest and the number of Welsh speakers is ever declining, despite strong local efforts. As a part of the United Kingdom, Wales is not in as strong a position as is Ireland when it comes to implementing local development and language retention policies. There seems little prospect of a significant increase in Welsh speakers; a continuing decline is more probable.

possessive. Commemorative place names commemorate an important place or person; there were sixteen such names. Descriptive place names identify an easily recognizable characteristic of a location, such as marshland; there were nineteen such names. Only five place names could not be easily accommodated by this classification.

Analyses of place names thus provide information about the spatial and social origins of settlers. They may also tell us something about their aspirations: witness such a name as Toivola (Hopeville) in Minnesota or Paradise in California. Place names result typically from the first effective settlement in an area; because the Finns in Minnesota were later European settlers, their opportunity to name places was restricted to a local scale. Other groups in North America, most notably the French and Spanish, have successfully placed their language on the landscape at the regional scale; witness the profusion of French names in Quebec and Louisiana and of Spanish names throughout the American southwest.

Place names are often an extremely valuable route to understanding the cultural history of an area. This is especially clear in areas of relatively recent first effective settlement, but is also the case in older settled areas. In many parts of Europe, for example, former cultural boundaries can be identified through place name analysis. Jordan (1988:98) provides an example dating from before 800 when the Germanic-Slavic linguistic border roughly approximated the line of the Elbe and Saale rivers in what is now Germany. German place names are evident west of the line and Slavic place names to the east. After 800, the movement east of German speakers has not removed the evidence of the earlier boundary. Throughout much of Britain, it is similarly possible to identify areas settled by different language groups by a study of place names; in northeast England, for example, evidence of Viking settlement remains in such place name endings as -by.

The 'Great American Desert'

Clearly, language is everywhere in landscape, but language can also make landscape. A striking example is the nineteenth-century use of the term 'desert' — as in the Great American Desert — to describe much of what is now the Great Plains region. The plains were little known by European Americans until a series of early nineteenth-century explorations gradually introduced the desert concept, which became a 'fact' following an 1820 expedition. The first American explorers further predicted that the desert would restrict settlement. The misconception arose because individual explorers recorded what impressed them most, because the small areas of desert were impressive, and because the region lacked trees and was a desert relative to the east. One clear consequence was delayed settlement and increased movement to the west coast. Thus language, in the form of a specific environmental type, affected human geographic changes in the landscape.

Not only is language in landscape, landscape is in language. This is apparent in various ways. First there is a general relationship between language distributions and physical regions because physical barriers tend to limit movement (this is evident on a world scale in Figure 7.8). Second, specific languages reflect the physical environment and the need for people to understand and cope with those environments, hence there are many Spanish words for features of desert landscapes and few comparable English words. Similarly, as languages move to new environments, it is necessary to add new words; the Australian 'bush' and 'outback' are examples of additions to English. Finally, the human landscape is in language, for example, as it reflects class and gender. Within a given language, word choices and pronunciations say a great deal about social origins. This area of research is now labelled sociolinguistics.

RELIGION

Religion is a second important cultural trait. The origins of religious beliefs are as complex as those of language, but the universality of religion suggests that it serves a basic human need. Essentially, a religion involves a set of beliefs and associated activities that are in some way designed to facilitate appreciation of our human place in the world. In many instances, religious beliefs generate sets of moral and ethical rules that can significantly condition many aspects of behaviour. Thus religion is an important cultural variable.

CLASSIFICATION AND REGIONS

A simple but useful classification of religions distinguishes between universalizing and ethnic types.

Universalizing religions

Universalizing religions incorporate mechanisms such as missionaries and conversion procedures designed to transmit their beliefs to others. There are three principal religions of this type.

Buddhism originated about 500 BC in northern India and about 100 BC began expanding to China, Japan, Korea, and southeast Asia. Christianity began some 500 years after Buddhism as an offshoot of Judaism, spread throughout Europe, and subsequently moved across the globe as Europe expanded. Today, Christianity is the most widespread religion spatially and most significant numerically. There are obvious links between language and religious migration. The third universalizing religion is Islam, which began some 600 years after Christianity (Muhammad was born in Mecca in 570) and subsequently spread throughout the Middle East, North Africa, and parts of Asia.

Each of these three religions might be better regarded as a religious group. Buddhism has two principal versions: Theravada Buddhism in southeast Asia and Mahayana

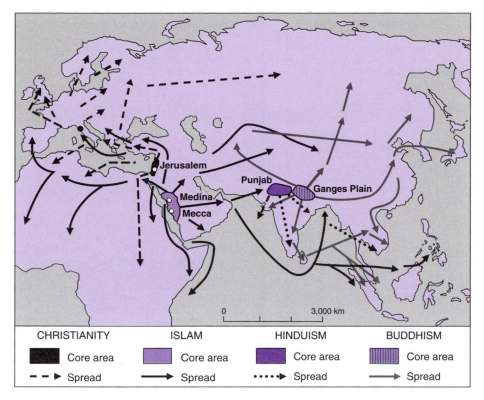

Figure 7.14 Origin areas and diffusion of four major religions.

Buddhism in east Asia. Christianity experienced a major east/west division in 1054 and a Protestant breakaway in the west in the 1500s. Islam is divided into two principal groups, Shiite and Sunni. Shiites are the majority in Iran and Iraq, while Sunnis dominate in Arabic-speaking areas.

Ethnic religions

Ethnic religions have an identity with a particular group of people and do not actively convert others. There are several hundred religions in this category. Numerically, the most significant is Hinduism, which evolved in northern India about 2000 BC. Hinduism has no dogma and only a loosely defined philosophy by religious standards. It is a polytheistic (belief in more than one god) religion and has close ties to the rigid social stratification of the caste system. A second ethnic religion is Judaism, the oldest monotheistic (belief in one god) religion, originating about 2000 BC in the Near East. Following the destruction of Jerusalem in AD 70, Jews dispersed throughout Europe, a process known as the Diaspora. Only in 1948 was the long-term goal of a Jewish homeland achieved with the creation of the state of Israel. Other ethnic religions include Shinto, the indigenous Japanese religion, and Taoism and Confucianism, religions primarily associated with China.

Spatial distribution of religions

Table 7.4 provides detailed data on the distribution of religious groups by selected major world regions, while Table 7.5 shows the distribution of Christianity's five subgroups.

Christianity is the leading religion numerically with 32.9 per cent of the world population nominally Christian, followed by Muslims (followers of Islam) at 17.4 per cent, Hindus at 13.1 per cent, and Buddhists at 6.1 per cent. A high 17.1 per cent are classed as non-religious and another 4.5 per cent as atheist. Clearly, then, the twofold classification excludes literally hundreds of religions. Many cultures favoured **animism** prior to the diffusion of universalizing beliefs. Animism has many versions, but essentially involves the belief that specific inanimate objects (such as the sun, moon, rivers, or mountains) are of value for spiritual reasons.

Figure 7.14 locates the origin areas of four major religions: Buddhism, Christianity, Hinduism, and Islam, while Figure 7.15 indicates the contemporary distribution of these and other religions. Especially clear from these two maps is the remarkable expansion of Christianity and Islam, two religions that developed in close proximity, both spatially and temporally. From their semidesert hearth areas, both sets of beliefs spread as a result of the more general spread of culture via military activity, exploration, and economic expansion. Buddhism and Hinduism have not experienced comparable spread; the former has effectively relocated to China, while the latter is restricted to India.

RELIGION, IDENTITY, AND CONFLICT

The importance of religion for individual and group identity varies considerably. For some, religion is irrelevant or marginal. In much of the more developed world, especially in urban areas, religion is not central to human activity.

Table 7.4: An overview of religious data, 1992

Religionists	Africa	Asia	Europe	Latin America	Northern America	Former Oceania	USSR	World
Christians	327,204,000	285,365,000	413,756,000	435,811,000	239,004,000	22,628,000	109,254,000	1,833,022,000
Muslims	278,250,800	636,976,000	12,574,500	1,350,500	2,847,000	100,500	39,229,400	971,328,700
Non-religious	1,896,000	691,144,000	52,411,000	17,159,000	25,265,000	3,291,000	85,066,000	876,232,000
Hindus	1,475,000	728,118,000	704,000	884,000	1,269,000	360,000	2,000	732,812,000
Buddhists	21,000	313,114,000	272,000	541,000	558,000	26,000	407,000	314,939,000
Atheists	316,000	161,414,000	17,604,000	3,224,000	1,319,000	535,000	55,898,000	240,310,000
Chinese folk religionists	13,000	186,817,000	60,000	73,000	122,000	21,000	1,000	187,107,000
New Religionists	21,000	141,382,000	50,000	530,000	1,421,000	10,000	1,000	143,415,000
Tribal religionists	70,588,000	24,948,000	1,000	936,000	41,000	67,000	0	96,581,000
Sikhs	26,000	18,272,000	231,000	8,000	254,000	9,000	500	18,800,500
Jews	337,000	5,587,000	1,469,000	1,092,000	7,003,000	98,000	2,236,000	17,822,000
Shamanists	1,000	10,233,000	2,000	1,000	1,000	1,000	254,000	10,493,000
Confucians	1,000	5,994,000	2,000	2,000	26,000	1,000	2,000	6,028,000
Bahais	1,496,000	2,680,000	91,000	801,000	365,000	77,000	7,000	5,517,000
Jains	53,000	3,717,000	15,000	4,000	4,000	1,000	0	3,794,000
Shintoists	200	3,220,000	500	500	1,000	500	100	3,222,800
Other religionists	433,000	12,292,000	1,469,000	3,570,000	485,000	4,000	333,000	18,586,000

Source: 1993 Encyclopedia Britannica Book of the Year (Chicago: Encyclopedia Britannica Inc., 1993).

TABLE 7.5: Distribution of Christian populations, 1992

Religionists	Africa	Asia	Europe	Latin America	Northern America	Former Oceania	USSR	World
Christians	327,204,000	285,365,000	413,756,000	435,811,000	239,004,000	22,628,000	109,254,000	1,833,022,000
Roman Catholics	122,907,000	123,597,000	262,638,000	405,623,000	97,022,000	8,208,000	5,590,000	1,025,585,000
Protestants	87,332,000	81,476,000	73,939,000	17,263,000	96,312,000	7,518,000	9,858,000	373,698,000
Orthodox	28,549,000	3,655,000	36,165,000	1,764,000	6,008,000	576,000	93,705,000	170,422,000
Anglicans	26,863,000	707,000	32,956,000	1,300,000	7,338,000	5,719,000	400	74,883,400
Other Christians	61,553,000	75,930,000	8,058,000	9,861,000	32,324,000	607,000	100,600	188,433,600

Source: 1993 Encyclopedia Britannica Book of the Year (Chicago: Encyclopedia Britannica Inc., 1993).

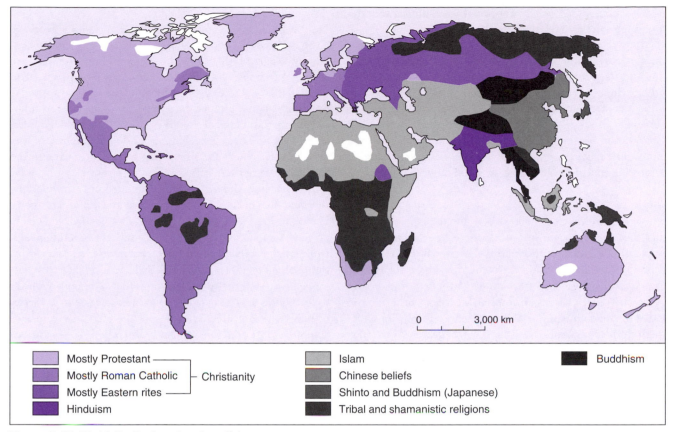

Figure 7.15 World distribution of major religions.

Some countries have actively rejected religion in favour of a political belief, such as communism, as an appropriate guide to beliefs and behaviour.

Despite such examples of the declining relevance of religion, in reality most humans are unable to separate themselves from religion, some because they choose to be actively religious and others because religion is typically one part of state identity. Indeed, for many people, religion is the basis of life and their religion is regarded as the one true religion. Not only do religions divide our world, they also often encourage people to behave in what may be seen, from a detached perspective, to be inappropriate behaviour; recall the views of James outlined at the beginning of this chapter.

Our human tendency to identify with a specific religion has prompted much military activity such as the Crusades and the European religious wars of the sixteenth and seventeenth centuries. Recent and ongoing conflicts in Pakistan (Islam/Hindu), Lebanon (Christianity/Islam), and Northern Ireland (Catholic/Protestant versions of Christianity) are but three examples. Such conflicts may simply use religious differences as an excuse or an easy label for other differences. But there is no doubt that, in many instances, religion promotes hostility to non-believers, a hostility that, combined with a general lack of understanding, leads to mistrust and conflict.

In some cases, religion is an even more potent unifying force than language for a group. Even more so than language, religion can resist outside influence. The North American experience involving the in-movement of many and varied groups suggests that religion is often the most lasting feature of a culture, retained long after language is lost.

RELIGIOUS LANDSCAPES

Religion and landscape are often inextricably interwoven. There are three principal reasons why this is so. First, many religions incorporate beliefs about and attitudes towards the physical environment — nature — and the human environment. Second, religion often influences land use. Third, many religions explicitly choose to display their identity in landscape.

Religious perspectives on humans and nature

A principal function of many religions is to serve as an intermediary between humans and nature, although the type of relationship favoured varies. Judaism and Christianity place God above humans but humans above nature. Hence, these religions incorporate an attitude of human dominance over the physical environment, an attitude that is reflected in numerous ways. Christians in particular see themselves as fulfilling an obligation to tame and control the land. An alternative type of relationship sees humans as a part of, not apart from, nature, both therefore having equivalent status under God. Such a view is generally characteristic of eastern religions, especially Taoism. The result is a very different human/land relationship (Box 7.7).

Many religions ascribe a special status to certain types of physical environment. Rivers, such as the Ganges for Hindus and the Jordan for Christians, are sacred places, while other religions similarly value mountains, as with Mount Fujiyama in Shintoism. Human environments may also achieve sacred status. Mecca for Islam and Lourdes for Catholicism are but two major examples. More generally, almost any religious addition to landscape is sacred whether it is a church, cemetery, shrine, or some other feature.

Religion and use of land

There are also religious beliefs about the use of land, plants, and animals. In some instances, these beliefs have major impacts on regional economies. Pigs are a common domesticated animal in Christian areas, but are absent in Islamic and Jewish areas. The traditional Catholic avoidance of meat one day per week prompted European fishermen to move to Newfoundland waters long before European settlement in the New World. Hindus regard cows as sacred (Box 7.8). These few brief examples merely scratch the surface of what is a very varied and complex issue.

Religion, then, influences human evaluation of space. Different religions incorporate different beliefs and attitudes and what is important to one group may not be important to another. When different groups value similar places, as in the case of sites in Jerusalem regarded as sacred by Judaism, Islam, and Christianity, the results are often unfortunate. When religion affects the way we use land, it can be a powerful cultural factor operating against economic logic.

Religious symbols in landscape

Landscape is also a natural repository for religious creations, a vehicle for displaying religion. Sacred structures, in particular, are a part of the visible landscape. Hindu temples are intended to house gods, not large numbers of people, and are designed accordingly; Buddhist temples serve a similar function. Islamic temples (mosques) are built to accommodate large numbers of people, as are Christian temples, cathedrals, and churches. The size and elaborateness of any temple often reflect the prosperity of the local area at the time of construction.

In Calcutta, Hindu temple architecture traditionally reflected the belief that temples are the homes of gods. Hence, because mountains are the traditional dwelling places of gods, temple towers were built to reflect mountain peaks, while small rooms inside the temples resembled caverns. In the late eighteenth century, this style was replaced by flat-topped two- or three-storey buildings that were built adjacent to the homes of the very rich. Finally, in this century, a series of new types appeared as a result of the pressures of urbanization and related institutional processes. These various types are described in detail by Biswas (1984) and are a clear example of the role played by changing power relations.

Sacred structures are important religious and tourist locations. The Golden Temple at Amritsar, India, is sacred to Sikhs but also attracts many others, as do the Taj Mahal, an Islamic mausoleum near Agra, India, and Westminster Abbey in London.

Places to house gods or to gather for worship are typical features of most landscapes. In many Christian communities, the church is a religious and social centre serving many

---

Box 7.7: The Mithila cultural region

An excellent discussion of the landscape implications of the human/land relationship implicit in Hinduism is contained in an article by Karan (1984) discussing the folk art of the Mithila region in north India, between the Ganges and the Himalayan foothills. Explaining the distinctive folk art requires an account of the people and the place they have created.

The Mithila region is an alluvial plain and is settled by some 15 million people at the high density of 722 per square kilometre. The population speak Maithili and are Vedic Hindu. Religion is central to all life, the central tenet being the unity of humans and all nature. The Maithilis believe in awareness of life as manifested in all things. Plants, animals, clouds, the sky, and water share the life-force equally with humans. This concept of oneness of all things thus affects attitudes to land and to others. The key social unit is the extended family and ultimate responsibility is to the group, not to oneself.

Cultural landscapes in the region reflect religious attitudes. Buildings are of local clays so that human impact is minimized.

The rice fields in the alluvial plain blend imperceptibly with the physical environment. This is not a landscape made by humans but rather one that humans share with all other aspects of nature. The folk art of the people is a profound expression of their being a part of nature; the colours used are made from natural materials, while the subjects of paintings are largely religious. The vast scope and seeming timelessness of the forces of nature are also fundamental themes in the folk art of the region.

In addition to providing an example of a type of human/land relationship that is atypical in a Christian area, this brief discussion also highlights the relevance of folk art as a part of a group's cultural record. This is relatively new territory for the human geographer.

Source: P. Karan, 'Landscape, Religion and Folk Art in Mithila: An Indian Cultural Region', *Journal of Cultural Geography* 5 (1984):85–102.

extrareligious functions. Some religious groups actively reject such external expressions of religion as churches and create landscapes devoid of religious expression (Box 7.9). Such cases are unusual.

Other ways in which religion creates a distinctive landscape include the construction of shrines (usually along routes) and the use of land for burying the dead. Hindus and Buddhists cremate the dead, while Muslims and Christians typically bury the dead, a practice that clearly requires much space.

Box 7.8: Religion and irrational lifestyles

Why do Hindus refuse to eat beef even when no other 'food' is available to them? The simple answer is that cows are sacred to Hindus because they are a symbol of everything alive, the mother of life. But why is this so? Is the sacred cow a belief that leads to irrational economic behaviour or is there a logical reason for the sacred cow?

To western eyes, the veneration of cows is nonsensical. Stray animals invade private property, government agencies provide homes for old cows, and when the Indian constitution was prepared, it included a bill of rights for cows — to western eyes an inappropriate economic state of affairs.

A careful analysis of the larger economic system in India leads to a different conclusion. As Harris (1974) has argued, there is a shortage of oxen, which are used as draft animals, and cows give birth to oxen. Oxen, and hence cows, are critical substitutes for tractors. Cows also provide dung used as fuel and as a household flooring material. In short, cows are an invaluable component in Indian agriculture and life. Thus, Hindu India makes very efficient use of cows. The fact that it does not appear that way to the Christian economist in the developed world does not alter the facts. Hindus venerate cows because cows are so important. They are not killed because such killing would, in the long run, disturb the larger pattern of Indian agriculture.

This is one example. Other seemingly perplexing questions can also be logically explained. Cultural riddles and irrational behaviours associated with particular religions do make sense once we place then in their larger context.

Box 7.9: Religious landscapes: Mormons and Doukhobors

Most landscapes include evidence of religious occupation. Places of worship are especially evident even in an increasingly secular developed world, but religious impact on landscape is not always quite so visible.

In the American and Canadian west, areas of Mormon settlement are distinctively different from other landscapes because of the control exercised by the church over many aspects of human activity. Thus, these Mormon landscapes are characterized by wide streets, roadside irrigation ditches, barns and granaries in towns, unpainted farm buildings, open fields around town, hay derricks, 'Mormon' fences, distinct architec-

tural use of brick, and Mormon ward chapels (Francaviglia 1978). The Mormons were able to impose these landscape features because they were the first effective settlers in their areas and because they are a cohesive, highly centralized group. These landscape features reflect the aims of the leaders rather than any particular ideology; most are pragmatic.

In some parts of western Canada, a very different situation prevails. The Doukhobors, like the Mormons, are a Christian sect, but with a very different set of core beliefs. One belief involves rejection of the 'externalities' of religion and thus there are no houses of worship, no crosses or spires in the landscape. In short, the Doukhobor areas are devoid of religious symbolism. A second belief involves the equality of life, which has led to a communistic form of settlement with all work and financial matters being group, not individual, focused. Communal living arrangements required the construction of distinct double houses that accommodate up to 100 people (Gale and Koroscil 1977).

These two examples emphasize the diversity of religious expression in landscape. Typical western landscapes show evidence of religious symbolism and beliefs (houses of worship, cemeteries, roadside shrines) while other areas, settled by distinctive religious groups, result in landscapes that are significant reflections of group organization. Such group landscapes may (Mormons) or may not (Doukhobors) include symbolic features.

7.3 Doukhobor women. (T. Wakayama)

SUMMARY

Our divided world

Our world is divided physically and, more crucially, culturally. Culture refers to our ability to knowingly change physical environments in directions suggested by experience.

Culture and society

Both are difficult terms to define. Culture can be divided into non-material and material; geographers study both categories. Society refers to a cluster of institutionalized ways of doing things. Society thus includes non-material culture such as language and religion. North American geographers have typically studied culture, not society, while British and European geographers have focused on society. This and the following chapter reflect the need to integrate both concepts and deal respectively with the geographical expression and spatial constitution of culture.

Human scale of analysis

Geographers can study humans at any scale from all humans to single individuals. It is typical, however, to use a cultural scale, a group of people with some recognizably common set of operating rules, hence geographers study language groups and religious groups.

Cultural evolution

Preagricultural groups probably numbered ten to thirty individuals and used language, fire, and tools. There was limited cultural variation prior to the development of agriculture except that tied to the physical environment. Agriculture and civilization were closely related; possible causes of both include irrigation, social change, class conflict, population pressure, and the availability of leisure time. Once in place, agriculture is typically permanent, whereas civilization has not proved to be. Civilizations collapse as a result of natural disasters or conflict.

Cultural regions

Regions can be delimited on various scales. World regions are useful overviews, but lack precision. North America can be usefully regionalized using the concept of first effective settlement. Europe is usefully defined by using specific traits.

Cultural landscapes

Cultural regions have distinct cultural landscapes because of the impact of culture on land and the varying human/land relationships. Cultural adaptation involves cultural changes in response to environmental and cultural challenges. Geographers have studied cultural adaptation using a variety of concepts, including those of culturally habituated predisposition, cultural preadaptation, duplication, deviation, fusion, and the core, domain, and sphere model.

Languages

Language is probably the single most important human achievement. In terms of understanding human groups, their attitudes, beliefs, and behaviours, language is of prime consideration. Indeed, it is not far-fetched to assert that language is the most appropriate single surrogate for culture.

The basis for group communication, language is the earliest source of group unity and the means by which cultures continue through time. For many groups, language is culture. Today there are perhaps 3,000 languages in the world. Individual languages can be grouped into families — languages that share a common origin. The numerically largest language family is Indo-European. Mandarin has more speakers than any other language, but English is the most wide-spread and the nearest to a world language.

Language as identity

Languages create barriers between groups and facilitate group identity. For many groups, language is the principal basis for a national identity. Multilingual states are characteristically less stable than unilingual states.

Language and landscape

Place names — toponyms — are the clearest expression of language on landscape. Place name studies help us understand early settlement. Language also helps make landscape; a place may become what it is named — as in the case of the Great American Desert. Both physical and human landscapes are in language; languages incorporate words for important physical features and language reflects class and gender distinctions.

Religions

For many, religion is the basis of life. Thus, religion is a useful variable for regionalizing and may be more potent than language for reinforcing group identity. Religions vary greatly in their ability to affect people and promote group cohesiveness.

Religion and landscape

Many religions function as an intermediary between humans and nature. Often, specific physical environments are ascribed a special status. There are also religious beliefs concerning the use of land that can affect regional economies. Landscape is also a natural vehicle for displaying religion.

WRITINGS TO PERUSE

BRONOWSKI, J. 1973. *The Ascent of Man*. London: BBC.
An outstanding study of cultural evolution from prehistory to the present.

BUTLIN, R.A. 1993. *Historical Geography: Through the Gates of Space and Time*. New York: Arnold.

A survey of the many different approaches to reconstructing the geographies of the past.

EDWARDS, J. 1985. *Language, Society and Identity*. Oxford: Blackwell.

A non-geographic study of the role played by language in larger cultural issues.

FRANCAVIGLIA, R.V. 1978. *The Mormon Landscape*. New York: AMS Press

An unusually detailed description of one of the most visually distinctive cultural landscapes in North America.

GASTIL, R.D. 1975. *Cultural Regions of the United States*. Seattle: University of Washington Press.

A thoughtful regionalization that can be interestingly compared to that of Zelinsky.

LEIGHLY, J. 1978. 'Town Names of Colonial New England in the West'. *Annals, Association of American Geographers* 68:233–48.

One example of a study of toponyms used to analyse settlement history.

MEINIG, D.W. 1969. *Imperial Texas: An Interpretive Essay in Cultural Geography*. Austin: University of Texas Press.

One example of several fine regional studies by this author. Includes several original approaches to region and landscape analysis.

PARK, C. 1994. *Sacred Worlds: An Introduction to Geography and Religion*. New York: Routledge.

Focuses on the spatial distribution of religion, the processes by which religion and religious ideas spread through space and time, and the visible manifestations of religion in the cultural landscape.

ROONEY JR, J.R., W. ZELINSKY, and D.R. LOUDON, eds. 1982. *This Remarkable Continent: An Atlas of United States and Canadian Society and Cultures*. College Station, Texas: Texas A and M University Press.

An original attempt to map and discuss a wide variety of cultural traits.

SALTER, C., ed. 1971. *The Cultural Landscape*. Belmont: Duxbury.

A collection of readings organized in a novel and stimulating fashion.

SHORTRIDGE, J.R. 1977. 'A New Regionalization of American Religion'. *Journal of the Scientific Study of Religion* 16:143–53.

A comprehensive statistical regionalization of religious differences.

SOPHER, D.E. 1967. *Geography of Religions*. Englewood Cliffs: Prentice-Hall.

The first book-length study of this topic, full of useful ideas and facts.

SPENCER, J.E. 1978. 'The Growth of Cultural Geography'. *American Behavioral Scientist* 22:79–92.

One of many overviews, but especially useful to the new student of geography.

WAGNER, P.L. 1974. 'Cultural Landscapes and Regions: Aspects of Communication'. *Geoscience and Man* 5:133–42.

An early statement regarding the importance of communication to our understanding of culture.

ZELINSKY, W., and C.H. WILLIAMS. 1988. 'The Mapping of Language in North America and the British Isles'. *Progress in Human Geography* 12:337–68.

A detailed review article.

8

Cultures: Symbolic and social landscapes

SOME OF THE PREVIOUS CHAPTER clearly contained material that had conceptual, social, or symbolic overtones. Most notably, the discussions pertaining to cultural adaptation provided examples of conceptual work, while the accounts of linguistic and religious identity emphasized a view of landscapes as places. Overall, however, such content was not distinguished by being theoretically informed. In this chapter the interest in cultural landscapes is continued, but the thrust is now strongly theoretical, not empirical; in other words, the underlying philosophy is typically something other than the empiricism that dominated Chapter 7. A useful way to smooth the transition from the empiricism of the previous chapter to the social theory introduced in this chapter is to consider the concept of vernacular regions.

VERNACULAR REGIONS

The cultural regions discussed in the previous chapter were typically examples of formal regions, areas with one or more cultural traits in common; they were the geographic expressions of culture. The concept of a vernacular region is rather different for it is a region that is perceived to exist by those living in the region and/or by those living elsewhere; such regions introduce the idea of the spatial constitution of culture. Clearly, there is overlap. A formal region may also be a vernacular region. The Mormon landscape can be defined formally (Francaviglia 1978) but it is also generally perceived to exist as a distinct entity by both Mormons and others.

CREATING A VERNACULAR REGION

In other areas, the link is not so clear. The area of French Louisiana, which is the only remaining remnant of the vast French Mississippi valley empire, is a distinctive vernacular region, but, according to Trépanier (1991), this perception is not an accurate reflection of cultural reality. The area has a French population that, although diverse in origin, is now associated with one group identity, namely Cajun. There are in fact four important French subcultures in the area; White Creoles, Black Creoles, French-speaking Indians, and the descendants of the Acadians who fled Nova Scotia after 1755 (now known as Cajuns). Furthermore, until recently, the Cajun group identity was perceived in largely negative terms by others and it is only since the late 1960s that a process of image modification, what Trépanier (1991:161) called 'beautification', has been taking place. The Creole identity, on the other hand, carried a positive image for both Black and White French speakers. How are we to explain these two anomalous developments?

Identification of the area with the Cajun group and beautification of the term 'Cajun' have both resulted from decisions made by government. In 1968, the Louisiana legislature created the Council for the Development of French in Louisiana (CODOFIL) to gain political benefits by cultivating a French image, but this did not garner any real popular support. Once created, CODOFIL was determined to unify French Louisiana using the then-negative Cajun label rather than the more positive Creole label. Trépanier (1991:164) interpreted this as a way to guarantee a White identity for the French-speaking area. Once this decision was made, it became necessary to improve the popular image of the Cajun identity, which was achieved through publicity campaigns, including the organization of Cajun festivals. The fact that a new governor was elected under the Cajun banner helped enormously.

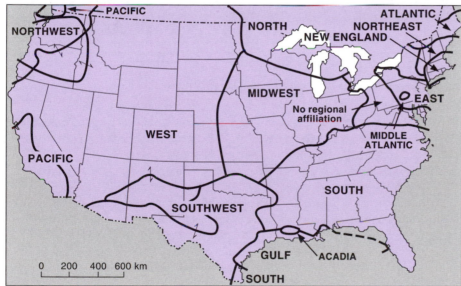

8.1 *(above)* Northern New England farming village, Vermont, USA. (Vermont Department of Travel and Tourism)

Figure 8.1 *(left)* North American vernacular regions.
Source: W. Zelinsky, 'North America's Vernacular Regions', *Annals, Association of American Geographers* 70 (1980):14.

For many, French Louisiana now exists as a vernacular region with a Cajun identity, a situation that ignores the variety of French-speaking people and the fact that both the Black Creoles and the French-speaking Indians reject the Cajun identity. As this example vividly illustrates, vernacular regions are in fact much more than areas perceived to possess regional characteristics; they are also regions to which specific meanings and values are attached — they have social and symbolic identity.

DELIMITING VERNACULAR REGIONS

North American geographers have paid much attention to the task of delimiting vernacular regions and have typically done so by collecting information on individual perceptions. Probably the most elaborate survey was conducted by Hale (1984); it involved 6,800 responses from such people as local newspaper editors. All such endeavours have to ensure that the responses received in some way represent those of average people, given the aim of identifying vernacular regions.

Zelinsky (1980) avoided this problem by studying the frequency of metropolitan business use of a specific regional term and a more general national term; for example, if the region being delimited is that of the American south, the incidence of the term 'Dixie' might be compared to the term 'American'. Figure 8.1 depicts the resulting fourteen vernacular regions, some areas that lack regional affiliation, and two areas that belong to either one or both of the two regions. It is most instructive to compare this map to that of cultural regions based on the concept of first effective settlement; the south, middle west, and New England are similarly located on both maps. Clearly the culture region and the vernacular region are related to similar underlying factors. From a Canadian perspective, it is notable that the western region extends deep into the prairie provinces.

VERNACULAR REGIONS AND SENSE OF PLACE

If an area has an identity such that it is named, then this clearly suggests that the identity is meaningful to those using the name. Expressed another way, vernacular regions are not simply locations, they are places. The name of the region conveys a meaning, or possibly more than one meaning. There are parts of the world that possess a powerful regional identity such that the mere mention of the name conjures up vivid mental images. Most North Americans, regardless of religious affiliation, recognize that the term 'Holy Land' refers to the land bordering on the western Mediterranean. Other places have one meaning for one group and a quite different meaning for another; for lovers of country music, Nashville is likely to be a meaningful place, while for others it may be little more than another American city.

Indeed, vernacular regions are often viewed more positively by those living within the region than by those outside. For many living in the American Bible belt, the name is a source of great pride, while for some outsiders, it is employed as a term of derision. In other cases, the name of the vernacular region may have little meaning and limited identity as a place to both residents and outsiders as it is merely an institutional creation employed for tourism or promotional reasons, but with no real roots in the region.

PSYCHOGEOGRAPHY

The term 'psychogeography' is used to refer to 'how people feel about, experience, paint themselves into the world and take that portrait back into themselves as literal parts of who they are and hence their well-being' (Stein and Thompson 1992:63). The aim in a psychogeographic study is to identify an internal self-image of a region rather than an outside image as is usually considered in studies of vernacular regions, although there is considerable overlap between the two (Box 8.1).

HOMELANDS

Another way to think about psychogeography, vernacular regions that are clearly perceived by those living within, is by means of the concept of a **homeland**. According to

---

Box 8.1: Psychogeography — the sense of Oklahomaness

Understanding a regional cultural landscape requires consideration of regional identity, which includes popular attitudes that have grown through long experience in a landscape. To determine regional self-awareness, Stein and Thompson (1992:65) used the state of Oklahoma 'as an example of a community of meaning within a politically-defined territory', what they call a cultural identity system. '"Oklahomaness" connotes what is distinctively Oklahoma; that is the boundaries of the identity and the contents within it' such that 'Oklahoma is first and foremost a state of mind which springs from a common pool of self images, a community of meanings which lends the state much of its regional character' (Stein and Thompson 1992:66).

What are these various self-images, the popular attitudes that shape the identity of Oklahoma today? First, there are images associated with Native Americans: of the Trail of Tears resulting from the forced migration of the Five Civilized Tribes from the southeastern United States to Oklahoma; and of reservation lands taken from them by the United States government to give to White Americans. Second, Oklahoma was the scene of five famous land rushes to the 'unassigned' land taken from the Native Americans (1889–1901); from these land rushes emerged an image of unbridled enthusiasm and opportunism as suggested by the labels 'Boomers' and 'Sooners'. These land rushes result in Oklahoma occupying a special

place in the American psyche; it is no accident that the chosen location and title of a musical show and film is 'Oklahoma!' and that it was so successful on Broadway from 1943 to 1948 at a time of exuberant American nationalism.

A third component of the regional identity is the tragic and haunting image of the 'Dust Bowl' of the 1930s, associated with the folk songs of Woody Guthrie and so movingly described in John Steinbeck's novel, *The Grapes of Wrath*. Cowboys, horses, cattle, and open spaces combine to produce a fourth image, one that today implies an 'ideological statement of moral superiority over those with whom Oklahomans have compared themselves and felt inferior' (Stein and Thompson 1992:73). Both of these images are linked to a fifth — the symbolism of land and sky.

Together, these images, along with a number of others (such as the football mentality), are central to an appreciation of the state of Oklahoma as a meaningful cultural identity system. There are boundaries delimiting an area that has real meaning for those inside. The images are not limited to a single group of people and are in fact closely tied to relations between those inside and those outside the region. Indeed, the identity of Oklahoma is continually changing as relationships with others change.

Nostrand and Estaville (1993:1), there are four basic ingredients of a homeland: people, place, sense of place, and control of place. Sense of place refers to the emotional feelings of attachment that people have for a place, while control of place refers to the requirement of a sufficient population to allow a group to claim an area as their homeland. Most homelands are associated with groups defined on the basis of common language, religion, or ethnicity.

RETHINKING CULTURE: A SYMBOLIC VIEW

In Chapter 7 we restricted discussion to only one view of culture, albeit an important one with strong roots in anthropology. The discussions so far in this chapter of vernacular regions, psychogeography, and homelands are clearly moving in a rather different direction. There are other interpretations of culture that recently have influenced human geography and that are in general accord with the type of empirical work described earlier.

With the rise of humanistic geography beginning in the 1970s, a symbolic interpretation of culture became prominent. This view broadens the culture concept to embrace more fully non-material culture and to emphasize that individuals create groups via communication; the focus is on a broader view of culture that allows human geographers to consider topics other than landscape, specifically topics that fall under the general rubric of what we have labelled the spatial constitution of culture. Jackson and Smith wrote:

> Culture in the sense of a system of shared meanings, is dynamic and negotiable, not fixed or immutable. Moreover the emergent qualities of culture often have a spatial character, not merely because proximity can encourage communication and the sharing of individual life worlds, but also because, from an interactionist perspective, social groups may actively create a sense of place, investing the material environment with symbolic qualities such that the very fabric of landscape is permeated by, and caught up in, the active social world (Jackson and Smith 1984:209).

Such an interpretation of culture was anticipated by Wagner (1975:11) when he wrote, 'The fact is that culture has to be seen as carried in specific, located, purposeful, rule-following and rule-making groups of people communicating and interacting with one another.'

These observations by geographers are in accord with a major school of thought in sociology, **symbolic interactionism**. This refers to a group of closely related theories derived from the ideas of the American philosopher, Mead (1934). Essentially, it is argued that humans learn the meanings of things as a result of their social interactions and, further, that human behaviour arises from reactions to perceived environments. Interactions with other people provide us with meanings for things that we are then able to use to understand those things. Once we have an understanding, we are able to define situations that we encounter and act accordingly.

With this rather different view of culture in mind, a view that is clearly in accord with the opening discussions in this chapter, we can now evaluate different approaches to the subject matter of this chapter. We include brief comments on two of the philosophies introduced in Chapter 2 — Marxism and humanism — and we introduce three additional sets of ideas: structuration, postmodernism, and feminist theory.

SOCIAL THEORY

DIVERSITY OF CURRENT APPROACHES

Social theory is not the property of any one discipline for questions about ways of life and about human behaviour are asked in all of the social sciences. Not surprisingly, there has long been a wide range of social theory, although during the middle part of this century, the social sciences were dominated by a positivistic interpretation based on theoretical ideas dominant in the physical sciences. This view centred on the creation of theory, the testing of hypotheses, and the creation of laws (see the discussion of positivism in Chapter 2). Questions of interpretation were not considered.

There have, however, been dramatic changes in our theoretical preferences since the late 1960s. The weakening of the dominant influence exerted by positivism has resulted in the flowering of a multitude of alternative approaches. Other versions of social theory that have come to the forefront are linked by a shared rejection of the claim that social science can be value neutral; by a shared rejection of the claim that a theory/hypothesis/law procedure is the highest level of explanation; and by the shared belief that social science is an interpretative endeavour such that questions of meaning and communication are relevant.

Many of the approaches that came forward with the partial decline of the previously dominant positivism trace their intellectual origins to earlier writers; for example, Marxism to Marx and humanism, in the form of phenomenology, to Schutz. Other approaches are more clearly of recent origin, notably structuration and postmodernism.

The diversity of current approaches generates two divergent responses among practising social scientists. Some contend that if social theorists are unable to agree among themselves, then what possible use is social theory for those actively engaged in research? Others see the matter differently, arguing that a proliferation of theoretical approaches is an invaluable way to avoid the dogmatism that results from a single dominant approach. This is an important matter to give thought to, but not one for which there is a single correct answer.

HUMANISM AND MARXISM REVISITED

One of the interests of contemporary human geography is with the social significance of space and place. From a humanistic perspective, there are concerns with the elements that combine to produce a sense of place, with the symbolism of landscape, and with **iconography**. From a more Marxist perspective, there is a concern with social inequality related to such variables as language, religion, ethnicity, class, gender, and sexuality, especially as they are reflected in space. Both groups recognize that space is important in the constitution of social life. But human geographers have not restricted their theoretical interests and today structuration, postmodernism, and feminism are proving to be important additions to our body of available approaches.

STRUCTURATION

The sociologist Anthony Giddens has developed a social theory that is concerned with the links between human agents (people) and the social structures within which they function. **Structuration** theory focuses on the capacities that permit people to institute, maintain, and alter social life. It is a body of ideas derived from numerous earlier contributions to social theory, including Marxism, and it has evolved in several publications — for example, Giddens (1984).

Individuals are viewed as agents operating within the contexts of local social systems, sometimes called **locales**, and the larger social structures of which they are a part. Capitalism is one example of a larger social structure, being a series of rules created by humans to facilitate human survival. But perhaps the key characteristic of structuration is the identification of the dualities associated with social structure and human agency. First, social structure enables human behaviour while at the same time behaviour can also influence and reconstitute culture. Second, the rules of any social structure are both constraining and enabling — they are constraining because they limit the actions available to individuals, but they are also enabling as the rules do not determine behaviour. There is an intriguing conceptual parallel here with environmental determinism.

Human geographers have responded favourably to the fundamental logic of structuration theory (see Gregson 1986). This is because Giddens has explicitly incorporated human geographic ideas in structuration theory and because it has been persuasively argued by a number of influential human geographers who have made significant theoretical and empirical contributions — for example, Gregory (1981). The extent to which the theory will have a substantial and long-lasting impact on human geography, however, remains to be seen. Box 8.2 provides one example of a structuration approach to a human geographic problem.

POSTMODERNISM

Few contemporary human geographers would disagree with the statement that **postmodernism** is an especially difficult body of ideas to understand. This is at least partly because postmodernism is, by definition, neither structured nor unambiguous and because there are multiple versions of postmodernism. Further, postmodernism is best represented in disciplines other than human geography, most notably in architecture, literature, and other expressions of culture. Despite these very real difficulties, it is important to attempt to convey the essence of postmodernism as it is assuming an increasingly important role in human geographic research.

Modernism

As the name implies, postmodernism is a reaction to **modernism**, a general term in culture history that refers to a varied set of mid-nineteenth century onwards breaks with earlier traditions. Modernism developed most fully in art and architecture, but is also full of implications for social science methodology. It assumes that reality can be studied objectively and validly represented by theories (as in positivism), and that scientific knowledge is practical and desirable. Modernism is linked to the rise of capitalism and the Industrial Revolution. It centres around traditionally liberal themes such as the rationality of humans, the privileged position of science, human control over the physical environment, the inevitability of human progress, and a search for universal truths.

The alternative of postmodernism

Postmodernism rejects the assumptions of modernism: reality cannot be studied objectively because it is based on language and therefore needs to be thought of as a **text**; further, all aspects of this text are related (it is intertextual) and therefore it is not possible for it to be accurately represented. Taken to the extreme, this means that truth is relative and, for practical purposes, non-existent. Causality does not exist and theory construction has no meaning.

So how does a postmodernist approach work? The emphasis is on the **deconstruction** of texts and the construction of narratives that do not make claims about truthfulness; often these narratives will focus on differences, uniqueness, irrationality, and marginal populations. Deconstruction questions authority, the established readings of a text, and the highlighting of alternative readings. Overall, postmodernism considers modernist claims to be arrogant, even authoritarian. 'Postmodernism and deconstruction question the implicit or explicit rationality of all academic discourse' (Dear 1988:271).

One of the principal attractions of postmodernism for contemporary human geography is the emphasis on cultural otherness; it is open to previously repressed experiences such as those of women, homosexuals, and those lacking power and authority generally. Later sections of this chapter embrace this diversity of experience.

Diverse postmodernisms

Not all postmodernism is quite as described above. Indeed, the concept varies considerably between disciplines and

Box 8.2: An application of structuration theory

It is noteworthy that the aims of the research summarized in this box are not significantly different to much previous work. The concern is once again with region and place, which we know are central to human geography. Indeed, there is a concern with the changing landscape and the changing character of place, which are central concerns of much traditional cultural and historical geography. What is different in this work is the basic position taken, which is derived from structuration theory; this means that the language used to express ideas and the concepts employed are different from those typically employed.

The key idea is that place is a human product: 'it always involves an appropriation and transformation of space and nature that is inseparable from the reproduction and transformation of society in time and space' (Pred 1985:337). The result of applying structuration concepts to the Swedish province of Skåne is, according to Pred, a greater understanding of people and place. Unlike more traditional approaches, structuration allows the researcher to discern the multitude of meaningful local variations in the evolution of places as these result from local variations in, for example, the timing of land enclosure, agricultural production, diets, and language.

The central concern is with the 'becoming' of a place and with place as a 'historically contingent process'. These terms are most easily explained by reference to Figure 8.2, which shows that:

> any place or region expresses a process whereby the reproduction of social and cultural forms, the formation of biographies, and the transformation of nature and space ceaselessly become one another at the same time that power relations and time-specific path-project intersections continuously become one another in ways that are not subject to universal laws, but vary with historical circumstances (Pred 1985:344).

There are seven propositions that further clarify how structuration theory is used in this example. These propositions are general in character and not limited to the specific region studied by Pred (1985).

1. Structuration is a continuous process; there are in fact the two interconnecting processes of social reproduction (structure) and individual socialization (agency).
2. What occurs in any given place and the meanings attached to that place are tied to the structuration process in item 1 above. Indeed, they are tied to the structuration process in that place and elsewhere.
3. Power relations are central to the social structure of a place as it is becoming. In accord with the idea of duality of structure, however, they can be transformed.
4. Power relations also influence what people know and perceive.
5. The manner in which power relations are tied to the becoming of a place depends on the degree to which local institutions are controlled locally or non-locally.
6. The becoming of a place may be dominated by institutional projects relating to production and distribution as these imply spatial and social divisions of labour.
7. The becoming of a place is also tied to individual biographies.

Notwithstanding the success of this and some other applications of structuration, relatively few authors have applied structuration as a research methodology, not least because the theory itself is complex. Indeed, in recent years Pred has favoured alternative conceptual schema to structuration, notably aspects of postmodernism — see, for example, Pred and Watts (1992).

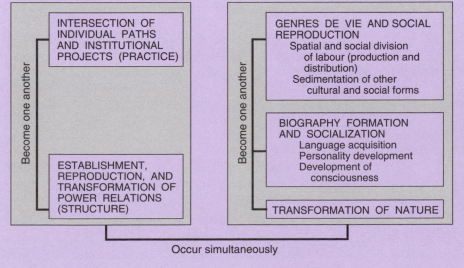

Figure 8.2 Components of place as historically contingent process.
Source: A. Pred, 'The Social Becomes the Spatial, the Spatial Becomes the Social: Enclosures, Social Change and the Becoming of Place in the Swedish Province of Skåne', in *Social Relations and Spatial Structures*, edited by D. Gregory and J. Urry (London: Macmillan, 1985):336–75.

even within disciplines. Some postmodernists do make concessions to modernity by becoming involved in social movements or by breaking down the barriers between researchers and subjects so that people are allowed to speak for themselves. These versions are relatively close to some earlier concepts of culture, such as symbolic interactionism, and some other social theories, such as humanism. There is then no one clear and unequivocal version of postmodernism in human geography, which is not surprising given that the central message of the concept is one that acknowledges, indeed welcomes, diversity. For two critical reactions to postmodernism in general, see boxes 8.3 and 8.4.

FEMINIST THEORY

There is no one body of feminist theory; rather, there are several sets of ideas with each tied to some larger body of theoretical thought, such as liberalism, Marxism, socialism, or postmodernism. Regardless, all such ideas are united by a common concern.

Women are systematically disadvantaged in most areas of twentieth-century life and **feminism**, regardless of the specific theoretical bent, is the advocacy of equal rights for women and men, along with a commitment to improve the social status of women. There are several explanations for the inequality, most notably the prevalence of societies characterized by **patriarchy**: men dominate women through various social relations. The *household* is the most analysed institution, but the wide variety of types makes generalizations difficult. *Employment* differences include the lower wages paid to women and their lesser involvement in paid labour. *Culture* is seen as a key factor explaining the construction of **gender** differences through socialization processes. *Sexuality and violence* are both seen as forms of social control over women. Finally, the *state* is seen as typically reinforcing traditional households and failing to intervene in cases of violence against women.

As suggested, there are several traditions of feminist thought and action; three are noted. The oldest tradition,

---

Box 8.3: A positivist reaction to postmodernism

It is hardly surprising that an approach such as postmodernism, which effectively undermines much previous academic work because of its insistence that there are multiple readings of any reality, will be subject to opposing views, nor is it surprising that human geographers who have worked primarily in a positivist tradition will be at the forefront of such criticism. In this box, the focus is on concerns expressed by Brian Berry, one of North America's most distinguished human geographers and, according to the Social Science Citation Index, the world's most frequently cited geographer since the 1960s.

In the course of reviewing a recently published book edited by Abler, Marcus, and Olson (1992) and titled, *Geography's Inner Worlds: Pervasive Themes in Contemporary American Geography*, Berry (1992) noted that a postmodern emphasis was advocated by several chapter authors and took exception to the implications of such an emphasis for human geography.

Berry expressed specific concern with the implications of the following two claims:

1. 'While scientific method remains accepted in physical geography, critical social theorists have raised challenges to the use of formal modeling in human geography ... Human geographers continue to be engaged in lively epistemological debate in which there is little consensus except for considerable negativism towards logical positivism' (quoted in Berry 1992:491–2).
2. 'Contemporary human geography reflects ... a postparadigm condition in which disciplinary practices and concepts appear, for good or bad, to have broken loose from any notion of disciplinary closure and unity' (quoted in Berry 1992:492).

Most importantly, however, Berry identified a fundamental contradiction between, on the one hand, the claim made by the editors of the book that geography needs to be a 'coherent, synthetic, global discipline focused on human use of the earth' (quoted in Berry 1992:493) and, on the other hand, the postmodern diversity contained in the book and which the editors applaud. For Berry (1992:493), the two claims are incompatible:

The success of *disciplines* is that they think in ways *disciplined* by theory; for them, paradigm shifts occur because older theories are found wanting and are pushed aside by theories that offer better explanation. But postmodernism carries with it a New Age social vision that embodies not only ideas of holism (the interdependence of all systems, spiritual as well as material) and earth awareness (the interdependence of all things on earth, including humankind) but also a particular view of human rights (the rights of all individuals to chase their *sadhana* and live transformative lives) that, in academic practice, translates into a concept of inner meaning (geographer's inner worlds) that is fundamentally antitheoretic ... If this be geography's *karma* so be it, but the implications are not pleasant. Detached from the world of science, a generation of geographers will, like the regionalists, have little accumulative wisdom to share with others, simply disparate fragments — some insightful, some beautiful, but none coming together into the whole cloth of theories that provide order to inquiry and help frame practice.

dating back to the late eighteenth century, is one of democratic, *liberal feminism* aiming to obtain equal rights and opportunities for women. Two more recent traditions developed in the 1970s, arguing that the oppression of women cannot be corrected by superficial change because it is embedded in deep psychic and cultural processes that need to be fundamentally changed. *Radical feminism* contends that gender differentiation results from gender inequality with the subordination of women being separate from other forms of social inequality, such as those based on class. *Socialist feminism* similarly emphasizes gender inequality, but links that inequality to class; what this means is that men and capital both benefit from the subordination of women.

TYPES OF SOCIETY

Our contemporary cultures reflect a long evolutionary process that has followed different routes in different parts of the world. This section summarizes the European transition from feudalism to capitalism and the experience of socialism. Using Marxist logic, each of these societal types is an example of a mode of production (refer to the discussion of Marxism in Chapter 2).

FEUDALISM

In northern Europe Roman rule disappeared by the sixth century and was replaced by a relationship between near equals, namely, kings and warriors (vassals). With the temporary collapse of the power of kings in the ninth century, each warrior assumed authority to govern all those who lived in his area of control. This arrangement evolved into **feudalism**, a non-centralized system of governance and social and economic organization involving two groups; a group of direct producers (peasants) and a group of feudal lords headed by a renewed monarchy. The peasants were socially inferior to the lords and subject to their legal and political domination. All land was owned by the monarch, but this ownership was effectively decentralized to the lords in return for their military and political support.

Feudalism has several important implications for both place and people. In a feudal society, people were defined by their social class and social mobility was limited. Further, peasants were not free to make choices about whom they worked for, a situation that Marx described as exploitation.

CAPITALISM

A new type of culture, society, and economy emerged in the eighteenth century in association with the Industrial Revolution. The term used to describe this type, **capitalism**, was first popularized by Marxists in the late nineteenth century. Capitalism is characterized by a separation of the producer from the means of production, a separation that is achieved through the transformation of labour into a commodity that can be bought and sold. The principal means of production are owned by a particular class. In the western world, our contemporary cultures are dominated by this particular form of economic and social organization.

Box 8.4: A Marxist reaction to postmodernism

Of our more established geographic approaches, it is not only positivism that is under attack from postmodernism but also Marxism. Although these two are very different, they share a common concern with the claims of postmodernism. Consequently, just as Berry, a leading human geographer in the positivist tradition, has taken exception to postmodernism, so also has David Harvey, the most influential Marxist human geographer since the early 1970s.

Harvey (1989) has expressed reservations in detail in a book titled, *The Condition of Postmodernity*. The greatest concern for Marxists is the postmodernist claim that Marxism is unable to cope with the growth of **disorganized capitalism** (as opposed to **organized capitalism**) or what is also known as the transition from **Fordism to post-Fordism**. Harvey firmly believes in the power of Marxist theory to explain these processes and events. The arguments here are very complex and there are numerous internal differences within the Marxist camp, but the opposition to the far-reaching claims of postmodernism is not difficult to understand.

From a Marxist perspective, Harvey argues that postmodernism reflects:

a particular kind of crisis within ... [modernism], ... one that emphasizes the fragmentary, the ephemeral, and the chaotic side (that side which Marx so admirably dissects as integral to the capitalist mode of production) while expressing a deep scepticism as to any particular prescriptions as to how the eternal and the immutable should be conceived of, represented, or expressed. But postmodernism, ... with its concentration on the text rather than the work, its penchant for deconstruction bordering on nihilism, its preference for aesthetics over ethics, takes matters too far (Harvey 1989:116).

Most critically, Harvey (1989:117) sees the rhetoric of postmodernism as dangerous because it ignores 'the realities of political economy and the circumstances of global power'.

We cannot resolve the debates introduced in this and the previous box, of course, but there is great value in beginning to appreciate some of the tensions that prevail in contemporary human geography just as in chapters 1 and 2 we identified some of the controversies of the past.

According to Marx, capitalism is distinctive because of its capacity for self-expansion through ceaseless centralization and concentration of capital, because of its continual technological changes to the production process, because of the cyclical nature of the associated process of development, and because of the increased division between classes and resultant class conflict. Like feudalism, capitalism was regarded by Marx as an exploitation of the peasant or working class, but because all people under capitalism are not fully free, they are alienated (Box 8.5).

Another important social theorist, Max Weber, constructed a quite different model of capitalism as an **ideal type**, emphasizing the origins of capitalism in sociological terms by reference to the religious ethic of Protestantism, the growth of cities, and the legal and political framework provided by the rise of a new type of nation state.

After the end of the Second World War, capitalism changed considerably with increased growth of major (often multinational) corporations, and with increased involvement by the state (often in the form of public ownership) in the economy. This is the organized capitalism referred to in Box 8.3. Most recently, there is evidence to suggest that a further transformation into a disorganized capitalism is occurring — this term refers to a new form characterized by a process of disorganization and industrial restructuring (again, see Box 8.3). The transition from organized to disorganized capitalism is partially a reflection of the uncertain success of capitalism as measured by economic instability, social injustice, poverty, and unemployment.

An understanding of **class** is central to our understanding of capitalist economies and societies. Unfortunately, class has not been a clearly defined and effectively used concept in geography. The reasons for the lack of interest in class are not difficult to understand. Class is a variable employed by sociologists either as a status structural category or as an expression of self-identity. The latter use ought to be of interest to geographers as there is clearly a tendency for different classes to locate in different areas; Blaut (1980) argued for class as the appropriate variable for delimiting cultures. Class is not simply a possibly useful cultural variable; it also implies links between class groups; specifically it implies a set of power relations where some people exert power over others. Seen in this way, culture involves a whole set of patterns of dominance and subordination that are reflected in behaviour and landscape.

From a Marxist perspective, class and class struggle are inseparable. Until the 1980s, human geographers considered class largely devoid of theoretical content, focusing instead on socio-economic status, but explicitly conceptual uses of class that integrate the concept with other bases for group delimitation are now becoming more evident (Pratt 1989).

SOCIALISM

Some contemporary cultures have rejected capitalism in favour of **socialism**, a form of social and economic organization that involves common ownership of production and distribution of products. Socialism first developed in the nineteenth century in a variety of forms, largely as a rejection of capitalism. The common feature of all socialist endeavours is the opposition to capitalist individualism, hence the term 'socialism', which focuses on community, the well-being of society as a whole, equality, and the vision of a classless society.

Multiple versions of socialism appeared in the nineteenth century with the major distinction being between communism and social democracy. It is notable, however, that the

---

Box 8.5: The concept of alienation

According to Marx, individuals in a capitalist system experience **alienation**: they lack control over their own lives. This alienation results from the labour market and the very real maldistribution of economic and political power in a state. When those without power have protested, a state has typically responded by legitimizing capitalism via democracy and guarantees of individual human rights. Capitalism, then, subordinates all individual members to the state; for example, individuals lack control over such basic needs as food and shelter, and are required to behave in particular ways. Simply put, according to Marx, capitalism dehumanizes humans. People do not control the means of production; rather, it controls them.

Whether a state is in the core, semiperiphery, or periphery (using the terminology of world systems), individuals are alienated. This alienation is most detrimental in the periphery and least detrimental in the core, in the sense that living standards are highest in the latter, but the principle is unchanged.

The concept of alienation is central to many discussions of global problems. Once alienated, a person no longer interacts closely with the natural world with the only influences being the person and nature; rather, each alienated person has a relationship with the natural world that is in some way organized by forces that are not part of the person or of the natural world.

As capitalism spread across the globe, essentially linked to European activity, alienation (in the capitalist form) became more and more apparent. Our contemporary global political and economic system continues to encourage the spread of alienation. Some authors contend that alienation is such an integral part of capitalism that the institution of the state evolved to legitimize capitalism and prevent any substantial popular opposition to the circumstances of alienation (Johnston 1986:176).

development of communist societies in Russia and elsewhere was not in accord with the ideas of Marx, and the recent rejection of communism in the former USSR and throughout eastern Europe has resulted in an expansion of the capitalist world. The possibility of a transition from the now-dominant capitalism to socialism seems unlikely. Indeed, for many, socialism is now seen as a welfare state within a basically capitalist economy.

LANDSCAPES AND POWER RELATIONS

Our discussions of social theory and related contemporary concerns in cultural geography make it clear that peoples and places are now being viewed in a variety of conceptually sophisticated ways. Human geographers are increasingly cognizant of the diversity of peoples and places as a result of the insights suggested by humanism and Marxism and, more recently, by structuration and postmodernism. Combined, these various and varied approaches invite explorations of more than cultures as ways of life and their related visible landscapes, the geographic expression of culture; they invite explorations of culture as a process in which people are actively involved and of landscapes as places constructed by people, the spatial constitution of culture.

POWER AND AUTHORITY

As suggested in the accounts of structuration theory and feminist theory, and as described in the discussion of types of society, there are significant social and spatial variations in the distribution of **power** and **authority**. Further, it will be recalled that the logic of world systems theory (introduced in Chapter 6) relied on the argument that there is a dominant core and a subordinate periphery in the contemporary world. It is not surprising, then, that an understanding of peoples and places is facilitated by the recognition of the importance of the distribution of power and authority.

A compelling example of the expression of power in landscape was the **apartheid** landscape of South Africa, formally set in place beginning in 1948 and finally dismantled in 1994 (Box 8.6 and Figure 8.3).

WELL-BEING

This term refers generally to the overall condition of a group of people, and thus we can have economic, social, psychological, and physical **well-being**. Landscapes vary all too clearly in terms of the well-being of their occupants. Such variations apply at various spatial and social scales. Many landscapes in areas of European overseas expansion originated and have continued to evolve in response to the needs of the imperial power. Also on the global scale, we have already identified that there are areas of feast and areas of famine; some children are born into poverty and others into affluence. On a more local scale, other landscapes (of disadvantaged groups such as the homeless, for example) similarly reflect the fact that there are dominant and subordinate groups in any culture. Human geographers are paying increased attention to these landscapes, at a variety of scales, and to the social and power relations that lie behind them.

In the 1960s, concern about social problems in the United States prompted the development of territorial social indicators as measures of well-being. In a pioneering study, Smith (1973) identified seven sets of indicators to represent seven different components of well-being: (1) income, wealth, and employment; (2) the living environment, including housing; (3) physical and mental health; (4) education; (5) social order; (6) social belonging; (7) and recreation and leisure. The purpose of the indicators is to reveal the extent to which groups of people in different places have different experiences that involve a different quality of life. Indeed, there are extreme inequalities at all spatial levels. Knox (1975) has conducted similar investigations for England and Wales. The question of the causes of social and spatial inequalities has been addressed by Herbert and Smith (1989).

Much of the work done in these general areas is labelled **welfare geography**; this is a general approach to a wide range of issues, but with an emphasis on questions of social justice and equality. Concern with the geography of well-being and with welfare geography has encouraged research into such important issues as the geography of education (focusing on the often inequitable spatial distribution of services and facilities) and the geography of justice (focusing on the variations in the spatial availability of social benefits). But the two principal areas of concern are the geography of crime and the geography of health and health care.

CRIME

Human geographers have addressed the study of criminal behaviour and activities from a range of philosophical perspectives, but with the central concern of relating crime to relevant spatial and social contexts (Evans and Herbert 1989). Mapping the incidence of crime has helped identify high-incidence areas such as areas of the inner city, of poverty, of low-quality housing, of high population mobility, and of social heterogeneity. These essentially spatial, and often empiricist or positivistic, studies are accompanied by studies of the social environments of criminals, the victims of crime, and the geography of fear, all of which are more clearly Marxist or humanist in focus.

Two areas of particular concern at present are the relationship between areas of criminal activity and urban decline, and the way in which our use of space is affected by our concerns about the likelihood of a criminal act (Box 8.7).

HEALTH AND HEALTH CARE

Susceptibility to disease, morbidity (illness), mortality, and health care provision are matters that relate closely to questions of well-being and the quality of life, all of which have been analysed by human geographers; in the text discussions of population in chapters 5 and 6, these topics were

Box 8.6: The origins of apartheid

The first permanent European settlement in South Africa, designed to supply Dutch ships with food produce, began in 1652; population growth was slow until the nineteenth century. The colony became British in 1806, and from 1806 until 1948 South Africa was predominantly British in political character. From the election of the Nationalist party in 1948 until the political victory of the African National Congress in 1994, the dominant political influence was Afrikaaner (Dutch). The Union of South Africa, founded in 1910, is now fully independent.

The history of South Africa since 1652 has been characterized by a series of conflicts between several relatively distinct societies. There have been, and continue to be, conflict between White and Black Africans, particularly. There is little doubt that one way to view the apartheid system is to place it in this long-term perspective of social conflict.

Segregation of different groups was characteristic of the South African landscape prior to 1948, but after that date, a series of laws were passed that politically institutionalized the system known as apartheid or separate development. Thus, apartheid has roots in both ongoing conflicts and in a long tradition of relatively informal segregation. Understanding apartheid, however, clearly requires further consideration of the social group that formally institutionalized apartheid, the Afrikaaners.

The Afrikaaners, as their very name suggests, have separated themselves from their European heritage. They identify themselves fully with Africa and are often called 'the White tribe'. Their language is derived from (but is different from) Dutch. Their religion is the Dutch Reformed Church. In 1948, the Afrikaaners gained political control of a state that comprised Whites, Blacks, Coloureds, and Asians. The Blacks include a variety of linguistic and tribal groups. The Afrikaaner reaction to attaining power was to institute apartheid. Afrikaaners agrued that apartheid was the best way to allow a socially fragmented country to evolve. The Afrikaaner position was that integration of different groups leads to moral decay and racial pollution, whereas segregation leads to political and economic independence for each group. Ten African homelands were created, four of which were theoretically independent (their independence was never recognized by the United Nations or by any government outside South Africa).

According to the Afrikaaners, then, apartheid provided Blacks with independent states and rights comparable to those of Whites. In practice, this was not the case. What apartheid enabled the Whites to maintain was their own cultural identity, political power, and exploitation of Black labour. Interestingly, the dismantling of the apartheid system, which began in the late 1980s and was concluded with the 1994 election of the first democratic government, proceeded relatively peacefully.

The apartheid system was unique, but this does not mean that it could not have developed elsewhere had the demographics been comparable. In the United States, White supremacy emerged in the nineteenth century and, had the Native Americans not been significantly reduced in numbers, a version of apartheid may well have emerged. In discussing apartheid, we must acknowledge that one specific cause was the Whites' failure to establish a majority (as the Whites did in the United States).

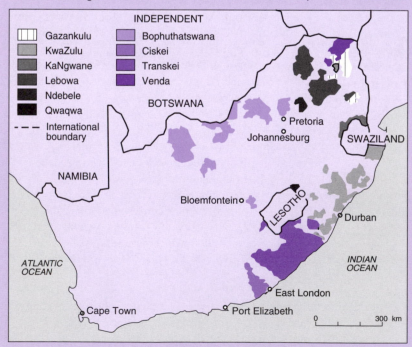

Figure 8.3 Apartheid on the national scale in 1975 — the 'homelands' planned by the South African government.

raised in a global context. Regardless of spatial scale, there has been a tendency to focus research on the less privileged members of society to demonstrate how health issues relate to a wide range of social and economic conditions. It is clear that health problems are related to the physical environment, to a variety of social factors such as housing, living conditions, and to access to health services. Several studies have emphasized the inequality of availability of services (see Jones and Moon 1987).

Much of the discussion in this chapter, in addition to other discussions in this text, focuses on the different types of landscapes. There are landscapes for the privileged and the less privileged, the advantaged and the disadvantaged, and they are very different places lived in by very different peoples. We now use the general labels of élitist landscapes and landscapes of stigma. There are obvious links between theoretical questions of class and gender, as already discussed, and these élitist and other types of landscapes.

ÉLITIST LANDSCAPES

The existence of élitist landscapes is evident, although they have been seldom analysed. Geographers studying cities have traditionally recognized that all cities include a range of social class areas; it is not unusual for the higher-class areas to be at a higher elevation than other areas, to be accessible to a lake, river, or ocean, or to be so located that they are not affected by the pollution generated by industry.

In British cities, it is usual for higher-status families to be spatially segregated and to generate such land uses as golf courses and specialty retail areas. In Melbourne, Australia, an élite residential area has been a consistent feature of the urban landscape, a feature that moved steadily south as the city grew. In an important sense, identity and landscape are very closely interwoven in these élite areas.

Élite regions can also be identified. In Britain, there is a long-standing distinction between north and south, a distinction that is one between non-privilege and privilege. More obviously, perhaps, many tourist areas, especially in less developed countries, are clearly landscapes for the privileged.

LANDSCAPES OF STIGMA

Other landscapes do not have a good name. In recent years, human geographers have identified pariah landscapes, landscapes of despair, and landscapes of fear. Examples of pariah landscapes include many ghetto areas in cities, Indian reserve lands in Canada and the United States, areas of Black residences in South Africa and South African homelands, especially prior to the changes following the election of the first democratic government in 1994 (see Box 8.6 and Figure 8.3).

The experiences of discharged mental patients prompted introduction of the term 'landscapes of despair', a vivid descriptor that can be equally applied to pariah landscapes. One of the most publicized of such landscapes in recent

---

Box 8.7: The geography of fear

Geographers have made considerable advances into the analysis of social and spatial variations in the distribution of fear, and there is compelling evidence to suggest that the fear of crime is a major problem affecting our perceptions of areas and our behaviours in areas (Pain 1992; Smith 1987). The problem is especially acute for women, the elderly, and other groups within society who are perceived as relatively powerless and vulnerable.

Fear of domestic violence, sexual harassment, and rape limits women's access to and control of space within and outside the home and also imposes limitations on social and economic activities. A study of the geography of rape and fear in one New Zealand city, Christchurch, demonstrates the extent of the problem. Christchurch has a population of 292,000 of whom 91 per cent are of European origin.

Pawson and Banks (1993) had two goals: first, to determine the spatial and social incidence of rapes that were publicly reported in the press; second, to map the spatial and social distribution of the fear of crime. The results of their research, based on study of their data and some national statistics, showed a spatial distribution that emphasized the inner-city area, and a general correlation with patterns of other criminal activity such as burglary, with areas of youthful population and

rental homes. Socially, it was demonstrated that younger women under 25 years are especially at risk from rapists; that about 50 per cent of the rapes occurred in the victim's home, not a public place; that most occurred at night during or immediately after the hours of social contact and at a time of minimal community surveillance; and that there is a marked seasonality, with fewer rapes in the winter months.

To map the spatial and social distribution of the fear of crime, a survey of about 400 persons was conducted. The results showed that, although fear is widespread, there is a distinct spatial pattern with the low-income, least stable, and high-rental areas reporting the greatest level of fear. Socially, women and the elderly displayed the greatest levels of concern, while fear at night was much greater than during the day. The evidence clearly showed that women and the elderly are often reluctant to use either their neighbourhoods or the city centre at night.

The conclusions of this research are clear and generally in accord with a variety of other studies. As this text has already emphasized in other contexts, we live in an unequal world. For women, the city at night may be seen as one dominated by males.

8.2 Urban slum, Brazil. (P. St Jacques, ACDI/CIDA)

years is the inner-city landscape of homeless people. All such landscapes are expressions of exclusion, spatial reflections of social injustices (Box 8.8).

As already noted, violence is closely identified with specific locations and many people must cope on a regular basis with landscapes of fear, with elderly people and women especially susceptible to violence, usually by males. Most people have a mental map that regards certain areas as safe, others as unsafe. Safe areas may only be safe for groups and/or during daylight hours. Large, open spaces, such as parks, are often perceived as unsafe, as are closed areas with limited exits, such as trains. The recent emergence of neighbourhood watch schemes is a clear reflection of increased fear and decreased social interaction in many city areas. Landscapes of fear result in a restricted use of public space.

GENDER IN THE LANDSCAPE

Gender, like class, implies a distinction between power groups — in this case, dominant males and subordinate females. Only in recent years have human geographers begun to pay serious attention to gender differences. This is unfortunate as gender differences clearly do matter; discussing humans as a whole effectively avoids acknowledging that throughout the world the lives and experiences of women and men are different. For example, women typically perform domestic work, which is unpaid, repetitious, and often boring, or are in low-paid and low-skilled jobs. Thus, women and men tend to do different work, in different places, and hence lead different lives, which include different visions of the world and of themselves. These differences are not explained in terms of biological differences — sex — but in terms of cultural differences — gender.

There is little doubt that most of our human geography is male-oriented, a result of the near universality of patriarchal, male-dominated cultures. Interestingly, however, this is not always easy to see in the landscape in quite the same way as are class or ethnic differences. But landscapes are both shaped by gender and also provide the contexts for the reproduction of gender roles and relationships (Monk 1992).

There are numerous examples of how landscapes, both visible and symbolic, reflect the power inequalities between women and men in their embodiment of patriarchal cultural values. In urban areas, *statues and monuments* typically reinforce the idea of male power by commemorating male military and political leaders: 'conveyed to us in the urban landscapes of Western societies is a heritage of masculine power, accomplishment, and heroism; women are largely invisible, present occasionally if they enter the male sphere of politics or militarism' (Monk 1992:126).

The design of *domestic space* is strongly influenced by ideas about gender roles. In many cases, including traditional Chinese and Islamic societies, areas for women and others for men are separated, with those for women isolated from the larger world. In western societies, the favoured domestic design has centred on the home as the domain of women and as a retreat from the larger world.

This same patriarchal logic has influenced the expansion and form of city *suburbs* with men seen as commuters and women as home-makers with ready access to schools and shops. It is now argued that this arrangement further disadvantages women who may become isolated because of the distance from town, or who struggle to cope with work at home and limited opportunities in paid employment; it is therefore both a reinforcement of traditional gender roles and identities and an obstacle to change.

GENDER AND HUMAN DEVELOPMENT

The evidence is compelling. There are many and varied expressions of gender in landscape, expressions that typically disadvantage women, but perhaps the clearest evidence is

---

Box 8.8: Consigned to the shadows

The title, 'Consigned to the shadows', is borrowed from an article in a geographical magazine (Evans 1989), and could refer to many disadvantaged groups. It does in fact refer to those who are mentally ill. Those of us living in larger urban centres are probably not unaccustomed to the occasional horrific news story about deplorable conditions in some home for the mentally ill. However, you may not have heard about the Greek island of Léros, where some 1,300 mentally ill people have little food, minimal medical care, and inefficient sanitation; when the odour becomes too unpleasant, these people are hosed down. People with various disorders are grouped together with no thought of proper care or cure.

Conditions on Léros may be appalling today, but the island is not an extreme case historically. Mentally ill people have typically been separated from the majority with no real attempt to cure or understand. Indeed, there is no real understanding of mental illness; in the United States until recently, homosexuality was regarded as a mental disorder. In Japan, private mental hospitals now house about 350,000 people (one of the highest per capita rates in the world) at least partly because it is relatively easy to have a person so detained.

The problem of properly housing the mentally ill is increasing. The World Health Organization predicts a dramatic rise in the number of people with serious mental disorders in the less developed world. This is not surprising given that such illness can be a result of a wide range of factors, including medical and dietary factors. A recent World Health Organization estimate places the number of mentally ill worldwide at 100 million.

that relating to data on human development. Table 6.10 detailed the Human Development Index (HDI) for a number of countries and Table 8.1 provides details of a more sophisticated gender-sensitive HDI that uses separate female and male estimates of life expectancy, adult literacy, mean years of schooling, employment levels, and wage rates. The results are most illuminating.

As seen in Table 8.1 Canada, which ranked first in Table 6.10 with an HDI of 0.982, falls to fourth place when the gender-sensitive measure is employed; Japan, Switzerland, and the United States (respectively second, fourth, and sixth in Table 6.10) also fare poorly in this revised ranking. Perhaps the key information in Table 8.1 is that concerning the female HDI as a percentage of the male HDI; in no case is the female HDI higher than the male HDI. The three countries with the highest values in this category are, respectively, Sweden, Finland, and Norway. Of those countries for which gender-based data are available, Kenya ranks lowest in this category with the female HDI being only 58.6 per cent of that for males.

ETHNICITY

DEFINITION

Ethnicity is often poorly defined or not defined by those who use the term; an examination of sixty-five studies of ethnicity noted that fifty-two of them offered no explicit definition (Isajiw 1974). The geographer, Raitz, advocated that **ethnic** should refer to any group that has a common cultural tradition, that identifies itself as a group, and that is a minority group: 'ethnics are custodians of distinct cultural traditions ... the organization of social interaction is often based on ethnicity' (Raitz 1979:79). Thus the group may be delimited according to one or more cultural criteria, but it is necessary that the group not be living in their national territory — Swedes in Sweden are not an ethnic group, whereas Swedes in the United States, given that they identify as a group, are an ethnic group.

TABLE 8.1: *Gender-sensitive measures of human development: Selected countries, 1992*

Country	Gender-sensitive HDI	Female HDI as % of male HDI
Sweden	0.938	96.16
Norway	0.914	93.48
Finland	0.900	94.47
Canada	0.842	85.73
United States	0.842	86.26
Switzerland	0.790	80.92
Japan	0.761	77.56
Kenya	0.215	58.60

Source: United Nations Development Programme, *Human Development Report, 1992.* (Toronto: Oxford University Press, 1992):21.

Although this definition is reasonable and clear, it is not universally employed. Indeed, the term 'ethnicity' is generally regarded as one of the least clear labels in social science and it is important to acknowledge these confusions. The greatest confusion occurs when terms such as 'race' or 'minority' are used interchangeably with 'ethnic'.

We learned in Box 3.4 that, biologically speaking, there are no such things as races (subspecies) within the human species and we learned in Box 3.5 of the historical popularity of racist thought as a means of identifying groups typically perceived to be inferior. It is common today to use the term 'race' to set apart outsiders whose physical appearance does not accord with some generally accepted norm; the key divide used is the most visible one, skin colour. It is important for us to acknowledge that the fact that races do not exist does not prevent the label from being applied.

Some groups are generally regarded as minorities, a term that refers to a state of exclusion, because they are in some way different from the majority. Common bases for delimiting a minority are language, religion, ethnicity, perceived racial identity, recent immigrant status, or any combination of two or more of these.

SHARED IDENTITIES: CHANGING IDENTITIES

Ethnicity refers to some sense of an enduring collective identity; time as well as space is shared because of the notion of common descent. Most groups identified as ethnic base their ethnicity on one or both of the two principal cultural variables discussed in the previous chapter — language and religion. In common with both of these identifying labels, ethnicity is also both inclusionary and exclusionary. Some people are defined as insiders because they share the common identity of the group, while others are seen as outsiders because they are different. By definition then, people belong to at most one ethnic group, but it is quite possible for an ethnic group to gradually change its identity and behaviour.

'Ethnic' is a convenient term to employ partly because it defies explicit definition. Human geographers often utilize the ethnic label in identifying and discussing cultural regions in rural and urban areas. Generally speaking, an ethnic region or neighbourhood is an area occupied by people of common cultural heritage who are voluntarily choosing to live in close spatial proximity.

Ethnic areas

It is usual for immigrant ethnic groups, especially those moving into urban areas, to experience assimilation or acculturation eventually. This is notwithstanding an initial period of social and spatial isolation that may lead to a low level of well-being, relative deprivation, and an ethnic colony, enclave, or **ghetto**. We can interpret these common initial experiences of an immigrant ethnic group — deprivation and residential segregation — as social and spatial expressions of outsider status. Often a local group identity is continually reinforced by **chain migration**, the process whereby migrants from a particular area follow the same paths as friends and relatives who migrated before them.

Despite the often negative experiences for a new immigrant ethnic group, **assimilation** is a common later experience. Most European groups that move into cities in the United States steadily lose ethnic traits such that the group eventually becomes a part of the larger culture. If **acculturation** occurs, the ethnic group functions in the larger culture, but retains a distinctive identity. In the United States, acculturation has been a characteristic process.

One key factor determining whether or not assimilation occurs is the degree of residential propinquity: if group members live in close spatial proximity, then social interaction with the larger culture is limited and assimilation unlikely. Once again we see an intertwining of space and culture. A second factor is the impact of state policies. Since 1971, in Canada there have been policies actively encouraging **multiculturalism**. The political logic for the introduction of multiculturalism was succinctly stated by then Prime Minister, P.E. Trudeau: 'Although there are two official languages, there is no official culture, nor does any ethnic group take precedence over any other' (quoted in Kobayashi 1993:205).

Immigrants moving into rural areas, especially if the movement is of a group and/or involves chain migration, tend to retain aspects of their ethnic identity longer than do those in moving into urban areas. Many of the regional landscapes of areas of European expansion in particular are characterized by distinct ethnic imprints. This is especially so if a group has a formal structure, as in the case of the Mormon group discussed in Chapter 7, but is typical regardless because of the relative isolation of rural areas. Historical and cultural geographers conduct studies of the extent of cultural change that accompanies and follows settlement and the landscapes created (see McQuillan 1993).

LANDSCAPE AS PLACE

At the beginning of this chapter, we highlighted the value of perceiving landscapes or regions as places with social and symbolic content. The discussions so far have pursued this idea using a wide variety of theoretical underpinnings and focusing on how a number of important variables — such as class, gender, and ethnicity — interact to form groups and related landscapes. A recurring feature of these accounts has been the reality of inequality — unequal social groups living in unequal landscapes.

We have provided enough evidence to demonstrate that societies, landscapes, and places are unequal. Groups in particular places have very different experiences; income levels vary, as do health, education, and overall environmental quality. The importance that the human geographer attaches to these evident irregularities is clearly related to philosophical persuasion. All human geographers are concerned about these issues: the empiricist might emphasize the need to describe facts accurately; the positivist might focus on a precise statistical statement of facts and development of theoretical explanations; the humanist might centre attention on the experiences of those living in such places; the Marxist might strive to explain issues, identify causes, and advocate change; the structurationist might emphasize the process by which the place has 'become' through time in relation to the distribution of power; the postmodernist might focus on the writing of alternative geographies inspired by the experiences of previously excluded groups; and the feminist might uncover the inequalities associated with the gendering of landscape. Each approach has merit and it is hoped that the combined results will promote solutions to the uneven distribution of well-being. It is appropriate now to return briefly to our underlying theme concerning the idea of landscape as place.

Landscapes are not simply locations; they are places, places in the sense that they convey meaning. This idea was central to our opening account of vernacular regions and is reflected in the increasing number and sophistication of iconographic analyses (Box 8.9). The meaning that a place has is a human meaning and is dependent on social matters. There is an important circularity here. Humans, as members of groups, create places; in turn, each place created develops a character that affects human behaviour. This circularity applies to all landscapes, but is perhaps most relevant in the extreme cases — landscapes of the advantaged and the disadvantaged, the privileged and the underprivileged, insiders and outsiders, rich and poor, and men and women.

PLACES IN GEOGRAPHY

Much of this chapter deals with ideas and subject matter that are relatively new in human geography and that are continually unfolding and placed into appropriate contexts. One distinguished human geographer argues that our focus should be on places, alternatively called regions, localities, or locales (Johnston 1991:67–8). The thrust of the argument, using the term favoured by Johnston — 'region' — is as follows.

1. Region creation is a social act and therefore regions differ because people made them that way; differences in physical environment are relevant, but are not a cause.
2. Regions are self-reproducing entities because they are the contexts within which people learn. Thus people are made by the places that they create.
3. Regional cultures do not exist separately from the individual members of the culture.
4. Regions are not autonomous units; they interact with other regions.
5. Regions are often the deliberate creations of those who are able to control.
6. Regions are potential sources of conflict.

Each of these six points has been incorporated in the discussions and examples presented in this chapter.

POPULAR CULTURE AND FOLK CULTURE

We now introduce two new terms, 'popular culture' and 'folk culture', which add an additional dimension to this account of the social and symbolic aspects of landscape. Aspects of popular culture and folk culture relate clearly to our interests in culture and landscape, in peoples and places.

Popular culture can be defined as the activities of those groups that conform to ever-changing attitudes and behaviours. In contrast, **folk culture** refers to those groups that prefer to retain long-standing attitudes and behaviours. There is a useful link between both popular cultures/folk cultures and ethnicity. Most folk culture groups are rural ethnic groups, often linked by a distinctive religion and sometimes by a distinctive language. Many popular culture groups are urban ethnic groups that often display specific ethnic attitudes and behaviours, such as a Caribbean carnival in London, England. Indeed many ethnic groups exhibit some of the characteristics of a popular culture and some of the characteristics of a folk culture; the popular/folk distinction is not always straightforward.

POPULAR CULTURE

Our use of this term centres on those cultures characterized by change and involve relatively large numbers of people. The activities of popular culture groups tend to diffuse rapidly, especially in more developed areas where people have the time, income, and inclination to indulge in popular culture activities.

Jackson (1989) offers an account of the rise of popular culture in nineteenth-century Britain, showing how the social activities of the working-class populations were constrained by more powerful classes. New places of entertainment (such as music halls) and activities (such as prostitution) became major issues of class struggle. Many issues today similarly involve the control of space, and Jackson (1989:101) argues that the 'domain of popular culture is a key area in which subordinate groups can contest their domination'. One of the better-known instances of social inequality that has expression in landscape relates to matters of sexuality. For example, in most large cities, there are gay/lesbian neighbourhoods that provide powerful examples of the spatial constitution of culture.

Other geographic analyses of popular culture include studies of landscapes and regions with relatively little emphasis on the culture itself. Thus there are studies of the landscapes of shopping malls, sports, gardens, urban commercial strips, and such regions as those of a particular type of music or recreation. Two brief examples are noted.

Shopping malls provide a vivid instance of the impact of popular culture on landscape. They are characterized by a sameness in that they are created as artificial landscapes; they are prime examples of landscapes of consumption. Malls are located in most urban areas in the more developed world and are usually enclosed, climate-controlled, lack windows, and cater to homogenized consumer tastes. As a product of popular culture, they encourage the further acceptance of popular cultures. When completed in 1986, the West Edmonton Mall in Edmonton, Canada, was the largest in the world, with 836 stores, 110 restaurants, twenty movie theatres, a hotel, and a host of recreational features (see Jackson and Johnson 1991).

Box 8.9: The iconography of landscape

Iconography is the description and interpretation of visual images to uncover their symbolic meanings. Because landscapes can be regarded as depositories of cultural meanings, it is possible to subject them to iconographic analysis. If successful, an iconographic analysis uncovers a landscape as being shaped by and at the same time shaping the regional culture. Two Canadian examples are briefly summarized.

Osborne (1988) focused on the development of a distinctive national Canadian iconography in both the artistic images of lands and peoples and the various responses to these images. Most notable of the various images identified by Osborne was the work of Tom Thomson and the Group of Seven; their declared aim was the enhancement of a national identity, which they attempted to achieve largely through their paintings of the Canadian north. During the first half of the twentieth century, these painters helped to create a distinctive image of Canada: 'rock, rolling topography, expansive skies, water in all its forms, trees and forests, and the symbolic white snow and ice of the "strong North"' (Osborne 1988:172). Although this image is clearly a limited one and is now complemented by other images, it remains a compelling example of an iconography.

A different application of iconographic logic involved an examination of the images of the city of Hamilton, Ontario, in their societal context, asking why these images appear as they do (Eyles and Peace 1990). The images derived primarily from newspaper accounts, although varied, were dominated by a single and negative theme--Hamilton as steeltown. Aspects of this image arose from frequent comparisons with neighbouring Toronto, with Hamilton seen as a blue-collar city of production, while Toronto is seen as a city of consumption. Because Hamilton is an industrial city, it is seen as the past rather than the future, as an area of economic decline, and as a polluted landscape. With the continued rise of Toronto as a major world city, the image of Hamilton is unlikely to change as comparisons will continue to be made.

Sport is an important part of popular culture and popular culture regions can be identified by sporting preferences. Organized sporting events are associated with the social changes that occurred during the Industrial Revolution. Major sports, such as baseball and soccer, have their organized origins in the late nineteenth century. Regions can be delimited in terms of spectator preferences and participant preferences; in the United States, Rooney (1974) has discovered clear regional variations. Thus, professional American football players come primarily from a cluster of southern states (Texas, Louisiana, Mississippi, Alabama, and Georgia). Geographers are often inclined to link such preferences to the concept of a vernacular region; certainly, the five states noted above can be identified as the American deep south.

FOLK CULTURE

In general, folk cultures are more traditional, less subject to change, and, in principle, more homogeneous than popular culture. This third identifying characteristic is of limited value as there is much evidence to assert that popular culture landscapes can also be homogeneous. Folk cultures have preindustrial origins and are less clearly related to class. Religion and/or ethnicity are most likely to be key unifying variables with traditional family and other social traditions as paramount. We may justly suggest that folk cultures are characterized by a lack of change.

Folk culture regions, such as those of Amish or Mennonite groups, display uniform landscapes that are relatively unchanging, a reflection of both central control and a desire by individuals to conform to group norms. Domestic architecture, fence styles, and barn styles vary from one folk culture to another. Other folk culture traditions relate to matters of food and drink with a rejection of mass-produced, popular-culture, fast foods for example, and a preference for established habits (Box 8.10). Music preferences similarly include a rejection of the products of popular culture and a preference for local styles. In North America, rock-and-roll music is a product of popular culture, while a style such as western swing is associated with north Texas and Oklahoma.

Geographic studies of folk culture are plentiful in North America and Europe and, at present, tend to be closely associated with the traditional landscape school of Sauer rather than with the more theoretically informed methods outlined and employed in this chapter, although the latter are likely to become increasingly prominent. Studies of popular culture are currently being conducted in the traditional landscape school and in the newer arena of cultural analyses enriched by social theory.

GROUP ENGINEERING

This brief concluding section is a provocative postscript to both the present and the previous chapter. We can argue that the real significance of culture is that it conditions our attitudes and behaviour. Do we then want to engineer ourselves? Is there value in attempts to change our cultural beliefs and practices? These are difficult and sensitive questions that provide a challenging conclusion to our discussions of culture.

The previous chapter began with a powerful metaphor — 'mountains of hate, rivers of inflexible tradition, oceans of ignorance' — using physical geographic terms to describe aspects of our human condition. Language, religion, ethnicity, class, gender, and sexuality have all at some time in some place been a justification for war, cruelty, hypocrisy, or dogma. The goal of a single universal language remains elusive; the prospects for uniting diverse religious beliefs are negligible; ethnic groups will continue to value traditions; class distinctions are possibly being increased, not decreased; and the likelihood of eliminating the cultural implications of gender and sexuality seem slight. All of these variables are human constructions and they are inevitable consequences of cultural evolution. This comment applies to gender in that gender-divided cultures are human creations, albeit with a biological basis. It is unfortunate that each variable not only involves divisions within the human population but also that divisive attitudes and behaviour are encouraged. For many members of cultural groups, especially religious groups, dogma is fact. Hence, the metaphor used by James is most appropriate.

Box 8.10: Geophagy

Geophagy, the intentional eating of dirt, is a widespread folk custom that is most commonly practised in sub-Saharan Africa. Specific preferences for types of dirt, usually clays but sometimes sand, are clear. The dominant reason for this custom is the belief that it provides essential minerals and it is practised especially by pregnant women and the ill.

Geophagy is a folk custom assuming different forms in different places and not universally practised. In the southern United States, some young Black children and pregnant Black women continue the custom. Geophagy is a good example of a tradition of questionable value that continues to be adhered to because it is a custom, despite being condemned by many.

This is but one example of a custom that appears to be largely unaffected by culture change. There are many other folk foods and folk medicines, but they are not typically as 'unusual' as geophagy.

Human diversity clearly has a very unfortunate dimension. These comments lead us to ask whether or not we ought to actively attempt to engineer ourselves — to devise and accept new values and new moral standards. We are now at a point in human cultural evolution when we can give serious thought to designing ourselves and hence our future. Given this development, we must consider the directions that human life needs to take. These are heady issues. Today, the more developed world is actively questioning traditional value systems. Only 100 years ago, racial hierarchies were proposed by serious scholars and accepted by other serious scholars; even more recently, most White people took the racial inferiority of others as fact. Today such theories are largely discredited and such attitudes are greatly reduced.

A world without human diversity at both group and individual levels may be unimaginable. Perhaps what we really need to achieve is a diversity that involves mutual respect between groups. Only then will our divided world be void of 'hate ... inflexible tradition ... [and] ignorance'.

SUMMARY

Vernacular regions

Vernacular regions are perceived to exist by those living inside and/or living outside the region. They may be created institutionally and typically possess a strong sense of place. A delimitation of such regions in North America closely parallels the regions delimited using the concept of first effective settlement. A related area of interest is known as psychogeography. Some vernacular regions may qualify as homelands if they are especially closely identified with a distinctive cultural group.

Symbolic interactionism

Provides a view of cultural groups as being created and maintained through communication; closely associated with humanistic concepts.

Current social theory

There is a variety of social theory to which human geographers are able to turn to facilitate their analyses of people and place. These include the humanistic and Marxist approaches first introduced in Chapter 2 and the approaches known as structuration and postmodernism. These two additional emphases are advances that are currently enriching our analyses, although both are complex and, especially in the case of postmodernism, highly diverse. In addition, human geographers are actively using radical and socialist versions of feminist theory.

Types of society

Three types of society, modes of production using Marxist terminology, are the feudal, capitalist, and socialist versions. The feudal and capitalist types share the characteristic of explicitly dividing people into unequal groups or classes. Our contemporary social and economic world is dominated by capitalism.

Landscapes and power relations

Human well-being is unequally distributed. Geographers now recognize élitist landscapes and landscapes of stigma with such landscapes as examples of places. Three useful variables in this context are those of class, gender, and ethnicity. Analyses include topics such as crime and health.

Studies focusing on landscape and power relations display a dramatic break with the traditional focus initiated by Vidal and Sauer especially. In brief we are seeing an increased use of social theory (especially Marxism), a lesser concern with the landscape per se, and an acknowledgement of the role of human agency as it relates to social relations and power relations. This newer focus first emerged in the 1980s. It is still in its infancy and the direction it will take is uncertain.

Gender

Landscapes are gendered as they reflect the dominance of patriarchal cultures — dominant men and subordinate women. Gender differences in well-being and the quality of life in general are especially evident at the aggregate scale using measures of human development.

Ethnicity

Ethnic groups are often loosely defined; the key linking variable may be language, religion, or common ancestry. They are minorities and may be seen as outside of larger society; ethnic groups may experience acculturation or assimilation. Canada is an officially multicultural society.

Place

We can be confident that there will be continued recognition of landscapes as signifying systems that have meanings; we used the term 'place' (Johnston prefers 'region') in this context. Further, we can be confident that landscapes will be interpreted as the ongoing outcome of struggles between groups and between different attitudes and behaviours.

Popular culture

This refers to the activities of those groups that conform to ever-changing attitudes and behaviours. Geographic studies of popular culture employ traditional and more recent conceptual backgrounds and include analyses of landscapes and regions.

Folk culture

This refers to those groups that prefer to retain long-standing attitudes and behaviours. Typically studied in the manner of the Sauer landscape school are a wide range of landscape features and folk activities.

Future cultural identity

Cultural variables are human constructions. They promote divisions between people and landscape. Unfortunately, such divisions have often resulted in disagreements between groups. Geographers now ask questions about our future cultural identity and the possibility of human engineering — deliberately creating new culture.

WRITINGS TO PERUSE

AGNEW J.A., and J.S. DUNCAN, eds. 1989. *The Power of Place: Bringing Together Sociological and Geographical Imaginations.* Boston: Unwin Hyman.

An edited volume, including both conceptual and empirical studies; focuses on the intellectual history of concepts of place and the linking of power and place.

CATER, J., and T. JONES. 1989. *Social Geography: An Introduction to Contemporary Issues.* London: Arnold.

An excellent overview of a wide range of subject matter, including ethnicity, crime, and gender.

DUNCAN, J., and D. LEY. 1993. *Place/Culture/Representation.* London: Routledge.

A series of essays aimed at representing cultural geography using diverse social theories; includes several examples with a Canadian focus.

KATZ, C., and J. MONK. 1993. *Full Circles: Geographies of Women over the Life Course.* New York: Routledge.

Analyses the lives and expectations of women of all ages in both more and less developed countries with reference to class, ethnicity, national identity, and individual values.

KEITH, M., and S. PILE, eds. 1993. *Place and the Politics of Identity.* New York: Routledge.

A radical text that critically examines the traditional power relations of male/female, heterosexual/homosexual, and White/Black.

LEY, D. 1983. *A Social Geography of the City.* New York: Harper and Row.

Excellent text that includes discussion of ethnic neighbourhoods, ghettoes, and landscapes of stigma.

MASSEY, D. 1993. 'Questions of Locality'. *Geography* 78:142–9.

An argument in favour of locality studies as the heart of geography.

MCDOWELL, L. 1989. 'Women, Gender and the Organization of Space'. In *Horizons in Human Geography*, edited by D. Gregory and R. Walford. London: Macmillan, 136–51.

An overview of the need for and implications of incorporating gender in human geography; 'add women and stir?'.

MOMSEN, J., and V. KINNAIRD, eds. 1993. *Different Places, Different Voices: Gender and Development in Africa, Asia and Latin America.* New York: Routledge.

Twenty case-studies of the living conditions of women in the less developed world; a radical challenge to western feminist approaches.

NOBLE, A.G., ed. 1992. *To Build in a New Land: Ethnic Landscapes in North America.* Baltimore: Johns Hopkins University Press.

An excellent first book-length study of North American ethnic groups and their landscapes; largely traditional cultural geography in content, but with some humanistic overtones.

NOSTRAND, R.L. 1992. *The Hispano Homeland.* Norman: University of Oklahoma Press.

A seminal contribution to homeland studies that identifies a distinctive group and their landscape through a reconstruction of their history.

ROSENAU, P.M. 1992. *Post-Modernism and the Social Sciences: Insights, Inroads and Intrusions.* Princeton: Princeton University Press.

A clearly written account on a complex subject; distinguishes between two principal versions of postmodernism and defines the key terms and issues.

SHIELDS, R. 1991. *Places on the Margin: Alternative Geographies of Modernity.* New York: Routledge.

A series of four case-studies, including Niagara Falls and the Canadian north, that emphasize the social construction of space.

SHORT, J.R. 1991. *Imagined Country: Environment, Culture and Society.* New York: Routledge.

An exploration of the relationship between people and the physical world, focusing on how images are reflected in the North American cinema, the British novel, and Australian painting.

SPAIN, D. 1992. *Gendered Spaces.* London: University of North Carolina Press.

An exploration of the spatial construction of gender cross-culturally; demonstrates that segregation, power, and gender are integral to one another in many cultures.

VALENTINE, G. 1989. 'The Geography of Women's Fear'. *Area* 21:385–90.

A brief factual account of the impact of violence on women's use of space.

WINCHESTER, H. 1992. 'The Construction and Deconstruction of Women's Roles in the Urban Landscape'. In *Inventing Places: Studies in Cultural Geography*, edited by K. Anderson and F. Gale. Melbourne: Longman Cheshire, 139–56.

All of the contributions to this edited volume make worthwhile reading. The piece by Winchester is especially helpful as an indication of how radical feminism can be combined with postmodernism in an empirical analysis.

9

The political world

We live in a divided world, and not the least evident and formidable of the divisions with which we are familiar are the many states which make up the jigsaw pattern of the world map. Professor Toynbee rightly reminds us that states exist to serve communities of individuals and that they have been accorded excessive importance. Doubtless if states were subordinated to a world authority, much of the violence, danger and fear implicit in everyday life would disappear. World government, however, is at present very much a hope, difficult of achievement; in a world of remarkably rapid change, men's minds are, in contrast, slow to throw off old attachments, attitudes and beliefs. The real world of today is organised into over 150 states and, although they carry out service functions as do public utilities, they behave very much like Greek gods, so much so that, given the great powers which they wield, they need to be taken very seriously (East and Prescott 1975:1)

Like other species, humans have always been concerned about partitioning space to separate themselves from other human groups. The creation of specific territories is the basis for political organization and political action. Political partitioning of space results in the most fundamental of geographic divisions — sovereign states. Thus, our long-standing predilection for delimiting territory — an area in which members of a group feel secure against outsiders and within the group — has gradually resulted in the creation of states. Most states are recognized as such by other states; their territorial rights are typically respected by others; they are governed by some recognized body, and they have an administration that operates the state. Individual members of states are thus 'spatially tethered and subject to rules' (Johnston 1982:4).

STATE CREATION

For each of us, as inhabitants of the earth, our most immediate and accurately perceived global map is probably that of the world divided into states. The familiarity with areas and shapes and sharp lines of international boundaries reflects our recognition that in the world security, territorial integrity and political power lie enmeshed in this pattern of states. Each state, large or small, is a sovereign unit (Douglas 1985:77).

DEFINING THE NATION STATE

'Nation' is not an easy word to define. As we have discussed in chapters 7 and 8, humans are divided into numerous cultures based on such variables as language, religion, and ethnicity; cultural affiliation is important not least because it provides people with an identity and a sense of community. The term **nation** is potentially applicable to these cultural groups. But not all cultural groups aspire to be nations and a key question is why some cultures regard themselves as possessing national identity while others do not.

Humans are also divided into much more formally spatially demarcated political units known as states. A **state** is a set of institutions, the most important being the potential means of violence and coercion; a state also determines rule making within its territory thus encouraging cultural homogenization.

A **nation state** is a clearly defined cultural group (a nation) occupying a defined territory (a state). Each nation state is, in principle, a political territory including all members of one national group and excluding members of other groups. In practice, there are few such nation states.

PREDECESSORS OF THE NATION STATE

Our current political divisions are the outcome of a long evolutionary process that began with claims to territory and that is now characterized by independent states, most of which aspire to be nation states — that is, political groupings of people with a common identity. In Europe before *c.*1600, the sovereignty of individual rulers focused on the allegiance of people rather than a rule over a territory; major exceptions to this generalization are the classical Greek and medieval Italian city states. The emergence of territorial sovereignty involved practical needs and moral discussions.

Classical Greece was not one political unit but an amalgam of city states. Both Plato and Aristotle gave thought to the size, location, and population of the ideal state. Plato favoured a self-sufficient and inward-looking state while Aristotle favoured a self-sufficient but outward-looking state. The principal state in Greek times was the Persian empire, a large and loosely structured unit that was built through conquest and dependent on efficient leadership. One of Aristotle's pupils, Alexander the Great, actively pursued the expansionist (outward-looking) policies favoured by his teacher to destroy the Persian empire and create an even larger but short-lived empire. From this time through to the sixteenth century, Europe experienced a succession of empires with major divisions typically based on religious grounds (for example, between Christian and Islamic areas and between Christian and pagan areas). Even after the rise of nation states empires continued to grow and decline.

The most characteristic state in medieval Europe was the feudal variety. Such states had uncertain boundaries and continually changed in spatial extent. Political control was usually exercised by a few individuals and there was a clear social hierarchy. Such feudal states were gradually replaced by absolutist states that concentrated power in a monarchy; these states in turn were replaced by nation states.

Historically, then, the fact that a political unit exists does not imply that the boundaries delimit a national identity. Many states (such as city states or tribal units) have been too small to qualify as nations, while other states (such as empires) have been so large as to be multinational.

THE RISE OF THE NATION STATE

Nationalism

Understanding the rise of nation states requires an appreciation of the concept of nationalism, a term formally introduced and defined in Chapter 7 during the discussion of language (see glossary). Nationalism is the belief that a nation (cultural group) and a state (political unit) should be congruent. Further, it is the assumption that the nation state is the natural political unit and that any other basis for state delimitation is inappropriate. This also implies that an aspiring nation state will argue:

- that all members of the national group have the right to live within the borders of the state
- that it is not especially appropriate for members of other national groups to be resident in the state
- that the government of the state must be in the hands of the dominant cultural group

It is not difficult to understand how an application of these arguments might lead to conflict with other states or to internal difficulties. Although this principle of nationalism is taken for granted in the modern world, as our brief historical account of states has indicated, it is a recent idea.

Explaining nationalism

The map of Europe that was redrawn in 1815 following the Napoleonic wars was one of states demarcated on the basis of dynastic or religious criteria; no real attempt was made at this time to correlate the national identities of populations with their states or their rulers. Indeed, such an exercise would have proved most difficult because the cultural map of Europe was exceedingly complex and because cultural groups included many social divisions. Such an exercise would also have been inappropriate because nationality was not of great importance at the time; people did not question boundaries based on dynastic, religious, or community differences. This raises the important question as to why national identity, despite obvious problems of definition, emerged as the definitive criterion for state delimitation during the nineteenth century. Five possible explanations are noted:

1. Nation states emerged in Europe in response to the rise of nationalist political philosophies during the eighteenth century.
2. Humans want to be close to those of similar cultural background.
3. The creation of nation states was a necessary and logical component of the transition from feudalism to capitalism, as those who controlled production benefited from the existence of a stable state (a Marxist argument).
4. Nationalism is a logical accompaniment of economic growth based on expanding technologies.
5. The principle of one state/one culture arises from the collapse of local communities and the need for effective communication within a larger group.

Regardless of the specific causes, nation states evolved during the last two centuries; in Europe, Germany and Italy evolved in response to some centrally organized attempt to politically unite a national group. The transition to a nation state in Britain occurred gradually throughout the eighteenth century, while in France monarchical absolutism was intact until the 1789 revolution; in both of these cases there was an early general correspondence between culture and state.

Nation states in the contemporary world
Despite the transition to nation states, our contemporary political map continues to include many examples of multinational or binational states. Good examples of multinational states are many African countries whose boundaries were drawn by Europeans without reference to African national identities. Nationalistic ideology has been most clearly articulated in African countries in the form of anticolonial movements, but without any real changes to the European-imposed boundaries to date.

Typically, such multinational states are politically unstable, prone to changes of government and/or instances of 'minority national' discontent. Canada and Belgium are examples of politically uncertain binational states. States that approach the nation state ideal, such as Italy and Denmark, are found primarily in Europe. States that incorporate members of more than one national group are not necessarily unstable; the United States has largely succeeded in creating a single nation out of disparate groups, while Switzerland is probably the prime example of a stable and genuinely multinational state.

EXPLORATION AND COLONIALISM

Clearly, not all states in recent centuries have attempted to accord with the ideal of the nation state. European countries in particular attempted to expand their territories overseas to create empires similar to those that existed in earlier periods in Europe and elsewhere. Such empires are, by definition, multinational and were typically regarded as additions to rather than replacements for nation states.

European countries began developing world empires about 1500, with most such empires achieving their maximum extent in the late nineteenth century. During the twentieth century, these empires have typically been dissolved. Figure 9.1 shows the late nineteenth-century British empire; large areas of the world were under British control at this time. Today, some 100 years later, the British empire is limited to a few island locations mostly in the Caribbean and south Atlantic (Box 9.1).

Exploration
Most empires began as a result of exploratory activity (but see Box 1.1 for an assessment of the contentiousness of the term 'exploration'). Overall geographers have paid little attention to the study of exploration (which is interpreted as the expansion of knowledge that a given state has about the world) other than to list and describe events. One geographer, however, has derived a conceptual framework for the

---

Box 9.1: Remnants of empire

There remain about 10 million people living in about forty political units still best described as colonies. Three general reasons can be cited as to why these forty continue to remain in a politically dependent situation.

First, several colonies are scheduled to change their political status in the near future in accord with some agreed-upon timetable. The best known example is probably Hong Kong, which is scheduled to revert from Britain to China in 1997. Similarly, Macao is to be transferred from Portugal to China in 1999. There are a number of other colonies currently controlled by France, Denmark, the Netherlands, and New Zealand that are liable to become independent as economic circumstances change.

Second, there are other areas currently remaining as colonies because of their perceived value to the colonial power. Both France and the United States use islands in the Pacific for military purposes, some of which are judged to be of great value; the United States, for example, is reluctant to relinquish control over Guam and Diego Garcia.

A third reason for continuing colonial status is that some areas may not be viable as independent states — some colonies appear destined to remain colonies. St Helena is a prime example.

St Helena is an island of 122 km² (47 square miles) with a population of 5,700. Located in the south Atlantic Ocean, it was reached by the Portugese in 1502. Subsequently the Dutch and British competed for the island and today, the island remains as one part of the much-reduced British empire. The competition that occurred reflected the location of the island as a stopover on the Europe–India route. The current population is ethnically diverse as a result of the movements of slaves from east Africa, indentured labour from China, and British and some other European settlement. Once St Helena ceased to function as a stopover because of the technological advances of steamships, refrigeration, and the creation of a new route using the Suez Canal in 1869, economic problems were soon evident. St Helena lacks a variety of resources and suffers from its small size and small population. For much of this century, the agricultural industry was dominated by flax and, at times, half of the working population was employed in this one activity. Today, agriculture is limited to livestock and vegetables produced for subsistence purposes only.

The future of St Helena is uncertain. It is isolated, reliant on financial support from Britain, and unlikely to prove a tourist attraction. From the British perspective, it is no longer a prestigious overseas possession but a financial liability (Royle 1991).

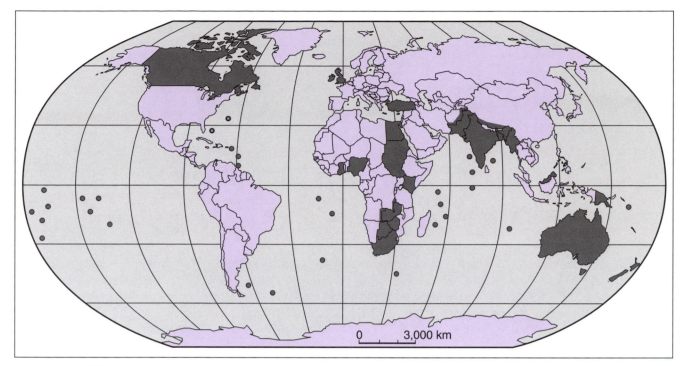

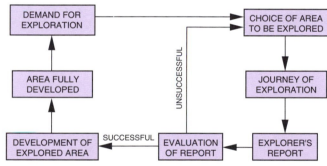

Figure 9.1 *(above)* The British empire in the late nineteenth century.

Figure 9.2 *(left)* Principal elements in the process of exploration.

Source: Adapted from J.D. Overton, 'A Theory of Exploration', *Journal of Historical Geography* 7 (1981):57.

process of exploration, a framework that focuses on the links between events (Overton 1981). As depicted in Figure 9.2, the process commences with a demand for exploration and may, in due course, lead to the 'development' of the new area. This 'development' is likely to mean exploitation of the explored area as a source of raw materials needed by the exploring country. Thus Britain, for example, viewed Canada as a source of fish, fur, lumber, and wheat, and Australia as a source of wool, gold, and wheat.

Colonialism

The way a new area is used by the exploring country is typically a function of perception and need. Economic, social, and political activity in explored areas that became colonies is typically determined by and for the exploring power. This is true of European empires and the acquisition of territory by the United States from 1783 onwards (Figure 9.3). In most cases, Aboriginal populations are eliminated or moved if their numbers represent a military or spatial threat. The economic activity encouraged is that which benefits the cen-

tral power. This is the process introduced in Chapter 6 known as colonialism (see glossary).

In recent European history, colonialism was a form of territorial conquest that involved expansion throughout the Americas, much of Asia (including Siberia), and most of Africa. Competition for colonies was a major feature of the world political scene from the fifteenth to the early twentieth century. The earliest colonial powers were Spain and Portugal, followed by other west European states, the United States, and Russia.

Explaining colonialism

The explanation for colonial expansion is complex; although often summarized as being for 'god, glory, and greed', it can also be explained in terms of various changes in Europe involving the prevailing feudal system and economic competition at the time. It is worth recalling that, shortly before the onset of European colonialism, China and the Islamic world were the two leading world areas; indeed, inventions such as gunpowder, the mariner's compass, printing, paper, the horse

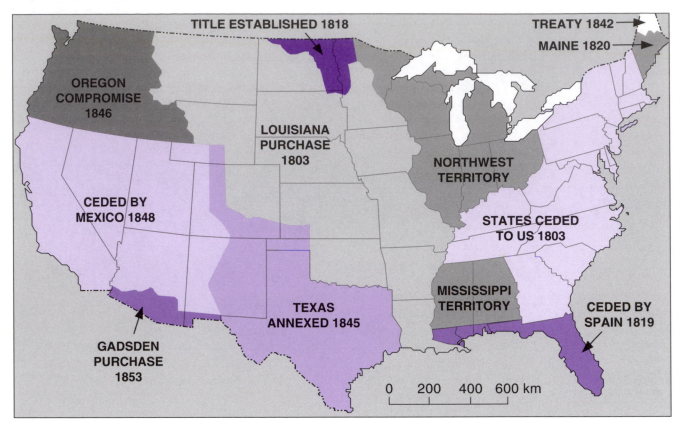

Figure 9.3 The phrase 'manifest destiny' reflects the belief that the new United States had an exclusive right to occupy North America. In 1803, President Jefferson purchased the Louisiana territory from France. Spain ceded the Florida peninsula in 1819, while war with Mexico resulted in the acquisition of Texas and the southeast. The Gadsden purchase of 1853 completed the southern boundary of the US. The northern boundary along the 49th parallel was established in 1848. Subsequently American interests expanded north to Alaska and across the Pacific to include Hawaii.

harness, and the water-mill first appeared in China. The fact that Europe was the region that initiated global movements appears to be related to the demands of economic growth that began in the fifteenth century, as well as to internal social and political complexity, turmoil, and competition. Reasons for the colonial fever that swept Europe, especially in the nineteenth century, included the ambitions of individual officials, special business interests, the value of territory for strategic reasons, and national prestige. But perhaps the most compelling motive was economic; raw materials were needed for domestic industries and the colonial areas provided outlets for industrial products.

Colonialism effectively ended following the Second World War with the removal of Japan and Italy as colonial powers and the failure of Britain to maintain a diverse empire. Three European powers — France, the Netherlands, and Portugal — insisted on trying to retain colonies, but all were unsuccessful.

Decolonization

In recent years there has been an increase in the number of states, from seventy in 1930 to about 187 in 1994. Most of the new states are areas achieving independence from a colonial power. Many of them are the result of either the recognition of a state on national grounds or are areas around which Europeans drew boundaries for their own reasons.

As we have seen, the idea of a national identity originated in Europe, but it has been welcomed by colonial areas that were discontented with colonial rule, had related aspirations for independence, and were offended by the psychological implications of foreign rule. The transition to independence was sometimes violent and sometimes peaceful.

Effects of colonialism

It is possible that most colonies resulted in net losses to the national economies of the colonial powers with profits going primarily to the stock exchange and to selected business and other leaders. For Britain, colonies were a means of reducing population pressure as the colonial areas included vast expanses of Canada, Australia, and South Africa considered suitable for European settlement. The effects on the areas colonized are complex. As discussed in chapters 5 and 6, many areas have experienced massive population growth and various associated economic problems because of the colonial experience and related death rate reductions. Many feel that **imperialism**, a process by which one group of people exerts power via their state over another group in another territory, has had great negative consequences for the less powerful area. This is one of the central arguments of world systems theory as introduced in Chapter 6. A related set of ideas, **dependency theory**, contends that African and Asian countries became poor as a result of being colonies, while the colonial powers were able to advance at their expense.

STATE CREATION: SOME CONCEPTUAL DISCUSSIONS

Early philosophical discussions that described or explained state creation and expansion included the writings of Plato and Aristotle, of Strabo who viewed the Roman empire favourably, of ibn-Khaldun who distinguished between the territories of nomadic and sedentary peoples, and of Bodin and Montesquieu who both favoured the emerging nation state.

Ratzel

The great early nineteenth-century German geographer, Ritter, viewed states as organisms. He saw both cultures and states as progressing through a life cycle, an idea that was to be elaborated on by Ratzel. Ratzel's seven laws concerning the spatial growth of states (Box 9.2) are not examples of rigorous scientific logic; rather, they are generalizations based on observations of a supposed ideal world. The notion of the state as a living organism was central to Ratzel's thinking and clearly subject to some critical appraisal as it was a reification (assumption of independent existence) of something that was actually a human creation. Ratzel's ideas have been highly influential both conceptually and practically.

Jones

A rather different view of states was developed by Jones (1954) based on some earlier ideas by Gottman, Hartshorne, and the political scientist, Deutsch. Jones proposed a chain of events as depicted in Figure 9.4. The chain begins with some political idea and concludes with the creation of a political area. Thus the idea of state creation may lead in due course to actual state creation. Between idea and area are decision, movement, and field. A positive decision is necessary to implement the initial idea. Movement of some variety follows any political decision and a field of activity is in turn created by the movement. This field leads ultimately to the delimitation of a political area. Jones briefly applies these ideas to the creation of the state of Israel. Zionism is the political idea; the Balfour Declaration is the political decision; the principal movement is the immigration of Jews; the resultant field is one of settlement and government activity, and the political area is that of Israel (Box 9.3).

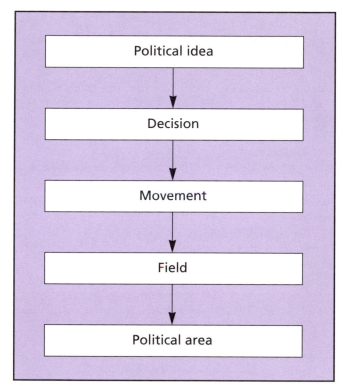

Figure 9.4 Jones field theory.

Box 9.2: Laws of the spatial growth of states

Ratzel's laws, published in 1896, are in fact a series of generalizations:

1. The size of a state increases with its culture.
2. The growth of a state is subsequent to other manifestations of the growth of people.
3. States grow through a process of annexing smaller members. As this occurs, the human and land relationships become more intimate.
4. State boundaries are peripheral organs that take part in all transformations of the organism of the state.
5. As a state grows, it strives to occupy politically valuable locations.
6. The initial stimulus for state growth is external.
7. States' tendency to grow continually increases in intensity.

The key idea is that a state grows as its level of civilization increases.

These laws present an interesting parallel with the Ravenstein laws (Box 6.2). In neither case is strict deterministic logic evident.

Deutsch

A third conceptual contribution is that of Deutsch (see Kasperson and Minghi 1969:211–20), who sees a process of state creation that can involve as many as eight stages:

1. transition from subsistence to exchange economy
2. increased mobility leads to formation of core areas
3. development of urban centres
4. growth of a network of communications
5. spatial concentration of capital
6. increasing group identity
7. rise of national identity
8. creation of a state

Both Jones and Deutsch emphasize evolution and focus primarily on human actions. Neither of them regard states as organisms. Indeed, as already detailed, it is important to appreciate that nations do not exist as conveniently packaged bundles of people waiting for states to be drawn around them (Taylor 1989). Nations and states are both human creations; explaining the existence of nations and nationalistic identity requires a consideration of ethnic distributions and the political construction of nations. This is evident, for example, in contemporary Quebec where aspirations for a separate political identity appear on occasion to be politically orchestrated. Nationalist tendencies may or may not directly involve the populations affected, but nations can be human creations.

Wolfe (1962) tested some of Deutsch's ideas in a Canadian context, while Box 9.4 presents a discussion of the process by which the Canadian state came into being, a discussion that

includes use of a thought-provoking — if somewhat intellectually dangerous — process known as the counterfactual method.

These conceptual discussions lead to one of the most exciting and yet provocative aspects of the study of our political world — geopolitics.

GEOPOLITICS (AND *GEOPOLITIK*)

Geopolitics is the consideration of the relevance of space and distance to questions of international relations. Geopolitical discussions originated in the late nineteenth century, but manipulation of the concept by some German geographers and Nazi leaders in the 1930s and 1940s resulted in the demise of these discussions. Only since the 1970s has geopolitics revived and once again become a legitimate area of interest.

The intellectual origins of **geopolitics** lie in the work of Ratzel, specifically in his seven laws of state growth. A Swedish political scientist, Kjellen, expanded on Ratzel to argue that territorial expansion was a legitimate state goal. He coined the term 'geopolitics' for the study of this particular and narrow field. Clearly, this definition of geopolitics focuses on only one aspect of the larger field, the relevance of space and distance to questions of international relations. It was, however, the narrow interest that dominated geopolitics until it was largely discredited in the 1940s.

GEOPOLITICAL THEORIES

A leading British geographer, Mackinder, writing in 1904 (see Kasperson and Minghi 1969:161–9), was the first to

---

Box 9.3: The Jewish state

The Jews are a religious group with roots in the area that is now Israel. Jerusalem, in particular, is an important symbolic location for Jews — as indeed it is also for Christians and Muslims. During the time of Jesus, this area (known as Palestine) was under Roman rule. In 636 it was invaded by Muslims and from then until 1917 (with a brief Christian interlude during the Crusades), it was a Muslim country, specifically Turkish after 1517. Over the centuries, Jewish populations moved away from this area, in a process known as the Diaspora, to areas throughout much of Europe and overseas. At the end of the First World War (1914–18), Syria and Lebanon were taken over by France, and Palestine and Transjordan (now Jordan) were taken over by Britain, under the League of Nations (the predecessor of the United Nations) mandates. The Palestine mandate — specifically the Balfour Declaration of 1917 — allowed for the creation of a Jewish national home but without damaging Palestinian interests. An impossible goal was being set.

In 1920, Palestine included about 60,000 Jews and 600,000 Arabs. Most of the Jews had arrived since the 1890s.

Jewish immigration increased under the mandate, especially as Jewish persecution became intense in Hitler's Germany. After the Second World War, Jewish survivors struggled to reach Palestine, and by 1947, there were 600,000 Jews, 1.1 million Muslim Arabs, and about 150,000 Christians (mostly Arab). Arab-Jewish conflicts were by now commonplace and Britain, unable to cope, announced that it intended to withdraw.

The United Nations then produced a plan to partition Palestine into a Jewish and an Arab state; Jerusalem was to be an international city. The Arabs did not accept the plan and the Jews then declared the independent state of Israel (1948). Conflict immediately flared between Israel and nearby Arab states and between Jews and Palestinian Arabs, and has continued to the present. Israel established relations with Egypt in 1979, with the Palestine Liberation Organization in 1993, and with Jordan in 1994. As of 1994, the principle obstacle to peace in the region is the lack of relations between Israel and Syria.

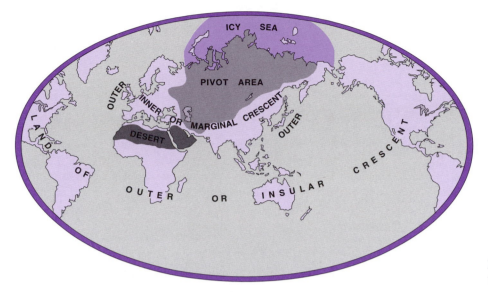

Figure 9.5 Mackinder heartland theory.

formulate a geopolitical theory — the **heartland theory**. Mackinder's work has strong environmental determinist overtones, reflecting British concerns about perceived Russian threats to British colonies in Asia, especially India. Mackinder contended that the Europe–Asia land mass was the 'world island' and that it comprised two regions — an interior 'heartland' (pivot area) and a surrounding 'inner or marginal crescent' (Figure 9.5). He wrote:

Who rules East Europe commands the Heartland;
Who rules the Heartland commands the World Island;
Who rules the World Island commands the World (Mackinder 1919:150).

Mackinder was arguing that location and physical environment were key variables in any explanation of world power distribution. The theory is flawed because of overemphasis on one region (east Europe) and because Mackinder could not anticipate the rise of air power. Overall, the theory has not proved to be an adequate explanation, but it did influence other writers in general and may well have exercised a continuing influence on the United States' policy in particular.

Certainly the earlier work of Ratzel, Kjellen, and Mackinder influenced the rise of *geopolitik* in the 1920s. **Geopolitik** is a specific interpretation of more general geopolitical ideas. It focuses on the state as an organism, on the subordinate role played by individual members of a state, and on

---

Box 9.4: A different Canada?

A renowned historical geographer has posed an intriguing question about Canada (Clark 1975): what would have happened if the French explorer, Samuel de Champlain, had not established a French empire on the St Lawrence in 1608?

His 1603 voyage up the St Lawrence as far as Montreal island left Champlain, an excellent geographer, unimpressed with the St Lawrence as an entry point to North America. Hence he spent the summers of 1604, 1605, and 1606 exploring the North American coastline south of his winter base in Port Royal (Nova Scotia), seeking a superior entry point. These three explorations south were time-consuming, painstaking travels characterized by misfortunes and delays. In each case Champlain was unable to leave Port Royal as planned because supply ships from France had not arrived and because of the after-effects of disease-ridden winters. These facts, combined with his meticulous surveys, prevented him from reaching as far south as the mouth of the Hudson River. Clark argues convincingly, then, that it was only because of a series of unfortunate circumstances that Champlain did not reach the Hudson River. Such a discovery was a legitimate possibility.

Consider the possible consequences. The first crucial consequence is that Champlain would have recognized the value of the Hudson as an entry point and committed the French to that location and not to the St Lawrence. Allow fantasy to rule for a few moments. Champlain could have travelled the Hudson to the Hudson-Mohawk confluence (present-day Albany); the French would then have allied with the local Five Nations Indians with the result being a powerful French core area, a better environment for the favoured seigneurial land-holding system, and a generally more attractive region for French settlement. Feasibly, such a French empire would have been indestructible.

Canada could be a very different place; indeed, it could be in a different place. This intriguing argument should encourage us to appreciate the value (and dangers!) of creative thinking. In human geography the actual and the possible both merit consideration.

the right of a state to expand to acquire sufficient *lebensraum* (living space). The individual most responsible for popularizing these ideas was Haushofer, a German geographer in Nazi Germany who was bitterly disappointed by the territorial losses Germany experienced as a result of the First World War. Haushofer's academic justification for German expansion coincided with Nazi ambitions. The extent of Haushofer's influence on actual events is uncertain but it is the case that German *geopolitik* resulted in a general disillusionment with all geopolitical issues.

Despite this disillusionment, several scholars continued to present their views on the global distribution of power. Spykman, writing in the 1940s (see Kasperson and Minghi 1969:170–7), stressed the importance of Mackinder's 'inner or marginal crescent' (rimland) to produce a **rimland theory**, arguing that the power controlling the rimland could control all Europe and Asia and therefore the world. As an American, Spykman thus saw considerable advantage to having the rimland fragmented, a view that has certainly influenced post-Second World War American foreign policy in Asia (Korea and Vietnam) and in eastern Europe. Also during the 1940s, de Seversky stressed that the United States needed to be dominant in the air to ensure state security (see Kasperson and Minghi 1969:84).

Geopolitics underwent a renaissance in the 1970s and is once again a respected aspect of political geographic study. The most dramatic geopolitical development since the Second World War is the end of the cold war.

THE CONTEMPORARY GEOPOLITICAL WORLD

During the cold war period, from the end of the Second World War until the early 1990s, the world included an ideologically based bipolar division into states belonging to the United States-dominated North Atlantic Treaty Organization (democratic and capitalist) and states belonging to the USSR-dominated Warsaw Pact (communist). This division no longer exists — a geopolitical transition has taken place.

This transition was first in evidence with the installation of a non-communist government in Poland, with the approval of the USSR, in 1989. This was followed with remarkable rapidity by the collapse of communist governments elsewhere in eastern Europe, the 9 November 1989 symbolic breaching of the Berlin Wall, the reunification of Germany in 1990, and the collapse of communism in the USSR in 1991. We are now entering a new and uncertain geopolitical world order.

THE STABILITY OF STATES

Maps of our political world are subject to frequent and often dramatic change. Changes may occur as a result of war or, in rather different fashion, as a result of the establishment of economic groupings of states. Many factors may affect the stability of a given state.

CENTRIFUGAL AND CENTRIPETAL FORCES

The geographer Hartshorne (1950) distinguished between centrifugal and centripetal forces. **Centrifugal** forces tear a state apart; **centripetal** forces bind a state together. When the former exceed the latter, a state is unstable; when the latter exceed the former, a state is stable. The most common centrifugal forces are those involving internal divisions in language and religion that lead to a weak *raison d'être* or perception of identity. Other centrifugal forces include the lack of a long, common history, as is the case in many former colonies, and state boundaries that are subject to dispute. The most

---

Box 9.5: The plight of the Kurds

According to some commentators, the biggest losers in the Middle East during the past 100 years have been the Kurds. Following the collapse of the Ottoman (Turkish) empire at the end of the First World War, other groups — Turks, Arabs, Jews, and Persians — have consolidated or even created their own states. The Kurds remain stateless despite totalling over 20 million dispersed in Turkey (10.3 million), Iran (4.6 million), Iraq (3.6 million), Syria (1 million) and in smaller numbers in other neighbouring states. Their failure to achieve state identity is also despite the fact that in 1920, Britain, France, and the United States acknowledged the need for a Turkish state: 'No objection shall be raised by the main allied powers should the Kurds ... seek to become citizens of the newly independent Kurdish state' (Article 64, Treaty of Sèvres, signed 20 August 1920, quoted in Evans 1991:34).

Why have the Kurds not succeeded in imposing their identity on a region to create a state? Part of the answer is that their identity is rather unclear. Most (but not all) are Sunni Muslims; their society in a mountain environment is typically poor and divided tribally, their dialects are varied, and there is no one political party that speaks for the Kurds. It can be argued that they are united by persecution rather than by their own identity.

Ideally, the Kurds favour a state in the mountainous border regions of Turkey, Iraq, and Iran. Unfortunately for the Kurds, this is an area valued by existing states as a resource-rich environment, including the oil supplies at Kirkuk. Iraq has used chemical warfare against the Kurds and, following the end of the 1991 Gulf War, an extensive campaign to slaughter the Kurds in Iraq, which caused massive population movements into Turkey. Ironically, Turkey is far from a safe haven: in 1979, the then-prime minister of Turkey stated that 'the government will defeat the disease (of Kurdish separatism) and heads will be crushed' (quoted in Evans 1991:35). The future is not promising.

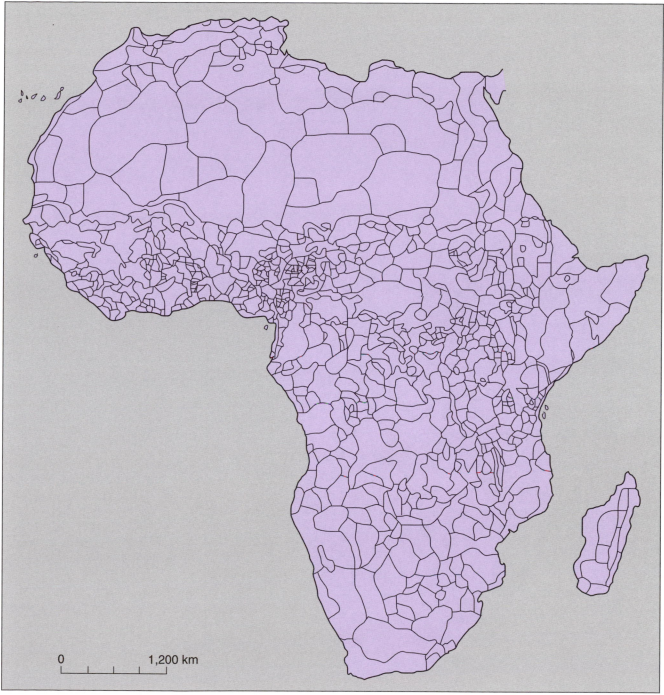

Figure 9.6 African ethnic regions.

Source: L.D. Stamp and W.T.W. Morgan, *Africa: A Study in Tropical Development*, 3rd ed. (New York: Wiley, 1972):41.

common centripetal force is the presence of a powerful *raison d'être*, a clear and well-accepted state identity. Other centripetal forces include a long state history and boundaries that are clearly delimited and accepted by others. The various centrifugal or centripetal forces are often closely related; thus, any internal divisions reduce the likelihood of a long, common history and agreed-upon boundaries. There are also close links between stability and peace and between instability and conflict within and between states.

INTERNAL DIVISIONS

States may face secessionist movements, situations where nations within states are actively operating against the state. One common aim is that of creating a new and separate state; possible examples include the Quebecois in Canada, the Flemish and Walloons in Belgium, and the Welsh and Scottish nationalist movements in the United Kingdom. An alternative aim is to link with members of the same nation in other states to create a new state; the Basques in Spain

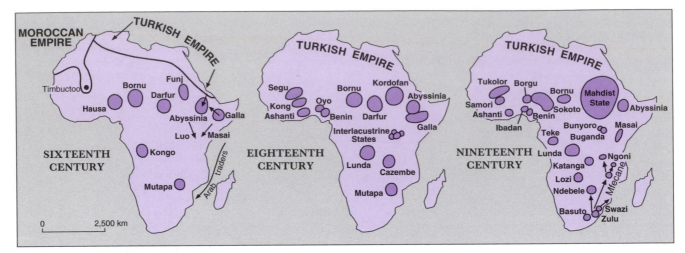

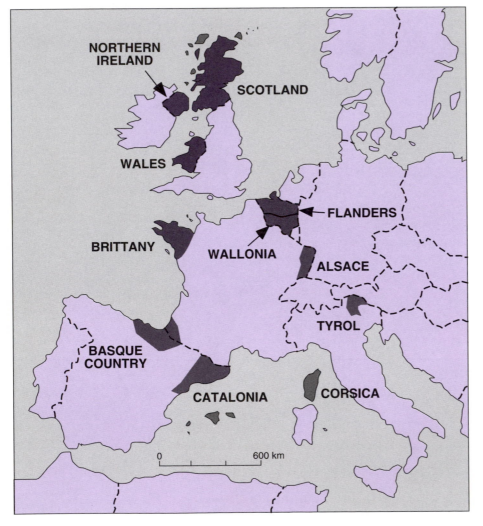

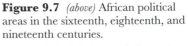

Figure 9.7 *(above)* African political areas in the sixteenth, eighteenth, and nineteenth centuries.

Figure 9.8 *(left)* Areas of dissent in Europe.

and France, and the Kurds in Iraq, Iran, Syria, Turkey, Armenia, and Azerbaijan are two examples (Box 9.5). A third aim is known as *irredentism* (see glossary). This is the situation that arises when a group in one state seeks to belong to another (usually neighbouring) state; Somali irredentism is one example.

Nation and state discordance in Africa

Discordance between nation and state is especially evident in Africa (Figure 9.6). There are only a few instances where a distinct national group correlates with a state; the southern African microstates of Lesotho and Swaziland are examples. African nations that had related states but no longer do so are more numerous and include Hausa and Fulani nations in Nigeria, Fon in Benin, and Buganda in Uganda. The states of contemporary Africa are not a reflection of a long African history but are the creations of colonialism that subsequently achieved independence. State boundaries, shapes, and sizes are all colonial creations, a reflection of past European rather than African interests. In consequence, Africa is characterized by political fragmentation. Some states have a high degree of contiguity (when there are many states, most states have many neighbours); some are very small states or are states with a small population; some have awkward shapes and long, often environmentally difficult boundaries, and some lack access to the sea. Each of these difficulties is related to the colonial past, a past that ignored national identity in the process of creating states.

National identity was evident in many of the precolonial African states. Early states in west Africa were Ghana and Mali (names that are those of modern states, but refer to different areas); both of these states succumbed to Islamic incursions in the eleventh century. States continued to grow and decline. Figure 9.7 offers maps for the sixteenth, eighteenth, and nineteenth centuries. The colonial impact is especially clear when we compare figures 9.6 and 9.7; European neglect of African national identities is a prime cause of African nationalism and contemporary instability in many African states.

Nations and states in Europe

In Africa present-day postcolonialist states rarely accord with nationalist aspirations, whereas in Europe present-day states correspond more closely with national identity. There are, however, many European instances of internal divisions, most of which have a linguistic and/or religious base. The framework proposed by Rokkan (1980) sees four functional prerequisites for states — economy, political power, law, and culture. These four are tied to what Rokkan sees as the constant tension between a European core and its periphery. Further, he identifies two axes: a north-south cultural axis from the Baltic to Italy, and an east-west economic axis. The north-south cultural axis was Protestant (a religion that favoured national aspirations) in the north and Catholic (a religion that cut across national boundaries) in the south. The east-west economic axis had key commercial centres in the west that were good partners in state formation in contrast to the eastern European cities that were unable to offer the resource base needed for the building of states. Rokkan therefore argued that the north-south axis differentiated the conditions of nation building while the east-west axis differentiated the conditions of state building. Ideal conditions for nation state building thus occurred in the northwest.

Contemporary nationalist movements in western Europe can thus be seen as relics of a Rokkan-type core/periphery cleavage in areas not yet homogenized into an industrial society with its related class divisions. In our present context this view is certainly most useful.

Separatist movements in Europe

Regionalism and associated striving for separate identities remains strong in western Europe. Figure 9.8 indicates a series of centres of dissent. In all of these areas there are political parties whose expressed policies are explicitly regional and nationalist. In some cases separation is favoured; in other cases self-government within a federal state structure is favoured, perhaps by means of **devolution**. In most of these European cases language is a key identifying factor, but it is also notable that areas asserting a distinct identity are typically peripheral and economically depressed.

Northern Ireland: Since partition in 1921, a large Catholic minority has sought to separate from the United Kingdom and integrate with predominantly Catholic Ireland.

Scotland: The Scottish National Party was formed in 1934 and achieved notable political success in 1974; in 1978, 33 per cent of voters favoured a devolved assembly with 31 per cent opposed. Today the Scottish National Party is focusing attention on representation in the European Community.

Wales: Plaid Cymru, the Welsh national party, was formed in 1925 to revive the Welsh language and culture; it favours self-government for Wales. The devolution referendum of 1978 saw 12 per cent in favour and 47 per cent against a devolved assembly.

Flanders and Wallonia: The two linguistically distinct regions of Belgium combined as a state in 1830 and have never achieved national unity.

Brittany: Breton (a form of Celtic) is the original language and is still spoken by 10 per cent of the population. Various groups agree that culture and language in the region need to be defended; separation is not typically favoured.

Alsace: A German-speaking area of France that changed between German and French rule in 1871, 1918, 1940, and 1945.

Basque country: A distinctive language region divided by the French-Spanish border; one organization practises terrorism while others strive for separate identity through democratic means. A Basque assembly was convened in 1977.

Catalonia: The Catalan language is being revived with 17 per cent of the population being Catalan speakers; a Catalan assembly was convened in 1977.

Corsica: A militant group that seeks full independence from France; today the French government perceives Corsica as a real threat to national unity; teaching Corsican is now allowed in schools.

Tyrol: An area of German speakers in northern Italy; total autonomy has sometimes been sought.

Other areas that have seen recent movement for autonomy or independence in western Europe include Galicia (northwest Spain), Andalusia (southeast Spain), German speakers in southeast Belgium, the Jura region (France–Switzerland), and Sardinia.

Thus, although language is the principal centripetal force in many western European states, there are numerous specific centrifugal forces. A similar comment applies also to pre-1989 eastern Europe and the former USSR where most nationalist sentiments were little publicized. Events since 1989 have highlighted the national problems of such states as Romania, where a Hungarian minority has been repressed; Bulgaria, where the majority of Bulgars attempted to assimilate Turks in their strivings to create a united Bulgaria; and Yugoslavia, where Serbian nationalism has proved to be highly destructive (Box 9.6).

After 1988, the former USSR experienced considerable separatist pressure from the three Baltic republics — Latvia, Lithuania, and Estonia — and from other southern republics with significant Islamic populations. In 1990, conflict flared once again between Christian Armenians and Muslim Azerbaijanis in the then Soviet Republic of Azerbaijan. It is important to appreciate that the former USSR was not a typical state; rather it was the last great world empire. In 1918, areas such as the three Baltic republics, Ukraine, Byelorussia, Georgia, Azerbaijan, and Armenia, all fought for independence from this empire (Box 9.7).

Cores and peripheries

One of the principal causes of social unrest in general is the presence of a **core/periphery** economic spatial structure. Rich cores and poor peripheries are common features of modern states and often exacerbate any separatist tendencies in the peripheral areas. Thus most of the areas of unrest shown in Figure 9.8 may feel disadvantaged not only in terms of national identity but also in terms of economics.

Box 9.6: Conflicts in the former Yugoslavia

From the mid-fifteenth to the mid-nineteenth century, the area that was to become the federal state of Yugoslavia was a part of the Ottoman empire. By the beginning of this century, some parts were independent, notably Serbia, while other parts were within the Austro-Hungarian empire. At the end of the First World War, Yugoslavia was founded as a union of south (Yugo) slavic peoples, although the name was not adopted until 1929. Following the Second World War, a socialist regime was established under the leadership of Tito, who dominated the political scene until his death in 1980, after which a form of collective presidency was established. The ethnically diverse area of Yugoslavia was organized as six socialist republics and two autonomous provinces (Table 9.1).

The ethnic diversity evident in Table 9.1 is further complicated by the major European cultural division between eastern and western Christianity that runs through former Yugoslavia as well as through Romania, Ukraine, and Belarus (Figure 9.9). It is therefore not surprising that this federal state was dissolved in 1991 following Serbia's attempts in 1988 to exert a greater influence in the federation by assuming direct control of the two autonomous provinces (Kosovo and Vojvodina).

The specifics of dissolution involved the 1991 secession of Slovenia, followed quickly by that of Croatia. In Bosnia, the dissolution has continued to involve considerable conflict between Croatian Muslims and Serbs, including that terrible euphemism, 'ethnic cleansing'.

Ethnic tensions have surfaced within the larger region of the Balkans, not only in the former Yugoslavia. Since the collapse of communist regimes, Bulgaria has felt threatened by its Turkish population and Romania by its Hungarian population.

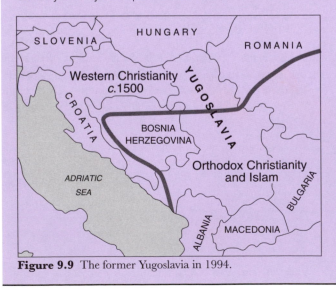

Figure 9.9 The former Yugoslavia in 1994.

TABLE 9.1: *Ethnic groups in the former Yugoslavia*

Republic	
Serbia	85% Serbs
Croatia	75% Croats, 12% Serbs
Slovenia	91% Slovenes
Bosnia-Hercegovina	40% Slavic Muslims, 32% Serbs, 18% Croats
Montenegro	69% Montenegrins, 13% Slavic Muslims, 6% Albanians
Macedonia	67% Macedonians, 20% Albanians, 5% Turks
Province	
Kosovo	77% Albanians, 13% Serbs
Vojvodina	54% Serbs, 19% Hungarians

Box 9.7: The collapse of the USSR

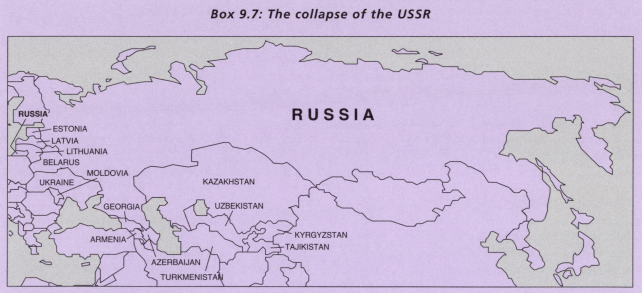

Figure 9.10 The former USSR in 1994.

The Union of Soviet Socialist Republics was never really a union; rather it was an empire tightly controlled by one of the republics, Russia (Figure 9.10). In this sense it was little different from its predecessor, the pre-1917 Russian empire.

Russian expansion from the small core area around Moscow is typically explained in terms of a continuing search for good agricultural land, seaports, and easily defended borders because of regular invasions by groups such as Vikings, Poles, Germans, and Mongols. Initial expansion of Russia was to the north and the White Sea port of Archangel was built by 1584, while to the west, St Petersburg was built in 1703. To the east, Russia expanded across Siberia; the Urals were crossed in 1582, the Sea of Okhotsk at the edge of the Pacific reached by 1639, Alaska claimed in 1741, and fur trade settlements located as far south as California during the nineteenth century. To the south, Russia moved down the rivers Don, Dnieper, and Volga, acquiring Ukraine and gaining access to the Black Sea by 1800. During the nineteenth century, Russia moved into the area between the Black Sea and the Caspian Sea.

Russia retreated from the American area in 1867 with the sale of Alaska to the United States, but in 1917, the new communist government inherited a massive Russian empire. Indeed, the empire expanded with the suppression of local independence movements and the reannexation of several areas that broke away from the Russian empire after the First World War, including Ukraine, Moldova, Estonia, Latvia, and Lithuania. From 1917 until the collapse of the USSR in 1991, there was considerable movement of ethnic Russians into other republics of the Union and policies favouring the Russian language and opposing religious beliefs. The collapse of the USSR was one part of the sweeping changes that occurred in Europe in the late 1980s and early 1990s, leaving a difficult and uncertain legacy of ethnic tensions and often ethnic conflict. Table 9.2 provides data on ethnic populations in the former republics of the USSR, which are now fifteen independent states, and highlights the high percentages of Russians in most of these states.

TABLE 9.2: Ethnic groups in the former USSR

Country (former republic)	Population (millions)	Titular nationality (%)	Principal minorities
Russia	147.0	82	Tatars 4%
Ukraine	51.5	73	Russians 22%
Uzbekistan	19.8	71	Russians 8%
Kazakhstan	16.5	40	Russians 38%, Ukrainians 5%
Belarus	10.2	78	Russians 13%
Azerbaijan	7.0	83	Russians 6%, Armenians 6%
Georgia	5.4	70	Armenians 8%, Russians 6%, Azeris 6%
Tajikistan	5.1	62	Uzbeks 23%, Russians 7%
Moldova	4.3	65	Ukrainians 14%, Russians 13%
Kyrgyzstan	4.3	52	Russians 22%, Uzbeks 12%
Lithuania	3.7	80	Russians 9%, Poles 7%
Turkmenistan	3.5	72	Russians 9%, Uzbeks 9%
Armenia	3.3	93	Azeris 2%
Latvia	2.7	52	Russians 35%
Estonia	1.6	62	Russians 30%

Note: Population data are for 1989.

Many peripheral areas have a specialized or short-lived economic base that is particularly subject to economic problems such as unemployment.

Internal divisions in Canada

Are there internal divisions in Canada? It is a huge country of 9.8 million km³ (3.8 million square miles), 26 million people, and large unsettled areas. Much of it is environmentally difficult — Cartier described the southern coast of Labrador as the land that God gave Cain. Further, Canada comprises ten provinces and two territories, has marked variations between regions and people living in traditional ways and areas and people living in modern settings. It is a plural society comprising Aboriginal peoples, the two founding nations of English and French, and a host of other national groups. Combined, these physical and human factors result in a high degree of regionalism — a major cause of instability. The unity of contemporary Canada continues to be threatened by Quebec, French language, separatism, and various western Canadian separatist movements.

BOUNDARIES

Boundaries mark the limits of a state's sovereignty. They are 'lines' drawn where states meet or where state's territorial waters end. A state's stability is often related to the character and location of its boundaries.

The characteristics that give identity to a nation, such as language, are very rarely as abruptly defined as its boundaries. For this reason, there are many examples of minority situations. In principle, such problems are lessened where the boundaries are *antecedent*; that is, the boundary preceded significant close settlement. Settlers moving into areas close to the boundary must acknowledge the boundary. Such boundaries are also often geometric — as in the case of the United States–Canada boundary west of the Great Lakes, which follows the 49°N parallel. The boundaries of Antarctica are a second example; in this case they were actually defined before significant exploration to minimize conflict between competing states over such matters as mineral rights.

Other boundaries are *subsequent*; that is, they are defined after an area has been settled and the basic form of the human landscape is established. Such boundaries may attempt to reflect national identities — for example, the present boundary of France is an approximation of the *limites naturelles* of the French nation — or may totally ignore such distinctions — for example, most colonial boundaries. Many subsequent boundaries are continually redefined, as in the case of western Europe.

All boundaries are artificial in the sense that what is meaningful in one context may be meaningless in another. Rivers are popular boundaries as they are easily demarcated and surveyed, but they are really areas of contact and not of separation and so often make poor boundaries. They may also be poor boundaries if the river is wide and there are island areas, as in the Mekong River in southeast Asia, or if the river changes course, as in the Rio Grande.

Divided states

In some instances a state is divided into two or more separate parts, a situation that increases the likelihood of boundary problems. A classic example occurred in the decolonization of Britain's Indian empire, beginning in 1947. Following widespread Hindu/Muslim conflict, the two states of India and Pakistan were created with India predominantly Hindu and Pakistan predominantly Muslim. But Pakistan itself was in two parts separated by 1,609 km (1,000 miles) of Indian territory, each part having roughly the same number of people. Perhaps inevitably, relationships between the two parts were poor. The capital was located in West Pakistan, first at Karachi, then at Islamabad. By 1971, a large part of the Pakistan army was actually fighting Bengali guerrillas in East Pakistan, and the independent state of Bangladesh was proclaimed.

Some states have experienced partition, in which case the boundaries between the newly formed states may be especially inappropriate. Examples of such partitioning include Germany (into East and West from 1945 until 1990) and Korea (into North and South since 1945). In both cases, one nation was divided into two states for reasons related to the political wishes of other states, and there are sound reasons to argue that such externally imposed and overly artificial boundaries will not be long-lasting.

GROUPINGS OF STATES

Our contemporary world is characterized by two divergent trends. First, as already discussed, many regions and peoples are actively seeking to create their own independent states; this is a continuation of the processes of nationalism and decolonialism (see boxes 9.5, 9.6, 9.7). Second, some groups of states are actively choosing to unite, perhaps even sacrificing aspects of their sovereignty.

European integration

The principal example of such a voluntary grouping began following the end of the Second World War in 1945, a date that marked the end of an old Europe. Various moves towards European unity culminated in 1957 with the creation of the European Economic Community (EEC) involving six countries: France, Belgium, the Netherlands, Luxembourg, Italy, and West Germany. These six states had already created a European Coal and Steel Community and a European Atomic Energy Community, and all three of these communities merged into a single Commission of the European Communities in 1967. A common agricultural policy was adopted. Other European countries that did not favour such close integration formed the European Free Trade Association in 1960. Gradually, however, the EEC has assumed dominance with Britain, Denmark, and Ireland joining in 1973, Greece in 1981, and Portugal and Spain in 1986 to increase the original six members to twelve. Austria, Sweden, and Finland are scheduled to join in 1995. Norway voted against joining the EEC in November 1994. It seems reasonable to assume that states in central and eastern Europe will be members in the foreseeable future.

What reasons are there for the apparent willingness of European sovereign states to sacrifice some components of independence? Probably most important is the inherent attractiveness of the idea of a united Europe in the context of a world dominated by the United States and, before 1991, the USSR. A fifteen-member EEC (the twelve of 1994 plus Austria, Sweden, and Finland) would have a population of 375 million, larger than the North American Free Trade Agreement (NAFTA) that links Canada, the United States, and Mexico. During the formative years of the EEC, the United States was seen as a real economic threat and the former USSR as a real military threat. Thus, it seemed that there was a place for Europe in the world — but not for some patchwork of European states. The evidence for this, from the European perspective, was an understanding of how the United States and the former USSR rose to superpower status. Both superpowers were large; they comprised compact blocks of territory with a low ratio of frontier to total area; they had substantial east-west extent that increased the area of comparable environment. Both were mid-latitude countries with large populations and densely settled core areas, and both had a wide variety of natural resources. A united Europe, it can be argued, shares some of the above characteristics. Unlike the United States but somewhat similar to the former USSR, however, a united Europe is a multinational creation. Perhaps a voluntary multinational state might be more stable than an involuntary one such as the empire of the former USSR.

Kohr (1957) offers a radically different view of Europe. According to Kohr, aggression is a result of great size and power, and a large number of small states would make Europe a more peaceful place. A Europe divided into states as shown in Figure 9.11 would minimize border and national minority problems. It is possible to see these divisions as a return to an original Europe. Current events in Europe appear to be moving in a different direction, but there is as well much evidence of increased regional authority within existing states.

Other groupings of states
There are many other examples of groupings of states in the world, typically prompted by economic circumstances. Comecon (the Council for Mutual Economic Assistance), an economic grouping of east European states that began secretly in 1949, began to lose power in 1989 and was terminated in 1991. Other groupings include the Commonwealth (an association of states that were formerly parts of

the British empire), NAFTA, the Economic Community of West African States, the Latin American Integration Association, the Organization of American States, the Organization of Petroleum Exporting Countries, the Organization of African Unity, the South East Asia Treaty Organization, and, most recently, the Arab Maghreb Union (Box 9.8). Finally, there is the prospect of a single world government. Since 1945 the United Nations has striven to provide member states with a means of limiting conflict. Few states remain non-members, and developing world states now comprise the majority.

THE ROLE OF THE STATE

Most of the world's population are citizens of specific states and, as such, are subject to the laws of that state, are unable to significantly change the state, and are spatially tethered. Our everyday life is irrevocably involved with government, for all states 'are active elements within society, providing services which are consumed by the public' (Johnston 1982:5). Political units are related to cultural, social, and economic units and it is often useful to define our human geographic regions on a political basis. The state is a factor affecting human life; it has evolved with society, and is a key element in the mode of production. Because of the state's crucial role in the human geography of people and place, it is important to appreciate that there are many different ways in which a state can be governed.

FORMS OF GOVERNMENT

The two principal political philosophies that are accepted today are those of capitalism and socialism (see glossary), but there are related or alternative ideas that are of importance to us. Fundamental to the capitalist form of government is **democracy** — rule by the people. Democracy implies four features: regular free and fair elections of the principal political offices by means of universal suffrage; a government that is open and accountable to the public; the freedom of state citizens to organize and communicate with each other, and a just society offering equal opportunity to all citizens. Democracy was important for a period in classical Greece, but it is only since the nineteenth century that it has reappeared as an important idea and only during the present century that it has been generally approved and commonly practised.

In some countries constitutional monarchies survive, although typically without any real power. Britain is an example of a country where the **monarchy** legitimizes a hierarchical social order, while the Netherlands, Denmark, and Norway have more democratic monarchies. While democracy is rule by the people and monarchy is rule by a single person, **oligarchy** is rule by a few, usually those in possession of wealth. The term 'oligarchy' was introduced by Plato and Aristotle; the favoured term today is 'élite'.

Figure 9.11 European ethnic regions.
Source: L. Kohr, *The Breakdown of Nations* (Swansea: Christopher Davies, 1957).

Box 9.8: The Arab Maghreb union

'Maghreb' is an Arabic word meaning 'west' and has long been used to designate Africa west of Egypt and north of the Sahara. The Maghreb union, involving Morocco, Algeria, Tunisia, Libya, and Mauritania, was formally established in 1989. In general, these five states comprise a geographic region in both physical and cultural terms, although the state of Western Sahara is not involved in the economic union.

Given the course of African history, this region has never been a single political unit and the contemporary boundaries are, as elsewhere in Africa, a consequence of European colonialism. Libya was colonized by Italy, and the other four states

by France. Libya achieved independence in 1951, Tunisia and Morocco in 1956, Mauritania in 1960 and Algeria (after eight years of war) in 1962.

The principal reason for the recent union appears to be a common desire to improve economies. There is a general assumption that integration will benefit all five states. There is also a link with recent developments in Europe, for when Spain and Portugal joined the European Economic Community in 1986, Morocco and Tunisia lost major markets for agricultural products. Finally, the union has occurred as internal rivalries appear to be diminishing (Arkell and Davenport 1989).

Government by **dictatorship** implies an oppressive and arbitrary form of rule that is established and maintained by force and intimidation; military dictatorships are common features in the contemporary world in such countries as Chile. **Fascism** is an extreme form of nationalism that, especially in Europe between 1918 and 1945, provided the intellectual basis for the rise of political movements to combat governments whether they were capitalist or socialist. In Italy and Germany, fascist parties achieved power through a combination of legality and violence.

The political philosophy of **anarchism** may emphasize either individualism or socialism and was advocated by two notable nineteenth-century geographers, Kropotkin and Réclus. It rejects the concept of the state and the associated division into rulers and ruled.

SOCIALIST LESS-DEVELOPED STATES

Socialism is an imprecise term, but we can identify two general characteristics. First, socialist regimes aim to remove any and all capitalist structures, especially private ownership of resources, the role of the market place in allocating resources, and related class structure. Second, the socialist state is able, in principle, to make substantial changes to society. Recent experiences in the socialist-developed world of eastern Europe have made it clear that neither of these characteristics has met with popular approval.

Socialism has had a major impact in many countries in the less developed world. Several African countries, notably Tanzania, have attempted to combine tradition with socialist ideals. In Latin America, Cuba has maintained a socialist system since 1959 in defiance of the dominant capitalist model, while socialist movements have played an important role in mobilizing the poor in various other countries, but it is in Asia that socialism has been able to exert a significant and continuing influence (Hirsch 1993). Today there are socialist or communist governments in four major Asian countries, namely China, Laos, Vietnam, and North Korea. Of these, only North Korea maintains a traditional hardline, secret, closed system. Indeed, North Korea today is run by a communist royal family; before his death in mid-1994, Kim Il Sung groomed his son, Kim Jong Il, as heir.

One particular version of socialism is **Maoism**, which is the revolutionary thought and practice of Mao Zedong who was the leader of the peasant revolution in China that resulted in the 1949 creation of the People's Republic. Maoism has two components: the strategy of protracted revolution to achieve power and the socialist policies practised after the revolution was successful. Both aspects continue to be influential in other parts of the world.

In many less developed countries, socialism has a strong anticolonial, nationalist content. Unlike former socialist states in the developed world, most of these states are firmly rural, not urban and industrial, in character. Although the details vary from state to state, it is fair to say that, even more than in capitalist states, individual behaviour is often determined by larger state considerations. Fertility policies, for example, are likely to be much more rigorous, as in China. Perhaps most importantly, applications of central planning result in considerable state involvement in people's everyday lives. The principal difference between capitalist and socialist states is that in capitalism state intervention in everyday life is tied to the concept of alienation, whereas in socialism it is directly tied to the central government. Whether capitalist or socialist, the state is a major factor affecting individual lives.

SUBSTATE GOVERNMENTS

It has been argued that capitalist states are a necessary part of the capitalist mode of production, while socialist states explicitly organize people and land via mechanisms of central planning. These comments have ignored the fact that, in many states, political authority is decentralized. The political geographer Paddison (1983) has distinguished three levels of decentralization (Figure 9.12).

- *Federal*: The most decentralized form of government; examples include Australia, Canada, and the United States; the purpose of **federalism** is to prevent one level of government from dictating to another.
- *Unitary*: The most centralized; local governments are a means by which the central state organizes the political hinterland.
- *Compound unitary*: This is intermediate between the federal and unitary types, and devolves substantial powers to subnational governments. There are two types: Type I has established regional governments, and Type II involves small (usually peripheral) areas that maintain some distinct identity while retaining ties to the state.

This useful classification demonstrates that most of the large population states (except China) are federal, and that the most common type of state is unitary. Any understanding of the state's role must therefore consider all appropriate levels of government.

Australia is a good example of a state in which regional units, called states, have considerable authority. The present subnational states were previously separate British colonies. Like many national boundaries, these subnational boundaries do not necessarily reflect contemporary circumstances. In the Australian case, the boundaries reflect decisions by the British government and various intercolonial rivalries. One impact of the colonial boundaries was recorded by Thomas Knox, an American traveller of 1888:

Beyond Goulbourn the railway carried our friends through the district of Riverina ... At Albury they crossed the Murray River and entered the colony of Victoria; a change of gauge rendered a change of train necessary, and Fred remarked that it seemed like crossing a frontier in Europe, the resemblance being increased by the presence of the custom-house officials, who seek to prevent

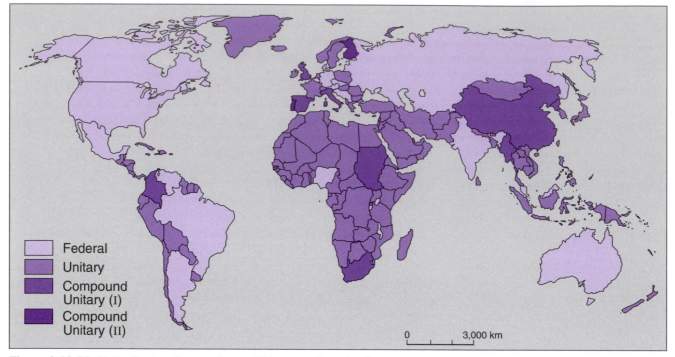

Figure 9.12 World distribution of types of states. This map reflects the distribution of states in the early 1980s; the principal change since that time is, of course, the collapse of the Soviet Union.

Source: R. Paddison, *The Fragmented State: The Political Geography of Power* (New York: St Martin's Press, 1983):32.

the admission of foreign goods into Victoria until they had paid the duties assessed by law (in Larkins and Parish 1982:49).

Although the transition from colonies to states, which occurred in 1901, has reduced the importance of the subnational units, there is no doubt that any discussion of Australian geography and landscape must consider the role played by the component parts of the federation both past and present.

EXERCISING STATE POWER

Capitalist states have a set of institutions and organizations through which the state's power is exercised (Clark and Dear 1984). These are important to the effective functioning of the state and include political and legal systems, the means to enforce the state's power, and several mechanisms to influence economic matters. There are typically significant spatial variations in the management of this **state apparatus** as there are also in the incidence of public-sector income and expenditure and the provision of **public goods**, including services. Studies of these matters by human geographers often focus on the unequal availability of, for example, health care or education, a general theme that was introduced in the previous chapter. Public goods are often deliberately distributed in an unequal manner to influence an electorate prior to an election (a popular American term for this behaviour is 'pork barrelling'); human geographers often think of social justice in terms of territorial justice. It

is clear that the form of government and the particular political philosophy favoured directly affect the manner in which state power is exercised.

One of the most critical aspects of the power exercised by individual states concerns the need for international cooperation in solving global environment problems. Recall 'The Tragedy of the Commons' (Box 4.2) to appreciate that state power is needed to ensure that industrial firms do not harm the environment. Also, some international authority is needed to ensure that individual states also behave appropriately. But there is not as yet any such international state, only occasional groupings of states for particular problems. There are at least two dilemmas here (Johnston 1993). First, governments in developed countries may be unwilling to protect the global environment when such actions will result in a loss of jobs and wealth (and therefore votes) in their own state. Second, governments in less developed countries typically contend that they cannot afford to implement environmentally appropriate policies and practices, and that they are not the principal causes of environmental problems.

ELECTIONS: GEOGRAPHY MATTERS

Much attention has been paid to the question of voting since Krebheil's (1916) analyses of geographic influences in British elections. The typical focus has been on spatial and temporal variations in voting patterns and on such causal variables as environment, economy, and society. Clearly, in any analysis of elections, geography matters. Geography matters with regard to the boundaries of voting districts, voting behaviour, government activity, and larger world issues.

CREATING ELECTORAL BIAS

In 1812, the governor of Massachusetts, Elbridge Gerry, rearranged voting districts (Figure 9.13) to favour his party. Thus, **gerrymandering**, a word coined by Gerry's political opponents, refers to any spatial reorganization designed to favour a particular party. It is a procedure that aims to produce electoral bias in one of two ways: by placing supporters of the opposition party in one electoral district, or by scattering those supporters so that they cannot have a majority anywhere. Racial gerrymandering has been especially common, as in congressional district boundaries in Mississippi (Figure 9.14). Prior to 1966, Blacks formed a majority in one of the five districts; the redrawing of boundaries was designed to ensure that Blacks were a minority in all districts.

Gerrymanders are deliberate creations that produce an electoral bias, but it is only recently that they have been regarded as violations of the Constitution in the United States. A second method, also widely used, involves **malapportionment**. In this case, boundaries are drawn so that small electoral districts are created containing voters of one's own party, while large electoral districts are created containing a majority of opposition voters. In many countries there is a typical density distinction between urban and rural areas, and rural voters can exert a greater impact than the equivalent number of urban voters. Given the possibility of a distinct attitudinal difference between rural and urban residents, this situation can be easily manipulated.

Clearly, electoral bias is not difficult to produce, but it can also be produced quite unintentionally. What comprises a fair procedure for drawing electoral boundaries? Objectively, the number of voters in each district ought to be as nearly equal as possible and districts should be contiguous. Other than these two considerations, there is probably little else that can be agreed upon. Some observers argue that a district ought to be a meaningful spatial unit in a human geographic context; others argue that districts ought to be as diverse as possible.

VOTING AND PLACE

It is not uncommon to assert that class is a dominant influence on voting behaviour. Thus, in Britain there is a distinction between the Labour party, representing the views of workers against those of the capitalist employers, and the Conservative party, representing the employers. A similar distinction can be made in the United States between the Democrats (who are worker-based) and Republicans (who are employer-based). In Canada, although the situation is less clear, the Liberals are more working-class oriented than are the Conservatives, although ethnic divisions can be a key factor here. Recent election results in Northern Ireland, Scotland, and Wales also exhibit a tendency for voting to be linked to ethnic or nationalist interests.

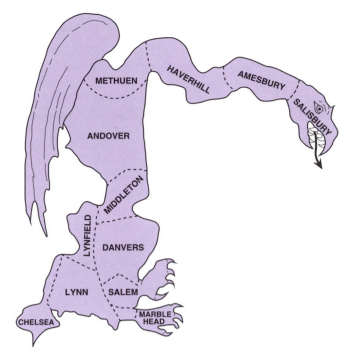

Figure 9.13 *(above)* The original gerrymander.

The intent of the first gerrymander (*Boston Gazette*, 26 March 1812) was to concentrate the vote for a particular party in a few districts. The term results from the salamander-like appearance of the districts and from Governor Gerry who signed the districting law.

Source: R. Silva, 'Reapportionment and Redistricting', *Scientific American* 213, 5 (1965):21.

Figure 9.14 *(below)* Congressional districts in 1960s Mississippi. (a) shows the pre-1966 districts; (b) the post-1966 districts that guaranteed a minority of Blacks in all districts — a deliberate gerrymandering.

Source: Adapted from J. O'Loughlin, 'The Identification and Evolution of Racial Gerrymandering', *Annals, Association of American Geographers* 72 (1982):180.

BLACK POPULATION

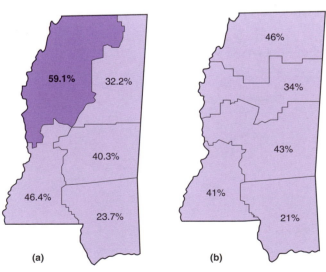

TABLE 9.3: *Disputes and conflicts involving the UN, 1945–90*

Operations by UN forces or military observer groups

Greek frontiers, 1946–9
Indonesia, 1947–51
Palestine/Israel, Egypt, Lebanon, 1947–
Kashmir, 1948–
Korea, 1950–3
Congo/Zaire, 1960–4
West Irian, 1962–3
(North) Yemen, 1963–4
Cyprus, 1964–
Dominican Republic, 1965
India–Pakistan, 1965–6
Iran–Iraq, 1988–
Afghanistan, 1989–
Angola, 1989–
Namibia, 1989–90
Central America, 1989–
Haiti, 1990

Former UN trust territories (now independent or merged with a neighbouring state)

Togolands
Cameroons
Rwanda
Burundi
Tanganyika
Somalia
NE New Guinea
Nauru
Western Samoa

Other disputes and conflicts that have involved the UN

West Berlin, 1948–9
Taiwan, 1949–71
South Africa, 1952–
Hungary, 1956
Algeria, 1956–61
Indochina, 1958–66
Tibet, 1959–61
Tunisia, 1961
Kuwait, 1961–3
Mozambique, 1961–75
Cuba, 1962
Rhodesia–Zimbabwe, 1962–79
Borneo, 1963–5
Gibraltar, 1965–
Falklands, 1965–
Czechoslovakia, 1968
Western Sahara, 1974–
East Timor, 1975–
Cambodia, 1979–
Iran–United States 1979–80
Afghanistan–USSR, 1979–
Malta–Libya, 1980
Belize–Guatemala, 1981
Libya–United States, 1986
Venezuela–Guyana, 1986–7
Iraq–Kuwait, 1990–

Source: A. Boyd, *An Atlas of World Affairs*, 9th ed. (New York: Methuen, 1991):30–1.

But is class really a key explanation of voting? Do people from different regions of a country but from the same class vote similarly? We have already noted that the special circumstance of nationalism can disturb this relationship. Place also matters. Recent geographic research has clearly demonstrated that where a voter lives is crucial. This is especially well documented by Johnston (1985) using data from the 1983 British election: national trends cannot simply be transferred to the local scale. Four local influences on voting can be identified.

The first of these involves sectional effects: 'a long-standing geographical element of voting whereby differences in local and regional political culture produce spatial variations in the support given to the various political parties' (Johnston

---

Box 9.9: Some traditional enmities

The contemporary world includes many examples of long-standing rivalries involving states and nations. Although these rivalries may not be evident at any particular time, they represent potential conflict areas. The following highly selective list does not include rivalries involving minority groups within states.

In Europe there are traditional enmities involving:

Poland–Germany
Poland–Russia
Romania–Hungary
Turkey–Russia
Turkey–Greece
Turkey–Bulgaria

In the Middle East there is a traditional enmity between Turkey and Syria and a whole set of post-1948 enmities between Israel and Arab states.

In Asia there are traditional enmities involving:

Thailand–Cambodia
Thailand–Burma
China–Russia
China–Vietnam
China–Mongolia
India–Pakistan
Japan–Korea

In Africa there are numerous ethnic rivalries, but they rarely assume the form of state rivalries. In southern Africa there are rivalries between the Zulu and other groups in South Africa and between Shona and Ndebele in Zimbabwe. In eastern Africa there is an Ethiopia–Somalia rivalry, and throughout the Sahel region, there are rivalries between Arabs to the north and Black Africans to the south.

In South America there are rivalries between Brazil and Argentina and, in general, between Indian and Hispanic groups.

1985:279). Sectional effects have also been clearly demonstrated in presidential elections in the United States. Long-standing geographic cleavages require any successful candidate to build an appropriate geographic coalition; these cleavages are so factored into the electoral college system that to be elected president, votes are not enough — votes in the right place are required.

A second local influence involves environmental effects such as level of unemployment — specifically, in the 1983 British election, the higher the level of unemployment in an area, the more successful the Labour party candidate — and the impact of an incumbent candidate — specifically, incumbency results in increased votes.

A third local influence involves campaign effects: 'Vote-switching was more common in the safer seats and the longer-established parties (Conservative and Labour) did better in the marginal constituencies' (Johnston 1985:287).

A fourth local influence concerns social contacts; individuals are influenced in their voting (as they are in other behaviours) by friends, neighbours, relatives, and fellow employees with whom they are in contact. Known as the **contextual effect**, this argument is in general accord with the symbolic interactionist interpretation of culture noted in Chapter 8 and it implies that analyses of elections need to consider social contexts.

Voting is class-based and place-based. Any successful political party needs to develop a strong social and spatial base, and any meaningful analysis of elections needs to consider class and place.

THE GEOGRAPHY OF PEACE AND WAR

'In geographical terms, this planet is not too small for peace but it is too small for war.' So wrote the geographer, Bunge (1988), drawing a distinction between all past wars and any future world war. Past wars have not been global in impact. A future world war will be on a global scale.

CONFLICTS

Conflict is typically between states and it is therefore important to understand who is in control of the state apparatus and how that control is exercised. The concepts of war and peace were only clearly distinguished in European thought beginning in the eighteenth century with the emergence of capitalism and nation states. During this time, war became a state activity with the creation of professional armies and the elimination of private armies; also at this time, war became a temporary state of affairs with long periods of peace between conflicts. Between the French Revolution and the First World War, wars were between nation states. After 1918 and until 1991, conflicts were ideological, variously between communism, fascism, and liberal democracy. These comments apply to the western world and it is only since the end of the cold war that other cultures have begun to play a major role in global politics.

The contemporary world contains many traditional enmities (Box 9.9). Most of these relate to ethnic rivalries and/or to competition for territory. Conflicts between peoples, a feature of human affairs for eons, have become increasingly formalized and structured over time with the increasing number of independent states. Indeed, some authors believe that aggression is a necessary human behaviour (Box 9.10). Since 1945, the United Nations has offered member states the opportunity to work together to avoid conflict, and in Korea many countries fought together under the UN flag to resist aggression. Table 9.3 lists a wide array of disputes and conflicts that have involved the UN.

Conflicts may be categorized as (1) traditional conflicts between states, (2) independence movements against foreign domination or occupation, (3) secession conflicts, or (4) civil wars that aim to change regimes.

Examples of category 1 since 1945 include three Indian-Pakistan wars, four wars involving Arab states and Israel, and the Vietnam War. Examples of category 2 arose primarily as a consequence of decolonization and include the

---

Box 9.10: Naturally aggressive?

A fundamental debate in social science concerns a particular human characteristic: aggression. Are we naturally (genetically) aggressive or do we learn (through culture) to be aggressive?

Certainly aggression is a normal part of the world we live in. It operates at many spatial and social scales and assumes a variety of forms. But there are variations between cultures — some exhibit considerable aggression, others relatively little. So, are we naturally aggressive?

Desmond Morris believes we are. In his popular book, *The Naked Ape* (1967), he argues that humans fight one another to establish dominance over each other, to defend territory and defend family; such fighting is, therefore, biologically pro-

grammed. Another popular writer, Konrad Lorenz, produces a similar argument in *On Aggression* (1967). Lorenz develops his views about human aggression only after detailed consideration of animal behaviour. Ashley Montague sees things very differently. In *The Nature of Human Aggression* (1976), Montague acknowledges that humans have become increasingly more violent as technology advances, but he does not see such behaviour as genetically controlled. Indeed, early human societies, far from being aggressive, are argued to be characterized by cooperation. For Montague, humans are neither 'naked apes' nor 'fallen angels'.

Belgian Congo (1958–60), Mozambique (1964–74), and Indonesia (1946–9). Examples of category 3 include Tibet (1955–9), Biafra (1967–70), and Philippine Muslims (1977–). Examples of category 4 include China (1945–9), Cuba (1956–9), Bolivia (1967), and Iran (1978–9).

The conflicts between states (category 1) have the greatest potential for further disruption of our human world. Geographers have focused on various theories of international relations to explain such conflicts between states. Principal variables in such theories include power, environment, and culture. Much of the related empirical work is strongly quantitative.

THE COSTS OF WAR

It is estimated that, between 1945 and 1990, the world spent perhaps $16 trillion for defensive or offensive reasons. Today the developed world spends twice as much as it did in 1960 and the less developed world spends six times as much (both figures allowing for inflation). There are some 29 million people employed by the armed forces and some 11 million employed in various weapons industries. Furthermore, our capacity to destroy is greater than ever and always increasing. Possibly as many as 25 million people have died in conflict since 1945 — this is the same total as those who died in the Second World War. Most of these deaths have been in the less developed world; most of the weapons used originated in the developed world. Military expenditures are a great drain on many economies; Israel spends some 27 per cent of gross national product in this way and many poor countries spend upwards of 10 per cent compared to a world mean of 5.6 per cent.

The world may not be at war, but it is most certainly not at peace with itself.

THE GEOGRAPHY OF NUCLEAR WAR

Geographers have contributed significantly to our understanding of the causes of conflict, but — possibly most important — they have helped clarify the consequences of war, especially nuclear war. In Britain, Openshaw, Steadman, and Greene (1983) have published an extensive series of estimates of probable casualties following any nuclear attack; Bunge (1988) has published a provocative yet penetrating *Nuclear War Atlas*, while other geographers have focused on possible climatological effects of a nuclear exchange (Elsom 1985). Nuclear war means national suicide for any country involved, while a large-scale nuclear war will affect all environments. Large areas of the Northern hemisphere are likely to experience subzero temperatures for several months regardless of the season. These temperatures and reductions in sunlight would adversely affect agricultural productivity. A nuclear winter would result in many deaths from hypothermia and starvation. It is not necessary to belabour these points. Let us consider instead the role that geographers can play in influencing public awareness

and alerting public policy to the follies of nuclear war.

Geographers have two important responsibilities: to teach people to love the land and the peoples of their state, and to teach people not to hate and fear other states. Undoubtedly, human geographers are in an enviable position to achieve such goals. They have studied the environmental and human consequences of nuclear war. They have used cartography (long a most effective propaganda tool) to teach and so are able to identify any deliberate territorial biases. And their study of landscapes, places, and their interrelationships can help produce the global understanding that is so desperately lacking in our contemporary world.

PERPETUAL PEACE OR A CLASH OF CIVILIZATIONS?

Despite the end of the cold war that began in 1945, the beginnings of significant disarmament, and the transition to democracy in South Africa, the early and mid-1990s have not seen the beginnings of what Kant called 'perpetual peace'. There has been a rise of nationalism, populism, and fundamentalism with conflicts erupting in numerous areas, including the Persian Gulf, the former Yugoslavia, Somalia, and Rwanda.

It seems possible that the next pattern of conflict will be what Huntington (1993:22) describes as a clash of civilizations or cultures:

It is my hypothesis that the fundamental source of conflict in this new world will not be primarily ideological or primarily economic. The great divisions among humankind and the dominating source of conflict will be cultural. Nation states will remain the most powerful actors in world affairs, but the principal conflicts of global politics will occur between nations and groups of different civilizations. The clash of civilizations will dominate global politics. The fault lines between civilizations will be the battle lines of the future.

Huntington (1993:25) identifies seven or eight major cultures: western, Confucian, Japanese, Islamic, Hindu, Slavic-Orthodox, Latin American, and possibly African. It is instructive to compare these with the regions proposed by Toynbee as detailed in Chapter 7 and the cultural regions delimited in Figure 7.2 and also link these basic ideas with those so well expressed by James and included as opening remarks to Chapter 7. It can be argued that conflict will be based on these cultural divisions for six basic reasons:

1. Differences between cultures are more fundamental than those between political ideologies; they include those of language, religion, and tradition. These are basic differences that imply different views of the world, different relationships with a god or gods, different social relations, and different understandings of individual rights and responsibilities.

9.1 Rwandan refugees arriving in Tanzania, 1994. (UNHCR)

2. The world is becoming smaller, which increases contacts and can intensify cultural differences.

3. The processes of ongoing modernization and social change result in people being separated from long-standing local identities and weaken the state as a source of identity. Religion, in various fundamentalist versions, is increasingly filling the resulting gap, providing a basis for identity.

4. A process of dewesternization and indigenization of élites is occurring in many parts of the less developed world, thus increasing the desire to shape the world in new ways.

5. Cultural characteristics, especially religion, are difficult to change.

6. Economic regionalism is increasing and is most likely to be successful when rooted in a common culture.

The argument for cultural differences as the principal basis for future global conflicts is a strong one. Indeed, we have already discussed numerous regional and local conflicts that are rooted in cultural differences, especially language and religion, in this and the two preceding chapters. We may not be naturally aggressive (see Box 9.10), but conflict at all social scales, much of which is not necessarily harmful, does seem to be fundamental to human existence.

SUMMARY

Dividing territory

Humans partition space, and the most fundamental division is the sovereign state. Almost all of us are subjects of a state. Loosely structured empires were typical of Europe before 1600. The link between sovereignty and territory emerged most clearly in Europe after 1600.

Nation states

In principle, a nation state is a political territory occupied by one national group. Despite the importance of nationalism and the territory-state-nation trilogy, there are many current examples of binational or multinational states. Numerous scholars have attempted to explain state creation: Ratzel viewed states as organisms; Jones developed a field theory progressing from idea to area, and Deutsch outlined an eight-stage sequence. We must not forget that nationalism — the idea that every nation should have its own state — is a complex notion that may or may not involve the populations affected. Nationalism is the belief that each nation has a right to a state; only in the twentieth century has nationalism become a prevalent view.

Empires: Rise and fall

World history includes many examples of empire creation and collapse. European-based empires evolved after 1500 and typically disintegrated after 1900, greatly increasing the number of states in the world. Colonies quickly became dependent on mother states in a process known as imperialism, in which one group exerts power, via their state, over another group located elsewhere. The most compelling motivation for colonialism was economic.

Geopolitics

This term refers to any discussion of the roles played by space and distance in international relations. Early geopolitical arguments by Ratzel and Kjellen included the idea that territorial expansion was a legitimate state goal. This idea dominated geopolitics until it was discredited in the 1940s as a consequence of being closely associated with a Nazi expansionist ideology. Other important geopolitical theories include Mackinder's heartland theory and Spykman's rimland theory. In the early 1990s, the cold war that began in 1945 ended and a geopolitical transition occurred.

States: Internal divisions

Whether or not a state is stable is closely related to the relative strength of centrifugal and centripetal forces. A key centripetal force is the presence of a powerful *raison d'être*. Thus a key centrifugal force is the lack of such and the existence of minority groups and hence secessionist movements. In

Africa examples of nation and state discordance abound as a result of state boundaries imposed by former colonial powers without any reference to national identities. European states typically have greater internal homogeneity, although there are many examples of minority issues, some of which might be explained by reference to core/periphery concepts.

The collapse of the former USSR, an ethnically diverse empire, is to be noted, as is the disintegration of Yugoslavia, a country that comprised very different republics and that is divided by a major religious boundary. Canada's instability is related to a wide variety of physical and human factors. Many areas that are asserting a distinct identity are peripherally located in a state and economically depressed.

Boundaries of states

These may be closely related to stability. State boundaries are lines, but national boundaries are not. Antecedent boundaries precede settlement; subsequent boundaries succeed settlement.

Groupings of states

There are two opposing trends in our political world: an increasing desire for minority groups to have their own states, thus further fragmenting our world, and an increasing desire for states to group together, usually for economic purposes. The European Economic Community is the principal instance of the second trend. As of 1994, it includes twelve countries and is scheduled to increase to fifteen (possibly sixteen) in 1995.

Substate governments

Political authority is decentralized in federal states. Many of the states with large populations are federal. As is the case with state boundaries, substate boundaries often do not reflect physical or human differences.

The power of the state

There are numerous types of states and numerous related political philosophies. States are important in many different arenas, both internally with regard to the distribution of public goods, and in terms of relationships with other states, especially regarding conflict and environmental concerns.

Elections

Election analyses require a geographic focus. Some electoral boundaries are deliberately drawn to create an electoral bias using gerrymandering or malapportionment. The supporters of political parties are often spatially segregated. Place is a significant factor related to voting behaviour.

Peace and war

Analyses of peace and war, with their obvious implications for both people and place, are inherently geographic. Geographers can contribute to an understanding of the causes and consequences of conflict. Even countries that are not at war are heavily committed to military and related expenditures. Geographers have indicated the consequences of any nuclear war and can also contribute to increased understanding between states. Peace education is a part of geography. There is a possibility that any future global conflicts will be based on differences in culture.

WRITINGS TO PERUSE

ABLER, R. 1987. 'What Shall We Say? To Whom Shall We Speak?' *Annals, Association of American Geographers* 77:511–24.
Focuses on the role of geography in our contemporary world, emphasizing the need to educate others on matters of key political concern.

ARCHER, J.C., et al. 1988. 'The Geography of US Presidential Elections'. *Scientific American* 259, no. 1:44–51.
This important article shows that the US electorate is divided by enduring geographic cleavages such that any winning candidate must build a geographic coalition.

BELL-FIALKOFF, A. 1993. 'A Brief History of Ethnic Cleansing'. *Foreign Affairs* 72, no. 3:110–21.
A summary article that helps place the current (1994) situation in Bosnia in a sound historical perspective; it also explains the importance of homelands in a political context.

BOYD, A. 1991. *An Atlas of World Affairs*, 9th ed. London: Methuen.
An indispensable factual account of our contemporary political world. There are companion atlases dealing with Africa, North America, and the European Economic Community.

BUNGE, W. 1973. 'The Geography of Human Survival'. *Annals, Association of American Geographers* 63:275–95.
A thought-provoking and deeply passionate piece by a most important geographer. Bunge is a challenging and distinctive writer who writes with deep concern in an often polemical fashion.

CUTTER, S.L. 1988. 'Geographers and Nuclear War: Why We Lack Influence in Public Policy'. *Annals, Association of American Geographers* 78:132–43.
An important statement addressing the actual and potential role of geography in peace education and nuclear awareness.

GOTTMAN, J. 1973. *The Significance of Territory*. Charlottesville: University Press of Virginia.
An original essay by a well-known political geographer with substantial conceptual and historical content.

KASPERSON, R.E., and J.V. MINGHI, eds. 1969. *The Structure of Political Geography*. Chicago: Aldine.
A pioneering collection of forty articles that includes works by Ratzel, Mackinder, Whittlesey, Hartshorne, and Wallerstein.

KLIOT, N., and S. WATERMAN, eds. 1983. *Pluralism and Political Geography: People, Territory and State*. London: Croom Helm.
A series of readings that includes good discussions of theory and a series of important examples.

KNIGHT, D.B. 1982. 'Identity and Territory: Geographical Perspectives on Nationalism and Regionalism'. *Annals, Association of American Geographers* 72:514–31.
A clear discussion of some of the difficulties of defining nation and state and the related problems of group delimitation.

LIND, M. 1994. 'In Defence of Liberal Nationalism'. *Foreign Affairs* 73, no. 3:87–99.
A succinct summary of the contemporary implications, especially in Europe, of the simple idea that every nation should have its own state.

MUIR, R., and R. PADDISON. 1981. *Politics, Geography and Behavior*. New York: Methuen.
This book demonstrates how political factors and processes influence and interact with spatial behaviour. The result is a very different type of political geography text with detailed focus on values, attitudes, images, and decision making.

O'SULLIVAN, P. 1986. *Geopolitics*. London: Croom Helm.
Focuses on geographic interpretations of the relationships between states. Includes substantive conceptual discussion.

PARKER, G. 1983. *A Political Geography of Community Europe*. Toronto: Butterworths.
Detailed analysis of one region that considers whether or not that region is acquiring distinct geopolitical characteristics; that is, is a 'new' region emerging?

PEPPER, D., and A. JENKINS, eds. 1985. *The Geography of Peace and War*. Oxford: Blackwell.
An excellent set of readings focusing on such issues as the cold war, the arms race, nuclear war, and peace education.

10

Agriculture

THIS CHAPTER FOCUSES on agricultural geography, typically regarded as one branch of economic geography; the other principal branches (settlement, industrial, and transportation geography) are discussed in chapters 11, 12, and 13 respectively. Agricultural geography has traditionally been regarded as the identification, measurement, and explanation of spatial variations in agricultural activities. Although agriculture has been viewed as primarily an economic activity, both cultural and political considerations are also causes of spatial variations in agricultural activity. For example, as discussed in Chapter 7, a range of cultural attributes, especially religious beliefs, affect crop and animal production. Similarly, as discussed in Chapter 9, the state intervenes in many aspects of our lives, including protecting farmers from fluctuations in prices, and it thus often negates the laws of supply and demand.

But it is the economic aspect that has dominated our human geographic interest in agriculture, and also our interests in settlement, industry, and transportation. It is because of this that Erickson (1989:223) could assert, 'No other academic discipline has had more influence on human geography than economics.' This may appear a surprising statement as in previous chapters our human geographic concerns have had less to do with economics than with such disciplines as anthropology, psychology, and sociology. But much of our human landscape-creating behaviour, especially with regard to the contents of this and the next three chapters, indeed has strong economic motivations.

As a social science, economics is concerned with matters relevant to human geography, namely with allocating scarce resources for production and consumption, resources such as the land and soil needed for agriculture and the minerals and power supplies needed for industry. Economics is also concerned with exchange transactions and with wealth accumulation and distribution, issues central to human geographic analyses of settlement and transportation. Box 10.1 details some of the relations between human geography and economics.

In the human landscape, economic factors have often been considered important and the discipline of economics, especially neoclassical economics, has offered us many methods and concepts (boxes 10.2 and 10.3). But as human geographers, we have become acutely aware of the dangers of any overly simplistic monocausal type of explanation; landscape creation is recognized as complex. Accordingly, this and the next three chapters reflect our diverse interests as human geographers; neoclassical economic content is evident but does not necessarily dominate. Indeed, today we are turning to a more socially and culturally aware human geography that is prompted by our discussions of regions and places in Chapter 8 and by the political economy and neo-Ricardian schools of economics.

THE AGRICULTURAL LOCATION PROBLEM

We begin our account of agriculture with a basic geographic question. Why are agricultural activities located where they are? Our answer centres on economic issues, but first considers a variety of other factors, all of which are interrelated. Indeed, for many geographers, the best way to tackle agricultural issues is with an ecosystems focus. Low technology agricultural systems have similar structures to natural ecosystems; higher technology systems introduce a concatenation of circumstances, including economic, cultural, and political interpretations of physical environment. Separating these variables is convenient but misleading.

SOME PHYSICAL CONSIDERATIONS

The agricultural landscape is typically comprised of individual farms and the details of the landscape result from the decisions of farm owners, managers, and workers. Such decisions may lie anywhere in the continuum already identified — from optimizing through satisfactive to minimal adaptation — and reflect a variety of factors.

Physical factors are important influences in agricultural decisions and hence in the creation of agricultural land-scapes. Animals and plants are living things and require appropriate physical environments to function efficiently. Farmers have two options: they can either ensure that there is a match between animal and plant requirements and the physical environment, or they can create artificial physical environments. Both options are common — irrigation and glasshouses are well-known examples of the latter. Some farmers are prepared to create artificial environments for specific agricultural activities, while other farmers choose not to practise activities that are environmentally inappropriate. Because many environments are suitable for more than one crop, the agricultural decision can be based on other factors; similarly, those other factors can play a role when the demand for a product is less than the amount that can be produced.

Box 10.1: Human geography and economics

Understanding the relationship between economics and human geography is facilitated by brief reference to the historical development of economics. Although it is not necessary to provide detailed descriptions, it is useful to note the following schools of economic thought.

- *Classical economics:* This school, which dominated from the mid-eighteenth century until about 1870, was founded by Adam Smith and exemplified by the work of David Ricardo. It explained the price of goods by reference to the labour theory of value that sees prices as reflecting the labour time that goes into the production of the good.

- *Marxian economics:* This school shares with classical economics an interest in the labour theory of value. It is a body of theory that is part of Marx's larger political economy and applies his theories of value and exploitation to price theory. It exposes the class relations that explain the inequalities of capitalist economies.

- *Neoclassical economics:* This school replaced the dominant classical view in the later nineteenth century and introduced the concept of use value, or utility, with the price of a good arising from the circumstances of supply and demand — the price that people are willing to pay. Neoclassical economics is closely associated with capitalism; once in place, the neoclassical view proceeded to dominate the discipline until the present.

- *Political economy:* Today, economics is also characterized by a political economy approach that has a strong policy orientation. It is partially derived from Marxian economics and rejects the basic logic of neoclassical economics.

- *Neo-Ricardian economics:* An important school since about 1960, this uses the labour theory of value and develops aspects of Marxian theory. It has been applied to problems of late twentieth-century capitalism.

What is the significance of these various developments for human geography? Following the institutionalization of geography in 1874, a branch of geography usually called **commercial geography** emerged, focusing on world and regional production and links to the physical environment; it was explicitly environmental determinist in content. This early economic geography had close ties to economics and, from the economics perspective, could be usefully seen as 'spatial economics'. As detailed in Chapter 1, the 1950s witnessed the decline of the descriptive regional focus, while the cultural geography school initiated by Sauer remained confident of its identity. In the broad field of economic geography, there was increasing dissatisfaction with regional and environmental determinist approaches and little apparent link with cultural geography. A fresh approach seemed necessary.

Economics, especially in the then-dominant and still very important form of neoclassical economics, offered that fresh approach. Geographers turned to positivism, the philosophy favoured by neoclassical economics, and to many of the contributions, especially theories, of economists, to test and further develop economic concepts in a spatial framework. Economic geographic problems related to agriculture, industry, settlement, planning, and transportation thus occupied an especially prominent place on the research agenda of human geographers from the early 1960s to some time in the 1970s. More recently, however, human geographers have also turned to both Marxian economics, in the form of a political economy approach, and to neo-Ricardian economics to support specific research endeavours. This and the following three chapters reflect both the dominant impact of neoclassical economics and the political economy and neo-Ricardian schools.

Climate

Climatic factors are the main physical variables affecting agriculture. Plants have specific temperature and moisture requirements. Optimum temperatures vary according to the activity, but growing season temperatures of between 18° C (64° F) and 25° C (77° F) are often required. Low temperatures result in slow plant growth, while short growing seasons may prevent a plant from reaching maturity. In many areas, frost may result in plant damage. Moisture is crucial to plant growth; too little or too much can damage plants. In many areas, especially semiarid zones, rainfall variability is a problem. Many of the major world wheat-growing areas are highly vulnerable in this respect — the Canadian prairies are one example. Figure 10.1 indicates the relationship between wheat yield and mean annual rainfall at a time when fertilizer was little used. Animals, especially dairy cattle, have water requirements; sheep are much more adaptable to water limitations.

Soils and relief

Other important physical variables are those of soils and relief. Soil depth, texture, acidity, and nutrient composition all need to be considered. Shallow soils typically inhibit root development; the ideal texture is one that is not dominated

---

Box 10.2: Location and interaction theories

Much of this and the following three chapters is concerned with the location of human activities and interactions between locations. Inspired by classical and neoclassical economics, this involves a concern with **location theory**, which explains geographic patterns at any given place and time, and with **interaction theory**, which centres on movements between locations.

As introduced in Chapter 2, theories are sets of interrelated statements that explain reality and that can be either static or dynamic. They are usually inductively generated from facts, and hypotheses are then deduced and tested in the real world (see Figure 2.1). Perhaps the greatest virtues of the typical theory are rigour and simplicity. Theoretical rigour results from the use of the scientific method; once a theory is derived, the process of hypothesis deduction and verification will not, in principle, vary according to a researcher's abilities. Verified hypotheses eventually assume the status of laws. Subsequently, these laws can become initial statements in other theories.

By definition, however, theories are simplifications of complex realities; indeed, it is often painfully apparent that a theory is not merely simplified but actually quite untrue. Lack of truth, however, is not a good criterion for criticizing a theory, for a good theory will offer insight and facilitate further work. Put simply, theories are not intended to be facts. Some would argue that any science — such as human geography — needs to be supported by fact and theory. The facts can be seen as the subject matter and the theories as the means of explaining the subject matter.

Incidentally, when positivist-type theory was introduced, critics of the theoretical approach argued that theories were not possible because all locations, all geographic facts, are unique. While such a statement is literally correct, it is highly misleading because locations are both general — that is, general types of location can be recognized — and unique.

Theories focusing on the explanation of economic locations have typically adopted the style of **normative theory**. Normative theory is concerned with what ought to be. By definition, then, such theory does not purport to explain reality but rather what reality would be like in a speculative situation. For example, our principal agricultural location theory is normative as it provides a description of agricultural land values and land uses under certain circumstances, and the most popular such circumstance — actually a simplification or omission — is that of the economic operator.

Box 10.3: The economic operator concept

The **economic operator** is a normative concept that assumes the maximization of profit as a result of perfect knowledge and a perfect ability to use such knowledge in a rational fashion. It is also more broadly known as **rational choice theory** and has its roots in classical economics, but is most closely associated with the neoclassical school. Each economic operator minimizes costs and maximizes profits. There is, of course, no such person as an economic operator in the real world — none of us is blessed with the required 'omniscient powers of perception' and 'perfect predictive abilities' (Wolpert 1964:537), but such an unrealistic concept is useful in economics and geography, for it allows those theories of which it is a part to generate hypotheses that are not encumbered by the complexities of human behaviour. Indeed, it is one of the basic propositions of neoclassical economics. The concept of the economic operator is really at one extreme on a continuum of possibilities that range from optimization to a minimum adaptation. We know that most human behaviour lies between the two extremes and can be described as **satisficing behaviour**, a concept that is a basic alternative to rational choice theory. Such behaviour satisfies the individual rather than resulting in an optimizing situation.

by either large particles (sand) or small particles (clay); most crops require neutral or slightly acidic soils, while the most crucial of the nutrients are nitrogen, phosphorus, and potassium. Soil fertility needs to be maintained if regular cropping is practised, and methods include the use of fallow, manuring, and chemical fertilizers. Relief (the shape of the land) affects agriculture through slope and altitude. Slopes are relevant in terms of their angle, direction, and related insolation in terms of their susceptibility to the use of machinery and soil erosion. Generally, the flatter the land the more suitable it is for agriculture. Altitude affects temperatures. In temperate areas, the mean annual temperature falls 6° C (11° F) for each 1,000 m (3,280 ft) above sea level.

SOME CULTURAL AND POLITICAL CONSIDERATIONS

We have advanced far beyond the environmental determinist type of reasoning:

> The Corn Belt is a gift of the gods — the rain god, the sun god, the ice god, and the gods of geology. In the middle of the North American continent the gods of geology made a wide expanse of land where the rock layers are nearly horizontal. The ice gods leveled the surface with glaciers, making it ready for the plow and also making it rich. The rain god gives summer showers. The sun god gives summer heat. All this is nature's conspiracy to make man grow corn (Smith 1925:290).

We have already introduced one important cultural factor, namely that farmers are not profit maximizers. Farmers in the developed world typically favour security and a relatively constant income rather than strive for the unlikely goal of profit maximization. Poor farmers in the developed and less

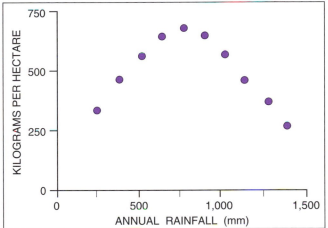

10.1 *(top)* Hutterite colony in southern Alberta, Canada. (Glenbow Museum)

Figure 10.1 *(bottom)* Relationship between the US mean annual rainfall and the wheat yield, 1909.
Source: Adapted from O.E. Baker, 'The Potential Supply of Wheat', *Economic Geography* 1 (1925):39.

developed world typically favour maximizing product output for subsistence rather than profit. These preferences can be interpreted as the consequence of cultural attitudes towards an essentially economic issue. More specific examples of cultural impacts are those relating to religion and ethnicity.

Religion and ethnicity
Group religious beliefs may emphasize specific agricultural activities because of the value placed on the activity or the product of the activity. Christianity, for example, values wine, which is used during the sacrament of Holy Communion. The resultant demand for wine encouraged the spread and

growth of viticulture, including, for example, its introduction by early Christian missionaries in California. Other agricultural activities are negatively affected by religious beliefs. Pigs are taboo in Islamic areas, while Hindus and Buddhists believe it to is wrong to kill animals, especially cattle.

Immigrants in a new land often bring a set of agricultural practices to which they continue to adhere after settlement. Thus, an area settled by a variety of ethnic groups often presents a patchwork appearance. North American examples abound. One geographer has explained the location of cigar tobacco production in the United States by ethnic variables (Raitz 1973). Another described two communities in south-central Illinois, one of German-Catholic ancestry and one of non-Catholic British ancestry (Salamon 1985). These communities are only 32 km (20 miles) apart and have similar soils, but there are substantial differences in farm size and organization. The German farms are small and diversified, including dairy, hogs, beef, and grain; the British farms are monocultural grain. It can be argued that these differences in farming practices stem from different attributes. Table 10.1 contrasts the yeoman Germans and the entrepreneurial British.

As Table 10.1 suggests, a variable such as landownership can affect farmer behaviour, especially innovative behaviour.

TABLE 10.1: *Contrasting farming types in Illinois, 1985*

Farming types

Yeoman	Entrepreneur
Goals	
Reproduce a viable farm and at least one farmer in each generation returns	Manage a well-run business that optimizes short-run financial returns
Strategy	
Ownership of land farmed preferred	Ownership plus rental land to best utilize equipment
Expansion limited to family capabilities	Ambitious expansion limited by available capital
Diversify to use land and family most creatively	Manage the most efficient operation possible
Farming organization	
Smaller than average operations	Larger than average operations
Animals plus grain, crop variety	Monoculture cash grain
Land fragmentation	Land consolidation
Landowners often operators	Landowners frequently absentee
Expansion of community territory	Community territory stable
Family characteristics	
Intergenerational cooperation	Intergenerational competition
Parents responsible for setting up son/heir	Incumbent upon son/heir to set up self
Many children, nonfarmers, live nearby	Often all children leave farming
Parents responsible for intergenerational transfer	Heirs responsible for intergenerational transfer
Early retirement geared to succession by children	Retirement geared to personal desires
Community structure	
Village central focus of community	Village declining
Community loyalty	Weak community attachment
Population relatively stable	Population diminishing
Strong church attachment	Church consolidations
Farmers involved in village	Farmers uninvolved in village

Source: S. Salamon, 'Ethnic Communities and the Structure of Agriculture', *Rural Sociology* 50 (1985):326.

Farms owned by their occupiers may be handled quite differently than rented farms or farms in a socialist economy where decision making is centralized.

The state

Political decisions affect much human behaviour, including farmers' behaviour. Governments have many reasons for choosing to influence farmer behaviour, but the most relevant in the contemporary developed world is the desire to support a relatively declining activity as measured by income. The characteristic government intervention involves fixing product prices and providing financial support for specific farm improvements such as land clearance (Box 10.4 and Figure 10.2). In the less developed world, governments provide assistance that enables farmers to adopt new methods and products.

The contemporary agricultural system is affected by state policies and national and international trade legislation. In Canada, the United States, Japan, and Europe, there are marketing boards, quota requirements, government credit policies, and extension services, all of which affect the agricultural landscape. Perhaps the best-known example of elaborate policies is that of the European Economic Community — in some cases, farmers have been paid not to grow crops (Bowler 1985).

SOME BASIC ECONOMIC CONSIDERATIONS

The spatial patterns of agricultural activities are the end product of complex physical, cultural, and political variables. But human geographers agree that the most important variables generally — not necessarily in a specific location — are those conveniently labelled economic. Agricultural products are the result of a market demand for them. Thus farming is subject to the basic laws of supply and demand. Figure 10.3 depicts characteristic supply and demand curves. Supply increases and demand decreases when the price received increases. The result is that an equilibrium price (P) can be identified at the intersection of the curves. The ideal economic world of **commercial agriculture** is occupied by profit maximizers (economic operators) who respond immediately to any price changes.

But this type of response is neither feasible nor desired by farmers in either the developed or less developed worlds. In the developed world, farmers also value stability and independence and may sacrifice profits accordingly. Further, such farmers may choose to respond to decreasing prices by increasing, not decreasing, supplies. Farmers in the less developed world are also not typically profit maximizers, although their behaviour may well be rational. **Subsistence agriculture** aims to produce the required amount of product to maintain family needs; profit maximization has no meaning for such farmers. The number of subsistence farmers has decreased significantly since about 1800 and continues to decrease. Available agricultural location concepts focus on commercial farmers.

COMPETITION FOR LAND

Theories concerned with the spatial pattern of farming activities provide a means of accounting for observed spatial variations. Geographers thus focus on the competition for use (between activities) of a particular location. Such competition arises because it is not possible to locate all activities at their economically optimal location, and different activities may have identical optimal locations. The obvious question is one of determining the most appropriate land use. Conventionally, land is assigned to the use that generates the greatest profits. Thus, in a general sense we can determine a hierarchy of land uses based on relative profits, but there is another way of viewing this situation. The greater the profits that a particular use generates, the more that use can afford to pay for the land. The maximum amount that a given use can pay is called the **ceiling rent**. Figure 10.4 indicates the relative rent-paying abilities of a variety of land uses. We can now identify the basic premise of much location theory: the competition among land uses, a competition fought according to rent-paying abilities, results in a spatial patterning of those land uses.

THE CONCEPT OF ECONOMIC RENT

Agricultural location theorists use this premise, combined with the concept of **economic rent**, which explains why land is or is not used for production. Land is used for production if a given land use has an economic rent above zero (Box 10.5). This economic rent concept can be related to one or more variables, but especially to measures of land fertility or distance from market. Land fertility was seen as the relevant cause in the formulation of the economic rent concept by the economist David Ricardo (1772–1823), while distance was used by the economist Johann Heinrich von Thünen (1783–1850).

THÜNEN AGRICULTURAL LOCATION THEORY

Thünen was a German economist and landowner who published two major works. A treatise entitled *The Frontier Wage* identified appropriate wages for agricultural workers. The second work, *The Isolated State*, published in 1826 (Hall 1966), was the result of Thünen's interest in economics (he was inspired by the work of Adam Smith) and of his forty years as manager of an estate on the north German plain.

The problem

When tackling the question of what agricultural activities should be practised where, Thünen conceived of a highly original method of analysis. He proceeded as follows:

Assume a very large city in the middle of a fertile plain which is not crossed by a navigable river or canal. The soil of the plain is uniformly fertile and everywhere cultivable. At a great distance from the city the plain shall end in an uncultivated wilderness by which the state is separate from the rest of the world (Johnson 1962:214).

Box 10.4: Government and the agricultural landscape

Governments are often interested in assisting agricultural areas because of a long-standing bias in favour of representation from such areas. Such assistance typically includes subsidized prices and tariff protection.

The Canada–United States border region between the Great Lakes and the Rocky Mountains provides an excellent example of an agricultural landscape's response to government policies. The landscapes differ despite similar climatic, soil, relief, and drainage conditions, although in fact the transboundary differences are lessening as the policies become less different. In the 1950s, government intervention was much greater in the United States, but by the 1970s, a decrease in American intervention and an increase in Canadian intervention meant that the landscape differences were greatly reduced.

The landscape differences for 1959–61 are clear in Figure 10.2; there are different crop combinations and different livestock combinations. The wheat allotment program in the United States meant that much wheat land was converted to raise barley, while the National Wool Act of 1954 encouraged sheep raising through a guaranteed price.

By the 1970s, the United States had moved away from support programs towards relatively market-oriented policies. On the other hand, Canada favoured greater intervention to ensure adequate incomes for farmers. As a result of these changes, the policies became more similar and the landscapes responded. The transboundary differences are significantly fewer today than in the 1950s and 1960s.

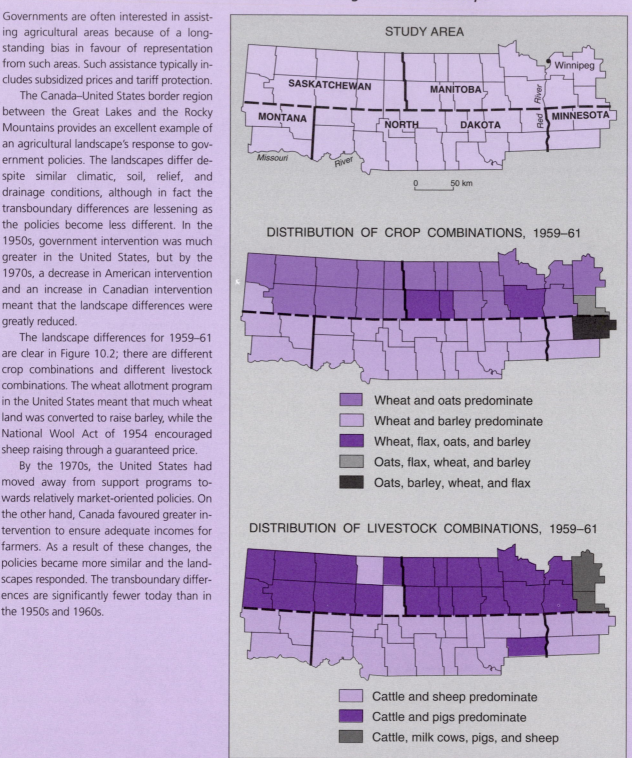

Figure 10.2 Crop and livestock associations, US–Canadian border.

Source: Adapted from H.J. Reitsma, 'Crop and Livestock Production in the Vicinity of the United States–Canada Border', *Professional Geographer* 23 (1971):217, 219, 221.

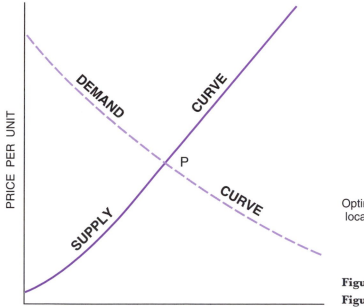

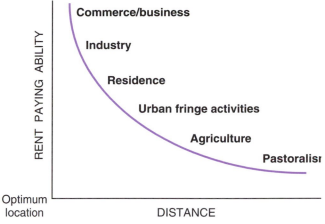

Figure 10.3 *(left)* Supply and demand curves.

Figure 10.4 *(above)* Rent-paying abilities of selected land uses.

Box 10.5: Calculating economic rent

It is important to clarify some details of the economic rent concept's calculation and spatial implications. Economic rent can be calculated as follows:

$$R = E(p - a) - Efk$$

where

R = rent per unit of land (dependent variable)
k = distance from market (independent variable)

E = output per unit of land
p = market price per unit of commodity
a = production cost per unit of commodity ⎫
f = transport rate per unit of distance per ⎬ (parameters)
  unit of commodity ⎭

This equation is a straightforward linear function between rent and distance. Rent is a function of distance with the details of the relationship determined by E (p – a), which gives the intercept, and Ef, which determines the slope of the line.

What is the spatial implication of economic rent as calculated above? Because economic rent declines with increasing distance from market, it eventually becomes zero. With more than one agricultural activity, it is likely that a series of economic rent lines appear, as in Figure 10.5. Where lines cross, one activity replaces another as the more profitable. The result is that agricultural activities are zoned around the central market and a series of concentric rings emerges.

The simple concept of economic rent as a function of distance results in a situation of spatial zonation.

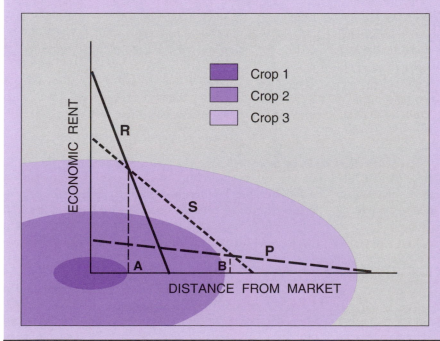

Figure 10.5 Economic rent lines for three crops and related zones of land use.

The assumptions

The complexity of the agricultural location problem prompted Thünen to devise a method that allowed for greatly simplified descriptions: the isolated state. Accordingly, he deliberately excluded several factors that are known to be relevant.

1. There is only one city, that is, one central market.
2. All farmers sell their products in this central market.
3. All farmers are profit maximizers — economic operators.
4. The agricultural land around the market is of uniform productive capacity.
5. There is only one mode of transportation by which farmers can transport products to market.

Thünen is thus investigating the agricultural location problem by excluding — or actually holding constant — such key variables as physical environment, humans, and transport. Only one key variable — distance from market — is allowed to vary. This is a distinctive and highly original contribution. It is, of course, in accord with his definition of economic rent that relies on distance from market.

The answer

Given the above, Thünen asked, 'How will agriculture develop under such conditions?' (Johnson 1962:214). The conditions are those of a controlled experiment isolating but one causal variable, distance from market. The answer is straightforward. Agricultural activities are located in a series of concentric rings around the central market, one zone for each product for which there is a market demand. Thünen calculated the location and size of each product zone by using data from his estate on production costs, market prices, transport costs, and other matters. The principle of concentric zones emerges from the concept of the isolated state, while the number, size, and content of zones is a function of particular places and times. Combining the isolated state concept, his own data, and the economic rent concept, Thünen identified two conclusions: (1) zones of land use develop around the market; (2) the intensity of each specific land use decreases with increasing distance from the market.

The crop theory

The first conclusion, the crop theory, is summarized in Figure 10.6. Remember that the specific activities evident in this figure reflect north Germany in the early nineteenth century, but that the principle of zones is general. In zone 1 are market gardening activities and milk production. Both these products are perishable, give high returns, and have high transport costs. Accordingly, they have a steep economic rent line as a result of a high Ef value, and a high intercept on the rent axis as a result of a high E (p – a) value. In zone 2 are forestry products used, at the time, for fuel and building. This activity could command a location close to market because of the high transport costs involved in moving such

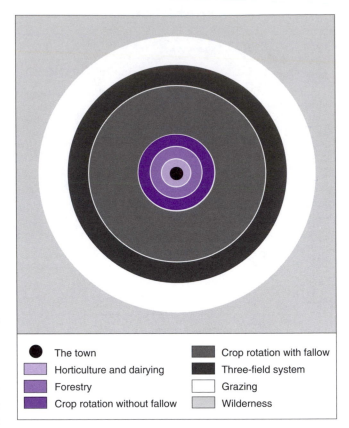

Figure 10.6 Agricultural land use in the isolated state.

Legend:
- ● The town
- ▨ Horticulture and dairying
- ▨ Forestry
- ▨ Crop rotation without fallow
- ▨ Crop rotation with fallow
- ▨ Three-field system
- ▨ Grazing
- ▨ Wilderness

bulky products. Rye, the principal commercial crop, is located in zones 3, 4, and 5. The difference between the zones is the intensity of rye cultivation, with zone 3 using a six-year rotation, zone 4 using a seven-year rotation and zone 5 using a three-field system. Livestock ranching with the production of butter, cheese, and live animals is in zone 6. Beyond zone 6 is wilderness that could be used if spatial expansion was needed, for example, as a result of increased product demand.

Figure 10.6 provides a summary of the crop theory and also demonstrates that product location is related to such issues as perishability and weight as they affect transport cost.

The intensity theory

The second conclusion is known as the intensity theory. In arriving at this conclusion, Thünen used his own production data to show that for any given product, the intensity of production decreases with increasing distance from market. The easiest example to identify is the cultivation of rye in zones 3, 4, and 5, but the principle also applies inside specific zones.

THÜNEN THEORY?

In light of our discussion of theory and scientific method in Chapter 3, we can ask if Thünen's work is indeed a theoretical contribution. The work does begin with a series of assumptions. Unfortunately, they are not valid assumptions, or patterns. They describe what ought to be — a normative description. As is the case with many economic models, a complex world is simplified to aid understanding; the spatial hypotheses produced are examples of ideal types (see glossary).

Thünen's work demonstrates the procedures of theory construction, but its designation as a theory is complicated by the inclusion of real world data into the procedures. It is nevertheless an excellent example of a type of theory to which geographers have been greatly attracted.

SOME MODIFICATIONS AND EXTENSIONS

Thünen was well aware that his work was a simplification. Accordingly, he modified several of the assumptions when elaborating on the basic hypotheses. The assumption of only one mode of transportation to market was relaxed with the recognition that a navigable waterway would serve to elongate zones (Figure 10.7). Relaxing the assumption of only one market and allowing a transportation network resulted in a more complex pattern of zones (Figure 10.8). It was also acknowledged that differential land quality would result in an emphasis on the better areas.

Despite its date of origin, and some obvious weaknesses, Thünen's work remains the fundamental contribution to the question of agricultural location for at least two reasons. First, it is a highly original piece of work that succeeds in simplifying the issue; as a result, the basic method has been used by many other location theorists, as we are to discover especially in chapters 11 and 12. Second, it focuses on the key geographic variable: distance as measured in terms of transport cost.

The principal premise behind Thüsen's economic rent and agricultural location concepts is that distance is the key causal variable. Human landscapes in general are clearly related to distance, regardless of how we choose to measure distance — straight-line distance, transport-cost distance, time, or any other format. Relationships between distance and spatial patterns, such that some spatial regularity is evident, lie at the heart of much theory and analysis.

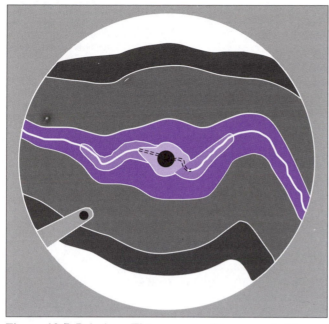

Figure 10.7 Relaxing a Thünen assumption — a navigable waterway.

DISTANCE, LAND VALUE, AND LAND USE

Numerous accounts of agricultural patterns represent geographers' formal testings of the Thünen hypotheses or testings of general relations with distance. The seminal work in this context is that by Chisholm (1962), a remarkably thorough pioneering survey of the scale ramifications of the concepts and an excellent survey of world evidence pertaining to the general issues.

Historical examples

Thünen's own work is a historical empirical study, and it is hardly surprising that it is applicable in a general historical context. Indeed, well before Thünen, there were examples of

Figure 10.8 Relaxing two Thünen assumptions — multiple markets and a transport network.

axioms, from which we can deduce hypotheses with confidence. Rather, they are simplifying assumptions that reduce the complexity of the problem. Given the character of the assumptions, we know that the hypotheses produced — the crop and intensity theories — are not literally correct. But we also know that they serve the purpose intended — that is, they describe an ideal state (interestingly, the original title of the work!). The two hypotheses describe a state of affairs against which we can compare real world agricultural land-use zonation. There are references to land use in Kent, England, regarding access to London, including a 1576 description and an 1811 account that identified four zones — clay pits, cattle pastures, market gardening, and hay.

For North America, many analyses of nineteenth-century agriculture highlight the Thünen principles in a spatial and temporal framework. One study of an area of Wisconsin that centred on Madison demonstrated that zonal change for any one period becomes temporal change as the urban centre increases in size (Conzen 1971). Thus, the area close to the expanding centre was dominated by wheat production from 1835 to 1870, by market gardening from 1870 to 1880, and by dairy farming after 1880. A study of western New York state focused on the related idea that a prolonged period of falling transport costs is associated with increasing regional specialization of agriculture (Leaman and Conkling 1975). It was shown that the expanding transport network of roads and canals allowed a frontier area, initially akin to Thünen's wilderness, to rapidly become, first, an area of wheat cultivation and, second, an area characterized by regional specialization related to access to market.

The southern Ontario frontier

Both of the American studies noted that areas experiencing increased settlement are transformed from non-agricultural areas to areas of subsistence agriculture and finally to a series of commercial agricultural types. Similar results were achieved in a study of nineteenth-century southern Ontario that focused explicitly on testing the Thünen hypotheses (Norton and Conkling 1974).

The area analysed lies north of the emerging centre of Toronto and can be regarded as one segment of the market area of that centre. Using data for 1861, the relationship between distance and land value is measured at $r = -0.6502$, suggesting that 42 per cent of the spatial variation in agricultural land values is explained by distance to Toronto (Box 10.6 and Figure 10.9). This analysis then follows the theoretical logic one step further by relating land values to a series

---

Box 10.6: Correlation and regression analysis

'A continuous theme in geographic research is that of analyzing the degree and direction of correspondence among two or more spatial patterns or locational arrangements' (King 1969:117). A traditional (and contemporary) approach to the problem is that of map overlays using geographic information systems. Another approach is to apply a set of quantitative procedures known as correlation and regression analysis.

By simple correlation and regression, we may determine the degree of association between two variables. The correlation coefficient, r, is a statistical measure of the relationship, where r may vary between –1 (a perfect inverse relationship) and +1 (a perfect direct relationship). Squaring the r value gives a coefficient of determination, r^2, which may be expressed as a percentage and which may be interpreted as the percentage variation in one variable explained by the other variable. A related set of procedures, known as multiple correlation and regression analysis, allow one variable to be related to two or more variables. In this case, the correlation coefficient is identified as R.

The procedure by which these results are obtained is as follows. The linear equation, $y = a + bx$, is used where:

y = dependent variable
x = independent variable
a = y intercept
b = slope of the line

The economic rent equation is of this form $[R = E (p - a) - Efk]$. This linear equation is the simplest equation for predictive purposes. A method known as least squares is used to calculate the best-fit line on the graph of x plotted against y. Once the best-fit line is calculated, we can measure how well it fits. If all points fall on the line, it is a perfect fit and $r = -1$ or $+1$; more typically points are not located on the line and r lies between –1 and +1. In cases where y does not vary with changes in x, r is close to 0 (Figure 10.9).

Procedures are also available to test the significance of r. The results reported in the text are all statistically significant.

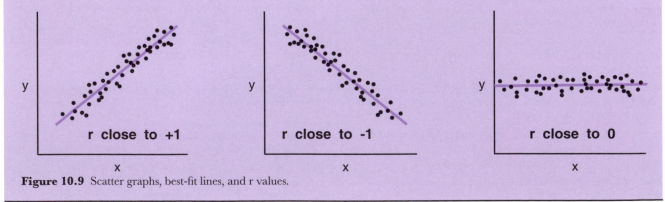

r close to +1   r close to -1   r close to 0

Figure 10.9 Scatter graphs, best-fit lines, and r values.

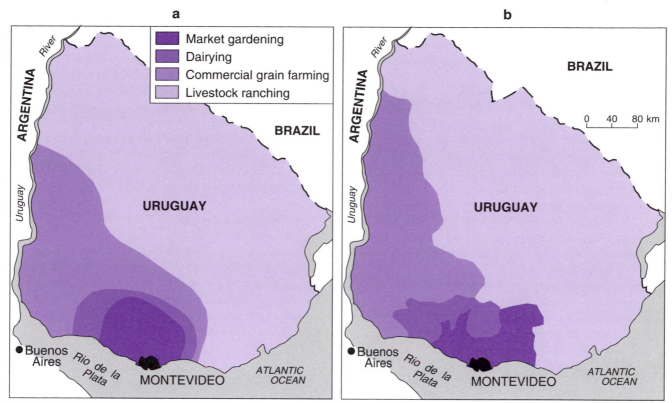

a

■	Market gardening
■	Dairying
■	Commercial grain farming
□	Livestock ranching

b

Figure 10.10 Agricultural land use in Uruguay: (a) as predicted by Thünen theory; (b) actual.

of variables — distance to Toronto, distance to smaller regional market centres, distance to major lines of communication, and a measure of land capability for agriculture. The statistical result is R = 0.7367, meaning that 54 per cent of the spatial variation in land values is explained by the set of independent variables.

A consideration of the 1861 pattern of land use showed an area that was divisible into two basic zones. First, close to Toronto was a zone of fall wheat, peas, and oats with production oriented for the Toronto market. Second, an outer zone was characterized by much unoccupied land and by spring wheat, potatoes, and turnips — an area anticipating commercial orientation. Thus, an incipient zonation was in place by 1861 and was to become fully realized in the later nineteenth century. This Ontario study also analysed the intensity hypothesis and showed that the intensity of the key commercial crop (fall wheat) decreased with increasing distance from Toronto.

Examples in the less developed world
More surprising than the historical examples is considerable evidence suggesting Thünen-type spatial patterns in less developed countries where much agricultural activity is subsistence, not commercial, in orientation. A regional pattern of land use is often in fact a mix resulting from the different agricultural decision making of relatively well-off commercial farmers and relatively less well-off subsistence farmers. Necessarily, these two groups have different economic out-

looks. The commercial farmers are concerned with profits and hence with rationalizing land use according to economic constraints, while the subsistence farmers are limited in their activities by lack of capital and inadequate size of holding. Despite these complications, Thünen-type patterns may emerge. Horvath (1969) analysed land use *c*.1964 in the vicinity of Addis Ababa, Ethiopia. This highland area is physically and culturally varied and yet had an area of eucalyptus forest around the city that was used for fuel and building material (an orientation outwards along roads was evident), vegetable cultivation within the forest and not as a separate zone, and a zone of mixed farming that included commercial and subsistence activity beyond the first forest and vegetable zone. Horvath concluded that this was an example of incipient zonation, similar to that noted for the southern Ontario example earlier. Unfortunately, the extreme political circumstances in contemporary Ethiopia make continuing analyses especially difficult.

A contemporary regional example
Analyses based on the Thünen theory are especially evident in contemporary circumstances and at the originally proposed regional scale. One example will suffice, that of Uruguay by Griffin (1973). Uruguay possesses many of the essential characteristics of the isolated state, and Montevideo has the key attributes of the theoretical central city. As a consequence, the Thünen model of ideal land use is very similar to the actual land use (Figure 10.10). The first zone is

one of horticulture and truck farming, followed by a zone of dairying, and finally the cereal zone. There are significant variations between model and reality related to such variables as soil fertility, transport efficiency, and ethnicity.

Examples on the continental and world scales
Thünen's logic is relevant even on the world and continental scales. Several geographers have proposed concentric patterns of land use for North America, Europe, and the world. Muller (1973) analysed the nineteenth-century United States and concluded that there was an expanding system of concentric rings. Jonasson (1925) described European agriculture in terms of an inner ring of horticulture and dairying, followed by zones of less and less market-oriented activity. At the world scale, Schlebecker (1960) traced the evolution of a world city from Athens to western Europe to western Europe/eastern United States.

The growth of such a world market and the related expansion of the supply area can be analysed by reference to import data. Over time, the mean distances agricultural imports have been moved has increased (Table 10.2). This spatial expansion of the export system is clearly zonal in character and has not occurred haphazardly. The changing details of the import system come from a modified Thünen system. Peet (1969) has detailed this situation specifically for wheat showing that, in many instances, the arrival of wheat was preceded by extensive animal rearing (the Canadian prairies are an exception). In all cases, wheat cultivation was associated with major population growth, transport expansion, and the rise of an urban network. Spatially, wheat imports to Britain came first from Britain itself, then the Baltic region, followed by the United States, western Canada, and Australia.

Examples on the local scale
There have also been tests of the theory on the scale of individual villages and farms. In such analyses, the concept of distance is equated less with transport cost than with movement minimization as related to time expended. Various investigations of villages have tested modified versions of

the crop and intensity theories and confirmed their relevance. Thünen-type patterns are also evident in the immediate vicinity of large cities. According to Sinclair (1967), the processes affecting land use close to cities are not those identified by Thünen, although the spatial consequence may be that of zonation. One focus is on the issue of urban expansion, specifically the anticipation of such expansion and hence of profits to landowners. The closer to the city, the greater the anticipation of urban expansion and hence the less capital investment in the land. This argument helps us understand why so many areas on the outskirts of a city are either vacant or are clearly in a situation of ephemeral land use.

Overall, the message is clear. The methods pioneered by Thünen are of enormous value to agricultural geographers and, as we shall see in later chapters, to industrial and urban geographers. Thünen's ideas have generated considerable research and, so long as there is an appreciation of the deliberate simplifications he employed, they will continue to stimulate work and provide an ideal world against which reality can be assessed.

WORLD AGRICULTURE: ORIGINS, EVOLUTION, AND REGIONS

The central focus of a discussion of world agriculture could be the links between physical environment and agriculture, or the links between regional population change and agriculture, or the links between distance from market and agriculture. But perhaps the most useful focus acknowledges that the contemporary pattern of world agriculture is the still-changing outcome of a long history: 'The imprint of the past is still clearly to be seen in the world pattern of agriculture. To understand the present, it is essential to know something of the evolution of the modern types of agriculture' (Grigg 1974:1).

ORIGINS

Agriculture has its origins in the **domestication** of plants and then animals. A domesticated plant is deliberately planted, raised, and harvested by humans; a domesticated animal is dependent on humans for food and shelter. As a consequence, domesticated plants and animals differ from the non-domesticated. From the human perspective, the domesticates are superior in that, for example, they bear more fruit or provide more milk. Domestication cannot occur rapidly; it is a long process involving improvement of the species through selective breeding.

When did the process of domestication first occur, and where, how, and why? In Chapter 7 we touched on these topics in a larger cultural context, but we lacked precise answers. In all likelihood, the rise of agriculture — actually a gradual transition from such activities as hunting and gathering — took place beginning some 12,000 years ago.

TABLE 10.2: *Average distances from London, England, to regions of import derivation (miles)*

	1831–5	1856–60	1871–5	1891–5	1909–13
Fruit and vegetables	0	521	861	1,850	3,025
Live animals	0	1,014	1,400	5,680	7,241
Butter, cheese, eggs	422	853	2,156	2,590	5,020
Feed grains	1,384	3,266	3,910	5,213	7,771
Flax and seed	2,446	5,229	4,457	6,565	6,275
Meat and tallow	3,218	4,666	6,018	8,093	10,056
Wheat and flour	3,910	3,492	6,758	8,286	9,574
Wool and hides	4,071	14,207	16,090	17,811	17,538

Source: After S. Leonard, 'Von Thünen in British Agriculture', *South Hampshire Geographer* 8 (1976):28.

From the initial centre (or more probably centres), agriculture diffused to other areas and today, preagricultural economic activities are marginal. As to where, the traditional interpretation focuses on a few Asian centres, most notably southwest Asia (present-day Iraq). More recently it has been argued that agriculture evolved in a relatively large number of centres independently — for example, southwest Asia, southeast Asia, east Asia, the North American midwest, Central America, western South America, southern Europe, northeast Africa, and Africa south of the Sahara. The origins of agriculture were quite possibly in response to different circumstances in different places.

The process of domestication

There is nothing difficult or complicated about the process of domesticating plants and animals; no knowledge of genetics is needed. It is simply a process that makes use of artificial selection as opposed to natural selection. Natural selection results in the survival and reproduction of those plants and animals that are best able to cope in a particular environment. Artificial selection involves humans allowing certain plants and animals to survive because they possess features judged desirable by humans, for example plants with bigger seeds or animals that are non-aggressive. The individual members of plant and animal species that humans favoured would reproduce and pass on the favoured characteristics; the individual members with less desirable traits would therefore be gradually eliminated.

Possible causes of domestication

Various explanations have been proposed to explain why humans became food producers. According to Sauer (1952), a lack of population pressure and the related available leisure time might have allowed for casual experimentation with plant and animal breeding in environments that were well endowed and that already encouraged a sedentary way of life. Southeast Asia, especially wooded hilly areas away from possible floods, is the suggested hearth area.

Perhaps the most satisfactory explanation combines a number of relevant variables, demographic, environmental, and cultural. The key idea, not yet proven, is that population pressure might have prompted a search for new supplies. This hypothesis argues that, although domestication is not difficult to practise, it is not especially attractive because it can require a great deal of work. Most societies did not need to increase food supplies because they had strategies to keep the population below carrying capacity (see glossary). But a climatic change at the end of the Pleistocene (see glossary) might have resulted in certain areas becoming very rich in food resources leading to an increase in a sedentary way of life and hence an increase in population. This increase in population caused movement into marginal areas and competition for space resulting in new strategies for food supply — plant and animal domestication. Various

authors, notably Boserup (1965) and Binford (1968), have proposed such a scenario.

Whatever the details of origin, by *c.*500 BC, agriculture was a major economic activity. Grigg (1974:21–3) identified five core areas and agricultural types evident by this date (Box 10.7).

EVOLUTION

Our current agricultural types and regions are the outcome of a long evolutionary process. Since domestication, plants and animals have diffused over large areas, much larger than those few areas where the wild varieties of the plants or animals were present. Several of the agricultural regional types to be discussed emerged in response to the process of European overseas movement and/or to the demands of

---

Box 10.7: Agricultural core areas

1. Southwest Asia
   Basic crops: Wheat, barley, flax, lentils, peas, beans, vetch.
   Basic animals: Sheep, cattle, pigs, goats.
   Subtypes: Irrigated farming in the Nile Valley, Tigris and Euphrates, Turkestan, and Indus Valley; dry farming was the basic southwest Asian complex; Mediterranean agriculture was a mix of cereals and tree crops such as figs, olives, and grapes; in northern Europe, oats and rye were added to basic crops.
2. Southeast Asia
   Two types originated on the southeast Asian mainland. Tropical vegeculture involves growing taro, greater yam, bananas, coconuts, and raising pigs and poultry; a form of shifting agriculture. Wet rice cultivation gradually displaced the first type.
3. Northern China
   Based on local domesticates (such as pigs, foxtail millet, soy beans, and mulberry) and imported domesticates (such as wheat, barley, sheep, goats, and cattle).
4. Africa
   Two agricultural types; both shifting.
   Tropical vegeculture based on yams.
   Cereal cultivation (millets and sorghum).
5. America
   Two agricultural types; no livestock.
   Shifting cultivation of root crops.
   Shifting cultivation of corn/squash/beans complex.

This summary is a reasonable picture of agriculture *c.*500 BC. Since that time, links between areas and types have occurred. Of particular relevance are those agricultural changes prompted by European expansion from the fifteenth century onwards.

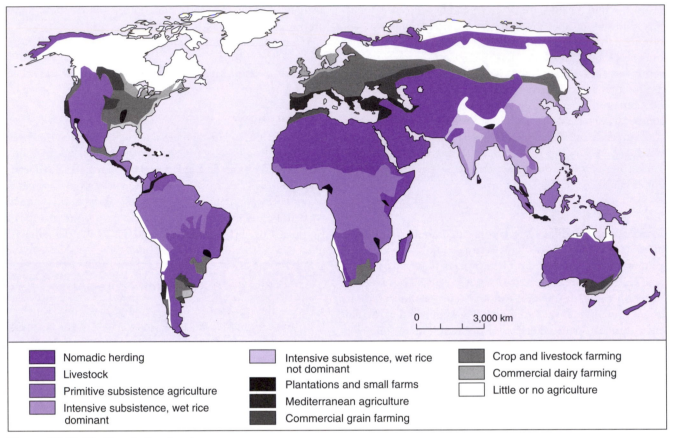

Figure 10.11 World agricultural regions.

Source: D.B. Grigg, *The Agricultural Systems of the World: An Evolutionary Approach* (New York: Cambridge University Press, 1974):4.

new population concentrations. In Europe, for example, agricultural activities prior to 1500 were the result of diffusion from the southwest Asian hearth areas; after 1500, there was widespread adoption of two American crops, potatoes and corn. Other agricultural changes resulted from increased market size; these included intensification, expansion of cultivated area, and increasing commercialization. Commercialization was also made more possible by techni-

cal advances in transportation, particularly since the mid-nineteenth century. Agricultural commercialization at this time was merely one component of substantial growth and change in the larger world economy (boxes 10.8 and 10.9).

Changes in English agriculture after 1750
A period of especially significant agricultural change began in England about 1750, possibly responding to population

---

Box 10.8: Frontier agriculture — subsistence or commercial?

To what extent were the early settlers' agricultural activities subsistence or commercial in orientation?

Many studies of agriculture in frontier North America have identified a brief subsistence phase followed by a dominant concern with commercial activities. Evidence of commercialization is provided by the settlers' demands for improved communications with available markets and by rapid transition from product diversification (characteristic of a subsistence economy) to product specialization (characteristic of a commercial economy). In South Africa, however, there is considerable evidence to suggest that there was a prolonged subsistence phase during

which farmers had limited contacts with markets. Why might this be the case? One answer is to contend that frontier settlement in South Africa differed from that in most other areas as it was not economically motivated. Rather, it is possible that movement to the frontier occurred for social and political reasons — the Dutch settlers' desire to escape the rule of the British in the established areas.

The message is clear. Frontier or other agricultural activity can often be best understood by reference to a wide range of processes. Economic activities, as we know, are often explained in terms other than economic ones.

increases, a social transformation from feudalism to capitalism, and industrial advances. The principal changes were therefore both institutional and technological.

Institutional changes included the process of enclosure — the subdivision of large, open, arable fields and large areas of pasture or wasteland into small fields divided by hedgerows, fences, or walls. This was part of a major social transformation from communal to private property rights. A second related institutional change was the creation of farms that were typically rented from a large landholder by a capitalist farmer employing labourers. Technological changes included the widespread adoption of mixed farming systems that included crops for animal and human consumption and the introduction of labour-saving agricultural machinery.

Clearly, the most radical changes in agricultural activities have occurred in the larger European world, creating 'the great gulf in productivity between the present agricultures of Asia, Africa and much of Latin America, and western Europe, North America and Australia' (Grigg 1974:53). Thus it was only in the mid-nineteenth century that the present character of commercial agricultural types was formed and the present differences in agricultural productivity realized. The link between industrial progress and agricultural progress seems clear.

REGIONS

This text recognizes nine major regional types of agriculture, a typology derived from the work of the noted American geographer Whittlesey and elaborated upon by Grigg (1974). Figure 10.11 shows the global distribution of these nine types, including a twofold division of one of the types, and identifies those areas with little or no agriculture.

Shifting agriculture

One of the earliest agricultural systems, shifting agriculture is today practised almost exclusively in tropical areas. It involves selecting a location, removing vegetation, and sowing crops on the cleared land. There is minimal land preparation and little care is given to the crops. Typically, agricultural implements are limited and livestock are not normally part of the system. After a few years, the land is abandoned and a new location is sought. In this farming system, land is not owned by an individual or family but rather by some larger social unit such as the village or tribe. Shifting agriculture has traditionally been for subsistence purposes, that is, for local consumption. Today, however, many such cultivators produce a cash crop, so that they have at least some market orientation.

Superficially, shifting agriculture appears to be wasteful and indicative of a low technology. There is, however, much evidence to suggest that it is a most appropriate method of maintaining soil fertility in humid tropical areas and is suited to circumstances of low population density and ample land. It provides adequate returns for minimal capital and labour inputs. Shifting agriculture is explained, then, by reference to a number of variables, including environment and population numbers.

Wet rice farming

Most of the rural population of east Asia is supported by wet rice farming, an intensive type of agriculture occupying only a small portion of the total land area. There are many versions of wet rice farming, but in all instances the crop is grown beneath slowly moving water for much of the growing period. This type is restricted to specific local environments, most notably to flat land adjacent to rivers. Low

---

Box 10.9: The pastoral frontier

The pastoral frontier was a new landscape in temperate grassland areas — including parts of North and South America, South Africa, Australia, and New Zealand — that experienced European overseas expansion. The Industrial Revolution and related population increases in the eighteenth and nineteenth centuries created huge demands for wool, hides, and meat. The newly settled areas' responses to these market demands were largely dependent on social and political considerations.

In South America, pastoral activities were viewed favourably by governments that saw no alternative uses for the land. Land-holding systems allowed individuals to control large areas. In the United States, government policies encouraged settlement by family farmers. This variance in attitude is particularly well exemplified in the history of southeastern Australia.

Between about 1820 and 1860, much of southeastern Australia's interior was settled by sheep farmers. These expansions occurred without any roads or security of tenure, and in circumstances that demanded self-sufficiency. State governments were unable to control this economically motivated settlement until the 1860s when there was a demand for arable land. Sheep farmers were then seen as occupying inappropriately large areas of land that could be put to better use. Hence, a series of land laws were passed to allow arable farmers onto land previously held by sheep farmers. The aim of the land laws was to create stable and prosperous agricultural societies, an aim not always easily achieved. Sheep farmers used a variety of tactics to prevent arable farmers from laying claim to land, such as building huts on wheels to be moved rapidly from site to site as inspectors appraised areas. Legal conflicts were common and the transition from pasture to arable was not really complete until about 1900.

walls delineate the fields and hold the water. Because this technology minimizes soil depletion, it is often possible to continuously crop areas, achieving multiple harvests each year. Wet rice farming involves large amounts of human labour and small areas of land — farms are small, perhaps only 1–2 ha (2.5–5 acres) and subdivided into fields. Cultivation techniques changed little until the 1960s when some areas were affected by the introduction of new rice varieties, fertilizers, and pesticides — a series of changes collectively known as the green revolution (Box 10.10). There are close links between wet rice farming, high population densities, and flat land environments (wet rice fields on terraced hillsides are characteristic only of south China).

Pastoral nomadism

This agricultural type is practised in the dry and cold areas of Africa, Arabia, and Asia, but is declining as a result of new technologies and changing social and economic circumstances. It remains an important type of agriculture in Somalia, Iran, and Afghanistan. Pastoral nomads are subsistence-oriented and rely on their herds for milk and wool. Meat is rarely eaten and the livestock — cattle, sheep, camels, or goats — are rarely sold. Usually one animal dominates, although mixed herds are not unknown. Pastoral nomadism likely evolved as an offshoot of sedentary agriculture, preferable in areas of climatic extremes where regular cropping was difficult. Continually moving in search of suitable pastures, these groups achieved considerable military importance until the rise of strong central governments brought them under control.

Pastoral nomads in the Atlantic Sahara region of northwest Africa are declining in numbers for a variety of interrelated reasons, including government persuasion, economic change, drought, and war (Arkell 1991). For more than 700 years the Sahrawis have survived in a difficult desert environment by migrating vast distances to locate water and pasture for their herds of camels and goats. The imposition of artificial colonial boundaries in northwest Africa divided the traditional territory of the Sahrawis, but did not prevent their movements. Only in the 1960s did economic changes, in the form of phosphate exploitation and related urbanization, prompt many of the herders to seek wage employment in urban areas.

More damaging changes occurred in the 1970s as a consequence of drought and war. The war resulted from Spain ceding its colonial interests in Western Sahara in 1975, leaving Morocco and Mauritania to compete for the territory and from the subsequent actions of a guerrilla group (backed by Algeria) claiming to represent the interests of the Sahrawis in Western Sahara. Many Sahrawis fled into Algeria and the ethnic group is now split into two — those in Algeria and those remaining in Western Sahara. Cross-border movements are now virtually nil as Morocco has constructed a series of defensive walls (sand and rubble parapets) around much of Western Sahara. The 1989 creation of the five-country Arab Maghreb union has not yet helped to resolve the conflict, and the future of the nomadic pastoralist Sahrawis remains uncertain.

Mediterranean agriculture

Mediterranean agriculture can be conveniently designated as a type because it is associated with a particular climate (mild, wet winters and hot, dry summers) and because it has played a major role in the spread and growth of agriculture in the western world. Traditionally, it has three components: wheat and barley, vine and tree crops (such as grape, olive, fig), and grazing land for sheep and goats. Both wheat cultivation and grazing activities are extensive land uses and low in productivity, while vine and tree crops are intensive. Typically, each of the three activities is evident on all farms. Spatially, this type evolved in the eastern Mediterranean, spread throughout the larger region by classical times, and then was exported to environmentally suitable areas overseas, such as California, Chile, southern South Africa, and southern south Australia. Population growth and technological changes mean that there are several versions of this type today with a general increase in irrigation and decline of extensive wheat cultivation.

Mixed farming

The four agricultural types discussed so far are subsistence in orientation; the remaining five are commercial. (Mediterranean agriculture might be described as mixed subsistence/commercial.) Mixed farming prevails throughout Europe, in much of eastern North America, and in other temperate areas of European overseas expansion. It is clearly a variant of the earliest farming in southwest Asia, involving both crops and livestock. The transition to modern mixed farming involved a series of changes, such as the use of heavier ploughs and a three-field system (in the early Middle Ages), the reduction of fallow, and the use of root crops and grasses for animal feed (circa seventeenth century), and general intensification (mid-nineteenth century). Most of these changes have been related to population and market pressures.

Contemporary mixed farming is intensive and commercial, being closely associated with large urban areas, and integrates crops and livestock. The principal cereal crop varies according to climate and soil — it may be corn, wheat, rye, or barley — while root crops are grown for animal and human consumption. Crop rotations are normal as they have environmental and economic advantages; they aid soil fertility and minimize the impact of price changes. As with the other farming types identified, there are many versions of mixed farming and there are, particularly, major differences between western Europe and the American midwest, with the latter having larger farms, especially high productivity, and advanced technology.

Dairying

This agricultural specialization is closely related to urban market advances in transportation and other technological changes that began in the nineteenth century. It is particularly associated with Europe and areas of European overseas expansion. Farms are relatively small and are capital intensive. The dairy products marketed in a given region are often related to distance from market, with those farms close to market specializing in fluid milk and those farther away producing butter, cheese, and processed milk. On the world scale, those dairy areas close to population concentrations — in western Europe and North America — focus on fluid milk while relatively isolated dairy areas — New Zealand, for example — focus on other less perishable products. Fluid milk producers are making increasing use of the feedlot system, which involves feeding cattle with purchased feed. The resulting agricultural activity begins to resemble a factory system as opposed to the traditional image of an agricultural way of life. This development is least evident outside of North America.

Plantation agriculture

A number of crops required for food and industrial uses can only be efficiently produced in tropical and subtropical areas. For this reason, as Europeans moved into the non-temperate world, they established an agricultural type known as plantation agriculture. Plantations produce such crops as coffee, tea, oil-palm, cacao, coconuts, bananas, jute, sisal, hemp, rubber, tobacco, groundnuts, sugar cane, and cotton. The production is for export to Europe and North America primarily. This type of agriculture operates on a large scale, using local labour with a European supervisor. A particular plantation produces only one crop.

Because plantations evolved within the larger context of colonialism, it is a type of agriculture with profound social and economic implications. It has been seen as land and labour exploitation. It is important to appreciate that plantation agriculture is not at fault here; rather it is the plantation as one component of colonial activity that can be criticized. The first plantations were established in the 1400s by the Portuguese in Brazil for sugar cane production and often used slave labour imported from west Africa. With the abolition of slavery, indentured labourers were brought in from India and China. The search for labour assumed global dimensions because the plantation economy was and is so labour intensive. For many areas today the outcome has been the introduction of alien peoples and eventual cultural conflict.

Given that plantations are located in one area but designed to serve the needs of a distant area, plantation companies are often multinational; that is, they operate in more than one country. Furthermore, such companies are often characterized by what is known as vertical integration — producing the crop, then refining, packing, and processing the by-products. Thus, although political colonialism is largely gone, it has in many instances been replaced by economic colonialism.

Ranching

Commercial grazing — ranching — is largely limited to areas of European overseas expansion and is again closely

---

Box 10.10: The green revolution

The term 'green revolution' refers to the creation of improved plant and animal strains and their introduction to the agricultural economies of the less developed world. Such improvements have been ongoing in the western world since the eighteenth century as one aspect of larger economic and technological changes. The less developed world, however, has been characterized by a lack of change. The initial breakthroughs came in the 1960s and resulted from scientific work in the more developed world.

Genetic improvements in rice and wheat have been the major advances. The new strains are higher yielding, respond well to fertilizer, are more disease resistant, and have a shorter growing season. In the 1960s, these strains were the means by which farmers could achieve subsistence levels and market a surplus. Impacts have been greatest in the Indian subcontinent and southeast Asia, but even in these areas the results do not match the optimistic expectations of the 1960s. As in attempts to introduce population control policies, a number of cultural and economic factors prevent (or at least delay) full acceptance.

Peasant farmers are usually conservative and unenthusiastic about abandoning traditional practices. For a new plant strain to be accepted by these farmers, governments found it necessary to offer subsidies and guaranteed prices. Furthermore, new strains have been most readily adopted by better-off farmers with capital and relatively large holdings, and in some cases such adoptions have involved the displacement of poor tenant farmers. Many of the poorest farmers are unable to adopt improved strains because of the costs of fertilizer and pesticides.

The green revolution is a mixed blessing. It can be argued that yield increases have averted some possibly catastrophic shortages, but, ironically, adoption of the new strains can actually result in an increasingly dependent status for the less developed country. The use of new technology increases imports of fuel, fertilizer, and pesticides and exposes the less developed country to such problems as variable oil prices. The need for less labour has led to increased unemployment. Another unfortunate aspect of adopting new strains is their lower protein content.

related to the needs of urban populations. Cattle and sheep are the major ranch animals, as beef and wool are the products in demand. Much European expansion into temperate overseas areas in North and South America, Australia, New Zealand, and South Africa was an expansion of a grazing economy. Ranching is usually a large-scale operation because of the necessarily low productivity involved, and hence it is associated with areas of low population density. In some areas the ranching economy has evolved into a livestock-fattening economy, most notably in the American corn belt where corn and soybeans are produced for animal consumption. This trend appears likely to continue; ranchers, especially in semiarid areas of the temperate world, are experiencing difficult economic times today.

Large-scale grain production

In large-scale grain production, the dominant crop is wheat grown for sale, often for export. Major grain producers are the United States, Canada, and Ukraine. Farms are large and highly mechanized. This type of production evolved in the nineteenth century to supply the growing urban markets of western Europe and eastern North America; it often displaced a ranching economy. In some areas wheat became the staple crop, as in mid-nineteenth century Ontario:

> A more classic case of a staple product would be difficult to imagine. More specialized in wheat production than the farmers of present-day Saskatchewan, Ontario farmers of the mid-nineteenth century exported at least four-fifths of their marketable surplus. Close to three-quarters of the cash income of Ontario farmers was derived from wheat and wheat and flour made up well over half of all exports from Ontario (McCallum 1980:4).

Such a staple was the engine of other economic growth in areas such as Ontario. Today, most grain areas are more diversified and some are now characterized more by mixed farming than by grain production; southern Manitoba in the Canadian prairies is a good example.

CONTEMPORARY AGRICULTURAL LANDSCAPES

AGRICULTURE AND THE WORLD ECONOMY

The relative importance of agriculture in the economy continuously decreases with economic growth (Grigg 1992). This can be demonstrated as follows:

1. For preindustrial economies, data for the World Bank category of low-income countries suggests that about 32 per cent of their gross domestic product (GDP) is derived from agriculture while about 68 per cent of their population is engaged in agriculture; similarly, the limited data

for contemporary industrial countries prior to the Industrial Revolution indicates that agriculture contributed about 40 per cent of the GDP and involved about 70 per cent of the workforce.

2. For industrial economies today, data for the World Bank category of high-income countries suggests that only 2.8 per cent of their GDP is derived from agriculture and only 5 per cent of their population is actively engaged in agriculture.

These figures exaggerate the extent of the relative decline of agriculture because the agricultural data for contemporary industrial countries do not include the food processing and related activities, but nevertheless the picture is a clear one. Industrial and service activities have replaced agriculture as the most important contributors to national economies.

There is considerable spatial variation in the timing of the relative decline of the agricultural sector. The decline occurred first in Britain, the home of the Industrial Revolution, where the agricultural percentage of the total labour force fell below 50 as early as the 1730s; in other European countries this decline to below 50 per cent occurred no later than the 1840s and has not yet occurred in much of Africa and Asia. Also, agricultural incomes are lower on average than are non-agricultural incomes, a gap that is widening all the time.

THE INDUSTRY OF AGRICULTURE

Today, agriculture in the developed world can be justifiably regarded as an industry, one in which the farmer is no longer the sole or primary decision maker. Friedmann (1991:65) asserted: 'the distinction between agriculture and industry is no longer viable, and should be replaced by the conception of an agri-food sector central to capital accumulation in the world economy.'

The Thünen theory might now be regarded as especially unrealistic because of its emphasis on horizontal relations on one level only (that of farmers) when there are horizontal relations on more than one level and also vertical relations between different levels. Five linked levels may apply — farmers, processors, wholesalers, retailers, and consumers. This is simply one way of acknowledging that the food supply system is becoming increasingly complicated. In recent years the greatest changes in this system have been the expansion of purchasing and selling power by the processing and retailing sectors.

Using the example of Quebec, Smith (1984:362) demonstrated that these changes have had a 'marked imprint on the landscape'. The processing level now involves fewer and larger plants and there is greater locational concentration. At the retailing level there is a similar tendency for a few large organizations to dominate the market. Further, there is increasing vertical integration with retailers assuming control of some processing and processors influencing farm operations. The consequence for the farmers themselves is a loss of independence in decision making.

Agriculture will continue to become more mechanized, with increasing use of chemical fertilizers and further replacement of labour by capital. The process is most evident in the developed world, but is also evident in the less developed world as a result of the green revolution. World grain production is increasing spectacularly — about 45 per cent between 1970 and 1990.

As is the case with industrial activity, much contemporary agricultural activity can only be understood when placed in the larger context of agricultural policies and trading agreements. In the less developed world, most policies aim to increase productivity and provide for higher farm incomes and better dietary standards. In the more developed world, policies aim to ensure secure food supplies, price and income stability, consumer interests, and regional development. Such policy objectives may be achieved through guaranteed prices, import controls, export subsidies, and various other amendments to the capitalist market system such as marketing boards. In Canada, for example, there are price support policies, income support policies, and supply-management policies, all subject to change as trading agreements become increasingly important.

A valuable concept to aid understanding many of the changes in the agricultural industry (and other industries as we will see in Chapter 12) is that of **restructuring**. In a capitalist economy, this can refer to changes in capital or changes in the movement of capital resulting from change in technology or labour relations. The process of restructuring as it applies to agriculture may involve spatial changes in agricultural activities, movements of capital from one level to another, and changes in the organization of production. Restructuring can be seen as a process of continual adjustments to changing circumstances.

GLOBAL TRADE IN FOOD

Two important causes of present agricultural change are the ever-increasing demand for food and the concentration of people in large urban centres. Food products need to move from areas of surplus to areas of deficit via sales and aid, movements that are bound to increase given different population growth rates. In recent years major food shortages have occurred in such countries as Bangladesh, China, India, Ethiopia, Nigeria, and Peru.

At least half of the world population relies on one of three cereal crops — wheat, corn, and rice. In many cases these are produced for local consumption, but there is also a substantial global trade in these cereals. Exports of grain are dominated by five countries — first and foremost by the United States, but also by Canada, Argentina, France, and Australia, while the distribution of the surpluses is concentrated in five private corporations. All five of these corporations are international giants with diverse economic interests. The principal importer of grain in recent years has been the former USSR, but several Asian countries are increasing their imports, notably China, South Korea, Indonesia, the Philippines, and Japan.

AGRICULTURAL MODERNIZATION IN THE LESS DEVELOPED WORLD

Throughout much of the less developed world, as suggested in Box 10.10, uncritical acceptance of green revolution innovations is proving to be problematic. As an example, Box 10.11 details some of the cultural and environmental problems involved in recent agricultural change in the Himalayan kingdom of Bhutan.

Box 10.11: Agricultural change in Bhutan

Bhutan is a mountainous land-locked kingdom between China and India that is inhabited by people of Tibetan or Hindu Nepalese origin. It is a less developed country that is only 46,500 km^2 (17,762 square miles) in size and with a population of 0.8 million in 1993; the CBR is 40, the CDR is 17, and the RNI is 2.3 per cent. It is a poor country with only 13 per cent urban population and is currently striving to emerge from geographic isolation.

Although agriculture is limited to about 3 per cent of the land, it is nevertheless the most important sector of the economy. Not surprisingly, agriculture is small scale and subsistence in focus, utilizing crops (especially rice) and animals. The government, with support from elsewhere, is attempting to change these circumstances. Before identifying the desired changes, it is worth noting that the current situation, based on self-sufficiency of households and sustainable agriculture, does not result in poverty at the individual or household level. The government is motivated to increase exports by the perceived value of modernizing.

Unfortunately, the changes being introduced are based on an uncritical acceptance of imported technology and ideas. Green revolution improvements, such as higher-yielding varieties of rice and wheat, are being diffused. It appears likely that these varieties with their associated needs for pesticides and fertilizers may not be well suited to the local environment — the cereal varieties being displaced, while lower yielding, may be much more appropriate. Similarly, the government has adopted new policies concerning two cash crops, cardamom (a spice) and potato, that appear to be environmentally and culturally inappropriate.

There is an important moral to this discussion. Blind, or at least shortsighted, acceptance of new technologies without careful evaluation of their consequences may be disastrous. Any change is best preceded by a trial period and a careful appraisal of the broader cultural implications (Young 1991).

The case of China

Liberalization of the Chinese economy began in 1978 with the opening up of the country to international trade and foreign investment. During the 1980s, the agricultural and industrial economies experienced reforms and modernization. Agriculture was the first sector of the economy to be affected (Morrish 1994). Between the 1950s and 1979, Chinese agriculture was collectivized to allow the communist government to exert maximum control over production. The rural commune, organized by officials, was the key social and economic unit and there were about 60,000 of them before 1979. The less rigorous system used since 1979 involves families working together on a voluntary basis; farmers use land on a contract basis, agree to deliver a specified amount of produce to the government, and are allowed to sell any surplus. The result of this reform was a steady increase in yields of about 6 per cent per year and in incomes. Many peasant farmers are now opting not to produce staple crops but are diversifying into new specialty crops.

But much remains to be achieved, for China continues to have about 65 per cent of the labour force involved in agriculture and yet this sector of the economy contributes only about 32 per cent of the GDP.

THE POLITICAL ECOLOGY OF AGRICULTURE

Political ecology merges two important areas of contemporary human geography, traditional human ecology and political economy as adapted from economics. This approach can be summarized as follows:

1. At the regional level, attempts are made to integrate the multiple social and ecological relations embodied in the organization of agriculture and land use.
2. This approach permits a focus on the politics of place, especially regarding differential power of groups and places.
3. It has clear implications for design and implementation of rural development policies.
4. There is recognition of the need to involve local populations as well as governments and organizations in the planning process.
5. The interplay of social practices (agency) and political-economic conditions (structure) is often considered (as outlined in the account of structuration theory in Chapter 8).
6. Thus a political ecology perspective adds significant new dimensions to more traditional human ecological approaches.

Box 10.12 discusses one application of this type of analysis that provides a good indication of the procedures and merits of the approach.

Other examples include those by Zimmerer (1991), which integrated regional political ecology concepts, structuration theory, a politics of place, and production ecology in an analysis of agricultural change in highland Peru, and by Grossman (1993:346), which used a political ecology perspective to 'highlight not only the impact of political-economic relationships on resource-use patterns but also the significance of environmental variables and how their interaction with political-economic forces influences human-environment relations'.

Gender relations and agricultural restructuring

There is increasing recognition, especially among those employing a political economy approach to the study of agriculture, of the significance of gender relations to the agricultural restructuring process. In the more developed world, there is interest in the growing discrepancy between traditional gender relations and identities in farming communities on the one hand and in society as a whole on the other hand. The trend for women to leave farming is generally considered a reflection of the conservative gender relations associated with farming communities.

In the less developed world, there is increasing recognition of women's historical and present roles. In the case of the west African country of Gambia, women have played important roles in the traditional subsistence cultivation of rice and in recent initiatives in horticultural production. Gambia is the smallest state in mainland Africa (11,300 km²/4,363 square miles) being a narrow strip of land bordering the Gambia River and encircled by Senegal except for a short Atlantic coast. It has a population of 0.9 million (in 1993), a CBR of 46, a CDR of 20, and an RNI of 2.5 per cent. The climate is subtropical and wooded savanna covers most of the area. There are five principal ethnic groups and some 85 per cent of the population are engaged in agriculture. Groundnuts are the one cash crop and account for about 90 per cent of exports.

Women play an important role in the subsistence cultivation of rice, but this activity does not generate any income. There have been recent government initiatives in dry season horticultural production and women have been enthusiastic in their involvement (Barrett and Browne 1991). If such initiatives increase women's incomes, they are clearly of great value; several of them yield well and generate the needed income, but typically have two deficiencies that are not dissimilar to those noted in Box 10.12.

First, the new schemes do not build upon the the local knowledge of horticulture as they are based on imported (or at least on non-local) knowledge. Second, they are not environmentally or economically sustainable. Once again, there is evidence of government attempts to improve agricultural circumstances, but such attempts are possibly misguided. In Bhutan and Gambia, long-established agricultural practices are ignored and new imported practices are favoured. It is not difficult to understand why governments are so tempted, but it is also important to appreciate that the new really needs to learn from the old. The agricultural practices that have survived, perhaps little changed, for long periods are usually grounded in economic and cultural reality; as such, they need to be improved, not replaced, and amended, not eradicated.

BIOTECHNOLOGY

It is appropriate to conclude this chapter by briefly identifying what might prove to be the most significant change in agricultural production since the first domestications of plants and animals. Biotechnology is a new field that permits alterations in the genetic composition of organisms, including food crops. General Foods, for example, has already made a caffeine-free coffee bean and other products are becoming available for consumers. From the perspective of food production and the world food problem, the greatest benefits are likely to be those associated with the creation of crops resistant to virus infections. Tickell (1992), however, outlined some of the problems of applying such scientific advances in practice and argued that the real problems of food production are not to be solved by technological advances but by solutions focusing on the problems of spatial inequalities that were addressed in Chapter 6.

SUMMARY

Economics and geography

Much of our human landscape-creating behaviour has strong economic motivations. Economics inspired much of the new human geography (often called spatial analysis) that surfaced and flourished in the 1960s. Contemporary studies

---

Box 10.12: Peasant-herder conflict in the Ivory Coast

A study of the current conflict between two different ethnic and agricultural groups in the west African state of the Ivory Coast provides a good example of the application of a political ecology approach to a traditional land-use problem (Bassett 1988).

1. What is the basic cause of land-use conflicts between peasant farmers and herders in the Ivory Coast? The traditional answer provided by human geographers is that of a declining resource base relative to population.

2. But this answer does not fully explain why some people lose while others gain land, nor why conflicts occur in such land-abundant areas as the northern Ivory Coast.

3. There is a need to focus on the politics of land use and the human ecology of agricultural systems.

4. In the northern Ivory Coast region, there are ethnic and economic tensions between Senufo farmers and Fulani herders regarding uncompensated crop damage by cattle in peasant fields. This problem began about 1970 and today there is widespread xenophobia towards the Fulani, plus acts of open hostility (eighty herders were killed in 1986). Some political candidates have inflamed the situation by making election promises of banishing the Fulani from the Ivory Coast.

5. But the official government policy is to encourage herders to reduce dependence on imported beef. This reflects the fact that the Fulani account for 33 per cent of beef production in the Ivory Coast.

6. The key factor is determining which of the two groups is able to exert greater political power. The answer to this question will likely influence future land-use patterns. The Fulani are new to the area, while the Senufo are indigenous, but current trends give rights to minorities as well as established majorities, which was not the case in colonial times.

7. The basic determinants of the conflict are outlined in Table 10.3. It is notable that both political ecological variables and traditional human ecological variables are included. Undoubtedly, the single biggest issue is crop damage.

TABLE 10.3: *Determinants of peasant-herder conflicts in northern Ivory Coast*

Ultimate causes	Proximate causes	Stressors	Counter-risks
Ivorian development model: • surplus appropriation by foreign agribusiness the state	Low incomes and beef consumption Insecure land rights	Uncompensated crop damage Political campaigns	Compensation Fulani expansion
• livestock development policies	Intersection of Senufo argriculture and Fulani	Theft of village cattle	Crop and cattle surveillance
Savanna ecology Fulani immigration	semitranshuman pastoralism		Corralling animals night

Source: T.J. Bassett, 'The Political Ecology of Peasant-Herder Conflicts in the Northern Ivory Coast', *Annals, Association of American Geographers* 78 (1988):456.

of agriculture, industry, settlement, planning, and transportation are varied and may incorporate economic, cultural, social, and political content.

The role of theory

Location theory explains geographic patterns, while interaction theory centres on movements between locations. Theories in general are valuable because of their rigour and simplicity. Many of the theories used by geographers are normative, that is, they describe what ought to be rather than what is. The economic operator concept assumes the maximization of profit as a result of perfect knowledge and a perfect ability to use such knowledge in a rational fashion.

The location of agricultural activities

Many factors need to be considered in any discussion of agricultural location. For some geographers, an ecological approach that considers all factors is most appropriate. Other geographers may focus on particular factors depending on the specific issue being tackled. Physical factors include climate, soils, and relief. As living things, animals and plants require appropriate physical environments, either natural or human constructed. Cultural, social, and political factors also play a role. A farmer's religious and ethnic background affects decisions, as do various levels of government. Economic factors include the basic principles of supply and demand.

Economic rent

The concept of economic rent is central to an understanding of agricultural location theory. Land uses compete for locations because it is not possible to have all activities at their economically optimal locations. There is, then, a spatial patterning of land uses, a pattern that results from their differing economic rents (itself a specific measure of rent-paying ability). Values of economic rent can be based on one or more variables; Ricardo based such values on land fertility while Thünen used distance from market.

Thünen theory

Thünen, a German economist, published a landmark book in 1826 that detailed a theory of agricultural location. It is a normative theory that holds all variables constant except distance from a central market. Two hypotheses are generated. First, zones of land use develop around the market. Second, the intensity of each specific use decreases with increasing distance from the market. Thünen proposed particular contents of the zones for his time and place (early nineteenth-century north Germany). Many modifications and extensions of this theory are possible and the theory and amendments to it have stimulated much geographic research.

Empirical analyses

Studies of the relationships between land value and distance, land use and distance, and land intensity and distance can be conveniently subdivided into those with a historical focus, those in less developed areas of the world, and those on various scales ranging from the world to individual farms. Overall, these analyses confirm the value of the Thünen theory as a basis for investigating agricultural patterns.

Agricultural origins

Agriculture had its origins in the domestication of plants and animals, a long process that began about 12,000 years ago. It seems probable that there were multiple areas of origin. Once the process began, these new technologies diffused to other areas. Today, preagricultural activities, such as hunting and gathering, are marginal.

Regions and types

There are nine principal types of contemporary agriculture, each occupying a particular region or regions. Three of these types are subsistence (shifting agriculture, wet rice farming, pastoral nomadism); a fourth type, Mediterranean agriculture, is mixed subsistence commercial, while the remaining five are commercial (mixed farming, dairying, plantations, ranching, large-scale grain production). These various types can be traced back thousands of years with the exception of plantations, which arose in relation to European overseas expansion; further, the modern versions of dairying, ranching, and large-scale grain production are very different from their antecedents.

Agriculture in the world economy

The importance of agriculture, relative to other economic activities such as industry, experiences a continuous decline through time as measured by labour force and contributions to GDP.

Agriculture as an industry

In many respects, agriculture today functions like an industrial activity with a number of levels — farmers, processors, wholesalers, retailers, and consumers — all of whom are related. In addition, national and international policies exert a considerable influence on agricultural activities at all levels. Partly because of this organizational complexity, traditional Thünen theory is of decreasing relevance as an explanation of the spatial distribution of agricultural activities.

Trade in food

Food is an important component of world trade. Much food is sent as aid to poor countries while other countries, notably the former USSR and several countries in Asia, are major importers of grain. The United States is the principal exporter.

Agriculture in the less developed world

China is currently experiencing major changes in the organization of agricultural activities as part of a transition from communism. In some other areas, such as Bhutan, green revolution technologies are misused, causing cultural and environmental problems. Recently, human geographers have employed a political ecology approach, which combines traditional human ecology with political economy, to address a variety of specific problems involving land-use change, land-use conflict, and the involvement of women in agriculture.

Biotechnology
Changing the genetic composition of food crops may be the greatest change in agricultural production since the first domestications.

WRITINGS TO PERUSE

BAKER, O.E. 1925. 'Geography and Wheat Production'. *Economic Geography* 4:389–434.
One of a series of articles by this early agricultural geographer, many of which exemplify a regional approach. See issues of *Economic Geography* between 1925 and 1933 for other examples.

BHATIA, S.S. 1965. 'Patterns of Crop Concentration and Diversification in India'. *Economic Geography* 41:39–56.
An early attempt to develop statistical measures for describing agricultural landscapes.

CONKLING, E.C., and M.H. YEATES. 1976. *Man's Economic Environment*. Toronto: McGraw-Hill.
An excellent text that discusses agricultural theory and landscapes within a larger economic context.

CONZEN, M.P. 1971. *Frontier Farming in an Urban Shadow*. Madison: State Historical Society of Wisconsin.
Detailed analysis of agriculture in an area experiencing the impacts of rapid settlement and commercialization.

FOUND, W.C. 1971. *A Theoretical Approach to Agricultural Land Use Patterns*. London: Arnold.
A sound overview of theoretical aspects of agricultural land use.

GILSON, J.C. 1989. *World Agricultural Changes: Implications for Canada*. Toronto: C.D. Howe Institute, Policy Study 7.
A valuable survey of current changes in world agriculture focusing on government policies and needed reforms.

GRIGG, D. 1984. *An Introduction to Agricultural Geography*. London: Hutchinson.
An excellent textbook dealing with factors affecting agricultural activities.

GROTEWORLD, A. 1959. 'Von Thünen in Retrospect'. *Economic Geography* 35:346–55.
A dated but insightful commentary that is easy to comprehend.

ILBERY, B.W. 1985. *Agriculture: A Social and Economic Analysis*. Toronto: Oxford University Press.
A well-written basic textbook.

LEAMAN, J.H., and E.C. CONKLING. 1975. 'Transport Change and Agricultural Specialization'. *Annals, Association of American Geographers* 65:425–37.
Statistical analysis of agriculture in nineteenth-century upstate New York.

PACIONE, M., ed. 1986. *Progress in Agricultural Geography*. London: Croom Helm.
Covers a wide range of topics such as theory, classification, diffusion, government policies, and marketing.

SAUER, C.O. 1952. *Agricultural Origins and Dispersals*. New York: American Geographical Society.
A classic study by the most prominent of twentieth-century cultural geographers.

SIMPSON, E.S. 1990. 'Plantations: Benefit or Burden?' *Geographical Magazine* 42, no. 1 (special supplement):103.
A brief but thought-provoking account of plantations as one component of colonialism.

11

Settlement

PERMANENT SETTLEMENTS BEGAN to form with the evolution of agriculture. Today, only a minority of humans engage in hunting and gathering or agricultural activities that involve regular movement from place to place. Permanent settlements are associated with specific economic activities, initially producing a food surplus and subsequently serving as a trade and industrial centre. Settlements that are today principally agricultural are usually called rural; those that are principally non-agricultural are called urban.

The distinction between rural and urban is not, however, quite so simple. Different countries use different criteria to determine whether or not a settlement is rural or urban; Table 11.1 gives samples of the many different definitions. Most of the settled surface of the earth is rural, although the world is rapidly becoming increasingly urbanized (Figure 11.1). Estimates for 1993 define 42 per cent of the world population as urban (using the different definitions used by each country); in the more developed world the figure is 72 per cent and in the less developed world it is 34 per cent.

In this chapter on settlement, we deal with two fundamental geographic issues — location and way of life — for convenience when discussing rural and urban issues. This division between rural and urban reflects a long-standing geographic tradition. Rural settlement geography is often thought to have originated with the early twentieth-century studies by Vidal and other European geographers focusing on settlement pattern and their links to agriculture. Urban settlement geography is more recent in origin; the first English language texts were published in the 1940s.

Our discussions of locational matters — rural, urban, and interurban — employ conceptual accounts based on historical experiences and theoretical accounts that are in a Thünen

TABLE 11.1: *Some definitions of urban centres*

France	Communes containing an agglomeration of more than 2,000 inhabitants living in contiguous houses or with not more than 200 m between houses, and communes of which the major part of the population is part of a multicommunal agglomeration of this nature.
Spain	Municipios of 10,000 or more inhabitants.
East Germany	Communes of 2,000 or more inhabitants.
Belgium	Communes of more than 5,000 inhabitants.
Denmark	Agglomerations of 200 or more inhabitants.
Canada	Cities, towns, and villages of 1,000 or more inhabitants, whether incorporated or unincorporated, including urbanized fringes of cities classed as metropolitan areas and other major urban areas. In 1961, also including urbanized fringes of certain smaller cities if the population of city and its urban fringe was 10,000 or more.
Japan	Urban municipalities (all *shi* and the *ku* of Tokyo-to) usually having 30,000 or more inhabitants and which may include some rural area as well as urban cluster.
Israel	All settlements of more than 2,000 inhabitants, except those where at least one-third of the heads of households, participating in the civilian labour force, earn their living from agriculture.
Mexico	Localities of 2,500 or more inhabitants.

Source: H. Carter, *The Study of Urban Geography*, 3rd ed. (London: Arnold, 1981):18.

This table confirms that there is no simple and agreed-upon definition of urban centre on the basis of population size, and emphasizes the danger of comparing data for different countries.

tradition. Our discussions of way of life, both rural and urban, are more conceptually diverse reflecting a wide range of economic and social theory, including Marxism, humanism, and postmodernism.

RURAL SETTLEMENT

Before about 1950 rural settlement issues were at the core of human geography. Today, rural issues receive much less attention than urban issues, perhaps because the urban experience is the dominant one for most people in the developed world. In this section we will focus on four topics: rural settlement patterns, especially as they relate to agriculture; rural settlement theory; changes in the rural way of life, and the analysis of land-use transition from rural to urban.

RURAL SETTLEMENT PATTERNS

Rural settlement locations assume many and varied patterns, depending on the physical environment, culture, social organization, political influences, and economic activities. Generally, patterns range between the extremes of dispersion (random or uniform) to nucleation (clustering) (see Figure 2.4). An important component of rural settlement landscapes is the pattern of fields, a landscape feature that often contributes greatly to the character of a place.

Dispersion

Dispersion of rural settlements, a relatively recent phenomenon, is especially prevalent in North America and parts of western Europe. In Europe, dispersion was prompted by the collapse of feudal social systems and the rise of an industrially based society. Between about 1750 and 1850, the English rural landscape was dramatically transformed by the process of enclosure (as discussed in Chapter 10), which included land consolidation, new field boundaries, and the construction of dispersed dwellings; governmental factors played a leading role in this transformation. As is the case with so many aspects of the human landscape, field patterns can be seen as spatial expressions of power relations; the preenclosure pattern reflected feudal social relations, while the enclosed pattern reflected individual ownership under a capitalist system.

In most areas of North America dispersed rural settlements were typical from the onset of European settlement. Governments' prevailing policy was to survey before establishing settlements, with the result that geometric patterns soon became the norm. Early French settlement in North America, for example, along the St Lawrence, introduced long, narrow lots at right angles to the river that gave all settlers access to water; these settlers did not own land but were responsible to seigneurs (feudal landlords).

The characteristic North American government survey system encouraged dispersed settlement; typically, each settler was granted a one-quarter section and required to construct a dwelling on the section. This type of survey began in 1785 when the United States land survey laid out a baseline

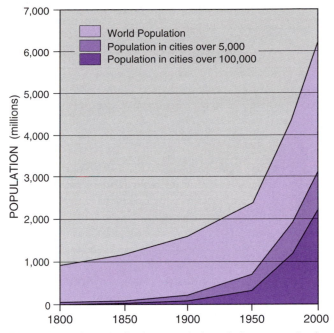

Figure 11.1 Growth of urban population relative to growth of world population.

west to the Pacific from the point where the Pennsylvania border meets the Ohio River. From the government's point of view, the big advantage of this type of survey was that it provided an orderly way to sell land. In Canada, the prairies provide one example of this system, although in this area the settlement landscape included both dispersed and nucleated forms (Figure 11.2 and Box 11.1).

Nucleation

Nucleated settlements dominate the rural landscape, at least partly in response to the basic human need to communicate and cooperate with others. Being with other people is important for security, social life, religious activities, and the regular exchange of goods and services. Reasons for the development of nucleated settlements include a scarcity of good building land (for example, in areas that experience regular flooding) a need to defend the group against others, the need for group labour to construct a particular agricultural feature, such as terraces on steep slopes or irrigation systems, and a political or religious imperative, as in the case of the Israeli kibbutz or the Mormon village.

Throughout much of the rural world, nucleation has been favoured over dispersion. In China the 1993 rural population is estimated at 74 per cent, most of whom live in nucleated village settlements. Although many of these villages have populations of several thousand, they are best described as rural settlements as the inhabitants are farmers. Often these settlements are surrounded by very small fields.

Nucleated settlements are not uncommon in some parts of North America. The New England landscape reflects early settlers' desire to live in close proximity to reinforce group cohesion and provide some protection against attack. Similarly, the landscape of the Mormon cultural region in Utah and parts of adjacent states is characterized by nucleation as a consequence of the central planning practised by the Mormon Church.

Nucleated rural settlement offers many different types of physical layout. Figure 11.3 identifies some examples. An irregularly clustered settlement suggests unplanned growth over time. Examples of regular settlements that may or may not result from some form of planning include elongated-street, green, and checkerboard settlements. Nucleated rural settlements vary in size from a few dwellings (a hamlet) through to villages with populations of perhaps 25,000 in some parts of the world (as already noted, the distinction between rural and urban is not a clear one).

RURAL SETTLEMENT THEORY

Most studies of rural settlement focus on particular settlements or regions and include substantial descriptive content, a tradition begun by the noted German researcher, Meitzen

(1822–1910). Beginning in the 1960s, however, increasing attention has been given to more general questions, such as why rural settlements are located where they are.

Bylund theory

A Swedish geographer, Bylund (1960), developed deterministic theoretical formulations for rural settlements' expansion into previously unsettled territory. The formulations are based on two principal variables: distance from the parent settlement, and relative land attractiveness (especially soils and climate). For an area in northern Sweden, Bylund assigned each possible location an attraction value and constructed a series of six settlement patterns. As with Thünen theory, the formulations allow for simplified descriptions. Comparing the simplified settlement patterns with maps of actual settlement in northern Sweden indicated that the two are similar — the essential features of the real world settlement are replicated. Thus, Bylund's two variables — distance from point of origin and land attractiveness — are confirmed as important.

Box 11.1: Rural settlement in the Canadian prairies

Rural settlement in the Canadian prairie region reflects central government policies and the wishes of some major ethnic groups. The result is a landscape that includes dispersed and nucleated farm settlements.

Settlement, which began in the 1870s, was preceded by a survey that allowed for sectional settlement and dispersed farmsteads (Figure 11.2). This arrangement or something similar prevailed in many other areas of European overseas expansion, such as southern Ontario, much of the United States, and much of Australia. Some land was assigned by Canada's central government to settlers while other land was granted to companies (especially railway companies) so that they could generate income by selling to settlers. Dispersed settlement was normal regardless of whether a settler received land directly from the government or indirectly from a company. Most prairie immigrants accepted this circumstance even though many had been used to village settlement. The major social implication of dispersion involved a new experience — isolation.

The principal exception to the prevailing pattern was the establishment of farm villages by specific ethnic groups. Doukhobors and Mennonites from Europe, Mormons from the western United States, German Catholics, and Jews opted for nucleation for religious reasons and/or because of a desire to continue a unified group, and the specific government requirements of sectional dispersed settlement were waived on request. These nucleated settlements are an infrequent but distinctive feature of prairie rural settlement. Most of the farm villages were of the elongated-street variety (see Figure 11.3), although the Mormons favoured a square grid plan.

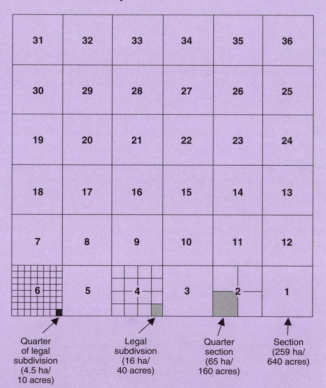

Figure 11.2 Township survey in the Canadian prairies. A typical township of 36 square miles (93 km²) is divided into thirty-six sections of 1 square mile (640 acres/259 ha) each, and subdivided into four quarter sections of 160 acres (64 ha) each. Settlers were usually assigned a quarter section. Close to urban centres, quarter sections might be divided into four legal subdivisions of 40 acres (16 ha) each.

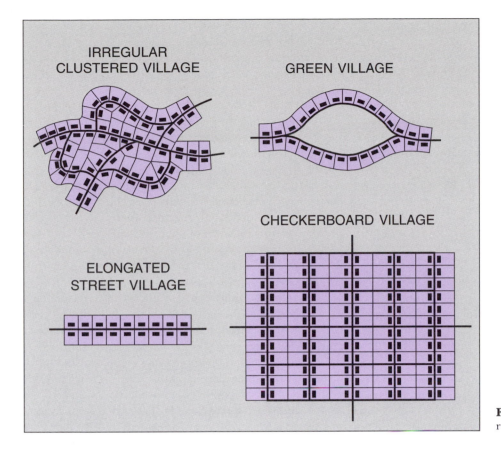

IRREGULAR CLUSTERED VILLAGE

GREEN VILLAGE

CHECKERBOARD VILLAGE

ELONGATED STREET VILLAGE

Figure 11.3 Examples of nucleated rural settlement patterns.

Hudson theory

An alternative approach to rural settlement theory is one by Hudson (1969). His work is deductive and based upon other geographic theory and ecological distribution concepts concerning the dispersal of seeds from parent plants. Hudson proposed three stages for the rural settlement of an area. *Colonization*, the first stage, is the movement of people into a hitherto unoccupied area and their settlement in the landscape. *Spread*, the second stage, is the subsequent filling up of the area by the offspring of the first colonizers. This second stage leads to the establishment of some nucleated rural settlements. The third stage is *competition* between the newly created nucleations as they vie for surrounding dispersed rural settlements to support their various activities (such as retail establishments). Hudson associates each stage with a characteristic pattern. The colonization pattern is one of concentric rings, the spread pattern is nebula-like with several distinct clusters, and the competition pattern is a regular lattice.

Some criticisms of rural settlement theory

Such theoretical discussions are of great value, but, as with all geographic theories, cautious use is crucial. Probably the single most important criticism of Bylund and Hudson is that their ideas fail — intentionally, of course — to incorporate the multitude of social, cultural, political, and other human variables. One critic, Grossman (1971), argued that such biologically derived principles as those used by Hudson are not applicable to rural settlements. Grossman's argument is that the process of settlement is largely dependent on human factors that vary from place to place, such as the interference of central planning, rather than on some unvarying human behaviour.

Rural settlement theory is not as well advanced as many other areas of geographic concern. These theories are, however, very useful reminders of the value of investigating geographic topics in a conceptual manner. Understanding specific cases is often greatly enhanced by referring to such conceptual efforts.

RURAL SETTLEMENTS IN TRANSITION

Gemeinschaft and Gesellschaft

Numerous studies of rural populations in diverse parts of the world have established that there are important differences between rural and urban ways of life. This was recognized in 1887 by the sociologist, Tonnies (see Cater and Jones 1989:170–1), who distinguished between *Gemeinschaft* (community), human association based on close personal contact as in village settings, and *Gesellschaft* (mass society), human association that is depersonalized as in urban settings. Rural settlements have historically been characterized by spatial isolation and limited spatial mobility, but the notion of a rural community with close informal social relations is somewhat misleading as the typical social framework in a rural setting was (and in many areas still is) one of a rigid social hierarchy.

The seminal work dealing with differences between rural and urban areas was accomplished by Wirth (1938), who contended that increases in size and density of populations result in increased anonymity, further division of labour, social heterogeneity, and enhanced mobility. Thus Wirth argued that urban areas were distinguished by a distinctive way of life. The lifestyle differences between rural and urban are generally well documented, although Pahl (1966) has argued that in some cases, the social differences between rural and urban areas are unclear.

Contemporary rural society

In most countries in the more developed world 'rural society has been subject to a process of cultural colonization in that the dominant images of rural life have been formulated by (middle class) urbanites and projected on to the countryside' (Cater and Jones 1989:194). The results are urban dominance and rural dependence that incorporate nostalgic myths of a rural idyll and better yesterdays.

The reality is that rural areas have been substantially modified by industrialization. The rural/urban difference, especially in the more developed world, is decreasing as a result of improved rural services and the proximity of urban employment opportunities for many rural dwellers. Even in those rural areas that are far from urban centres, contemporary technology ensures a certain uniformity in lifestyle as a consequence of access to similar mass media.

Depopulation or repopulation?

A principal feature of rural areas today is depopulation because of spatial concentration of economic activities. In 1961 there were 141 towns and villages in Saskatchewan; by 1993 the number was reduced to sixty-two. At the beginning of this century, the settlement process involved the growth of nucleated centres, which provided goods and services around the grain elevators that were established at 16-km intervals along railway lines, but this settlement system quickly became obsolete. The attraction of rural life diminished as agriculture technology changed, becoming capital intensive and requiring less labour, and as the recession of the 1980s lowered farm incomes.

For Canada, a detailed analysis by Keddie and Joseph (1991) suggests that the period 1981–6 was one of sustained urban growth comparable to that which occurred in the 1960s. During this five-year period, there was a 5 per cent rate of urban growth in Canada compared to a 0.8 per cent rate of rural growth. Table 11.2 summarizes the percentage change in regional rural population for the period 1981–6 and indicates substantial regional variations, from the prairie region with a rural population change of –3.1 per cent to the Atlantic region with a figure of +3.5 per cent. This is another timely example of the dangers of using data for large areas.

Urban growth continues to proceed apace throughout the world. Or does it? At least until about 1970 the data were clear — people moved from less to more densely populated areas — but in the 1970s, there was a decline in this movement in some parts of the more developed world and possibly even a countermovement. In some areas there is evidence of rural repopulation, a trend that prompted the term **counterurbanization** (Hall et al. 1973). According to some researchers, this is a trend that results from industrial changes in a less spatially concentrated economy. Other possible explanations include an increasing appreciation for rural lifestyles, especially by the growing number of seniors and the rapid proliferation of new communication systems that reduce the need to be spatially close to services. But there is a simpler explanation. Birth rates in the more developed world are low and hence there are fewer people to move from the less densely populated areas.

Significantly, the Canadian analysis noted earlier was prompted by some evidence that suggests rural population growth in the 1970s, but this possible trend was not evident, as we noted, for the first half of the 1980s. Any rural repopulation in Canada during the 1970s was a 'fleeting and regionally specific phenomenon' (Keddie and Joseph 1991:379).

THE RURAL-URBAN FRINGE

The transition zone between rural and urban areas often has multiple land uses and varied social and demographic characteristics. Indeed, the absence of a clear break between the two areas is a distinctive feature of contemporary land use and lifestyle. One way to interpret this situation is to focus on the centrifugal forces that encourage urban land uses to expand in the fringe belt. Urban development usually includes concentration and deconcentration. The term **urban sprawl** is often used to describe the deconcentration that involves low-density expansion of urban land uses into surrounding rural areas (Box 11.2).

TABLE 11.2: *Percentage change in rural population by region, Canada, 1981–6*

Region	Aggregate % change
Atlantic	3.5
Quebec	-0.1
Ontario	3.4
Prairies	-3.1
British Columbia	-1.1
Canada	0.8

Source: Adapted from P.D. Keddie and A.E. Joseph, 'The Turnaround of the Turnaround? Rural Population Change in Canada, 1976 to 1986', *Canadian Geographer* 35 (1991):371.

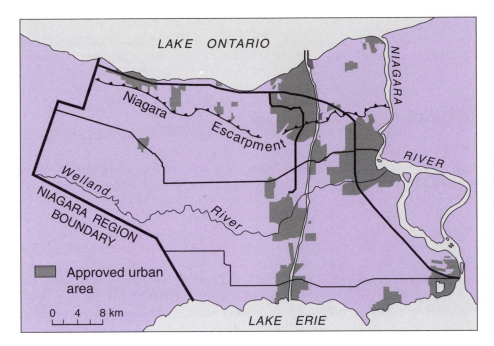

Figure 11.4 Approved urban areas in the regional municipality of Niagara, 1989. The Niagara region (population 370,000) is at the western end of an urban region containing 5 million people. North of the Niagara escarpment, a fertile plain — the Niagara fruit belt — edges Lake Ontario.

Source: Adapted from H.J. Gayler, 'Changing Aspects of Urban Containment in Canada: The Niagara Case in the 1980s and Beyond', *Urban Geography* 11 (1990):377.

The Niagara Peninsula

In much of the world the transition from rural to urban land uses can be a cause for concern. One particularly sensitive example is the Niagara Peninsula region in Canada (Figure 11.4). This region is one of only two important soft-fruit producing areas in Canada, but it is close to large urban areas while containing a number of urban centres. Concerns about urban encroachment were first raised in the 1950s, but there was little attempt to limit the growth of urban areas until the 1970s, as growth was considered good.

Box 11.2: Exurbanization in southwestern Ontario

In many areas deconcentration is occurring as residential movement to the rural/urban fringe. The question arises as to whether such movements are simply an extension of existing suburbs or represent a form of counterurbanization as already defined; another way to think of these different possibilities is in terms of the push and pull factors introduced in the discussion of migration in Chapter 6.

For Woodstock in southwestern Ontario, Davies and Yeates (1991) show that both considerations are relevant. In their study the focus is on **exurbanization** and a distinction is made between exurbanite households located in dispersed rural settlements (54 per cent of the total in the area around Woodstock) and exurbanite households located in villages (46 per cent of the total in the area around Woodstock). As shown in Table 11.3, 'the exurban-rural group is part of the new urban-to-rural migration pattern subsumed within the idea of counter-urbanization, while the exurban village group is part of the long-standing trend of suburbanization related to a search for more housing space and privacy at lower prices' (Davies and Yeates 1991:186). This conclusion is based on the relatively high valuation that the exurban-village group place on lower taxes and lower house prices and on the relatively high valuation that the exurban-rural group place on the amenity attractions of rural areas. Both groups place a high value on privacy, larger houses, more land, the attractive rural landscape, a better quality of life, and less crime.

TABLE 11.3: *Factors influential in the decision to move to exurban locations around Woodstock, southwestern Ontario*

Factor	Exurban-rural % important	Exurban-village % important
Better schools	34.4	42.4
Lower taxes	33.3	53.8
Less commuting	26.2	24.7
Friends and relatives	25.1	31.0
More privacy	95.1	82.3
Lower house prices	43.7	65.2
Larger house	63.9	58.9
Better services	15.3	19.0
More land	80.9	75.9
Proximity to urban area	48.6	51.3
Attractive landscape	79.2	72.8
Better quality of life	85.8	86.0
Less crime	56.8	72.2

Source: S. Davies and M.H. Yeates, 'Exurbanization as a Component of Migration: A Case Study in Oxford County, Ontario', *Canadian Geographer* 35 (1991):177–86.

As a result of community pressure and subsequent government action, rulings in the 1980s identified substantial areas that could only be used for agriculture or related purposes — a significant reversal of attitude.

ORIGINS AND GROWTH OF CITIES

Since about 1750, the world has experienced a number of transformations: the rise of capitalist societies and economies, a rapidly increasing population, a proliferation of states, the many effects of new technologies, and more as well as larger urban centres. By 1850, the major world cities were concentrated in the newly industrializing countries, a pattern that remained well into the twentieth century. Most of these major cities were European capitals or were located on the eastern seaboard of the United States. Recently, however, especially since *c*.1950, most cities (particularly the largest ones) are in former colonies in the less developed world. Before 1950 cities in the less developed world were typically transportation centres or colonial government centres that formally excluded most indigenous people. With independence, cities in the less developed countries acted as population magnets, offering employment and wealth to some, but only minimal benefits to the majority. In 1800, perhaps 3 per cent of the world population lived in cities; by 1900, the figure was 13 per cent, and, by 1993, 42 per cent. Increasingly, cities are the focal points of our modern civilizations.

Today, the majority of large cities are in the less developed world (Table 11.4), and tend to be located on the coast. It is also notable that many capitals are in the large-city class. The importance of transportation and of services is clear.

URBAN ORIGINS

The city/civilization link is an ancient one, the two words sharing the Latin root *civitas*. Civilizations create cities, but cities mould civilizations. James Vance (1990:4) quotes Winston Churchill: 'We shape our houses but then they shape us.' The earliest cities probably date from about 3500 BC and were preceded by large agricultural villages. It is no accident that the birth of cities coincided with major advances, such as the invention of writing. From the beginning, cities have been 'the chief repositories of social tradition, the points of contact between cultures, and the fountainheads of inspiration' (Smailes 1957:8).

The origins of **urbanization** and the urban way, **urbanism**, are debatable. Here are four possible explanations. In some areas the production of an *agricultural surplus*, possibly as a result of irrigation schemes, may have been sufficient; city life only became possible as the progress of agriculture freed some group members from the need to be producers of food. The earliest such cities were located in the Tigris–Euphrates region (present-day Iraq), the Nile Valley (Egypt), the Indus Valley (Pakistan), the Huang Valley (China), Mexico, and

Peru and were associated with the rise of agriculture in these regions (see Figure 7.1). These cities were the residential areas for those not directly involved in agriculture. They were small in terms of population and area. Populations ranging from 2,000 to 20,000 were normal, although those of Ur on the Euphrates and Thebes on the Nile may have been as large as 200,000.

Cities were first established in agricultural regions, although specific locations were often chosen on navigable waterways. Some cities are probably established as *market places* for the exchange of local products or on long-distance trading routes: Venice was a prime example of a port city, while Baghdad was a port in the desert.

Elsewhere cities may have started as *military, defence, or administrative centres*. The Greeks introduced cities to their colonial areas to exercise control, and in many cases planned the city in advance as a grid pattern with a central space and a series of routes intersecting at 90°. The Romans, who knew that urbanization was the key to controlling conquered areas, continued the Greek initiatives as the empire expanded. Many North African, Middle Eastern, and European cities can trace their origins to either Greek or Roman times.

A fourth suggested explanation proposes that cities arose as *ceremonial centres* for religious activity. Chinese urban locations were selected using geomancy, the practice of discovering hidden meaning through omens; once an auspicious site was located, the city was laid out in geometric fashion. Chinese cities were square in shape, in accord with the belief that the earth was square and the desire to be a part of nature rather than to dominate it (Wheatley 1971).

PREINDUSTRIAL CITIES

Use of the term 'preindustrial city' is somewhat misleading as there were notable differences between cities in different parts of the world prior to the Industrial Revolution — a

TABLE 11.4: *The ten largest urban areas, 1992, and projections for 2000 (population in millions)*

City	1992	2000
Tokyo	27.2	30.0
Mexico City	20.9	27.9
São Paulo	18.7	25.4
Seoul	16.8	22.0
New York City	14.6	14.6
Osaka	13.9	14.3
Bombay	12.1	15.4
Calcutta	11.9	14.1
Rio de Janeiro	11.7	14.2
Buenos Aires	11.7	12.9

Source: United States Bureau of the Census, International Data Base.

Note: For this table, a city is defined as population clusters of continuously built-up area with a population density of at least 12,950 per km² (5,000 square miles). Different definitions and criteria result in different detailed data.

reflection of cultural differences. It is a useful term in our introductory context, however, as a means of emphasizing the significance of the changes that accompanied industrialization. Before the advent of capitalism and the related Industrial Revolution, most cities were concerned with marketing, commercial activities, and craft industries, and also served as religious and administrative centres. As new features in the human landscape, Wirth (1938) argued that cities developed a new way of life — a new social system and a more diverse division of labour. Political, cultural, and social change followed the economic change from preagricultural to agricultural societies. Rural and urban settlements were complementary, the former producing a food surplus and the latter engaging in a series of new functions. From the beginning, then, cities have been functionally different from surrounding rural areas.

Inside the preindustrial city

Unlike modern industrial cities, however, the preindustrial city's homes, workshops, markets, and other functions were located in a relatively haphazard fashion. The principal evidence of any planning inside the city was a basic division between élite and other areas. According to Sjoberg (1960), the élite in all preindustrial cities occupied the central core in the economy, culture, and politics, regardless of time and place. Vance (1971) argued that, in Europe, the guild system was more crucial to urban differentiation than the feudal system; thus, there was land-use zoning in the sense that workers in a particular craft were spatially concentrated.

Preindustrial urban growth

In Europe the process of urbanization faltered after the collapse of the Roman empire in the fifth century and did not resume for some 600 years because of a decline in interarea movement and trade. Urban centres in Europe revived gradually as commerce revived, as political units gained increasing power, as population increased, and as agricultural technology advanced. These developments became evident in the eleventh century. The resulting urbanization revived old cities and established new ones throughout Europe, especially in areas well located for trade such as on the Baltic, the North Sea coast, and the Mediterranean. Urbanization was encouraged in the seventeenth century by the economic philosophy of **mercantilism**. This philosophy contended that governments should be actively involved in economic activities in order to help states attain their maximum economic potential. Thus, mercantilism encouraged colonial expansion and city growth.

While Europe went through these periods of decline, stagnation, and rebirth, other areas of the world had different experiences. In China and in much of the Islamic world, the earliest cities flourished throughout this time. Traditional cities outside Europe began to experience substantial change only with the arrival of European colonial activity.

Preindustrial cities followed the development of agriculture and soon assumed distinctive functions and characters. The typical city served an agricultural population and also functioned as a trading centre. In Europe the trading function was greatly enhanced by the acceptance of mercantilism, but all of this changed dramatically with the advent of industrialization.

INDUSTRIAL CITIES

The Industrial Revolution began in England about 1760 and combined a series of changes that altered human landscapes. The principal economic change was related to the new merchant and entrepreneurial middle class who opposed the state's interference in mercantilism and subsequent limits on profits. With the rise of the merchant class, mercantilism faded and the final remnants of feudal society and economy disappeared. The new economic force of capitalism (see glossary) emphasized growth, profits, and a market economy. Capitalism, combined with a series of technological advances pertaining particularly to new energy sources, thus led to the Industrial Revolution and the industrial city.

Industrial urban growth

Urbanization has been one of the key phenomena of the industrial age involving the spatial movement of large numbers of people and major changes in social life. Industrial cities grew with remarkable rapidity in key resource locations, especially on coalfields. One of the first industrial areas in Britain was Coalbrookdale, northwest of Birmingham, where Abraham Darby first smelted iron ore with coke instead of charcoal in 1709. After 1760, the steam engine became available and old industrial activities, especially textile and metal production, were transformed. The importance of coal for power and transportation resulted in major new concentrations of activity in the new industrial cities. Machinery, capital, and labour were located in the cities.

Urban growth proceeded apace in the new industrial areas — in Britain from 1760 onwards and then in western Europe, the United States, parts of southern Europe, Russia, and Japan by the end of the nineteenth century. The rapidity of growth is illustrated by the United Kingdom whose 24 per cent urban population in 1800 increased to 90 per cent in 1993. Since the Second World War, industrialization has occurred in many parts of the less developed world. Today, in many states most of the population is in cities, following the British lead, although most states continue to have a rural majority. Even in states that continue to specialize in agriculture, cities have grown as a result of the increasing importance of service activities.

EXPLANATIONS FOR URBAN GROWTH

The changes described earlier have economic and social causes. Generally, economic causes involve the financial benefits that accrue from city growth, namely reductions in

11.1 A planned capital during construction, Canberra, Australia. (National Capital Development Commission 3407/1)

assembly, production, and distribution costs. According to this view, cities are the most economic settlement decisions. **Economic base theory** reduces urban economies to two interdependent sectors, basic and non-basic. The basic sector comprises all those activities that produce goods and services for sale outside the city. The non-basic sector comprises all those activities that produce goods and services for sale inside the city. Detailed consideration of this distinction allows some important conclusions. First, the larger the city, the less dependent it is on the basic sector. Second, the larger the city, the more it is able to grow. Social cause explanations for city growth revolve around the fact that groups offer security and that cities result once the necessary social structures are in place. It seems reasonable to conclude that economic and social causes play a role.

URBAN PLANNING

Urban planning has a long history, but only in recent years have most cities grown in anything other than an unregulated and uncontrolled fashion. Generally, urban planning results from an awareness of problems. It implies a loss of individual rights and a search for the common good. Some

early cities had planning in their core areas resulting in central squares and a geometric street pattern; others clearly grew haphazardly.

Origins

Modern urban planning originated in Europe and the United States only in the late nineteenth century. Interestingly, the planning movement had origins in anarchism, with such notable figures as Patrick Geddes and the geographers Peter Kropotkin and Elisée Réclus at the fore, as well as in utopian socialism as personified by Robert Owen, a Scottish philanthropist. In 1898, Ebenezer Howard introduced the idea of the **garden city**; cities to be built according to a master plan to provide a spacious and high-quality environment for working and living. Each city was intended to house 32,000 people in an area of 2,429 ha (6,002 acres) and was characterized by a concentric pattern of land use, wide streets, a low-density housing, public open spaces, and a **green belt.** Several settlements have been built in accord with a modified version of the garden city concept, while the general ideas have been most influential in urban planning in general.

Le Corbusier and Wright

Perhaps the best known urban planner was the Swiss architect, Le Corbusier, who designed totally new cities, but who has had very few of his designs converted to reality. Le Corbusier favoured increasing circulation, the amount of open space, and density; this was to be achieved by rebuilding city centres. In the 1920s, these were revolutionary ideas. The new planned city was to have an organized spatial structure with residences located according to social class. Only two cities have been built on these lines — Chandigarh, the new capital of the Punjab, India, and Brasilia, the new capital of Brazil. In America, the urban designs of the celebrated architect, Frank Lloyd Wright, were never fulfilled, although his impact on architectural style has been substantial. Le Corbusier and Wright reflect a modernist (see modernism in glossary) approach that regards the best designs as pure and simple.

Applications

Both Britain and the United States have been affected by urban planning philosophies and practices with roots in a variety of reformist movements. The essential belief was that social conditions were a product of the built environment; low-quality cities were seen as a cause of disease and crime. Since the Second World War, planning has centred on physical design. In Britain three key acts — the Distribution of Industry Act (1945), the New Towns Act (1958), and the Town and Country Planning Act (1947) — set the scene. Since 1946, twenty-eight new towns have been created (Figure 11.5). In some cases the aim was to disperse population from overcrowded cities; in others the aim was to support specific regional industrial policies. Probably the most significant consequence of this British planning has been to limit the growth of large urban areas. In the United States there has been much less government intervention and a lack of a common national framework. Zoning is a common means of ensuring that urban land is appropriately used. Planning at the national scale has typically been concerned with public housing and slum clearance. Overall, in most parts of the more developed world and in some parts of the less developed world, planning may affect urban location and internal structure.

URBAN LOCATIONS: I. HISTORICAL EXPLANATIONS

Why are urban centres located where they are? Various answers to the question have already been noted. Urban centres may locate in agricultural areas to serve rural populations, they may occupy strategic locations, they may be key trading centres, or their location may be determined by governments with some overall plan in mind. Occasionally, the location of a set of centres can be explained by reference to, for example, government planning. Thus, the settlement pattern on the Nile *c.*1317–1070 BC has been explained in terms of government selection of sites to maximize control

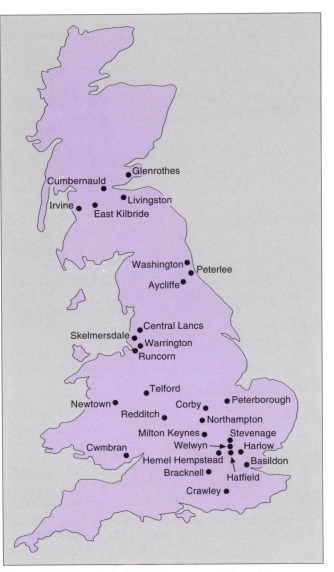

Figure 11.5 Location of British new towns. Most of the new towns in the south of England were located so as to relieve pressure on London; others were new industrial towns.

of populations (Church and Bell 1988). Such situations are hardly typical.

MERCANTILE MODEL: VANCE

According to Vance's mercantile model (1970), the initial growth of an urban centre results from external factors such as long-distance trade. The example of North America presents the process:

Stage I   An initial search of the North American coasts to gather information relevant to European interests.

Stage II   An initial testing of the quality of resources discovered and, in some cases, use of specific resources as staple products.

Stage III Settlement at key points, both coastal and inland, to facilitate export of products.

Stage IV   Expansion of the economic region by establishing urban locations, usually at key transport sites on waterways.

Stage V   Emergence of the key trading locations as cities.

Close links between this mercantile model and what is known as the *staple model* of economic growth indicate the former is effectively an offshoot of the latter. The staple model is often used by economic historians, especially for such areas as Canada and Australia, to describe situations where one or a few key resources are exploited and are responsible for much other economic growth. Examples of Canadian staples are, in an approximately chronological sequence, fish, fur, lumber, wheat, and minerals. In Australia the sequence is wool, gold, and wheat.

The use of staple concepts and related trading to explain urban network evolution is novel. Reaction to Vance (1970) has been mixed. Sargent's (1975) study of Arizona towns confirmed the model; Earle (1977) found it inadequate to explain the pattern of early colonial towns in North America, and Meyer (1980) considered it inadequate.

METROPOLITAN EVOLUTION: BORCHERT

A second historical model, again applied to North America, identifies four technological stages and relates urban growth to each phase. Borchert (1967) proposed the following:

Stage I   Sail wagon, 1790–1830. During this stage, urban locations were primarily related to availability of water transportation.

Stage I   Iron horse, 1830–70. Rail networks focused on ports, but also encouraged urban growth along railway lines in the new hinterlands. Many new urban centres were established during this stage.

Stage III   Steel rail, 1870–1920. Expansion into the large western region of North America and development of existing centres as industrial cities.

Stage IV   Auto-air-amenity, 1920 onwards. Creation of a dense road network, suburban sprawl, and new cities in the south.

These ideas are similar to the mercantile model as they explain urban growth in North America, but there is less emphasis on trade and greater emphasis on the technology of transportation.

SELECTIVE URBAN GROWTH: MULLER

The model devised by Muller (1976) to explain North American towns is based on the mercantile model. Three stages are envisaged:

Stage I   Pioneer periphery. Initial location of urban centres related to early agriculture, few trading contacts, little commercial agriculture.

Stage II   Specialized periphery. Dominant urban centres emerge at points of regional contact.

Stage III   Transitional periphery. The region is now fully linked to national networks and the urban pattern is a continuation of the past with the export of agricultural products and the establishment of new locations as manufacturing becomes important.

This third model argues that a sorting procedure for towns begins after the first stage, with early site advantages subject to change, especially with changes in transportation. The argument that towns were evident from the outset of settlement was advanced earlier by Wade (1959:1): 'The towns were the spearheads of the frontier. Planted far in advance of the line of settlement, they held the west for the approaching population.'

OTHER CONCEPTUAL CONTRIBUTIONS

Although the mercantile model and other related contributions appear relevant to nineteenth-century North America, they appear less appropriate for other times and places. This limitation is because they are partly linked to the mercantilist economic philosophy that played such an important role between the feudal period in Europe and later capitalism.

Attempts to explain the Australian and South African urban experiences have also utilized versions of the mercantile model. Conceptual contributions have focused on the importance of trade, related port development, and the role of distance. One additional concept applied to the Australian and South African cases is that of core-periphery. Cores or central areas dominated by incoming Europeans included ports, capitals, and mining centres. Peripheral areas occupied primarily by natives experienced less urbanization and largely remained outside of the growing economic region.

Rozman's model (1978) identified seven stages of premodern development (Table 11.5): (1) a preurban stage with unspecialized settlements; (2) cities locate separately from

TABLE 11.5: SEVEN STAGES OF PREMODERN URBAN DEVELOPMENT

Stage	Number of levels	Characteristic
1	0	Preurban
2	1	Tribute city
3	2	State city
4	2, 3, or 4	Imperial city
5	4 or 5	Standard marketing
6	5 or 6	Intermediate marketing
7	7	National marketing

Source: Adapted from G. Rozman, 'Urban Networks and Historical Stages', *Journal of Interdisciplinary History* 9 (1978):79.

one another and with only weak links to rural areas; (3) an administrative hierarchy becomes evident; (4) high levels of centralization based on administration prevail; (5) commercial centralization emerges, as do some periodic markets; (6) increased commercial activity results in five or six levels of cities; (7) a national marketing system is evident. This model identifies the number of settlement levels, which are groupings by population size, associated with each stage. For England, the second stage evolved in the second century BC, the third stage was bypassed, stage four evolved in the first century AD, stage five in the tenth century, stage six in the twelfth century, and stage seven in the sixteenth century. Although this model is extremely broad, it has the clear advantage of being less time- and place-specific than the earlier formulations.

Comparison of the various models identified earlier with the theory of agricultural location discussed in the previous chapter reveals substantial differences. Although Thünen's work had a specific factual stimulus, it largely succeeded in being a general explanation of agricultural location. Historical explanations of urban location clearly do not meet the expectation of generality. Fortunately, however, there is a theory available — central place theory — similar to that of Thünen in that it is of general applicability.

URBAN LOCATIONS:
II. CENTRAL PLACE THEORY

Thünen's most important contribution to geography is the concept of the isolated state, that is, the idea of setting up a grossly simplified world to observe the effects of a particular variable. In Thünen's case, the variable of distance expressed as transport cost was highlighted in a study of agricultural location. The German geographer, Walter Christaller (1893–1969), similarly set up an isolated state to observe the role played by distance for the location of urban centres.

Christaller's seminal work was clearly influenced by Thünen and possibly more directly by Alfred Weber, a German economist. This work was a theoretical contribution to a settlement geography that was descriptive at the time. Like Thünen, Christaller was a pioneer. The English title of the work, which was not translated until 1966, is *Central Places in Southern Germany* (Christaller 1966). It has proven highly influential for subsequent theoretical and empirical research. At the time of publication, there was little interest in the book and only after the Second World War did it become known; it was best received in the Scandinavian countries and the United States. During the 1960s, the work came to the fore as one component of the spatial analytic movement that swept English language geography at that time.

In 1940, a second German academic — the economist, August Lösch (1906–45) — published *The Economics of Location*, a book that included many of Christaller's ideas. Central place theory, as discussed in this section, integrates the work of both theorists. For sound heuristic reasons, however, it is largely couched in the language introduced by Christaller. Box 11.3 provides an interesting insight into the approach employed by this major theorist.

ASSUMPTIONS

Theories begin with assumptions that simplify the key issue under investigation. The assumptions of Christaller's central place theory are similar to those utilized by Thünen.

1. There is a flat, never-ending land surface. Christaller introduces this assumption to avoid the complications implied by a physically variable landscape. The plain is never-ending because Christaller did not want any boundary complications to arise.
2. This uniform plain has a uniform distribution of rural population; furthermore, each member of this population has identical purchasing power and behaves similarly. Christaller introduces this assumption to avoid the complications of a landscape with variable human character.
3. There is a homogeneous transport surface that allows for equal ease of movement for all members of the population in all directions. Christaller introduces this assumption to avoid the complications of a network of roads and other lines of communication.
4. Finally, it is assumed that the above three statements combine to describe some preliminary stage in landscape evolution with all subsequent evolution related to the growth of cities as service centres.

FOUR KEY CONCEPTS

Christaller developed his theory using four key concepts. First, the concept of a central place is introduced. A *central place* is an urban centre that evolves because it provides a series of functions for the surrounding rural population. Thus a central place originates because of the demand for functions generated by rural areas. Second, the *range* of a good is the maximum distance that people are prepared to travel to acquire a particular good or service. Third, the *threshold* is a measure of the minimum number of people required to support the existence of a particular function. Finally, we have the concept of *spatial competition*, the idea that central places compete with each other for customers.

These four concepts apply at various scales and can be most easily highlighted by considering the simplified example of one business, for example, a bakery. What conditions need to be satisfied for a bakery to operate profitably? First, there must be enough people wishing to purchase bakery products. For every central function, there is some minimum level of demand that varies according to the specific function. It is, for example, relatively low for a bakery and relatively high for a dentist. This is the threshold concept. The second condition to be satisfied relates to the distance people are prepared to travel to shop at the bakery given certain prices. There must be people located within the range of the business in question who are willing to purchase at the prices offered. Finally, the people within the range who represent the demand must not have access to other competitive bakeries. In this manner the concepts apply in the context of a single business and in the context of a collection of businesses clustered in an urban centre.

HEXAGONS

In principle, for the most efficient use of space, central places should be located as shown in Figure 11.6. Central places that compete with each other for the purchasing power of rural populations will tend to locate so as to produce a regular arrangement (Box 11.4). This triangular lattice is the most economical use of space. Within a central place system, the ideal shape of each market area should be a circle, but, as Figure 11.7 shows, circles either leave some areas unserved or some areas served twice. The ideal shape that completely covers an area without any overlap is the hexagon. This is because the hexagon is the nearest geometric figure to a circle; it has the greatest number of sides and provides total coverage without duplication.

Box 11.3: Theory construction — by Christaller

In a brief article originally published in German, Walter Christaller explained how he developed central place theory. The explicit references to Thünen and Weber are illuminating. Christaller wrote:

> Besides geography and statistics, I was also interested in sociology, which at that time (in 1913), when I was beginning my studies, had begun to exist as a new scholarly discipline ... At that time I witnessed the efforts of my teacher Alfred Weber in Heidelberg to create an industrial location theory.
>
> ... I continued my games with maps: I connected cities of equal size by straight lines, first of all, in order to determine if certain rules were recognizable in the railroad and road network, whether regular traffic networks existed, and, second of all, in order to measure the distances between cities of equal size.
>
> Thereby, the maps became filled with triangles, often equilateral triangles (the distances of cities of equal size from each other were thus approximately equal), which then crystallized as six-sided figures (hexagons). Furthermore, I determined that, in South Germany, the small rural towns very frequently and very precisely were 21 kilometres apart from each other. This fact has been recognized earlier, but had been explained as being due to the fact that these cities were stopover places for long distance trade traffic, and that in the Middle Ages the distance daily by a cart was about 20 kilometres.
>
> My goal was staked out for me: to find laws, according to which number, size and distribution of cities were determined.
>
> ... It was clear to me from the very beginning that I had to develop a theoretical schema for my regional investigation — a schema which, as is customary in national economics, is set up by isolating the essential and operative factors. It was, thus, as in the case of Thünen's Isolated State, abstracted from all natural and geographic factors, and, also, partly to be abstracted from all human geographic factors. It had to be accepted as a symmetrical plain, without obstructions such as rivers or mountain ranges, with a uniformly distributed population, in order to then determine where, under such conditions, the site of a central city or market could form. I thus followed exactly the opposite procedure that Thünen did: he accepted the central city as already having been furnished, and asked how the agricultural land was utilized in the surrounding area, whereas I accepted the inhabited area as already having been furnished, and subsequently asked where the city must be situated, or, more correctly, where should the cities be situated. Thus, I first of all, as is said today, developed an abstract economic model. This model is 'correct' in itself, even if it is never to be found in the reality of settlement landscape in pure form: mountain ranges, variable ground; but also variable density of population, variable income ratios and sociological structure of the population, historical developments and political realities bring about deviations from the pure model. In the theoretical portion of my investigation, I thus did not satisfy myself with setting up a model for an invariable and constant economic landscape (thus, for a static condition) — but instead I also tried to show how the number, size and distribution of the central places change, when the economic factors change: the number, the distribution and the structure of the population, the demands for central goods and services, the production costs, the technical progress, the traffic services, etc.
>
> ... I was able to find surprising concurrences between geographical reality and the abstract schema of the central places (the theoretical model) especially in the predominantly agrarian areas of North and South Bavaria.

Source: W. Christaller, 'How I Discovered the Theory of Central Places: A Report About the Origin of Central Places,' in *Man, Space and Environment*, edited by P.W. English and R.C. Mayfield (New York: Oxford University Press, 1972):601–10.

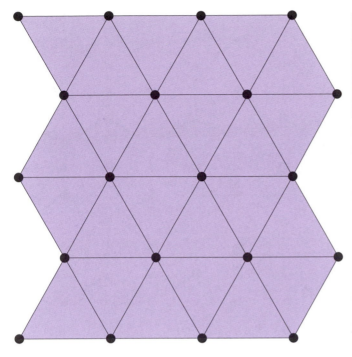

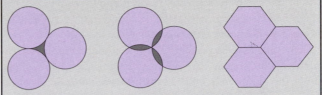

Figure 11.6 *(left)* A triangular lattice.
Figure 11.7 *(above)* Theoretical trading areas

equidistant from one another. For a two-order situation, the result is depicted in Figure 11.8. It appears that central place theory suggests a hierarchy of settlements and a nested hierarchy of trading areas.

THREE PRINCIPLES

We have just outlined what Christaller called the marketing principle. He also proposed the transportation and administrative principles. For all three principles, central places are located equidistant from each other, although in any given region one of the three may dominate or, more probably, the effects of all three will be evident.

The *marketing* principle invokes the principle of least effort: people will always go to the nearest centre that provides the required service. Christaller called this a k=3 arrangement, with k as the number of settlements at a given level in the hierarchy served by a central place at the next highest level. Figure 11.8 shows that each high-order centre serves three low-order centres (that is, one-third of each of the low-order centres that surround it, plus itself as a low-order centre). Note that higher order centres include all the functions of lower order centres.

The *transportation* principle allows for as many large centres as possible to be located on communication lines between centres. (Notice that we are now amending one of

HIERARCHIES

Central places, of course, contain more than one central function. Bakery products are not the only ones purchased by rural populations. Furthermore, different central functions have different threshold populations and different ranges. We find, then, that some central places have many functions and are correspondingly large, while other central places have relatively few functions and are correspondingly small. Christaller identified seven such levels of settlement and showed that the highest order settlements (the largest) will be few in number, while the lowest order settlements (the smallest) will be more numerous. But the spatial regularity is not disturbed because for any given order, settlements are

---

Box 11.4: Nearest neighbour analysis

Geographers may describe the spatial pattern of urban centres (or of any other phenomena represented on a dot map) by a statistical procedure, nearest neighbour analysis. This can be used to describe a point pattern as clustered, random, or uniform (see Figure 2.4).

This procedure indicates the degree to which an observed distribution of points deviates from what might be expected if the points were randomly distributed in the same area. A single statistic, R_n, is calculated.

R_n equals distance observed [average distance between each point and its nearest neighbour] divided by distance expected

$$\frac{1}{2\sqrt{\text{density of pattern}}}$$

If $R_n = 1.0$ the two distances are identical and the pattern is random. If $R_n < 1.0$ the observed distance is less than the expected distance, indicating that the points are relatively clustered. If $R_n > 1.0$ (to a maximum of 2.15) the points are relatively equally spaced — as proposed by central place theory.

Although a useful statistical procedure, nearest neighbour analysis needs to be employed at an approximate spatial scale (recall figures 2.6 and 2.7). When only the first nearest neighbour of each point is used, a repeated pattern of two or more closely spaced points, occurring at large spatial intervals, gives a low value of R_n, although the pattern may be dispersed. Thus, it is not uncommon to incorporate second- and third-order neighbours or to confirm the results with an alternative procedure.

the initial assumptions.) The result is shown in Figure 11.9; there is a k=4 (the centre itself plus one half of each of six lower order centres) system in place.

The *administration* principle assumes that a central government would not risk subdividing trading areas or regions. Thus the hierarchy is built up through the addition of whole regions. Figure 11.10 depicts the arrangement of centres and shows that a k=7 situation applies, that is, a centre serves itself plus six lower order centres.

Other authors, especially Lösch (1954), have contended that Christaller's three principles are but special cases of a more general situation. Extending this argument results in spatial arrangements that are much more complex than those proposed by Christaller. Box 11.5 provides some evidence to support the general hypotheses proposed in central place theory.

THE RANK SIZE RULE

Closely linked to central place theory, although separate in origin, is the rank size rule. Originally devised in 1913, the rule is limited to establishing a numerical size relationship between centres in a region. The rule is as follows:

$$P_r = P_1/R$$
where
P_r = population of centre r
P_1 = population of largest centre
R = rank size of centre r

Thus the population of a centre is inversely proportional to the rank of that centre. The largest centre in a region is named the **primate city**.

Geographers have noted that some regions conform to a rank size distribution with the second largest city being one half the size of the largest and so forth, while other regions conform to a primate distribution with one centre being more than twice the size of all other centres. This difference can be explained in various ways. Thus, rank size distributions tend to apply in large countries, those with a long history of urbanization, and those that are economically and politically complex. Primate distributions tend to apply in small countries, those with a short history of urbanization, those with simple economic and political structures, and those at the centre of colonial empires.

INSIDE THE CONTEMPORARY CITY

Perhaps no topic reflects the changing nature of geographic research quite as well as the geography of the urban area. In the 1960s, spatial analytic researchers used simple models to focus on the internal structure of cities. Since then there has been a concern with the city as it is perceived by its inhabitants and as a symbolic landscape, on the quality of life in the city, and on cities as places. There has also been additional emphasis on the structure of the city, including

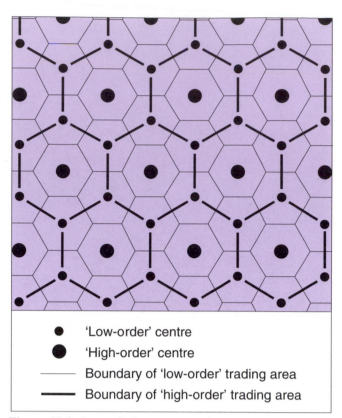

- ● 'Low-order' centre
- ⬤ 'High-order' centre
- ——— Boundary of 'low-order' trading area
- ▬▬▬ Boundary of 'high-order' trading area

Figure 11.8 A central place system — the marketing principle.

the rise of postsuburbia and the postindustrial city. The conceptual transition has been from positivism, to humanism and radical approaches, and most recently to other social theory. This transition in conceptual work is introduced in Box 11.6 and is exemplified in this discussion of the contemporary urban area.

THE INTERNAL STRUCTURE OF URBAN AREAS

A simple urban land-use model operates on the same principles as Thünen theory, with different potential land uses having different economic rent lines (see glossary). Thus, land uses with the greatest values of economic rent at the city centre will have steep rent lines and hence occupy land adjacent to the centre. In principle, a city has a series of concentric zones. A typical arrangement might have businesses at the centre with industrial and residential uses appearing at greater distances from the centre (see Figure 10.4). The value of land decreases with increasing distance from the centre. A popular diagram of the urban land value surface that allows for a non-homogeneous transport surface is shown in Figure 11.11.

There are three basic models of the internal structure of urban areas (Figure 11.12). The *concentric zone* model (Figure 11.12a) is from the pioneering work in 1925 of a group of Chicago sociologists who applied ecological concepts from biology and botany to urban areas; the principal figures associated with this movement were Park, Burgess, and McKenzie (Park and Burgess 1921). This Thünen-type model is an

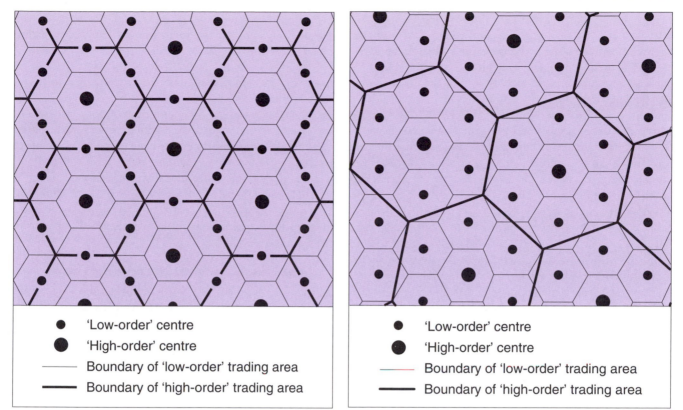

Figure 11.9 *(left)* A central place system — the transportation principle. In this instance, the low-order centres are located at the middle of each of the six sides and not at the corners of the hexagons of the high-order centres; hence each low-order centre is linked to two, not three, high-order centres.

Figure 11.10 *(right)* A central place system — the administration principle.

Box 11.5: Testing central place theory

The essential hypotheses of central place theory relating to the spacing and size of centres and to the shape and size of market areas have been evaluated many times and for many different regions. Combined, these studies tend to confirm the basic hypotheses. Many of the early classic studies are described in detail by Berry (1967).

In evaluating central place theory, it must not be forgotten that it considers urban centres as service centres. It does not take into account urban centres set up, for example, because of a specific transportation or resource need. Thus, Cape Town in South Africa began as a Dutch settlement on the route between Europe and Asia; the location decision had nothing to do with a surrounding rural population's demand for services. Similarly, the many resource towns of northern Canada, such as Sudbury, are located at a site for resource exploitation, again without any stimulus from a surrounding rural population. But analyses show that regional central place networks allow for such factors as they evolve; a modified version of a central place network can evolve from non-central place beginnings.

One of the most interesting criticisms of central place theory concerns the extent to which a recognition of hierarchies is legitimate or whether it merely results from the application of convenient but non-specific terms, such as hamlet, village, and town. There is no doubt that the relationship between number of functions and urban size varies, but it may be that there is a continuous functional relationship rather than a series of discrete stages.

One example of a formal testing of the hypothesis relating to urban centre spacing and size will suffice. Brush (1953) investigated southwest Wisconsin and identified three levels of settlement: hamlets, villages, and towns. There were 142 hamlets, seventy-three villages, and nineteen towns; hamlets averaged two functions, villages eighteen functions, and towns forty-two functions; furthermore, hamlets averaged an 8.8-km (5.5-mile) spacing, villages a 15.8-km (10-mile) spacing, and towns a 33.9-km (21-mile) spacing. This one example provides strong evidence to support Christaller's work.

It is difficult to exaggerate the importance of central place theory in the geography of urban centres and the spatial analytic school that briefly dominated geography in the 1960s and remains important today. Central place theory is a major conceptual contribution, helping to elevate urban geography in importance and stimulating other location theory.

extension of economic rent logic. The central business district is at the city centre; the second zone is a transitional area comprising original industries and older houses; the third, fourth, and fifth zones are all residential and become increasingly affluent with increasing distance. This model, like that of Thünen, is a combination of theoretical logic and first-hand experience of the real world. As a descriptive device, the model applies best to North American cities.

The *sector* model of urban land use assumes that internal structure is largely conditioned by the location of routes that radiate outwards from the centre (Figure 11.12b). This model was developed by Hoyt (1939) and is also a descriptive formulation. It is an improvement over the concentric zone model as it incorporates distance and direction. Sectors develop partly because once a specific use is attracted to a line of communication, other uses locate elsewhere; industrial uses, for example, tend to repel residential uses.

The *multiple nuclei* model incorporates an additional variable, namely that of several discrete centres in the urban area. Developed in 1945 by Harris and Ullman (1945), this model can take many forms in reality — one example is shown in Figure 11.12c. The number of nuclei varies according to city size and details of development. For North American cities, five general land uses were proposed: central business district, wholesaling and light manufacturing area, heavy industrial area, residential areas, and suburbs.

These three models are attempts to generalize about the internal structure of urban areas. None of them purport to be universally correct and, in general, all are limited to the North American and twentieth-century experience. Combined, they provide a very useful set of descriptions, valuable indicators against which real world structures can be assessed. They are all essentially positivistic as they focus on simplifications of a complex reality. There is a general failure to answer (or even ask) questions about processes, about the causes of particular spatial configurations. Valuable as these models are, they do not provide adequate accounts of the internal structure of urban areas.

Interestingly, the distribution of population densities is remarkably consistent for all urban areas and bears little relation to the details of internal structure. Typically, densities decline as a negative exponential function of distance from the centre of the city (see Figure 2.5).

URBAN EXPANSION IN THE LESS DEVELOPED WORLD

Much of the rapid urban growth in the less developed world is a result of the growth of **squatter settlements**. These are uncontrolled expansions on the periphery of a city and are comprised of low-quality, poorly serviced, temporary dwellings. Such squatter settlements have serious health, social, and economic problems. As they age, squatter settlements may become an integral part of the city, reducing pressure on rental areas inside the city and offering a base for new migrants. In many instances, squatter settlements have achieved a genuine integration with the larger city as unplanned but necessary growth (Box 11.7).

Box 11.6: Modern and postmodern urban theory

Twentieth-century urban theory has undergone a series of changes. The first dominant theoretical focus, which developed in the 1920s, was that of the Chicago school of *urban ecology*, which regarded the city as a social organism and interpreted change in terms of analogies with ecology. The most important process was the competition for urban space that resulted in spatial segregation of groups. Explanations also centred on a group's dominance of an area and on the invasion of an area by a group that then succeeded some other group in that area. A second related focus, which played a leading role in the 1960s, was that of the *neoclassical urban land rent* Thünen-type theorists who interpreted urban areas in terms of the rational economic operators' behaviour (see glossary).

Both of these approaches were criticized from a number of alternative perspectives as neither explained many of the changes taking place in the modern city, especially those associated with the larger processes of transition from Fordist to post-Fordist modes of production (as introduced in Chapter 8, see glossary) or of the related economic and social restructuring (as introduced in Chapter 10, see glossary). *Political economists* pay special attention to the conflicting interests, on the one hand, of financiers concerned with capital accumulation and, on the other hand, of inner city residential and small business interests. In human geography, this Marxist approach is associated particularly with the work of David Harvey and is discussed in Box 11.10.

Another focus, sometimes labelled *urban managerialism*, is derived from the work of the sociologist, Max Weber, and places a greater emphasis on the state and other institutions. *Humanistic* concepts have also been applied to the study of urban areas, especially in the form of interpretive social research.

Most recently, there has been a movement towards some form of *postmodern* account of the city. A basic contention of postmodernism, as discussed in Chapter 8, is that all theories are to some degree repressive as they privilege one explanation above all others. Postmodernists therefore question the legitimacy of all modern urban theories.

Two excellent accounts of urban theory that refer to appropriate examples of each of the above types are those of Bassett and Short (1989) and Cooke (1990).

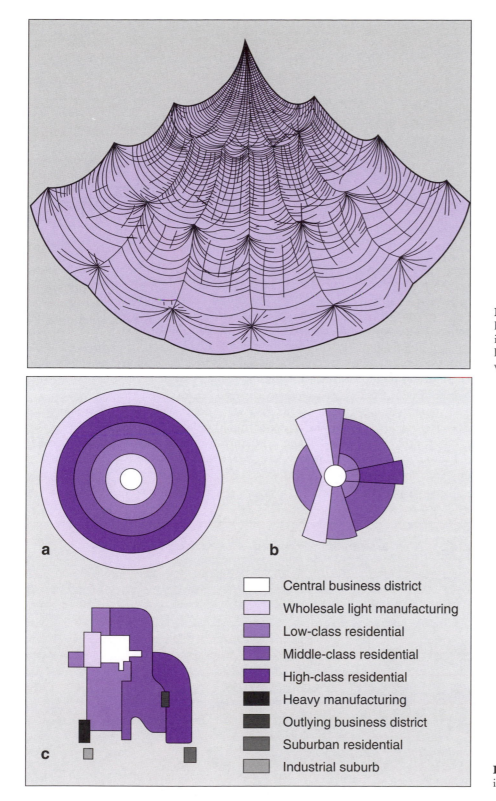

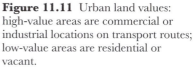

Figure 11.11 Urban land values: high-value areas are commercial or industrial locations on transport routes; low-value areas are residential or vacant.

Legend:

- ☐ Central business district
- ☐ Wholesale light manufacturing
- ☐ Low-class residential
- ☐ Middle-class residential
- ☐ High-class residential
- ☐ Heavy manufacturing
- ☐ Outlying business district
- ☐ Suburban residential
- ☐ Industrial suburb

Figure 11.12 Three models of the internal structure of urban areas.

Mexico City is engulfed by shanty towns. Furthermore, it is located in an unfavourable environment that continually suffers from flooding, subsidence, and shortages of fresh water. The urban explosion began about 1900 as a result of development policies favouring industrialization. The city has doubled in size since 1970 and some estimate that it will double again by about 2020. It is not an exaggeration to describe Mexico City as a social and environmental disaster — a disaster that is ever worsening because of continuing growth.

In general, cities in the less developed world are experiencing intense pressure to provide additional housing, but are incapable of responding adequately. Further, there are clear disparities in the rates at which different groups in the city are able to access improved services, disparities that are a direct reflection of relative power (Lowder 1994).

Another feature of urban expansion is the growth of **conurbations**. The original example is the region from Boston to Washington on the east coast of the US, but recent and future examples are predominantly in the less developed world. Regardless of location, many cities in the less developed world suffer from overcrowding, limited provision of services, traffic congestion, unemployment, damaged environments, and ethnic conflicts.

POSTINDUSTRIAL CITIES

One of the most distinctive features of twentieth-century urban change in the more developed world has been the growth of suburbs and, most recently, of postsuburban forms (Box 11.8). Indeed, the major current change, possibly exemplified by postsuburban expansion, is the rise of the postindustrial city. Such cities have origins not tied to any industrial considerations and have an employment profile that favours professional, management, administrative, and skilled labour employment as opposed to manufacturing employment. Physically, the city includes office buildings and public institutions. In addition to the new postsuburban forms, cities that serve as government and business centres are prime examples as are, possibly, major tourist and retirement centres. In many such cities, especially those changing from the industrial to the postindustrial form rather than those being newly created as postsuburbia, there is a high degree of social polarization between the new middle classes and the working poor or unemployed (Ley 1993).

Gentrification

Since the Second World War, there has been a partial reversal of the trend for high-status populations to live away from the city centre. Inner city convenience, especially commuting to work and the less tangible concept of quality of urban living, prompt many to choose to live close to city centres. The process of transforming an often derelict or low-quality housing area is known as **gentrification**, from the English word 'gentry', meaning middle class. Evidence of this process is abundant; most cities, large and small, in the more developed world are undergoing gentrification (Ley 1992).

The dogma of the previously dominant ecological models is that neighbourhoods decline after reaching maturity, but gentrification is a clear exception; it is a version of redevelopment prior to complete neighbourhood decline. This distinctive feature of many urban areas may be related to the concept of the postindustrial city; it appears to be one consequence of economic and social restructuring. As is the case with many changes, gentrification has positive and negative characteristics. For some middle-class populations, it provides an opportunity to live close to the city centre, but for the poor, it is a threat to their tenure in an area because of the increasing house prices that may result in displacement. The issue of displacement is an important one because it raises the important question of prevention —

Box 11.7: Urbanization in the less developed world — the case of Brazil

Brazil is typical of the less developed world as it has experienced explosive urban growth in recent years, especially since about 1945. Recent urban growth has accompanied an economic transition from agriculture to industry.

The earliest urban centres in Brazil were founded under Portuguese rule and were located on the coast to facilitate the export of sugar and other products. São Paulo, founded in 1554 by Jesuits, remained relatively unimportant until it became the key port for the export of coffee in the late nineteenth century. Growth has been accompanied by a distinct spatial expression of class. Working-class areas arose close to industrial sites, while élite areas developed on higher ground; in some cases, these areas are enclosed by security fences and patrolled by guards. Shanty towns, *favelas*, are common on the city's outskirts. Many new immigrants to cities such as São Paulo subsist by becoming involved in what is called the *informal sector*, for example, selling goods in the street or providing a service such as washing cars. Such activities are necessarily low income, but are viewed as temporary. For migrants, the attraction to cities is not the shanty town or informal sector employment but the perception that the city offers better schooling, better medical services, a better water supply, and the prospect of permanent employment. As shanty towns grow, there is enormous pressure on urban governments to provide the needed services, but inevitably extended urban services attract even more migrants. This may be unfortunate as shanty towns are usually decrepit and lack basic amenities, but it can be viewed positively as shanty towns can be seen as reception areas for migrants, providing a transition between rural and urban ways of life.

The contemporary Brazilian city is thus characterized by marked class segregation and decentralization. Urban problems abound and appear to be far from resolution (Godfrey 1991). It is, however, instructive to relate the text comments on counterurbanization and the discussion on declining fertility in Box 5.3 to this issue. If birth rates decline substantially in the less developed world, then the problems of shanty towns will at least be decreased somewhat.

should governments intervene in the process to protect the more vulnerable groups?

Postmodern processes

Contemporary cities in the more developed world are structured by twentieth-century urban planning and growth. They are the product of industrialization and modernity in general. But there are now new and different processes operating — postmodern processes. To date, postmodernism's principal impact is architectural, reproducing local vernacular building styles. Two typical examples are the refurbishing of old buildings in the numerous waterfront developments in North American cities and the construction of suburban office clusters as opposed to modernist office towers. Buildings are, of course, both artefacts and expressions of a way of life, and a failure to study architecture is also a failure to study the social meaning of place.

The global city

Some major world cities — notably New York, London, and Tokyo — are now so similar that they comprise a new type of city — the global city. The two forces prompting this convergence are postindustrial forces — the increasing dispersion of manufacturing on the global scale and the increasing concentration of the control of that production in the global city (Sassen 1991).

THE IMAGE OF THE CITY

Our discussion of the city's image relies heavily on a seminal work by Lynch (1960) in which the visual quality of urban areas is exemplified by case-studies of Boston, Jersey City, and Los Angeles. Visual quality is the ease with which residents are able to identify parts of the city and integrate these parts into the whole. What image do residents have of their cities? What is their mental map? These questions are important both practically (for getting around) and emotionally (for making sense of where we live).

Five key features of the urban landscape are identified as ordinarily perceived by city residents. *Paths* are various routes through the city, and many parts of the city are seen from paths. *Edges* are boundaries between perceived districts. *Districts* are parts of the city that are in some way perceived as different compared to surrounding areas; they are equivalent to geographic regions. *Nodes* are key points in the city such as major intersections. Finally, *landmarks* are distinctive physical objects such as buildings, statues, or hills.

Since approximately 1970, geographers have begun to expand their horizons substantially in their analyses of the nature of cities. Models of internal structure are valuable, but say little (if anything) about the human experience. Mental map research is intriguing and appears to offer a valuable approach to urban planning, but it has not generated a substantial body of subsequent work.

THE CITY AS IMAGE

Urban landscapes, in common with other human landscapes, can be interpreted symbolically, that is, as physical representations of cultural identities. Iconography (see glossary) is a popular technique for uncovering the meaning of cities by reference to their particular historical and cultural contexts. In a series of studies, Cosgrove (for example, 1984) has analysed the cities of Renaissance Italy by considering them in the context of humanistic culture, which was expressed in an urban landscape that displayed wealth through monuments and patronage of the arts.

Urban landscapes are constructed to represent and legitimate the power that some are able to exercise (Duncan 1990). Planned capital cities provide good examples. They are often

---

Box 11.8: Suburbia and postsuburbia

Suburbs are areas outside the central area of a city and separate from industrial areas comprising housing and basic services that are working class or middle class. Although there is evidence of suburbs in preindustrial times, the basic stimulus for suburbanization was industrial urbanization and the associated development of transportation. Suburbs are a compromise between the need to live close to places of employment in the central city or industrial areas and the desire to be close to open spaces and rural areas. Many twentieth-century suburbs, especially in Europe, are oriented to the working class and incorporate large areas of public housing. As such, they have often become areas of urban deprivation, social anonymity, residential congestion, and pollution. Further, in many cities public transportation systems have not been adequately modernized, adding an additional problem to working-class suburban life.

The most recent suburban trend (what is sometimes called postsuburbia) results from the increasing decentralization of offices and factories, the decline of manufacturing employment and the processes of economic restructuring. According to Knox 1992:26), 'Economic decentralisation has resulted in so much commercial and industrial development in the outer reaches of large US metropolitan areas that central cities and downtown Central Business Districts no longer dominate them the way they used to.' The concept of postsuburbia emphasizes that suburbs have matured into places with the qualities of the larger city itself; thus residences, offices, and light industry are all included in the 'edge cities' of postsuburbia. An excellent example of such a city is Tyson's Corner, a rural corner of northern Virginia in the mid-1960s, that now includes the ninth largest concentration of retail space in the United States.

designed to proclaim power and wealth and hence they incorporate grand avenues, large squares, public buildings, and monuments — all ways to symbolically assert mastery. But the reality is so often different to the plan. Both the capital of the United States, Washington, and the capital of Brazil, Brasilia, are planned cities, but both have grown into areas with massive problems of poverty, crime, and homelessness.

Indeed, many cities today have an especially distressing image. As with the larger human world of which they are a part, cities seem to have brought out the best and the worst of humans. Many cities are centres of civilization, others are unhealthy and unsightly creations that are a nightmare landscape for their inhabitants (Box 11.9).

URBAN AREAS AS PLACES

As we know, cities are internally complex, with different land uses, different realities for different people, and different districts — districts of high social status and others of low social status.

Perhaps an understanding of the urban environment can best be approached by considering the many and complex ways in which social structures and individual actions relate. We acknowledge that group identity is the mainspring of human society. If groups live together, then there is a shared understanding of place. When places have a meaning shared by many, there is a lack of conflict over land use. But in any urban area, there are locations used by many groups — such as roads — and these take on a multitude of meanings and

may become areas of conflict. Thus any urban area has two types of space: places occupied by groups whose members have a common image and a shared meaning, and locations used by diverse groups but belonging to none. The architect and planner, Greenbie (1981), distinguished between proxemic space, such as neighbourhoods, and distemic space, such as roads. Proxemic spaces reflect group identity, while distemic spaces have varied meanings according to the social background of the individual. Traditionally, the most successful and stable distemic space is the market.

Social theory that focuses on individuals as members of groups encourages us to acknowledge the meanings that different people attach to different locations. It becomes easier to understand the values that people ascribe to a place and the conflicts that can emerge over land use. Insights are offered and understanding is enhanced.

Residential areas and neighbourhoods
Urban areas are characterized not only by a number of different land uses but also by distinctive residential areas. Typically, these areas are distinguished on the basis of class, ethnicity, or some other cultural variable. In European and some other cities prior to the Industrial Revolution, the most distinctive urban residential district was the Jewish district, usually labelled as a ghetto (see glossary). Ghettos are apart from, rather than a part of, the larger city and are held together by the internal cohesion of the group and the desire of non-group members to resist spatial expansion of

---

Box 11.9: Some urban problems

Consider Mexico City. According to the United Nations Environmental Program and the World Health Organization, it is the most polluted of large world cities. Health guidelines are exceeded by a factor of two or more for levels of sulphur dioxide, suspended particulate matter, carbon monoxide, and ozone; levels of lead and nitrogen oxide are almost as bad. The principal source of these various pollutants is the burning of fossil fuels. They affect human health directly through breathing and indirectly through drinking water and contamination of food (Middleton 1994).

Consider Lagos, the capital of Nigeria. It combines serious physical site and human problems; much of the site consists of swamp, and the land is unstable. Drainage and sewage problems worsen daily. The population is increasing rapidly because of larger economic issues prompting rural out-migration. It is disease-ridden, congested, and virtually unlivable.

Cairo, the largest city in the Middle Eastern region, treats less than half of its sewage, with the remainder ending up in the Nile or in local lakes. Diarrhoea and dysentery are common, and there is a real danger of typhoid and cholera outbreaks. Bangui, in the Central African Republic, has a sewage system that was built for a population of 26,000 — a system that has

never been extended or improved; the current population is about 500,000. Addis Ababa, in Ethiopia, with a population of about 1.5 million, is the largest city in the world not to have a sewage system.

Urban problems are not limited to the less developed world. Los Angeles is dependent on water that is imported several hundreds of miles via aqueducts, but it also has an expensive system of channels and dams for protection from flooding that results from annual snow melt. It also sometimes experiences serious ozone pollution problems when hydrocarbons and nitrous oxides (emitted by automobile exhausts) react in oxygen. Health warnings are issued when the ozone level exceeds a certain figure: in 1987, warnings were issued on thirty-six days; at Glendora, about 30 km (18.5 miles) downwind, warnings were issued on 135 days.

A final example: the major world city of Tokyo is always in a precarious position because of the likelihood of a major earthquake. An earthquake, similar to that of 1923 (which caused 140,000 deaths), would effectively destroy the city, causing a collapse of the national economy and creating economic impacts throughout the world.

the group. Box 11.10 provides an account of such conflicts from a Marxist perspective.

During the Industrial Revolution, class divisions became more evident and spatial consequences were apparent. By the beginning of the nineteenth century, British commentators were acutely aware of the establishment of working-class districts close to the factories and the establishment of middle-class districts elsewhere, especially on the outskirts of the urban area. Distinct residential areas appeared most obviously in large immigrant-receiving North American cities in the nineteenth century, with distinctions based on ethnicity as well as class. Once a particular ethnic group was large enough, it settled as a group, usually in an inexpensive area close to employment opportunities.

Elsewhere in the world, such segregation might be based on religion, as in Belfast, Northern Ireland. In South Africa, even before the formal institutionalization of apartheid, cities were clearly divided on explicitly ethnic lines. Residence variation also results from lifestyle preferences and the proximity of major employers such as universities and hospitals. There is no doubt that society and space are irrevocably entwined.

The link between society and space is perhaps clearest when neighbourhoods are evident. The concept of neighbourhood is not formalized, but it does imply a district that reflects social values. Some geographers contend that the best indicator of neighbourhood identity is the presence of neighbourhood activism (Smith 1985). Unfortunately, some

---

Box 11.10: A Marxist interpretation of the urban experience in a capitalist world

David Harvey is a human geographer who, after producing a major methodological work on positivism in 1969, has focused primarily on arguing the merits of a Marxist philosophy for our understanding of the urban way of life (Harvey 1969, 1973, 1982, 1989). According to Harvey, Marxism offers a broad theoretical perspective (a metatheory or higher level theory) that allows geographers to approach such diverse issues as the built environment, urban economy, and local urban culture.

Thus, using Marxist theory and adding spatial factors, Harvey (1982) argued that uneven spatial development is a necessary accompaniment of capitalism. This argument applies on varied scales, including the urban scale. Thus, urbanization is one more component of the uneven spatial development that

occurs in a capitalist context. The spatial generality of these ideas justifies use of the term 'metatheory'.

Understanding urban places requires an appreciation of flows of capital and surplus value through systems of cities, and an appreciation of the class relations characteristic of a capitalist society. Explaining residential differentiation, for example, requires recognizing that the more powerful classes have resources to allow them to acquire land of their choosing. In Marxist terms, then, the explanation lies in the superstructure (see Box 2.6). It is not appropriate, from this perspective, to see residential differentiation as resulting from the preferences of all people but from the preferences of a powerful few.

Box 11.11: Slums and the cycle of poverty

The rapid expansion of urban industrial areas in the nineteenth century resulted in high-density, poorly serviced housing areas of poor quality. These areas contrast markedly with the neighbourhoods of preindustrial cities and with rural villages as there was a marked separation of the poor and the rich. In 1845, the slums of Nottingham, England, were described as follows:

... nowhere else shall we find so large a mass of inhabitants crowded into courts, alleys, and lanes as in Nottingham, and those, too, of the worst possible construction. Here they are so clustered upon each other; court within court, yard within yard, and lane within lane, in a manner to defy description. Some parts of Nottingham [are] so very bad as hardly to be surpassed in misery by anything to be found within the entire range of our manufacturing cities (Hoskins 1955:218).

As these nineteenth-century slums were cleared, they were replaced by equally poor forms of low-cost, often high-rise,

housing. Both types of slum have been associated not only with poverty but also with high levels of crime, vandalism, and substance abuse. Many commentators have proposed that there is a **cycle of poverty** that perpetuates slums as a way of life and limits the prospects for integration with the larger urban area. Slums are often associated with distinct ethnic groups, although the terms 'ghetto' and 'slum' are not synonyms.

The problem of poverty, of which slums are but one component, is perhaps the biggest urban problem. Knox writes:

... poverty comes as a package. Low income is at the core, but it is inextricably bound up with poor diets, poor physical environments, poor physical health, the psychological stress of continuously having to make ends meet, and the economic, social, and political disadvantages of being stigmatized as a "loser" in a highly competitive society (Knox 1994:299).

neighbourhoods have an identity that is negative — slum districts and service-dependent ghettos, for example, and any neighbourhood activism aims not to preserve identity but generate change (Box 11.11). In a challenging work titled *Landscapes of Despair*, Dear and Wolch (1987) describe areas dominated by a disadvantaged group, the mentally ill; dependent elderly, the physically disabled, and substance abusers are other examples of disadvantaged groups. Acknowledging that human landscapes reflect social processes, geographers recognize the need to eliminate such ghettos by eliminating their causes.

SUMMARY

Why settle?
Permanent settlements were initially associated with specific economic changes, specifically the production of an agricultural surplus; they also reflect social needs.

Rural or urban?
This distinction is not a clear one. For 1993, 42 per cent of the world is classed as urban; 72 per cent in the more developed world and 34 per cent in the less developed world.

Two key issues
A geographic concern with settlement focuses on why people settle where they do and on their way of life.

Patterns of rural settlement
Rural settlement locations assume patterns ranging from dispersed to nucleated. Dispersion is less characteristic; in Europe it dates from the early eighteenth century and is typical of many temperate areas of European overseas expansion. Nucleated rural settlements may be irregular or regular.

Rural settlement theory
The theoretical contributions of Bylund and Hudson provide simplified descriptions. For Bylund, locations are explained in terms of distance and land attractiveness. For Hudson, settlements are equated with seeds from plants, and a three-stage process, with each stage prompting a specific pattern, is proposed. Both theoretical contributions fail to consider a wide range of cultural and political variables.

Rural/urban differences and population movements
Especially in the more developed world, rural/urban differences (*Gemeinschaft* and *Gesellschaft*) are decreasing. A transition from rural to urban is commonplace (especially close to urban areas) as urban residents move to rural areas as a result of a counterurbanization process. Urban sprawl is also occurring. Today there is evidence of both rural depopulation and repopulation.

Urbanization and urbanism
The world experienced a series of profound changes beginning about 1750: the rise of capitalism, great population increases, the emergence of nationalism, technological advances, and more and larger cities. In 1800, 3 per cent of the world's population lived in cities; by 1900, the figure rose to 13 per cent, and in 1993, it was 42 per cent. This trend is increasing. The earliest cities date from about 3500 BC as successors to large agricultural villages. The earliest cities (those before about 1760) and many cities in less developed areas can be conveniently labelled preindustrial — they function primarily as marketing and commercial centres. Urbanization was greatly enhanced when a mercantilist philosophy prevailed in Europe. With industrialization and the philosophy of capitalism, cities increased in number and size at a rapid pace. The resultant industrial cities were closely related to resource locations and, in terms of their inhabitants' way of life, represented a marked departure from earlier experiences.

Economic base theory
The basic sector of an urban economy comprises all those activities that produce goods and services for sale outside the urban area; the non-basic sector comprises all those activities that produce goods and services for sale inside the city.

Urban planning
Most cities show the effects of some planning, especially those in the more developed world. The modern planning movement originated in Britain and the United States in the late nineteenth century. Major modernist figures include Le Corbusier and Frank Lloyd Wright.

Explaining urban locations
There are many attempts to explain why cities are located where they are. Traditionally, the concern was with specific cities, but there has been increasing concern with networks of cities. A series of historical explanations have been offered; these tend to be most applicable to areas of European overseas expansion and involve recognizing a series of stages. Examples include those by Vance, Borchert, and Muller. The most influential and persuasive explanation, however, is the central place theory expounded by Christaller and Lösch. This theory, conceptually similar to the work of Thünen, was published in 1933. A series of simplifying assumptions, combined with four key concepts, generate hypotheses concerning the size and spacing of cities. There are numerous testings of the hypotheses.

The rank size rule
This rule states that, in a given country, the population size of a city is inversely proportional to the rank of that city where the largest city is rank 1.

Changing urban theory

Major twentieth-century theories that focus on the internal structure and identity of cities include human ecology, neoclassical, Marxist political economy, Weberian, and humanistic emphases. Most recently, postmodernists have questioned all such attempts at explanation on the grounds that a theory privileges one explanation above all others.

Inside the city

Twentieth-century urban growth includes substantial squatter settlements in many less developed areas and the emergence of conurbations. There are three basic models that purport to explain the internal structure of cities — concentric zone, sector, and multiple nuclei models. Each appears to be valid to some degree, although they are of decreasing relevance in contemporary cities. Most recently, there is evidence of the rise of the postindustrial city and such distinctive features as suburban office areas and gentrified inner areas.

Global cities?

It is possible that the two global processes of deconcentration of manufacturing and concentration of economic control are contributing to the creation of similar global cities, such as London, New York, and Tokyo.

Urban images and urban places

What image do residents have of the cities they inhabit? Lynch's work suggests that the typical image includes recognition of five key features — paths, edges, districts, nodes, and landmarks. Most cities can be easily divided into areas of distinctive land use and even into areas of distinctive residential use. Residential areas that are different to other areas are most common in large immigrant-receiving cities with differences based on a variety of cultural, social, and economic variables. Sometimes, parts of cities can be clearly demarcated as neighbourhoods. Much contemporary research recognizes the value of integrating spatial and social matters and is enriched by considering various social theories. Human geographers also recognize that cities are symbolic features that are often created to reflect power relations; this is especially so with capital cities.

Urban problems

In many respects, cities are the high points and low points of civilization — the best and the worst human landscapes. Many urban areas have environmental and social problems that partly result from their great size.

WRITINGS TO PERUSE

BERRY, B.J.L. 1981. *Comparative Urbanization: Divergent Paths in the Twentieth Century*. London: Macmillan.

An insightful overview of numerous urban issues by a leading geographer.

BOURNE, L.S., and D.F. LEY, eds. 1993. *The Changing Social Geography of Canadian Cities*. Montreal and Kingston: McGill-Queen's University Press.

An edited collection that addresses a range of social urban topics in the Canadian context; clear emphasis on contemporary social theory.

BRYANT, C.R., and J.R.R. JOHNSTON. 1992. *Agriculture in the City's Countryside*. London: Frances Pinter.

A detailed survey that provides numerous insights into the changing land uses of the rural/urban fringe.

BUNTING, T., and P. FILION, eds. 1991. *Canadian Cities in Transition*. Toronto: Oxford University Press.

A substantial collection that considers Canadian cities from a wide range of urban geographic perspectives. Includes evolutionary analyses, central place analyses, links with agriculture, industry and transport, and a wide range of philosophical viewpoints.

BRUNN, S.D., and J.F. WILLIAMS. 1977. *Cities of the World: World Regional Urban Development*. New York: Harper and Row.

A study in comparative world urban development that includes discussion of city growth, specific cities, problems of city living, and possible solutions.

CARTER, H. 1983. *An Introduction to Urban Historical Geography*. London: Arnold.

Thorough review of urbanism and urbanization through time. Varied international content.

CLARK, D. 1982. *Urban Geography*. London: Croom Helm.

Well-written textbook covering all aspects of urban geography.

CLOUT, H.D. 1972. *Rural Geography: An Introductory Survey*. Toronto: Pergamon.

An introductory survey focusing on Europe that highlights the diversity of rural geographic research topics.

DAVIES, W.K.D., and D.T. HERBERT. 1993. *Communities Within Cities: An Urban Social Geography*. London: Belhaven Press.

A good urban geography textbook centred on the concept of community; includes an innovative account of community development.

GAD, G. 1991. 'Toronto's Financial District'. *Canadian Geographer* 35:203–7.

The first of a series on Canadian urban landscapes in this journal. Later examples include the Toronto downtown, gentrification in Vancouver, markets in Ottawa, wholesaling in Edmonton, working-class suburbs in Hamilton, and neighbourhoods in Quebec City and Winnipeg.

GAYLER, H.J. 1990. 'Changing Aspects of Urban Containment in Canada: The Niagara Case in the 1980s and Beyond'. *Urban Geography* 11:373–97.

A detailed article focusing on a specific case-study and the wider implications of urban expansion.

HALL, P. 1989. *Cities of Tomorrow: An Intellectual History of Urban Planning and Design in the Twentieth Century*. Oxford: Blackwell.

A scholarly and entertaining history of urban planning.

KNOX, P. 1994. *Urbanization: An Introduction to Urban Geography*. Englewood Cliffs: Prentice-Hall.

A substantial textbook that covers all aspects of urban geography emphasizing contemporary changes in urban areas that involve decentralization and recentralization.

LEY, D. 1983. *A Social Geography of the City*. New York: Harper and Row.

Excellent textbook with full discussions of such issues as neighbourhoods and urban social change.

SMITH, R.H.T., E.J. TAAFE, and L.J. KING, eds. 1968. *Readings in Economic Geography: The Location of Economic Activity*. Chicago: Rand McNally.

Although dated, this volume includes excellent statements of location theory and selection of important empirical studies. Relevant to chapters 10, 11, 12, and 13.

STELTER, G.A., ed. 1990. *Cities and Urbanization: Canadian Historical Perspectives*. Toronto: Copp Clark Pitman.

A useful collection of essays, including a good discussion of settlement system evolution.

12

Industry and regional development

IN THE STUDY OF spatial variations and changes in economic activity, it has been customary for human geographers to distinguish three levels of economic activity. *Primary* production includes farming, fishing, forestry, and mining and occurs where appropriate resources are available. *Secondary* production converts primary products into other items of greater value. This process is called manufacturing (literally, making by hand) and requires a labour force, energy supply, and a market. Most manufacturing occurs in factories and a uniform product results. *Tertiary* activity refers to those economic activities involved in moving, selling, and trading the goods produced at the first two levels and to service activities such as professional and financial services. Recently, it has been helpful to identify a fourth level: *quaternary* activity is economic activity that specializes in assembling, transmitting, and processing information and controlling other business enterprises. It includes professional and intellectual services, such as management consultants and education. Quaternary sector employment is the favoured growth area in the postindustrial city discussed in the previous chapter.

As with any classification, the divisions between the four levels are not absolute and the classification is useful, although not ideal from our geographic viewpoint. For the purposes of this chapter on industry, it is convenient to include primary production (except farming, which is discussed in Chapter 10) and all secondary, tertiary, and quaternary production (except transport-related activities, which are discussed in Chapter 13).

In this account of industry and regional development, we need to focus on several issues that human geographers have tackled from a variety of theoretical perspectives. Our first concern is finding out why industries are located where they are. The answer is a form of neoclassical economic theory that displays all the advantages and disadvantages of that approach. Second, we describe and discuss the origins and evolution of twentieth-century industrial landscapes; landscapes that owe their basic character to the Industrial Revolution. Third, the world industrial map is described with emphasis on the differences between the more and less developed worlds and also on the dramatic rise of Japan and the newly industrializing countries. Our fourth concern is with the industrial restructuring that has been apparent since the 1970s. This includes discussions of the transition from Fordism to post-Fordism and of the deindustrialization evident in the more developed world; a variety of theoretical perspectives, especially forms of Marxism, are included in this discussion. The fifth and final concern in this chapter is with the geography of uneven development and the related attempts by states to plan industrial growth to favour certain regions at the expense of other regions.

THE INDUSTRIAL LOCATION PROBLEM

Locational questions and relevant theories are central issues for geography. This is clear for agricultural and settlement issues and is equally important for industry. Once again, theories need to be considered in the context of particular social and economic systems, especially when we realize that industries strive to minimize costs, including labour costs, and maximize profits. Many different types of economic organizations play roles in the industrial process. Households are less important today than they were in the past as far as industry is concerned, but they do produce and reproduce; they are not usually directly concerned with profit.

Firms are a second type of economic organization and make up the commercial sector; they may be owned by one person, or be partnerships, cooperatives, or corporations. Corporations are most important in terms of control of the economy. Firms operate their business in factories. Industrial location theory explains why factories are located where they are (Box 12.1).

EARLY LOCATION THEORY

For economists such as Adam Smith, J.S. Mill, and David Ricardo, industrial location was related to the location of agricultural food surpluses that could feed industrial workers. Probably more important was Adam Smith's realization that under the new doctrine of capitalism, the goal of location decisions was purely financial. There is little doubt that this incentive remains paramount today. In some form it is at the heart of most industrial location theory since the writings of Adam Smith. One principal exception to this generalization is Marx's work, which saw industrial location as one of many issues that were best explained in terms of political inequalities. But we can now turn to the classic industrial location theory, the least cost theory of the economist, Alfred Weber.

LEAST COST THEORY

The first attempt to develop a general theory of the location of industry was that of the German economist, Alfred Weber (1868–1958), brother of the sociologist, Max Weber. This work, published in 1909 and translated into English in 1929 (Friedrich 1929), is another example of a theory that does not purport to summarize reality. Rather, it aims to say where industrial activities ought to be located; it is a normative model.

Weber follows the Thünen tradition of setting up a number of simplifying assumptions:

1. Some raw materials are ubiquitous; that is, they are found everywhere. Examples suggested by Weber are water, air, and sand.
2. Most raw materials are localized; that is, they are found only in certain locations. Sources of energy fall into this localized category.
3. Labour is available only in certain locations; it is not mobile.
4. Markets are fixed locations, not continuous areas.
5. The cost of transporting raw material, energy, or the finished product is a direct function of weight and distance.

Box 12.1: Factors related to industrial location

A number of factors play roles in the typical traditional industrial location decision.

Various measures of *distance* have been very important. Distance from sources of raw material is a consideration because such resources are unevenly distributed and vary in quality, quantity, cost of production, and perishability. With improvements in transport and changing industrial circumstances, this factor is less relevant today than it was in the nineteenth century. Similarly, distance from an energy supply is declining in importance as a factor related to location. During the nineteenth century, availability of energy meant location at a source of coal, but the emergence of such energy sources as electricity has freed industry from this locational constraint. Distance from market is another factor. As is the case with the other distance variables, it tends to decrease in relevance with improvements in transportation.

Availability of *labour* also shows spatial variation in quality, quantity, and cost. Although theoretically perfectly mobile, labour is often limited to specific locations because of political restrictions on movement or because of social reasons. *Capital*, in the form of tangible assets (such as machinery) or intangible assets (such as money), can be a key locational consideration. Capital may be mobile. Finally, in this identification of traditionally important variables, *transport* is a key factor that is linked to several of the variables already noted.

Each of these factors plays a role in the typical location decision. The proportions of each used in the production process may be varied; for example, capital might be substituted for labour. When an industrialist is faced with many options, the number of feasible locations is increased.

Also important for understanding industrial location are factors such as those related to the nature of the product and to various internal and external economies. With regard to *internal economies*, it is evident that each manufacturing activity has an optimum size and that benefits may be gained from either horizontal integration (enlarging facilities) or vertical integration (involvement in related activities at one location). *External economies* often result from similar industries locating in close proximity.

Finally, industrial location decisions often result from various human and institutional considerations. The key human factor is a crucial one — *uncertainty*. Simply put, decision makers do not have adequate information about all the various factors. They may wish to minimize their costs, but they are unlikely to be able to do so. We have already distinguished between optimizing and satisficing behaviour and in the current context, such a distinction is crucial. Location decisions, then, often reflect uncertainty, individual judgements, and chance factors. The key institutional factor is the role played by *government*. There is often a conflict between economic logic and what governments see as socially desirable.

Thus, the greater the weight the greater the cost, and the greater the distance the greater the cost.

6. Perfect economic competition exists. This means that the industry consists of many buyers and sellers; any one participant cannot affect product price.

7. Industrialists are economic operators (see glossary) who are interested in minimizing costs and maximizing sales.

8. It is assumed that the physical geography (climate and relief) and the human geography (cultural and political systems) are uniform.

These assumptions are unrealistic but necessary for Weber to achieve his goal of creating a least cost location format. Given these assumptions, industrialists will locate industries at least cost locations in response to four general factors — transport and labour (interregional factors) and agglomeration and deglomeration (intraregional factors).

Transport costs

Weber's first concern was with transport cost. To find the point of least transport cost, two factors play a role: distance and the weight being transported. Weber contended that the point of least transport cost is at a location where the combined weight movements involved in assembling finished products from their sources and distributing finished products to markets are a minimum.

Weber also introduced the **material index**: the weight of localized material inputs divided by the finished product weight. Thus, if an industry has a material index greater than 1, then the point of least transport cost, and therefore the selected location, will gravitate to the localized material sources; if an industry has a material index less than 1, then the point of least transport cost, and therefore the selected location, will gravitate to the market. Historically, the importance of orientation to material resources has declined because fewer industries use bulky and heavy materials than in the past and because of advances in transportation.

In general, Weber's predictions are correct. Webber (1984:57) gives the example of the soft drink industry where 1 ounce of syrup, a ubiquitous material (water), and a 4-ounce can combine to produce 12 ounces of soft drink and a 4-ounce can. Thus, the material index is 1 + 4/12 + 4 = 0.31. This material index is less than one and, indeed, the soft drink industry is typically market-oriented. Smelting primary metals represents a converse example, as do many agricultural processing industries; these industries remain predominantly material oriented because the cost of materials still forms a substantial share of the total cost.

Locational figures

Weber introduced a graphical procedure for tackling such problems. A simple locational figure is shown in Figure 12.1. In this case there are two raw material sources (R_1 and R_2) and a market (M) and the industry locates according to the

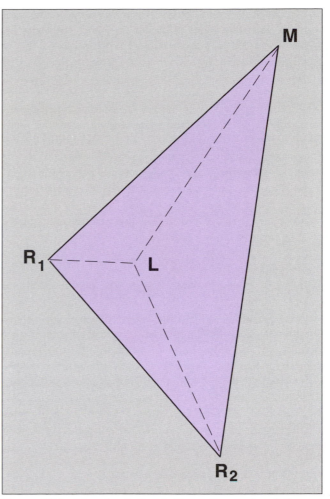

Figure 12.1 A locational triangle. Depending on the relative attractiveness of each of R_1, R_2, and M, there will be one location that is the least-cost location (L).

relative attraction of each of the three. If only ubiquities are used, then the locational figure is reduced to one point, the market. If only one raw material is used and that material loses no weight in the production process, then location can occur at the material source, at the market, or anywhere on a straight line between the two. Clearly, the locational figure is only needed in those cases where weight-losing materials are used. Use of such weight-losing materials is typical and the characteristic locational figure will have more than three sides, thus generating a complex problem. Geographers usually solve such locational problems algebraically rather than by referring to diagrams.

Labour costs

Although transport costs are the key to Weber's work, he also acknowledged the importance of labour costs. Such costs represent a first distortion of the basic situations already described and can be approached by mapping the spatial pattern of transport costs and then comparing this pattern to relevant labour costs.

To achieve this comparison, Weber introduced the concepts of **isotims** (lines of equal transport costs around material sources and markets) and **isodapanes** (lines of equal additional transport cost drawn about the point of minimum transport cost). Maps of isotims and isodapanes are examples of cost surfaces. Figure 12.2 shows isotims around two material sources and one market, while Figure 12.3 superimposes isodapanes on the isotims. Weber identified a critical isodapane, outside of which industrial location would not occur — industries can be attracted to low labour cost locations only if they lie inside the critical isodapane. A low labour cost location will attract industrial activity only if the savings in labour cost exceed the additional transportation costs of moving the raw materials to, and the finished products from, the low-cost labour site.

Agglomeration and deglomeration

Combined, transport and labour are the two interregional considerations. Agglomeration and deglomeration (see glossary) are intraregional considerations that, like labour, can potentially cause deviation from the least transport cost location. *Agglomeration* economies result from locating a production facility close to similar industrial plants, thus allowing a sharing of equipment and services and generating large market areas that aid the circulation of capital, commodities, labour, and information. *Deglomeration* economies are the reverse of agglomeration economies and result from location away from congested and high-rent areas.

Least cost theory: An evaluation

Weber's is a normative theory and does not attempt to describe or summarize reality. The major criticisms of Weber's work relate to the simplifying assumptions. Markets, for example, are not simple fixed points as Weber chooses to assume. Labour markets are characterized by discontinuities associated with age, gender, ethnicity, and skill. Despite such problems, Weber's basic logic is simple and sound (Box 12.2).

The focus on transport costs has been continued and substantially added to by Hoover (1948). The Weberian locational figure changes when transport costs are not directly proportional to distance. The cost per unit of distance is less for long hauls than for short hauls (Figure 12.4). Further, a step structure is often used (Figure 12.5).

MARKET AREA ANALYSIS

According to market area analysis thought, Weber's fundamental error is his assumption that industrialists seek the lowest cost location. Rather, for the market area analyst, industrialists are profit maximizers. Least cost theory is a form of variable cost analysis, that is, it is concerned with spatial variations in production costs, while the market area argument is a form of variable revenue analysis, that is, it is concerned with spatial variations in revenue.

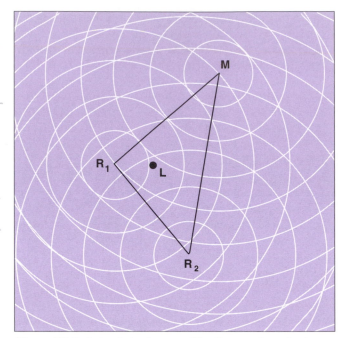

Figure 12.2 A simple isotim map. The diagram shows isotims around the two raw material locations (R_1 and R_2) and the market (M).

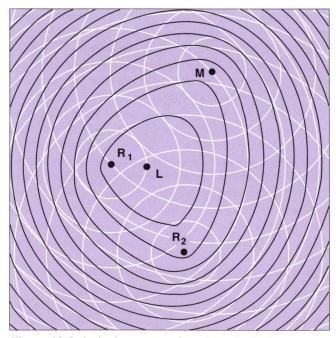

Figure 12.3 An isodapane map. Superimposing isodapanes on an isotim map indicates how far from L an industry may locate to take advantage of a particular low-cost labour location.

In market area analysis, the location that results in greatest profit can be determined by identifying production costs at various locations and then taking into account the size of market area that each location is able to control. The argument is that industries will attempt to monopolize as many

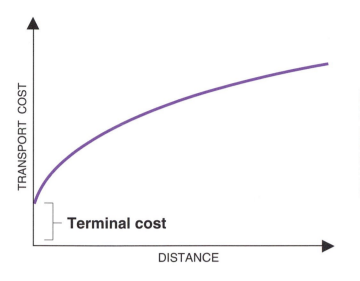

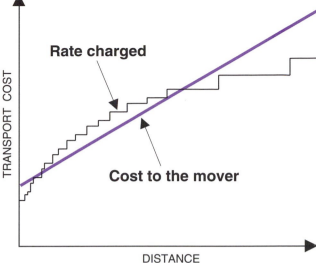

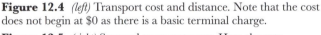

consumers as possible — they seek a **spatial monopoly and, in doing so,** exhibit **locational interdependence**. A classic example of this argument is that of the Hotelling (1929) model, which proposes that the agglomeration of industry is a logical outcome of certain demand conditions (Box 12.3).

This argument about seeking a spatial monopoly is relevant, but if least cost theory tends to de-emphasize demand, then market area analysis tends to de-emphasize factors other than demand. The most important market area theorist is Lösch, according to whom the ideal market area is hexagonal as per Christaller. Like least cost theory, market area analysis is a normative approach.

BEHAVIOURAL APPROACHES

Normative approaches allow us to identify rational or best locations and understand aspects of actual locations through comparison. But is such an approach satisfactory?

Figure 12.4 *(left)* Transport cost and distance. Note that the cost does not begin at $0 as there is a basic terminal charge.

Figure 12.5 *(right)* Stepped transport costs. Here the rate charged does not relate directly to distance because of the rate structures used by the movers.

Many geographers think not, preferring to focus on explicit analyses of why things are where they are rather than on abstract analyses of where they ought to be given certain assumptions. Normative approaches consider why things are where they are only in an indirect fashion. A set of approaches that can be labelled as behavioural attempt to correct this situation.

In brief, behavioural approaches centre on people's subjective views. The concept of the economic operator is rejected and the concept of satisficing behaviour (see glossary) is paramount. A simple example demonstrates the relevance of this approach. The beginning industrialist does

---

Box 12.2: Testing Weberian theory

There are many examples of tests of least cost industrial location theory. This box discusses a classic example of the application of Weberian concepts to the particular industrial locations of the Mexican steel industry in the 1950s (Kennelly 1968). The choice of an example that dates back more than forty years is quite deliberate as least cost theory is much less relevant today than it was in the past. The choice of the iron and steel industry is also quite deliberate as least cost theory can be appropriately applied to this type of heavy industrial activity.

In 1950, the Mexican steel industry included two principal steel plants and a number of other plants. Production covered a range of products and accounted for about 50 per cent of Mexican requirements. Location of plants, particularly the major ones, was related especially to the availability of raw

materials resources, such as iron ore, coke, oil, and market scrap. The location of labour appeared to be of minimal importance, as unskilled labour was quite mobile. Access to market was important, however, with major markets in Monterey and Mexico City. Overall, the 1950 Mexican steel industry was thus in accord with basic Weberian principles; Kennelly (1968) determined least cost locations to reach this conclusion.

Weberian theory is useful in explaining industrial locations of this type, but its logic weakens with technological change and is less relevant to most contemporary location decisions. Other (especially Marxist) approaches offer insights into industrial location decisions that the traditional theories do not attempt to consider.

not really debate where to locate; rather, the selected location is the existing location of the industrialist (or the nearest feasible location). Oxford, England, became the home of Morris, a major automobile manufacturer, for this non-least cost and non-market area reason. The location selected was the home area of the industrialist, William Morris. Admittedly, had that home area been a remote Scottish island, then some other location would have been selected. Specific locations are often chosen for subjective reasons.

Contemporary industrial firms, especially large corporations, are complex decision-making entities. Location decisions made by small, often single-owner, firms are less likely to be economically rational than those made by large firms, but other considerations compete with least cost and maximum profit logic for all firms. It is also clear that decision makers do not typically have access to all relevant information and, furthermore, that such information is subject to various interpretations. Geographers are well aware of the less than perfect mental maps (see glossary) held by decision makers. Although behavioural issues are not as easily theorized or quantified as the more directly economic issues, they are always factors to be considered in any analysis of industrial location.

THE INDUSTRIAL REVOLUTION

Before the Industrial Revolution, industrial activity was important but limited. Each major civilization was involved in such industrial activities as bread making, brick making, pottery, and cloth manufacture. These activities were located in the principal urban centres and close to raw materials.

In addition, each household was an industrial organization that produced clothing, furniture, and shelter. All industries involved minimal capital and equipment, were structured simply, and were small in size and output. Energy sources (such as wood) and raw materials (such as agricultural products and sand) were relatively ubiquitous, not localized. Given these characteristics, industries could be dispersed. As a result, there was little evidence of an industrial landscape. Transport systems were limited and markets for products were local rather than regional or national. All these circumstances and outcomes changed as a result of the series of developments that we call the Industrial Revolution.

ORIGINS

Between 1760 and 1860, the industrial geography of England was dramatically transformed. Elsewhere in western Europe the transformation occurred over shorter time spans during the nineteenth century. England, then, was the first industrial area.

Overall, the Industrial Revolution is well named as it was a period of rapid and cumulative change unlike anything that had occurred previously. Machines replaced hands, and inanimate energy sources were efficiently harnessed. On the other hand, the term fails to acknowledge that the process of industrialization was an uneven one and that other countries did not simply imitate the British case. It is also misleading to tie down the beginning of the Industrial Revolution too specifically. The English beginnings date back to the late sixteenth century and traditional craftsmen were still very important as late as the mid-nineteenth century. Belgium followed the English example most closely, but, unlike England,

---

Box 12.3: Locational interdependence — the Hotelling model

In 1929, an American economist, Harold Hotelling, considered the idea that industrial firms producing similar products might be locationally interdependent because of the shared demand. The basic logic was that any individual location decision must be tied to those of rival suppliers. With this in mind, Hotelling (1929) developed a model to show how one firm is attracted to the location of another firm. The classic, simple example is that of two ice-cream sellers on a beach. In this situation, two producers are competing to supply identical goods to consumers evenly spread along a linear market, therefore assuming that production costs are uniform, ice-cream selection is uniform, and demand is uniform. The two vendors will locate back to back at the centre of the beach, each serving half of the total market; such locational decisions allow each producer to supply consumers at the extremities of the market (the two ends of the beach) without any loss of marketing advantage.

How does this locational pattern come about? Initially each vendor may locate at the centre of one half of the beach. Such a pattern is socially optimum as it minimizes the average distance that consumers have to travel to purchase ice-cream and it also benefits each vendor equally — they each service half of the beach. But if one of the two vendors moves towards the centre, then that vendor has an advantage — selling to their original half and to a part of the other half — thus prompting the other vendor to also move towards the centre. The eventual result is an equilibrium pattern with both vendors back to back at the centre of the beach.

The fundamental advantage of this model is that it provides an alternative explanation of agglomeration to the model offered by Weber. It is, of course, a highly simplified situation that assumes identical products, identical production costs, the price of the product covering the cost of transport for the consumer, a linear market, and an unchanging demand. It also does not allow for locational inertia — not many industrial firms are as footloose as these two ice-cream vendors!

emphasized the metallurgical industries far more than the textile industries.

The major component of this change was the rise of factories, a form of large-scale production. Factories developed from earlier small establishments and domestic household activities. They were not only larger, they were also more mechanized and required more capital. Further, because of their reliance on newly used localized energy sources, they tended to agglomerate. The key energy source was coal and the first truly industrial landscapes arose in the coalfield areas of England. There is, however, much more than this to the Industrial Revolution.

The late eighteenth century witnessed the rise of a capitalist system that emphasized individual success and profits. The new capitalists seized the opportunity provided by a whole series of technological developments in the metals and clothing industries. Some examples: a furnace that burned coal to smelt iron was first used in 1709; a spinning jenny for multiple thread spinning was first used in 1767; a steam engine as an energy source was first used in 1769, and an energy loom for weaving was first used in 1785. These are only a few examples of a whole series of inventions, most of which were developed in England.

Recall our definition and discussion of energy in Chapter 4. Energy is the capacity to do work and the more successfully we can utilize other forms of energy, the more successfully we can achieve our wants. Also recall our definition of technology as the ability to convert energy into forms that are useful to us. Given the importance of coal for energy and the focus on metal and clothing industries, the first agglomerations occurred on the coalfields of northern and central England and in the traditional areas of cloth manufacture in northern England. These industrial agglomerations caused rapid increases in urban populations as workers moved closer to places of employment.

The basic logic of the Weberian least cost theory is useful in understanding these developments. Localized energy sources meant that industries were located at those sources and, if possible, at the sources of raw materials. The need to reduce costs meant that the Industrial Revolution was paralleled by a transport revolution. In England this involved first improved roads, then canals and then railways.

Many of the generalizations above are best clarified by brief reference to nineteenth-century industrialization, especially to the early British iron, steel, and wool industries.

EARLY INDUSTRIAL GEOGRAPHY

Analyses of these early industrial landscapes include traditional descriptive accounts that employ Weberian theory and other theoretical accounts that acknowledge the role played by economic power and organizations. Box 12.4 considers the case of the Tyneside coalfield in northeast England using concepts of power and organization.

Iron and steel

Prior to the Industrial Revolution, there were two principal centres of iron production — one in central England and one in southern England. Both had local ores and used wood (charcoal) as the energy source. The iron industry expanded rapidly as coal was substituted for charcoal from 1709 onward and as an efficient steam engine was developed. As iron production increased, new uses were found. The first

---

Box 12.4: Geographical change and industrial growth — the Tyneside coal industry

Understanding the changing geography of the Tyneside coalfield requires not only traditional industrial location concepts but also an awareness of the larger social, political, and economic framework.

In the second half of the eighteenth century, the Tyneside coalfield in northeast England dominated the British coal industry. Capital and control in the coal industry were concentrated in a merchant guild known as the Company of Hostmen. This group had a monopoly on coal trading and were able to use it to dominate local government. Between 1700 and 1726, there was significant spatial expansion of the coalfield as the shallow, easily worked seams became exhausted; sinking deeper pits was not a technological option at the time. Much of this expansion was accomplished by capitalists who realized the value of gaining control of the various factors of production, especially land. Those responsible for expansion formed a group known as the Grand Allies in 1726 and, by 1750, they controlled production in the region. Further, the Grand Allies were regularly in conflict with other capitalists in their attempts to introduce a price and output fixing arrangement.

The power exerted by the Grand Allies declined after 1770, by which time they controlled just less than half of the collieries. This decline in power was related to continued spatial expansion into areas that could be mined using post-1750 technology. The Grand Allies had not anticipated this technological change and had invested heavily in coal areas that were best exploited with older technology. Thus technical changes resulted in the Grand Allies owning inappropriate coalfield areas. In addition, changes in the economic environment, especially the banking system, contributed to their loss of power and destroyed their monopoly.

The monopoly was, however, quickly replaced by another capitalist organization, 'The Limitation of the Vend', which was a form of cartel. As this brief summary makes clear, an understanding of the Tyneside coal industry is best achieved by reference not only to Weberian and market area arguments but by reference also to social, political, and economic forces (Cromar 1979).

cast-iron bridge was built in 1779 in central England and most new industrial machinery was constructed of iron.

By the early nineteenth century, new iron areas dominated as technological advances continued to favour coalfield locations such as south Wales and northcentral England. A key factor emerged after 1825 when the first railway was constructed in northeast England. The railway boom was most evident in the 1840s and gradually allowed some movement away from coalfields. Railways, combined with the discovery of new iron ore sources, allowed a new industrial centre to develop in northeast England. Other major mid-nineteenth-century iron and steel areas were located in the English Midlands, south Wales, and central Scotland.

Textiles

Before the Industrial Revolution the woollen industry was specialized and spatially concentrated in three areas: southwest England, east-central England, and northern England. The manufacturing process remained unchanged from the fourteenth century to the late eighteenth century, when various inventions that were part of the larger Industrial Revolution prompted radical changes. Of the two major textile industries, cotton was subject to change first, followed by wool, as most of the new procedures were more easily applied to cotton manufacture. The major changes were as follows: a decline in domestic production and a rise in factory production; use of steam engines; rapid expansion in northern England related to availability of coal, and associated declines in the other two areas. In the case of the textile industry, the principal technological breakthroughs occurred in villages and small towns in northern England as mechanized factories were constructed near suitable sources of water.

In the case of the cotton industry, supplies of the raw material came initially from the East Indies, but supplies were irregular and a new source was essential. The Caribbean and southern United States were suitable source areas, but the task of separating cotton fibres from seeds was labour intensive and hence, even with a slave economy, supplies were limited. The invention of the cotton-gin in 1793 allowed American plantations to flourish and the English factories to be adequately supplied. The cotton mills of northern England were so dependent on imported cotton that mill workers were laid off because of reduced imports during the American Civil War. In the case of wool, the new area of supply was southeastern Australia.

Industrial landscapes

The Industrial Revolution and related rapid population and urban growth combined to create new landscapes. Factory towns arose, especially in northern England, and migrants from the south and rural areas poured into the new employment centres. In environmental and social terms, the result was often disastrous.

Before the onset of industrial change, the landscape was agricultural and predominantly rural, although before the use of steam energy, the early industrial landscapes became more congested and included some factories as opposed to domestic industrial premises. Probably the first real factory in Britain was a silk mill in northcentral England that was completed in 1722; it was five to six storeys high, employed 300 people, and used water energy. This indicates that coal was not necessary for industrialization, although it certainly accelerated the process of change. The first factory was followed by many others as the century progressed, but they were neither smoke nor dirt- producing. It was steam energy that resulted in major landscape change. More factories meant more workers crowded into small areas, and the use of coal meant that landscapes were soon polluted. Land was valuable and small terraced homes quickly became slums. In addition to poor-quality housing, these cities were characterized by smoke pollution — a situation that has been rectified in many cities only since about 1960.

Diffusion of industrialization

After about 1825, the technological advances evident in Britain diffused rapidly to mainland Europe (especially Belgium, Germany, and France) and to North America. The British did not favour diffusion of their industrial innovations. In some cases attempts were made to maintain secrecy about particular advances, but these attempts failed. In the United States, Pennsylvania and Ohio were the early industrial leaders because of their high-quality coal sources, while in the Russian empire, the driving force was the discovery of coal in Ukraine. Japan began to industrialize after establishing cultural contacts with the United States and rejecting feudalism in 1854.

As the nineteenth century progressed, one European area — the Ruhr region — offered coal, small local iron ore deposits, and an excellent location in terms of water transport. The demand for labour was so great that immigrants came from elsewhere in Europe, giving this region an unusually diverse ethnic character.

WORLD INDUSTRIAL PATTERNS

SOURCES OF ENERGY

Most energy today is produced by fossil fuels — coal, oil, and natural gas. Historically, coal is the most important fuel, but in the 1960s it was replaced by oil as the dominant energy source globally. The peak year for oil was 1973 when it provided 41 per cent of the world's energy; since then this has declined because of a combination of high prices, occasionally erratic supplies, more efficient use, and increasing use of hydroelectricity. Oil continues to be the chief source of industrial energy. A few countries tend to dominate world production of coal and oil. In 1960, the major sources of oil were the United States, Russia, and Venezuela, with only about 15 per cent in all of the Middle East and North Africa. Today, the Middle East and North Africa dominate.

Seven countries are now responsible for more than 80 per cent of world coal production: the United States, Ukraine, Russia, China, Poland, Great Britain, and Germany.

To date, there are few substitutes for gasoline. Countries such as Brazil and Canada are converting agricultural products to alcohol that can be blended with gasoline while some areas, such as Switzerland, are increasingly experimenting with electric cars. Different countries generate electricity in different ways. Nuclear reactors are heavily used in France, Belgium, and South Korea; Denmark is increasingly experimenting with wind, and many other countries such as Norway, India, New Zealand, Canada, as well as countries in South America and Africa, rely primarily on hydro generation (Box 12.5).

Energy supplies in Britain
In Britain the transition from coal to oil began in the 1960s. As we have seen, the Industrial Revolution began in Britain and relied heavily on coal. As local coal sources were depleted, imports were necessary but proved very expensive. As early as 1945, Britain began to import oil from the Middle East. Cost and supply uncertainties for political reasons prompted a search for oil and gas in the North Sea. Gas was discovered in 1965 and oil in 1969, and both were soon determined to be available in commercial quantities. By 1980 production was equal to domestic demand with thirty-six oilfields and twenty-five gas fields developed. What does the future hold? The current supply areas are being depleted and the key question is the extent to which declines can be countered by new discoveries. Some predictions suggest that 100 to 300 new fields will be developed by about 2020, and that the overall level of production will be maintained (Band 1991).

INDUSTRY IN THE MORE DEVELOPED WORLD

The more developed world is responsible for a disproportionate share of manufacturing employment and output. Most of this industrial activity continues to be framed within a capitalist social and economic system. Thus, decisions that affect the geography of industry and employment opportunities are taken by industrial firms. Small firms must cope with larger economic changes, while large firms tend to influence the form of the economic environment within which they operate. All firms must nevertheless take international trends into account. In some cases firms may find their home markets threatened by imports; in other cases firms may rely heavily on export markets. The global manufacturing system is dominated by the more developed world, especially North America, western Europe, and the west Pacific (especially Japan). These three regions control the export and import of manufactured goods. Many industries are part of this global manufacturing system through trading activities. Other industries are multinational as they own or control production in more than one country. **Multinational corporations** and other large firms tend to be less committed to a specific industrial location than small firms because of the spatial separation of production from organization (Dicken 1992).

Industrial firms today are still concerned with costs and profits and hence the basic logic of least cost theory and market area analysis continue to be relevant. While there is no doubt that many location (and other industrial) decisions may not be optimal, this may not be crucial in the long term. Thus, firms continue to make location decisions on the basis of the products, production technology, labour costs, sources of raw material and energy, capital availability, markets, and land costs — all considerations implicitly or explicitly contained within the traditional theories. An additional key consideration is the political context (Box 12.6).

Major industrial regions
There are four leading centres of manufacturing in the more developed world: North America, western Europe, the former USSR, and Japan. Together, the United States and Canada form the most productive industrial region in the

---

Box 12.5: An energy 'crisis' — New Zealand in 1992

Most of New Zealand's electricity is hydro-generated and thus dependent on adequate supplies of water. The principal source area is South Island, with both mountains over 3,000 m (9,843 ft) and high levels of precipitation. Electricorp, one company in which the state is a major shareholder, is responsible for generating most of the needed electricity and selling to regional boards that in turn sell to consumers. Electricorp is required by the government to make a profit. Because of this requirement, the company emphasized inexpensive hydro generation, closing down a coal-based generating station in 1990 and an oil-based station in early 1992. Unfortunately, hydro generation is dependent on weather conditions, and a drought that began in late 1991 posed a real threat to the generation of electricity.

By early 1992, lake levels were significantly lower than normal and a period of very cold weather increased demand for electricity. Electricorp was obliged to ask consumers to reduce consumption, to pay compensation to industrial users who reduced their use, to reopen the oil-fired station that was recently closed, and to lower storage lakes below normally permitted minimum levels. The problem was resolved in mid-1992 when the drought ended, but the consequences have been considerable. It is estimated that the cost to Electricorp was about NZ $300 million, and that the 1992–3 gross domestic product would be reduced by about 0.5 per cent because of the reduced manufacturing output (Murgatroyd 1993).

world. Within this large region, the key area is the northeastern United States and southeastern Canada. Its advantages include close ties to Europe, availability of energy and various raw materials, good internal transport facilities, and a considerable market. There are numerous local industrial regions within northeastern North America, including southern Ontario and southern Quebec, southern New England, the Mohawk Valley, the southern Lake Erie shore, and the western Great Lakes. Box 12.7 provides an account of the industrial geography of Canada.

Western Europe is second to North America in terms of manufacturing output and, like the North American region, consists of numerous local industrial areas, the principal of which are in central and northern Britain, the Ruhr and the mid-Rhine valleys in Germany, and northern Italy. Britain had the advantage of the earliest industrial start, which, by the mid-twentieth century, was all too clearly a disadvantage. This is also true of many major European coalfield areas, although the Ruhr Valley in northwestern Germany, which included the most important coalfield in Europe and developed a large iron and steel industry, has been much more successful in coping with recent industrial changes. In northern Italy, rapid diversification after the Second World War transformed a textile region into a textile, engineering, chemical, and iron and steel region using local gas and imported oil.

Until the dramatic political upheavals of the late 1980s and early 1990s, industrial location and production in the former USSR and the eastern European countries were determined by central planning agencies. Recent changes have not, of course, significantly altered the spatial pattern of industrial activities. In the former USSR there are five major industrial areas, four of which are now in Russia. Those based in Moscow and Ukraine were established in the nineteenth century. The Moscow area is market oriented and first developed as an industrial region that specialized in textiles; today it is diversified with a wide variety of metal and chemical industries using oil and gas. In Ukraine, industry is centred on a coalfield and there are also supplies of iron, manganese, salt, and gas. It is a major iron and steel and chemical industrial area. The other three regions were developed by the USSR government after the 1917 revolution and are located in southern Russia. The Volga area to the east of Ukraine was a relatively secure location, based on local oil and gas sources during the Second World War. Major industries are machinery, chemicals, and food processing. Further east, the Urals area is a source of many raw materials and as well as a major iron and steel and chemicals area. Industrial growth was promoted by the former USSR government because of the great distance to the western frontier region. Furthest east is the Kuznetsk area, which is similarly endowed with minerals and has a major coalfield.

Japan is relatively small in size; it has few industrial raw materials and almost all raw materials are imported; it is also a substantial distance from major world markets, but Japan has overcome these obstacles by making use of its large population. Since the Second World War, Japan has become a major exporter of industrial goods by keeping labour costs low. By 1970 the goods being produced included high-quality goods as a result of the highly skilled labour force. Japan is now a leading producer of such products as automobiles, radios, and cameras (Box 12.8).

NEWLY INDUSTRIALIZING COUNTRIES

Japan is no longer alone. The story told in Box 12.8 has encouraged many imitators. A group of other Asian countries aspire to achieve similar levels of industrial success. During the 1970s the economies of four Asian countries accelerated rapidly — South Korea, Taiwan, Hong Kong, and Singapore — and another four are accelerating today — Malaysia, Thailand, Indonesia, and the Philippines. Together, these are known as the newly industrializing countries (often called NICs). Japan is the model to be followed. In all cases labour costs are low and productivity is high. These various Pacific Rim countries are all high-growth rate areas experiencing labour shifts from agriculture to industry as occurred in much of nineteenth-century

---

Box 12.6: Explaining local industrial change

A concern with local industrial change (for example, the closure of a particular steel plant) needs to include local and global perspectives. There is no direct and predetermined link between local events and global circumstances, but global circumstances mediated through various levels of decision making are important.

In the case of the steel industry, major changes began in the early 1970s as the market for steel ceased to expand, and as production grew in some newly industrializing countries, such as Brazil, South Korea, and Japan. Thus the world steel trade increased, putting pressure on plants in North America and Europe to cut production.

These global changes were then mediated through to the local scale. The key mediation occurred at the level of the state. Thus the state effectively exploited divisions in the labour force by, for example, offering redundancy pay to older workers and alternative employment to younger workers. The state also weakened unified opposition to proposed plant closures by identifying specific plants for closure, rather than reducing production in all plants in a relatively uniform manner. These arguments are put forward in detail for the United States, France, West Germany, and Britain by Hudson and Sadler (1989).

These arguments, like those proposed in Box 12.4, are examples of Marxist thought. In both cases emphasis is on the relevance of social and political (rather than explicitly economic) frameworks.

Europe. These Asian countries are the principal but not the only NICs; other examples are Brazil, Mexico, Greece, Spain, and Portugal.

Most successful of the NICs is South Korea. Until 1950 this was a poor, less developed country characterized by subsistence rice production. Today South Korea is an industrial giant that began with heavy industry, then automobiles, and now high-technology products. It is the Japanese case repeated. For South Korea, this transformation took place in less than forty years.

Box 12.7: Industry in Canada

Plentiful natural resources mean that Canada has a well-developed primary (resource extraction) industrial sector. Its immense geographic extent and consequently dispersed national market poses a challenge to many firms because of high transportation costs. In addition, the primary sector is scattered throughout Canada, while the secondary sector is concentrated in the St Lawrence lowlands — southern Ontario and southern Quebec. Proximity to the United States has resulted in a manufacturing sector that includes many branch plants of American parent companies.

Following the in-movement of Europeans, there was a series of staple economies — fish, fur, lumber, wheat, and minerals. By the mid-nineteenth century, Canada was a major exporter of primary (resource) products and an importer of secondary (manufactured) products. Manufacturing in Canada at that time was largely limited to the processing of agricultural and other primary products for domestic and export markets. Following Confederation in 1867, the Canadian government pursued two policies with direct relevance to the infant industrial geography, subsidizing railway building and introducing tariff protection for manufactured products. The immediate beneficiaries were the established areas of the St Lawrence lowlands, and a factory system was soon evident. Montreal and Toronto especially grew as industrial centres.

The capital needed to develop manufacturing industry initially came from Britain, but as this source lessened, a branch plant economy owned by the United States developed. American firms were thus able to bypass the protective tariffs, while Canada benefited by the introduction of financial and other capital and technology. After the Second World War, the United States became less globally competitive (especially compared to Japan) and as Canada became increasingly nationalistic, the branch plant economy became less and less attractive.

Today, the Canadian manufacturing industry is highly regionalized and many geographers see Canada in terms of heartland and hinterland, or core and periphery. The heartland has the manufacturing industry, while the hinterland is the resource region; Toronto and Montreal have a fabrication economy, while the remainder of Canada has a resource-transforming economy. Each economy faces a different set of problems. The Toronto and Montreal industrial economies are suffering because of imports, especially from Japan and other Asian countries. The remainder of Canada, which is largely dependent on raw materials, is subject to fluctuations in world price and demand over which they have no control. Because Canada has many resources, problems for one economy are usually buffered by the fortunes of another at the national level. At the local level, however, many areas are subject to boom or bust. Single resource towns, of which there are many, are the most vulnerable.

Box 12.8: Industry in Japan

Japan comprises numerous islands scattered over about 2,500 km (1,553 miles). Cultural contact with Europe and North America was initiated only after 1854 with the rejection of feudalism, and Japan developed into an industrial power comparable to those other areas by 1939. This achievement included territorial expansion. Industrialization continued after the Second World War despite the loss of Korea and the considerable bomb damage. During the 1950s and 1960s, Japan excelled in heavy industry, especially shipbuilding, but by the late 1960s, a transformation had taken place with the new emphasis on automobiles and electronic products. Most recently, Japan has focused on computers and biotechnology. During each of these three phases since the Second World War, Japan has been a world industrial power.

To explain these remarkable industrial successes, it is usual to note the low labour costs, the high level of productivity, the emphasis on technical education, the minimal defence expenditures, the aid from the United States because of the perception that Japan served as a bulwark against Chinese communism, and the distinctive industrial structure. Japanese industry is characterized by small specialized firms linked to the major corporate giants, such as Nissan and Sony. This distinctive industrial structure facilitates the rapid acceptance of new technologies.

Another aspect of the Japanese success story is the considerable investment overseas. There are many Japanese companies operating in North America and Europe especially, and Japanese banks dominate international finance.

Current evidence suggests that these trends are to continue. One motivation for recent North American free trade agreements and continuing European integration is awareness of the need to react to the dramatic industrial growth in much of Asia.

Several NICs and some other less developed countries — including South Korea, Singapore, Taiwan, Hong Kong, the Philippines, and Mexico — have set up export-processing zones to attract transnational or multinational corporations. These are manufacturing areas where both the raw materials and the finished products are exported. Industries are attracted for three general reasons: first, because of inexpensive land, buildings, energy, water, and transport; second, because of a range of financial concessions in such areas as import and export duties; third, because of low workplace health and safety standards and inexpensive (usually young female) labour. It is commonly argued that women are less likely to be disruptive than men, and that they more willingly accept difficult working conditions and low wages. There are few advantages other than waged employment for the processing country, and even this advantage varies depending on larger international economic circumstances. The parallel between these export processing zones and agricultural plantations (discussed in Chapter 10) is a compelling one.

There are especially close links between these export processing zones and high-technology companies; in Mexico about 850 zones known as *maquiladoras* are set up within easy reach of the high-technology suppliers in such areas as the Santa Clara Valley in California (Silicon Valley) and the Dallas–Fort Worth area in Texas (Silicon Prairie). The high-technology materials, such as silicon chips, are manufactured in high-technology plants in the United States and transported to the *maquiladoras* for the simple but labour-intensive assembly into the finished product, which is then transported to the United States for sale.

INDUSTRY IN THE LESS DEVELOPED WORLD

The distinction between more and less developed is never straightforward. The NICs are in the less developed world, although their recent industrial history clearly provides compelling evidence of economic success. The NICs are frequently referred to as economic miracles. The typical less developed country has been much less successful industrially and miracles are not likely.

One obvious distinction between the more and the less developed worlds is the degree of industrialization, with many countries in the latter group not yet having passed through an Industrial Revolution stage. It is often hoped that industrialization might lead to employment for the unemployed or underemployed rural poor. Further, the less developed world favours industrialization to demonstrate economic independence, to encourage urbanization, to help build a better economic infrastructure, and to reduce dependence on overseas markets for primary products.

Problems

Potentially, industrialization has many advantages. Why, then, is it so difficult for many countries to achieve it? A major difficulty is the continuing effect of former colonial rule. Historically, European colonial countries saw their colonies as producers of needed raw materials for their domestic manufacturing industries and as markets for those industries. This is a difficult legacy to remove. Most countries in the less developed world do not have the necessary infrastructure nor the capital to develop that infrastructure. Furthermore, fundamental social problems resulting from limited educational facilities do not encourage the rise of a skilled labour force or the emergence of a domestic entrepreneurial class. Finally, these countries often have an insufficient domestic market in terms of spending capacity to make industrial production economically feasible.

Prospects

Despite these difficulties, there has been considerable industrialization in many less developed countries since the Second World War. Major successes have been in import substitution: that is, manufacturing goods that were previously imported, often with tariff protection policies in place. Examples include light industries that are not technologically advanced, such as food processing and textiles. Heavy industry has been more difficult to develop, even though much of the world's raw materials come from the less developed world. In general, the less developed world provides much evidence contradicting the standard primary-secondary-tertiary sequence. The more typical sequence is primary-basic tertiary-secondary; the quaternary sector will possibly rise in due course.

India

The example of India highlights many of the general points noted earlier. Following independence in 1947, India was a producer of agricultural products, but today has a much more diversified industrial structure. Industrial landscapes have been created and the government has attempted to reduce regional disparities. Since 1951, India has used a series of five-year plans to guide development. Initially, heavy industry was stressed, with self- reliance and greater social justice evident in later plans. Given that its industrial transformation began only in 1951, India has achieved remarkable success because of a substantial market, available resources, adequate labour, and relatively sound government planning.

China

As noted in our discussion of agriculture in Chapter 10, liberalization of the Chinese economy began in 1978 with the opening up of the country to international trade and foreign investment. Since the 1980s, agricultural and industrial economies have experienced reforms and modernization.

During the 1950s, industry development was based on the idea of large technology-intensive, state-funded factories and small labour-intensive, locally organized units (Morrish 1994). This strategy has proven unsuccessful because the state-owned factories were expensive to operate and highly inefficient. Shortages of consumer goods were commonplace.

A new regional development policy was introduced in 1980 with the establishment of a number of special economic zones that enjoy low taxes and other financial concessions to attract foreign investment. These are similar to the export processing zones already discussed. A reform program was initiated in 1984 that involved less state control, decreased subsidies, increased response to market forces, and encouraged local collective and private industrial enterprises. One result is that recent industrial production in China has outstripped that of South Korea, increasing at 12 per cent per annum between 1980 and 1990.

RECREATION AND TOURISM

Recreational activities are now a major component in most people's lives. They include time spent with family and friends, reading, or participating in or watching sports. It can be broadly defined to include tourism, an industry that is generated by the more developed world but that occurs in both more and less developed worlds. The tourism industry has experienced dramatic growth since about 1960 and, according to the World Travel and Tourism Corporation, it is now the largest industry in the world and is growing at a phenomenal rate — 23 per cent faster than the overall world economy (Wheat 1994:16). The number of tourists in the world today is about 500 million, but the importance of this industry cannot simply be measured in numerical terms for 'tourism is a significant means by which modern people assess their world, defining their own sense of identity in the process' (Jakle 1985:11).

Beginnings

In the more developed world, tourism and recreational activities are important features of modern life. The concept of tourist travel, particularly an annual holiday, is a product of the Industrial Revolution with annual holidays negotiated between employer and workforce. Other holidays were times of religious observance. In seventeenth-century Europe, travel holidays included visits to spas for medical purposes and then seaside resorts became popular because they supposedly had health benefits. By the nineteenth century, the annual holiday was established and seaside resorts in Europe, especially Britain, became playgrounds for the working classes.

Tourist attractions

There are six principal tourist attractions:

- good weather (usually meaning warm and dry)
- scenery (coastal locations are especially favoured)
- amenities for such activities as bathing, boating, or general amusement
- historical and cultural features (old buildings, symbolic sites, or birthplaces of important people)
- accessibility (increasingly tourist travel is by air; in general, costs increase as the distance travelled increases)
- accommodation

Areas with an appropriate mix of these attractions tend to be major tourist areas. In the more developed world, tourist areas are usually urban or coastal. In Europe, many of the principal coastal areas are peripheral, as in the cases of Spain and Greece.

One of the most distinctive types of contemporary tourist attraction is the tourist **spectacle**. These places are examples of popular culture that have often been created by developers and planners and willingly accepted by consumers (see Chapter 8 for a discussion of popular culture). Examples of such attractions include Disney World, major sporting events, world fairs (Ley and Olds 1988), and large shopping malls, such as the West Edmonton Mall (Hopkins 1990; Jackson and Johnson 1991; Shields 1989). Analyses of such places by human geographers cover a range of issues, including humanistic interpretations of their meanings.

Tourism in the less developed world

In the less developed world, tourist areas (mostly coastal) are growing in response to increasing demand in the more developed world, but there are several problems for the less developed country in this relationship. Most importantly, dependence on tourism makes a country vulnerable to the changing strategies of the tour companies in more developed countries; tourist areas need to compete with each other because they may offer little that is distinctive — there are many destinations for sun, sand, and surf that consumers can be directed to by tour companies. Nevertheless, less developed countries enthusiastically promote tourism to provide local employment, stimulate local economies, and earn foreign exchange. The tourist areas likely have improved communications and services and are atypical of the larger economy. Clearly, the tourism industry is one more example of the dominance of the more developed world over the less developed world.

Tourism's impact on local ways of life is contradictory. On the one hand, local crafts and ceremonies are encouraged, but on the other hand, tourism can destroy local cultures and dramatically change local environments. Increasing awareness of such problems is prompting the growth of environmentally aware tourism, sometimes called **ecotourism** or sustainable tourism. Belize, in Central America, is probably the best-known ecotourist destination, having hosted two major conferences on the topic. In Belize, tourism accounts for 26 per cent of the gross national product and is of ever-increasing importance in the economy because of declining prices for traditional cash crops, especially sugar cane. Awareness of the fragility of coastal ecosystems encouraged the concern with ecotourism, although several observers are unable to distinguish between the ecotourist developments in Belize and other more traditional developments (Wheat 1994:18).

Today, tourism and related recreational activities are a major growth industry in the tertiary sector. Current evidence suggests continued growth as travel costs decline and as surplus income in the more developed world increases (Box 12.9).

INDUSTRIAL RESTRUCTURING

Major changes in industrial geographies, changes that incorporate the transition from Fordism to post-Fordism (see glossary), are currently taking place in response to changes in three types of technology. Combined, these changes represent a transition to **flexible accumulation**, making it easier for companies to take advantage of spatial variations in land and labour costs and to serve larger markets. First, production technologies (such as electronically controlled assembly lines and automated tools) are increasing the separability and flexibility of the production process. Second, transaction technologies (such as computer-based, just-in-time inventory control systems) also increase locational and organizational flexibility. Third, circulation technologies (such as satellites and fibre optic networks) are enhancing the exchange of information and increasing market size. The subsequent industrial restructuring is assuming three principal forms:

- There is a changing relationship between corporate capital and labour as machines replace people, as manufacturing industry declines, and as multinationals seek low labour cost locations (as noted in the account of export processing zones)
- Both the state and the public sector are playing new roles with the shift from **collective consumption** (such as schools and hospitals) to cooperative public-private projects and to deregulation (as discussed in Chapter 11)
- There is a new division of labour at various spatial scales as the new technologies allow corporations to react rapidly to variations in labour costs (as exemplified in the account of export processing zones and by discussions in Chapter 11)

LOCATIONAL CONSIDERATIONS

Decision making, especially location decision making, by industrial firms today often differs markedly from that proposed by Weber. Increasing emphasis on technology and decreasing emphasis on materials and localized energy sources mean that, for many industries, transport costs are no longer the main criterion. Rather, many contemporary firms trade off two types of cost. First, there are labour and land costs as considered by Weber. Second, there are now a whole set of new costs associated with exchanging information between firms (Box 12.10).

Thus, a major debate today in studies of industrial activity, especially location decisions, concerns the implications of labour (and land to a lesser extent) costs on the one hand and information exchange costs on the other hand. For many industrial activities, information can be rapidly exchanged at low cost by electronic means, but there is a danger in such impersonal exchanges if the information is lacking in clarity. In principle, firms that make location decisions on the assumption that they are able to exchange information successfully are able to seek out low labour cost locations. A decentralized (deglomerated) industrial pattern results. A prime example is the successful expansion of many high-technology Japanese firms to other countries. In situations where firms lack confidence in their ability to exchange clear and unambiguous information, they will prefer to locate in close proximity (to agglomerate) in order to facilitate personal, face-to-face exchange of information.

Acknowledging the contemporary relevance of information exchange introduces two possible patterns of industrial location. Decentralization occurs if long-distance electronic information exchange is feasible, but centralization occurs if the information is lacking in clarity and requires personal

Box 12.9: Tourism in Sri Lanka

Sri Lanka is a characteristic tourist destination in the less developed world, as it includes attractive environments but also a problematic and uncertain industry. Long described as a paradise by Europeans, it is a logical area, from a European perspective, to be marketed as a tourist destination. There are scenic landscapes, including beaches, tropical lowlands, and mountains, as well as a rich and diverse cultural heritage. During the 1960s, the Sri Lankan government created the necessary infrastructure of hotels, roads, and airline facilities, and improvements and additions to this infrastructure have continued. There was spectacular growth of the tourist industry and, by 1982, tourism was second only to tea as an earner of foreign exchange.

A period of decline between 1983 and 1989 was related to the heightening ethnic conflict described in Box 6.6. Tourist numbers declined by more than 50 per cent because of the well-publicized internal conflicts, and northern and eastern Sri Lanka experienced the greatest losses in tourist numbers and hence the greatest economic damage. Since 1989, the industry has recovered and is increasingly attracting visitors from elsewhere in Asia. As of the mid-1990s, the future of the industry seems secure because of strong government support and careful regulation (O'Hare and Barrett 1993).

contact to be effective. Decentralization is typically associated with firms (often multinationals) that mass-produce a standard product and thus do not need to be concerned about possibly ambiguous information exchange, while centralization occurs when functionally related firms need to exchange information on a person-to-person basis because their products are linked but are not identical.

SERVICE INDUSTRIES

There is a generally accepted assumption that as an economy progresses, there is a transition from primary (or extractive) to secondary (or manufacturing) to tertiary (or service) activities. Figure 12.6 provides a simple description of the relative importance, in terms of employment, of the different sectors of industrial activity. Indeed, some geographers contend that, especially in the more developed world, there is an increasingly service-oriented postindustrial society. This may be so, but there have long been service industries.

According to Daniels (1985:1), a service 'is probably most easily expressed as the exchange of a commodity, which may either be marketable or provided by public agencies, and which often does not have a tangible form.' Clearly, service industries are diverse, so diverse that they encompass not only a tertiary sector but also quaternary and quinary sectors. Thus, transportation and utilities are tertiary; insurance and real estate are quaternary, while education, health, and government are quinary. Whatever their sector, the service industries are crucial components of any economy and merit the attention of geographic researchers.

Services have long been important, growing along with manufacturing activity during the Industrial Revolution, but have achieved their most rapid expansion since the Second World War. In the United States, for example, between 1980 and 1990, manufacturing employment decreased 3 per cent

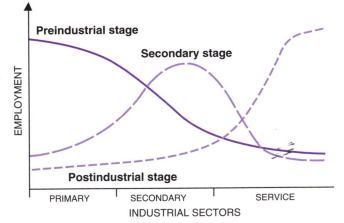

Figure 12.6 Simplified relationship between economic growth and distribution of employment.

while service employment increased 20 per cent. In the less developed world, the dominant service activities include retailing and distribution, while in the more developed world more specialized services such as banking and advertising are important.

Geographic attempts to explain the location of service industries have focused on the central place model described in Chapter 11. This work has made it clear that service locations can be determined by considering such standard causal variables as transport costs, the location of demand, and economies of scale. In addition, services have an especially strong predilection for agglomerating. They are located in accord with population and have clustering tendencies. Furthermore, location decisions appear to be especially vulnerable to such behavioural variables as the availability of information and the interpretation and use of that information. A consideration of behavioural variables helps explain

---

Box 12.10: Deindustrialization and reindustrialization

Both of these spatial trends result from the larger process of restructuring, specifically sectoral changes in capital, and can also be seen as one component of the transition to a postindustrial society. The term 'deindustrialization' refers to a reduction in manufacturing that is usually most easily measured by reference to employment data. It is most evident in the older industrial areas and such activities as iron and steel, textiles, engineering, and shipbuilding. The major social consequence is unemployment at unacceptably high levels.

A process of reindustrialization at least partially counters industrial decline. This process assumes various forms. First, there is an increasing tendency for small and/or new firms to be more competitive. This occurs particularly outside the traditional industrial areas and is related to the issue of information exchange by electronic means. Second, high-technology industrial

activities, especially micro-electronics, are expanding rapidly in terms of output, if not in terms of employment. Such industries are locating in environmentally attractive areas where skilled workers choose to live. A third aspect of reindustrialization is the expanding service industry. As a consequence of rising incomes and changing lifestyles, there is now significant growth in tourism and recreation industries, with accompanying impacts on environmentally attractive areas. Other service industries, especially banking and information services, are also expanding in major urban centres.

The late twentieth-century new industrial landscape is thus quite different to the nineteenth-century landscape. The landscape itself is visually different, the work experience is different, and the location of the landscape is changed. The need for coherent regional policies is compelling.

the clustering tendencies. Information diffusion takes place most rapidly and effectively in local networks. Indeed, some argue that recent changes in the technology of information diffusion have led not only to the clustering of certain services but also to spatial changes in settlement importance. Others, however, contend that new information technologies are leading to increased decentralization, possibly even to eventual domestication of service workplaces. Regardless of where service industries locate, there is little doubt that they are able to contribute significantly to increases in productivity and living standards. In Canada, for example, service sector growth is a crucial component of larger economic prosperity (Grubel and Walker 1989).

A geography of service industries also involves consideration of the availability of those services that are so much a part of human well-being, among them health care and educational facilities. Indeed, any study of service industries is virtually a microcosm of much contemporary geography because of their many links with other geographic topics and because of the importance of services in contemporary economies and societies.

INDUSTRY AND SOCIETY

The industrial geography discussed so far has said little about either humans or environment. Our concerns have been essentially spatial and economic, a fair reflection of the dominant traditions. Increasingly, however, as we have been seeing, human geographers have turned their interest to issues related to political and social circumstances. Since about 1970, many researchers have realized that much industrial geography can be best treated with reference to social theory. Their argument is that industrial activity does not take place in a politically and socially neutral background according to the principles of neoclassical economics. As a result, it is necessary to turn to contemporary social theory to gain real insights into the geography of industrial capitalism. Unfortunately, from an introductory standpoint, the range of theory is considerable, although the greatest emphasis is on versions of Marxist thought. Massey (1984), for example, provided a Marxist theoretical analysis of the changing industrial geography of the United Kingdom, while Graham et al. (1988) outlined a framework based on Marxism for investigating structural change in industries. Such industrial geographies represent significant departures from the usual Weberian or other type of economic theory and give priority to relating spatial and social issues by reference to such topics as class and gender as they have evolved in a capitalist framework.

It is now commonplace to argue that hypotheses about industrial (or other) locations lack completeness unless they are placed in some appropriate social, political, and institutional context. Massey (1984) argued that places vary in many ways, including workplace social relations, local political circumstances, and the position of places in the division of labour; thus it is necessary to consider each of these in any analysis of industrial location and the changing distribution of employment. The focus on workplace social relations often requires incorporating gender differences.

Gendered employment

Today in most of the more developed world, a majority of women of working age are in the labour force, which is a significant change. In Canada, for example, female participation rates in the labour force more than doubled between 1951 and 1981, but this change has not typically involved changes in the type of work done by females, as a majority are still employed in clerical, service, and low-skill jobs. Why is this so? One clear reason is discrimination by employers. In an account of the steel industry in Hamilton, Ontario, Pollard (1989) cited evidence of sexist hiring practices by the principal steel company, Stelco. Between 1961 and 1978 Stelco received between 10,000 and 30,000 applications from women for production jobs — none were hired. In the same time period, about 33,000 men were hired for production jobs. Following considerable publicity, intervention by the union and the Human Rights Commission, women began to be employed. But the evidence suggests that being employed was merely a beginning to the resolution of a larger social problem. While on the job, women were made to feel uncomfortable by men in the workplace and members of the community, especially wives of male employees. Clearly, there is a great deal involved in any attempt to challenge established gender divisions of labour; larger social traditions are at stake.

Environmental considerations

The topic of industrial impacts on environment was discussed in some detail in Chapter 4. The concern here is with the locational implications of the environmental consequences of industrial activity. In many countries new regulations concerning pollution are affecting industry in terms of the amount produced and the disposal of waste products. In principle, environmental considerations add a new cost factor that may affect location and production decisions. Interestingly, in a study of the effects of the National Environmental Policy Act in the United States, Stafford (1985) concluded that the policy would not lead to major locational changes. As regulations become increasingly more rigorous, however, locational effects seem likely.

THE GEOGRAPHY OF UNEVEN DEVELOPMENT

Much of our discussion has hinted at the relationship between industry, larger issues of economic change, regional disparities, and the possible value of regional planning. It is immediately evident that there are spatial inequalities at a variety of scales in terms of various criteria that we might label as developmental or quality of life. How do we explain these differences and deal with them?

EXPLAINING REGIONAL ECONOMIC GROWTH

Explaining spatial variations in economic development and quality of life is not easy. Traditionally, societies travel along a sequential path of development. Marx, for example, envisaged society passing from primitive culture, to feudalism, to capitalism, to communism. Rostow (1960) saw a transition through a series of stages from primitive society to mass-consumption society (Box 12.11). Others saw society experiencing changes in the dominant occupation from primary to secondary to tertiary. Geographers have considered these various generalizations and contributed to them. In general, the sequence is as follows. In the early stages of development, economy and society are fragmented, trade is limited, and

---

Box 12.11: The Rostow model of economic growth

This is a classic example of a developmental model that generalizes the economic development of capitalist states. There are five stages proposed for countries to follow based on the European experience.

1. *Traditional society:* Subsistence agriculture, domestic industry, and a hierarchical social system; a stable population-resource balance.
2. *Preconditions for take-off:* Localized resource development because of colonialism or activities of a multinational corporation; export-based economy; often a dual economy; the Canadian Shield and the Canadian north might be considered to be in this stage.
3. *Take-off to sustained growth:* Exploitation of major resource; possibly radical and rapid political change; the NICS may be in this stage or the next.
4. *Drive to maturity:* Creation of a diverse industrial base and increased trade.
5. *Age of high mass consumption:* Advanced development of an industrial economy; evident in the more developed world when the model was developed in the 1960s.

This is a useful model in an introductory human geographic context because it links easily to the demographic transition model (Chapter 5), the mobility transition model (Chapter 6), and the global diffusion of western cultures (Chapter 9 especially). As with these other models, however, the simplifications in the Rostow model can be dangerous — what happened in Europe and North America does not have to happen elsewhere (as Box 5.3 details with regard to fertility). It is also worth remembering that economic growth and development are not necessarily the same thing — recall the discussion of the Human Development Index in Chapter 6.

---

primary activities dominate. Subsequently, as transport improves and regions become more specialized and less subsistence oriented, manufacturing develops and trading becomes important. Finally, the service sector develops and regions are highly specialized. This idea of stages of development has been subjected to considerable criticism — recall the discussion of world systems theory in Chapter 6.

An alternative view is the *staple* theory of economic growth. A staple is a primary industrial product that can be extracted at low cost and for which there is a market demand. Originally developed by Canadian historians, the staple approach appears particularly applicable to areas of European overseas expansion. Staple success means economic growth in the areas of extraction and in the export centres, hence staple concepts have direct impacts on regional growth. Indeed, other scholars have employed staple concepts to build such models. Pred (1966) focused on the staple's multiplier effects.

A third general approach to the explanation of regional economic growth is that of Friedmann (1972), which utilizes *core and periphery* concepts. A core region is a dominant urban area with potential for further growth. Peripheries include areas of old established settlement characterized by stagnant, perhaps declining, economies, some of which may be former staple production areas; these are called downward transition areas. There are two other types of peripheral regions. Upward transition regions are linked to cores and continue to be important resource areas. Resource frontier regions are peripheral new settlement areas. 'Peripheral regions can be identified by their relations of dependency to a core area' (Friedmann 1972:93).

A fourth approach argues that growth does not appear everywhere at the same time; rather, it manifests itself in points of growth. This *growth pole* concept has resulted in a substantial but rather confused literature. Thus, a new area of resource exploitation will induce economic growth, but not necessarily at the source. Urban centres often serve as growth poles.

Each of the four approaches has both widespread application and serious deficiencies. As generalizations, they provide insights, but not explanations. Geographers investigating particular spatial and social issues tend to use the most appropriate model for their specific purposes. How, then, have geographers tackled the problems of unequal regional growth?

CORRECTING REGIONAL ECONOMIC INEQUALITIES

Spatial variations in economic development and quality of life are normal outcomes of any process of change. Central or core areas will usually experience innovations before outer or peripheral areas, and the uneven distribution of resources prompts spatial variations. Unfortunately, in many parts of the world these natural spatial variations have resulted in differences that society considers inappropriate. Hence, most countries have regional development policies.

The Canadian example

Governments have long interfered in economic change. In Canada, for example, companies have been granted monopolies, the progress of settlement has been dictated, and tariff barriers have been installed — all before 1900. In this century, however, such intervention has begun to play a major role in redressing spatial imbalances. The typical policy encourages growth in depressed, usually peripheral, areas.

A number of regions in Canada qualify as relatively underdeveloped. Indeed, some would contend that almost all of Canada outside of the St Lawrence lowlands qualifies for this description. In the case of the maritime provinces of New Brunswick, Nova Scotia, and Prince Edward Island, economic and related social problems stem from a peripheral location some 1,600 km (994 miles) from the major centres, a poor resource basis, national tariff policies, and increasingly centralized transport and production systems. Inequality is evidenced by low wages and high unemployment. The Canadian government has attempted to address these issues with policies aimed to stimulate growth, including offering low-interest loans to farmers and others and tax subsidies for industrialists. Results have been evident, but the basic problems have not been solved.

It is not uncommon for Canadians to consider some 75 per cent of their country, the Canadian Shield and the north, to be frontiers producing primary products for export that primarily benefit the Canadian core area. In many respects, these resource extraction areas are a part of the less developed world. Others, notably Native populations, see these areas as a homeland. These are two very different images that are not easy to reconcile (see Bone 1992).

The European example

The European Community is a second example of a political unit that contains core and peripheral regions with the resultant regional inequities. As of 1994, the European Community had twelve member states and the process of economic integration is accelerating. Enlargement and integration of the EC has, however, occurred within a context of significant regional disparities. The three most recent additions, Greece (1981), Spain (1986), and Portugal (1986), have increased the number of peripheral regions with low incomes and high unemployment. There are, therefore, very strong arguments for a European regional policy; to date, the principal policy has been to provide funds to depressed regions. An effective regional policy is clearly needed.

Decaying industrial areas

In addition to the problems of peripheral regions, there are also many areas depressed as a result of outdated industrial infrastructures (see Box 12.10). A useful way to consider industrial change is to recognize four stages:

1. Infancy refers to initial primary activities and domestic manufacturing

2. Growth refers to the beginnings of a factory system
3. Maturity refers to full-scale development of manufacturing and related infrastructure
4. Old age refers to evidence of decline and inappropriate industrial activity

Today, the typical depressed region is on a coalfield, lacks diversification, and is overly reliant on heavy industry. Unemployment is high and out-migration is normal. Depressed regions are similar in principle to peripheral regions that were formerly dependent on a staple as both are unable to diversify when circumstances change.

In a country such as the United Kingdom, which has both peripheral and depressed areas, there has been considerable government intervention since the 1930s. Areas are variously designated as special development areas, development areas, and intermediate areas, designations that change according to political circumstances. Government intervention in the United Kingdom takes the form of development controls, financial incentives, and creation of industrial estates.

Planning in socialist countries

The clearest examples of regional development are those involving socialist-planned economies. By definition, socialist states do not rely on the capitalistic dynamic of private ownership and entrepreneurship. Rather, they involve the rationalization of industry and the implementation of comprehensive planning. As of 1994, the majority of socialist states were in the less developed world. Regional policies have included attempts to industrialize large cities (as in China during the Great Leap Forward (1958–9) when perhaps 20 million rural residents moved into cities) and attempts to limit large city growth. The seventh Five-Year Plan in China (1986–90) focused on uneven development and rapid urbanization, a direct reflection of the consequences of earlier plans that had largely failed to address the issues of income differences between urban and rural areas and of urban coastal domination of industry. The constants evident in Chinese planning are those of state direction, collective ownership of the means of production, and controls over population movements — all features rare in capitalist countries.

We end this chapter with an intriguing observation suggesting the inevitability of uneven development. Deng Xiaoping (ninety-one in 1995), the political leader of China, once noted that it was perfectly in order for some parts of China to become rich before other parts; in fact, this was the way it had to be (Freeberne 1993:420).

SUMMARY

Levels of production

Traditionally three levels are recognized: primary (extractive), secondary (manufacturing), and tertiary (service). A fourth level (quaternary) involves information transmission.

Levels of economic organization

Two levels are recognized. Households produce and reproduce; they are of decreasing importance. Firms make up the commercial sector and operate business in factories.

Factors related to location

Industrial location theory explains why factories are located where they are. Traditionally, relevant factors include transport; distance from raw material sources, energy supplies, and the market; availability of labour and capital; the nature of the industrial product; internal and external economies; entrepreneurial uncertainty; and governmental considerations.

Least cost theory

The typical location decision aims to maximize profits by minimizing costs or maximizing sales or both. Least cost theory was developed by Weber and published in 1909. It is a normative theory that identifies where industries ought to be located. Following a series of simplifying assumptions, Weber concluded that industries locate at least cost sites determined by transport costs, labour costs, and agglomeration/deglomeration benefits. Solving locational problems required defining such concepts as material index, locational figure, isodapane, critical isodapane, and isotim. Others, especially Hoover, have improved Weber's formulation of the cost of transport variable.

Market area analysis

This is a second major theoretical approach that centres on profit maximization rather than cost minimization. Lösch is the principal theorist.

Behavioural approaches

These approaches to the industrial location problem focus on the subjective views held by people and the tendency to be a satisficer rather than an optimizer. Contemporary industrial firms are, typically, complex decision-making entities. Behavioural issues are important, albeit more difficult to quantify or theorize.

Industrial Revolution

Prior to the Industrial Revolution, industrial activity was domestic, small in scale, and dispersed. A revolution in industrial activity and related landscape occurred between 1760 and 1860, beginning in Britain. Factories replaced domestic production, mechanization proceeded, and localized energy sources were used. All this change was related to the rise of capitalism, which emphasized individual initiative and profits. New industrial landscapes appeared on the coalfields of Britain and were associated with city growth, rural to urban migration, and a series of transport innovations. Inner city slum landscapes were common features in the new agglomerations on coalfields and in the traditional textile areas in northern England.

Energy

Coal was the key fuel during the Industrial Revolution, but was replaced by oil in the 1960s. Some countries, such as New Zealand, are dependent on hydro-generated electricity supplies, which may be disrupted by drought conditions.

Industry in the more developed world

Industry is responsible for a disproportionate share of manufacturing employment and output. Most of this activity continues to take place in a capitalist economic and social framework. The global manufacturing system is dominated by North America, western Europe, and the west Pacific. Multinational corporations are of increasing importance.

Canadian industry

The dominant characteristics of Canada's industrial geography include abundant natural resources, a well-developed extraction industry, a branch plant economy because of proximity to the United States, a dispersed national market, a dispersed primary sector, and highly concentrated secondary and tertiary sectors.

Newly industrializing countries

Japan was the first of the NICs and has been followed by other Asian countries (notably South Korea), some southern European countries, and Brazil and Mexico. Export processing zones using inexpensive and relatively unskilled labour have been established in some countries, for example, the *maquiladoras* in Mexico.

Industry in the less developed world

Industrial growth is highly desired by many less developed countries to bolster weak economies, provide employment, demonstrate economic independence, encourage urbanization, help create a better economic infrastructure, and reduce dependence on overseas markets for their primary products. But industrialization is difficult to achieve, not least because of the effects of colonial circumstances. Many countries have experienced their greatest industrial successes in import substitution. India has achieved much success since 1951, while China opened the doors to international trade and foreign investment and also reduced state control of industries during the 1980s.

The tourism industry

The largest in the world, this industry is generated by the more developed world, although favoured destinations are in both more and less developed worlds. Major spectacles are a distinctive form of tourist attraction. Many less developed countries take advantage of attractive climates and landscapes to cultivate a tourist industry, although their success, as in the case of Sri Lanka, can be affected by volatile political circumstances. Tourist areas may benefit economically, but there may also be negative cultural and environmental consequences. Some countries, such as Belize, are encouraging ecotourism.

Industrial restructuring

Technological changes in production, transaction, and circulation are currently prompting some major changes in the global economy and in national and regional industrial geographies.

Information exchange

Industrial location decisions are increasingly made without reference to transport costs and with reference to the relative importance of information exchange and labour costs. Efficient electronic transfer of reliable information allows firms to decentralize by seeking out areas of low labour cost. Where information exchange requires personal contact, centralized patterns of location result.

Service industries

As one component of economic restructuring and the transition from Fordism to post-Fordism, industrial societies are moving to a postindustrial service stage. Some industries are especially diverse and hence relevant location theories are varied; central place theory is the major explanatory set of concepts, while behavioural approaches seem especially relevant. New information technologies are undoubtedly affecting the location of services.

Some social issues

Industrial geography, like other branches of economic geography, has neglected social issues. Contemporary industrial geography is becoming more concerned with such issues as gender division of labour and, accordingly, is increasingly turning to various forms of social theory.

Spatial inequalities

Whatever scale is employed, we see spatial variations in economic circumstances and quality of life. Some see this as a result of different societies being at different stages along some sequential path of development. Another view sees certain areas as producers or former producers of staples while other areas are consumer oriented. A third view contends that the world comprises a series of cores and peripheries as a logical consequence of economic change. Correcting spatial inequalities is seen as a government responsibility. Governments in capitalist countries frequently implement regional policies to improve the status of peripheral regions and regions that are depressed as a result of being older industrial areas. The most obvious instances of planning the spatial economy take place in socialist countries. Economic restructuring continues to contribute to uneven development.

WRITINGS TO PERUSE

BALE, J. 1981. *The Location of Manufacturing Activity*. Edinburgh: Oliver and Boyd.

An easy-to-read basic text that is full of examples of industrial location issues; focuses on British examples.

BATHELT, H., and A. HECHT. 1990. 'Key Technology Industries in the Waterloo Region: Canada's Technology Triangle (CTT)'. *Canadian Geographer* 34:225–34.

A detailed analysis of thirty-three high-technology firms showing that the smaller firms are the most research intensive, that all the firms are strongly linked to local economic activities, and that the strongest locational factors are the availability of skilled labour and local residence or education.

BLACKBOURN, A., and R.G. PUTNAM. 1985. *The Industrial Geography of Canada*. London: Croom Helm.

A very readable account that includes discussions of each Canadian region and raises many of the issues that face manufacturing industries in the more developed world.

BRITTON, S., R. LE HERON, and E. PAWSON, eds. 1992. *Changing Places in New Zealand: A Geography of Restructuring*. Christchurch: New Zealand Geographical Society.

An account of the experiences of New Zealand following ideologically driven government decisions after 1984 to favour economic deregulation as a means of coping with changes in the international economy.

DICKEN, P. 1993. 'The Changing Organization of the Global Economy'. In *The Challenge for Geography: A Changing World, A Changing Discipline*, edited by R.J. Johnston. Oxford: Blackwell, 31–53.

An eminently readable account of contemporary trends in the global economy with details of some changes in industrial geography.

EDGINGTON, D.W. 1993. 'The New Wave: Patterns of Japanese Direct Foreign Investment in Canada During the 1980's'. *Canadian Geographer* 38:28–36.

An analysis that demonstrates that Japanese investment during the 1980s helped diversify and strengthen the economic importance of the Canadian heartland in southern Ontario.

ENSTE, H., and C. JAEGER, eds. 1989. *Information Society and Spatial Structure*. London: Belhaven Press.

A series of readings with a focus on new information technologies and their relation to service industries.

HAMILTON, F.E.I., ed. 1974. *Spatial Perspectives on Industrial Organization and Decision Making*. London: Wiley.

An explicit reaction to neoclassical orthodoxy; an early indication of behaviourists' concern with how industrial location decisions are actually made as opposed to how they ought to be made.

HAY, A.M. 1976. 'A Simple Location Theory for Mining Activity'. *Geography* 61:65–76.

A useful article that contributes to a largely neglected area and that is illustrated by an educational game.

JONES, K. 1989. 'Editorial'. *The Operational Geographer* 7, no. 2:2.

An introduction to a series of useful articles on services and marketing activities in Canada, especially Ontario.

KEEBLE, D. 1989. 'Core-Periphery Disparities, Recession and New Regional Dynamisms in the European Community'. *Geography* 74:1–11.

A detailed account of regional issues with clear discussions of deindustrialization, reindustrialization, and tertiarization.

MATHER, C. 1993. 'Flexible Technology in the Clothing Industry: Some Evidence from Vancouver'. *Canadian Geographer* 37:40–7.

This account of the impact of flexible technology on the Vancouver textile industry shows that some old technologies continue to be important and that the impacts on labour are largely negative.

NORCLIFFE, G. 1993. 'Regional Labour Market Adjustments in a Period of Structural Transformation: An Assessment of the Canadian Case'. *Canadian Geographer* 38:2–17.

A detailed account of changes associated with the transformation from mass production to flexible production that demonstrates that the Canadian experience is quite distinctive because the transformation has been mediated by a specific staple regime of accumulation.

SANDBERG, L.A. 1989. 'Geographers' Perception of Canada in the World Economic Order'. *Progress in Human Geography* 13:157–75.

An interesting account of recent and current perceptions of Canada in the global economic context that aids understanding of regional disparities.

SMITH, D.M. 1981. *Industrial Location Theory: An Economic Geographical Analysis*, 2nd ed. New York: Wiley.

A comprehensive account of all aspects of industrial location theory; one of the best books available on the general theme of the location of economic activity; integrates cost models with demand factors over time.

WATTS, H.D. 1987. *Industrial Geography*. London: Longman.

An excellent survey of the field with a distinctive focus on the human consequences of contemporary industrial change.

WEBBER, M.J. 1972. *Impact of Uncertainty on Location*. Cambridge, Mass: MIT Press.

An excellent if rather advanced statement concerning decision making in industrial and other location decisions.

13

The geography of movement

THROUGHOUT much of this book, we have considered the question of why things are located where they are. The facts of human geography are sometimes located near one another and at other times they are separated by great distances. There is nevertheless overwhelming evidence to suggest that the spatial location of geographic facts is not random. The distinguished British geographer, Wreford Watson (1955), described geography as a discipline in distance — an explicit acknowledgement of the way humans use space and distribute their activities as an adjustment to distance.

More generally, an American sociologist, Zipf (1949:6), asserted that, 'an individual's entire behavior is subject to the minimizing of effort', and applied this **principle of least effort** to human movement to suggest that location decisions are made to minimize the effort required to overcome the **friction of distance**. Distance considerations have indeed been central to most locational decisions, as humans typically choose to minimize movement (Box 13.1). For example, as we have seen in earlier chapters, the choice of crop grown is related to market location and the choice of residential location is related to workplace location.

In this chapter we are concerned with the movements of people, goods, and information that are necessary as a consequence of location decisions; it is helpful to glance back at the discussions of space, location, distance, and diffusion in Chapter 2 as you begin reading this chapter. It is also helpful to acknowledge at the outset that the friction of distance decreases with improvements in communications. Weberian least cost theory is of decreasing relevance as an explanation for industrial location decisions, as is Thünen theory for

agricultural location decisions, while the opening up of areas such as Bhutan (Box 10.11) and South Korea (Chapter 12) hints at the effects of our shrinking globe. Further, as suggested during our discussions of current economic restructuring, especially in Chapter 12, the importance of physical distance for many locational decisions is lessening because of the new information technologies: 'What matters ... is not so much the physical distance between cities, between firms, between factories and shopping malls, but the degree to which they are hooked in to the flow of ideas, capital, and people' (Knox 1994:57).

CONCEPTS OF DISTANCE AND SPACE

Different interpretations of distance and space are appropriate for different activities. Thus distance may be best measured in terms of, for example, time or money and, by extension, sets of distances, spaces, may be similarly measured. Our discussions of distance and space are traditionally couched in terms of Euclidean geometry, a tradition that has been described as a 'container view of space' (Harvey 1969:208), as space is seen as a framework in which geographic facts are located and geographic events occur. It is useful to appreciate, however, that there are many different types of geometries. Riemann geometry, for example, is more appropriate than Euclidean geometry when we are concerned with a spherical surface, such as the surface of the earth, because in Riemann space, the shortest distance between two locations is a curved (not a straight) line. The characteristics of a particular geometry can be described in terms of the path of minimum distance between two points

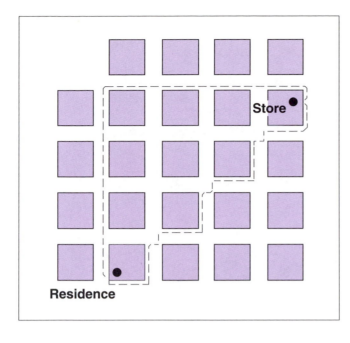

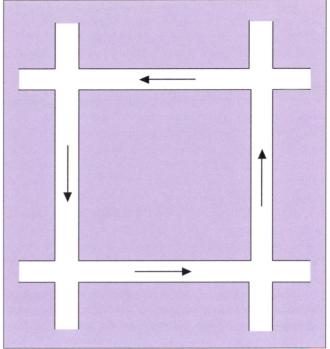

Figure 13.1 *(left)* A non-straight line, shortest distance route.

Figure 13.2 *(right)* A one-way system — impact on the distance travelled. To travel a block, the distance covered may be one block or three, depending on direction.

— geodesics. Thus, when we note that economic and time distance are different, we are saying that economic space has a different geodesic than time space.

This chapter will consider five different but related concepts of distance and space: physical, time, economic, cognitive, and social.

PHYSICAL DISTANCE

The spatial interval between points in space is the physical distance. It is often measured with reference to some standard system and, as such, is a precise measurement. Sometimes, however, physical distance is measured in less precise terms. In ancient India, the standard unit of distance measurement was the *yojana*, which was the distance that the royal mode of transportation, an elephant, could travel from dawn to dusk (Lowe and Moryadas 1975:22). Even in advanced technologies, imprecise measurements may be used; in North American cities it is common to measure distance by reference to the number of city blocks.

The shortest travel distance between points is often not a straight line. In a grid-pattern city, it is a series of differently oriented straight lines (Figure 13.1). In other instances the physical distance between two points may be related to the direction of travel, as in the case of vehicle movement in a one-way-street system (Figure 13.2).

Box 13.1: Ekistics — a science of human settlements

In the late 1950s, a Greek planner, Doxiadis, developed a science of human settlements, which he named *ekistics*. The central assertion was that all human settlements were a part of a complex system that included nature, humans, our built environment, and the networks that link humans. Within the context of this complex, all-encompassing system, Doxiadis further asserted that humans strive to:

1. Maximize contacts with all parts of the system.
2. Minimize the effort expended in achieving the first goal.
3. Optimize use of space so as to maintain contacts, but, at the same time, maintain an appropriate distance.

4. Optimize the quality of all contacts.
5. Optimize the integration of the first four goals.

Doxiadis concluded that settlements locate with these goals in mind. Thus, movement takes place along specific paths, which form the basic network of contacts. Settlement locations and transport routes are a direct consequence of the behavioural imperatives contained within the five goals. This stimulating contribution is a useful reminder of the important role played by movement and contacts in human life. The principal source for information on this topic is the journal *Ekistics*, which has been published since 1957.

TIME DISTANCE

For some movements, especially those of people rather than materials or products, time is important and thus the selected route may be the quickest rather than the shortest. Time distance is related to the mode of movement, traffic densities, and various regulations regarding movement. Figure 13.3 depicts isochrones, lines joining points of equal time distance from a single location, for the city of Edmonton. It is clear that travel time is not directly proportional to physical distance; if it were, the isochrones would be equally spaced concentric circles. Figure 13.4 represents an imaginative extension of these ideas showing Toronto in physical space and time space; the time space map has space stretching in the congested central area and shrinking in the outlying areas — a direct consequence of the greater time needed to travel a given distance on congested as opposed to freely flowing routes. Figure 13.4b is not, of course, fixed; the extent of stretching and shrinking varies according to the time of day and day of the week, with rush hour on a business day generating the most stretching. We might justifiably ask if there is 'a place for plastic space' (Forer 1978:230).

Converging locations

It is clear that travel times typically decrease with improvements in the technology of transport, a concept that Janelle (1969) called **time-space convergence**. We can conceive of locations converging on each other; these are locational changes in relative space — recall that in Chapter 2 we described relative space as subject to continuous change. It is possible to calculate a convergence rate as the average rate at which the time required to move from one location to another decreases over time. For example, Janelle (1968) calculated the convergence rate between London, England, and Edinburgh, Scotland, from 1776 to 1966 as 29.3 minutes per year. More generally, when the first global circumnavigation took place, it required three years (1519–22); today, we are able to circle the globe on regularly scheduled flights in less than two days, and information can be transmitted around the world almost instantaneously.

ECONOMIC DISTANCE

Movement from one location in space to another usually involves a monetary or some other economic cost; economic distance, then, can be viewed as the cost incurred in overcoming physical distance. As is the case in time distance, there is not necessarily a direct relationship between physical distance and other measures. If we consider the movement of commodities, for example, it is not uncommon to find that costs increase in a step-like fashion and that the cost curve is convex (see Figure 12.5). Similarly, in some cities taxi fares are determined not by physical distance but by the number of zones involved.

For many industries and other businesses, cost distance is of paramount importance. There is considerable logic to

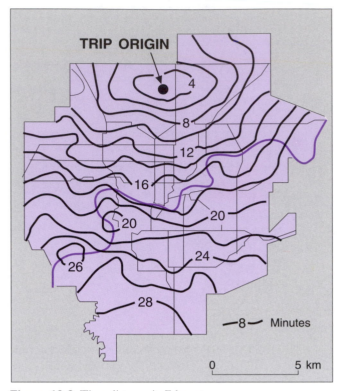

Figure 13.3 Time distance in Edmonton.
Source: Adapted from J.C. Muller, 'The Mapping of Travel Time in Alberta, Canada', *Canadian Geographer* 22 (1978):197–8.

the notion that economic activities should be mapped in economic space and not physical or container space.

COGNITIVE DISTANCE

Spatial cognition relates to our varying individual abilities to be precise in determining locations and distances. Each individual has a different interpretation of reality, that is, actual locations and actual physical distances. Understanding individual interpretations may be essential if we are to understand human spatial behaviour. Indeed, cognitive distance may be a better explanation for human movement than physical distance, but how is cognitive distance to be measured? Physical, time, and economic distance can all be measured precisely, but cognitive distance is quite different. Cognitive distances are determined by asking respondents about the relative proximity of a number of locations from a given location. The importance of cognitive distance to behaviour is then assessed by analysing the discrepancies between physical and cognitive distance. Also of interest are the reasons why cognitive distances differ from physical distances.

Most research has demonstrated that cognition affects our measures of distance and our larger spatial comprehension. Cognition itself is affected by the individual and social characteristics of humans. Most of us are able to be relatively precise about distances and locations in the area that we know well, but as physical distance increases, cognitive distance

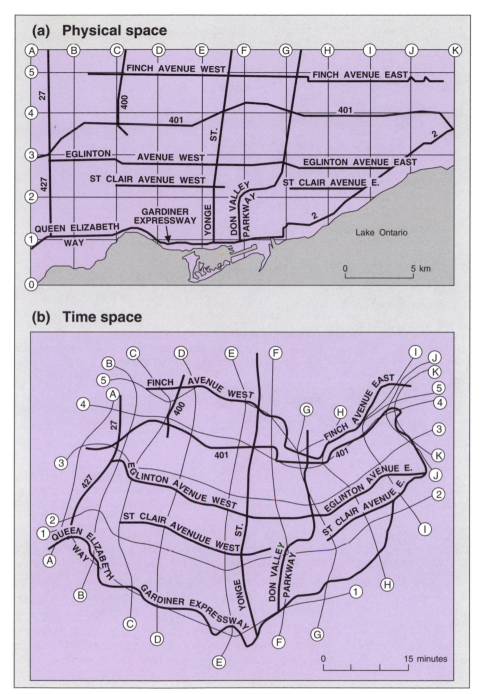

(a) Physical space

(b) Time space

Figure 13.4 Toronto in physical space and time space.
Source: G.O. Ewing and R. Wolfe, 'Surface Feature Interpolation on Two-Dimensional Time-Space Maps', *Environment and Planning* A9 (1977):430, 435.

becomes less precise. It is also clear that favoured locations are judged to be closer than locations that are disliked. In North America, for example, many people feel that the west coast is an attractive place to live and hence may judge it to be closer than it actually is in physical, time, or economic terms. Chapter 6 includes a brief account of mental maps (see Figure 6.2 and glossary).

Although it is clear that cognition affects behaviour in space, including movement, geographers remain uncertain of any specific cause and effect relationships. It remains to be determined whether or not cognitive distances and spaces can be used to predict spatial behaviour and spatial preferences.

SOCIAL DISTANCE

The concept of social distance is particularly complex with a variety of interpretations in the sociological literature. **Proxemics** is the study of social distances defined by individuals, as in circumstances of such social interactions as conversation, and is typically the interest of social psychologists, not geographers (Hall 1966). At the group scale of analysis, the correspondence between social and physical

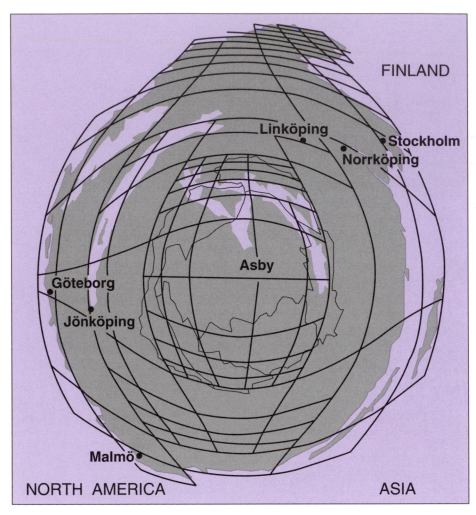

Figure 13.5 Logarithmic transformation of distances from Asby, Sweden.

distance has been studied with special reference to residential distributions. Following the lead of urban sociology, geographers have analysed the links between social status (often interpreted as class) and residence. Decisions about residence result in the creation of relatively distinct social spaces. One of the ways to measure social distance is by the amount of interaction between groups — for example, the degree of intermarriage. Social distance is most complete in cultures with rigid hierarchical **caste** systems.

TRANSFORMATIONS OF DISTANCE

In geographic studies the use of physical distance may often be inappropriate; the relevant measure may actually be time, economic, cognitive, or social. This is one reason (others are associated with the requirements of certain statistical procedures) why it may be useful to transform physical distance data. As an example, we turn to the classic work of the Swedish geographer, Hagerstrand (1967).

Recognizing that movement intensity decreased sharply with increasing distance from the focal point because potential movers were more unfamiliar with distant places than with nearer places, Hagerstrand mapped the logarithms of the physical distances (Figure 13.5). The focal point for Hagerstrand's analysis was the small town of Asby, which is located at the centre of a map that represents the information space of Asby residents. The cartographic result of this logarithmic transformation is that nearby distances appear exaggerated, while distances farther away seem to shrink. Compare this map to a more conventional map of Sweden to appreciate the displacement of the other locations identified. Many types of human interaction as well as human movement exhibit this negative relationship to distance. Figure 13.5 may well be the correct cognitive map. From a practical cartographic viewpoint, a logarithmic-transformed map creates room for most information where it is most needed, at the shorter distances.

A rather different type of transformation is the **topological map**: the correct spatial order of locations is maintained, but space itself is elastic as distance and direction can be altered. An excellent example of a topological transformation is a map of the London underground system (Figure 13.6). If we assume that most of us view distance

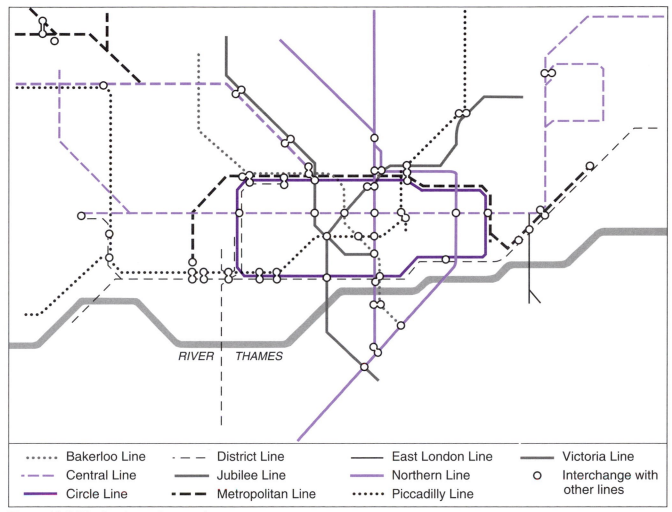

Figure 13.6 The London underground system.

..... Bakerloo Line	- - - District Line	—— East London Line	—— Victoria Line
- - Central Line	—— Jubilee Line	—— Northern Line	O Interchange with
—— Circle Line	••• Metropolitan Line	..... Piccadilly Line	other lines

not in precise quantitative terms but rather in relative terms (such as less than or greater than), it may well be the case that some topological transformations are highly relevant to the geography of movement — perhaps more relevant than the original untransformed physical distance data.

GRAVITY AND POTENTIAL MODELS

Many geographic studies of movement between locations have demonstrated a close link between the quantity of movement on the one hand and both the physical distance between locations and the sizes of locations on the other.

Gravity model
The gravity model describes the relationship between quantity of movement and distance as follows:

$$I_{ij} = \frac{M_i \times M_j}{D_{ij}^{\,b}}$$

where
I_{ij} is the interaction (movement) between locations i and j

M_i and M_j are the sizes (usually population sizes) of locations i and j
D_{ij} is the distance between locations i and j
b is a distance decay function

This formulation is a human equivalent to the gravity concept in physics. It proposes that the quantity of movement between two locations increases as their size increases; that is, the larger $M_i \times M_j$, the greater the movement. It also proposes that the quantity of movement between two locations decreases as D_{ij}, the distance between the locations, increases. Empirical evidence suggests that the impact of distance is not uniform and that D_{ij} needs to be raised to some power, thus $D_{ij}^{\,b}$, where b varies, depending upon circumstances, from as low as 0.5 to as high as 3.0. Note that a high value of b implies considerable distance friction, while a low value of b implies the reverse. As noted in the opening comments to this chapter, transportation and technological advances cause the value of b to decrease; that is, the frictional effects of distance decrease (Box 13.2).

This valuable model was introduced by the nineteenth-century social scientist, H.C. Carey. Carrothers quoted Carey as follows:

> Man, the molecule of society is the subject of Social Science ... The great law of *Molecular Gravitation* [is] the indispensable condition of the existence of the being known as man ... The greater the number collected in a given space the greater is the attractive space that is there exerted ... Gravitation is here, as everywhere, in the *direct* ratio of the mass, and the *inverse* one of distance (Carrothers 1956:94).

We have already encountered one of the first uses of this concept in our discussion of migration. In 1885, Ravenstein used a variation of the model when analysing migration flows between English cities. Other applications appeared from the 1920s onwards and included a 'Law of Retail Gravitation' (Reilly 1931).

The gravity model assumes that interaction — whether it is movement of people, movement of goods, telephone links, or whatever — takes place between locations. It also assumes that the greater the size, the greater the interaction, and it is thus treating all individuals as the same; we are well aware that this assumption is overly simplistic — the importance of individuals in terms of their ability to generate interaction is not equal.

A principal attraction of the gravity model is simplicity. The concept is attractive also because it has parallels in physical science and has provided good descriptions.

Potential model

The gravity model is concerned with relationships between pairs of points. The potential model is conceptually similar to the gravity model, but is concerned with the influence that many points have on one point; it can be expressed as follows:

$$V_i = \sum_{j=1}^{n} \frac{M_i}{D_{ij}^b}$$

where V_i = potential at location i
M_i = size of location i
D_{ij} = distance between locations i and j
b = a distance decay function

Thus, if a town, i, is located in a region with other towns, then town i has some potential for interaction with other towns. The potential at i is an aggregate measure of the influence of all other, $j = 1$ to n, places. As in the gravity model, distances may be raised to some power to accommodate the friction of distance. Again, this model has been extensively used by human geographers.

These applications of physics analogies to social phenomena are known as **social physics** and, as hinted above, we may not be too happy about treating humans as molecules and places as masses. Perhaps the value of these models results from the supposition that 'while it may not be possible to describe the actions and reactions of the individual

---

Box 13.2: The tyranny of distance

The Tyranny of Distance is the title of a book by the eminent Australian historian, Geoffrey Blainey, dealing with the history of Australia. Blainey (1968:2) began: 'in the eighteenth century the world was becoming one world but Australia was still a world of its own. It was untouched by Europe's customs and commerce. It was more isolated than the Himalayas or the heart of Siberia.' The twin ideas of distance and isolation frame the processes of Australian change in the European era. There are four principal 'distances' of importance:

1. Distance between Australia and Europe
2. Distance between Australia and nearer lands
3. Distance between Australian ports and the interior
4. Distance along the Australian coast

It is the first distance that is regarded as most important. The distance concept proposes new explanations as to why Britain sent settlers to Australia in 1788 and why the initial colony was at first weak but gradually succeeded. Similarly, the distance concept aids an understanding of the Chinese immigration in the 1850s and of later immigrations from southern Europe. Within Australia, the distances from coast to interior and along the coast are closely linked to economic and other change. The

first Australian staple, wool, was effectively able to overcome the distance from pasture land to port and the distance from port to Europe. Wool opened up much of southeastern Australia and was followed briefly by gold and then wheat. These products, along with technological innovations, combined to overcome the problems of distance.

Blainey is acutely aware that distance was not the only relevant variable — climate, resources, European ideas, and events were also important. Similarly, it is continually acknowledged that distance is not a simple concept; it is fluid, not static, and can be measured in many ways. In later writings Blainey has noted that distance is also a key to understanding the Aboriginal history of Australia.

Sensibly used, the distance concept is key to understanding the unfolding of life in many parts of the world. Blainey's work is eminently geographical.

Although the distance concept has not been explicitly applied by Canadian historians, it is central to a leading thesis in Canadian history — Laurentianism. This thesis, conceived by Harold Innis, places the Canadian historical experience in the context of staple exports and communications that, as in Australia, effectively overcame the tyranny of distance.

human in mathematical terms, it is quite conceivable that interactions of groups of people may be described this way' (Carrothers 1956:201). Certainly, not all human geographers are comfortable with such a supposition. You may be somewhat discouraged by the treatment of humans as objects!

Extensions of the basic models

The gravity and potential models are simplifications of complex phenomena and, quite appropriately, there have been numerous variations and extensions of the basic models. Two closely related approaches are noted. First, there is the approach of entropy-maximizing models introduced by Wilson (see Wilson and Bennett 1986). The term **entropy**, which originated in thermodynamics, refers to the disorganization in a system: entropy-maximizing models identify the most likely state of a system — that is, how many geographic facts are located in certain areas, given certain constraints. They have been extensively used for modelling spatial interaction. Second, there are location-allocation models that can be used to determine the optimal location — that is, the location that minimizes movement and other costs of facilities such as hospitals that satisfy consumers.

LINKS TO REGIONAL SCIENCE

In the late 1940s, the hybrid discipline of regional science developed in response to a dissatisfaction with the achievements of regional economic analysis. Largely the creation of the American economist, Walter Isard, it favoured the use of neoclassical economic theory and statistical techniques, and Isard wrote two major works (1956, 1960). This influential new discipline proved most attractive to the many human geographers who favoured spatial analysis as a dominant approach to their discipline. Regional science is important in the present context because it proved to be, quite understandably, one of the principal means by which social physics procedures were accepted and practised by geographers. The most notable of these geographers was Warntz (for example, 1957). Since about 1970, with the humanist and Marxist critiques of positivistic spatial analysis, geographers' interest in regional science and social physics has waned considerably. As has been evident in our discussions in chapters 10, 11, and 12, contemporary geographers are attracted to approaches such as political economy as opposed to neoclassical regional science.

DIFFUSION

Spatial patterns are rarely static and to understand present patterns, it is often helpful to analyse past patterns. Similarly, an understanding of the present may assist prediction of future circumstances. Much of this book has acknowledged this important point, at least implicitly, since the importance of time was first noted in the introduction.

Diffusion is best interpreted as the process of spread in geographic space and growth through time. Migration, the movement of people, can be regarded as a form of diffusion,

although, more typically, the concern is with a particular **innovation**, such as a new agricultural technique. Diffusion research has a rich heritage in geography; indeed, there are three specific approaches to this work — namely those associated with cultural geography, spatial analysis, and political economy.

DIFFUSION RESEARCH IN CULTURAL GEOGRAPHY

Until the 1960s, most diffusion research took place under the general rubric of historical and cultural geography as one component of attempts to understand cultural origins, cultural regions, and cultural landscapes. It is specifically associated with the landscape school initiated by Sauer. Typical studies focused on the diffusion of particular material landscape features such as housing types, agricultural fairs, covered bridges, place names, or grid-pattern towns. The usual approach was to identify an origin and then describe and map diffusion outwards from the origin; the work of Kniffen (for example, 1951) is exemplary in this context.

Various issues emerged from this research interest. In many instances, debate centred on the question of single or multiple invention and hence attention focused on the number and locations of hearth areas. The problem of agricultural origin source areas and subsequent diffusion is an example of this concern. The importance of ethnic or social characteristics was frequently acknowledged with different groups showing different degrees of receptivity to innovations. A detailed study of cigar tobacco production in the United States focused on the significance of ethnicity: 'In each of the tobacco producing districts, tobacco culture came to be identified with hard work, clever farming techniques, economic independence, and with an ethnic group' (Raitz 1973:305); in Wisconsin, for example, there is a very close relationship between people of Norwegian descent and tobacco production.

Diffusion continues to be a central concern in cultural geography as evident in the standard texts; indeed it can be regarded as one of five central themes (Jordan, Rowntree, and Domosh 1994).

DIFFUSION RESEARCH AND SPATIAL ANALYSIS

The character of diffusion research changed substantially following the work of Hagerstrand (1951, 1967), who pioneered the use of models and statistical procedures. This second emphasis was associated in North America with the rise of spatial analysis as a major approach in human geography after about 1955 and, although the specific emphases are quite different to those evident in the cultural geographic analysis of diffusion, the central concern is unchanged — to study diffusion as a process effecting changing human landscapes. The importance of chance factors was explicitly acknowledged by Hagerstrand in the use of a procedure known as Monte Carlo **simulation**, which allowed for the likelihood of any given acceptance of an innovation to be interpreted as a probability. Spatial analysis-oriented diffusion research also introduced a number of themes that we might call empirical regularities because they are consistently observable in geographic analyses. Four of these empirical regularities are now noted.

Neighbourhood effect
If some innovation, such as a new farming technology, is being diffused, then adoption might first occur close to the source of the innovation and, through time, occur at greater and greater distances from the source. This is the neighbourhood effect — a term that describes the situation where diffusion is distance biased. In general, the probability of new adoptions is higher for those who live near the existing adopters than it is for those who live farther away. The simplest description of this situation is, of course, a diagram of a series of concentric circles of decreasing intensity with increasing distance — similar to the ripple effect produced by throwing a pebble in a lake.

The neighbourhood effect occurs in those circumstances where an individual's behaviour is strongly conditioned by the local social environment. It is likely to be evident in small rural communities with limited mass media communication where the most important influences on behaviour are personal relationships — the *Gemeinschaft* society introduced in Chapter 11. The logarithmic transformation described in Figure 13.5 is descriptive of such circumstances. The neighbourhood effect is likely to be least evident in urban settings and in circumstances of improving transport and communication technologies that reduce distance friction.

Hierarchical effect
This second empirical regularity is an alternative to the first: the hierarchical effect is evident when larger centres adopt first and subsequent diffusion spreads not only spatially but also vertically down the urban hierarchy. Thus, the innovation jumps from town to town in a spatially selective fashion rather than spreading in a wave-like manner as proposed in the neighbourhood effect.

This effect is likely to be evident in circumstances where the receptive population is urban rather than rural and is increasingly evident as the technology of transport improves (Box 13.3).

Resistance
Reception of an innovation does not, of course, guarantee acceptance. Resistance to an innovation is a sociological attribute and the greater the resistance, the longer the time before adoption occurs. There is also a spatial pattern of resistance with urban dwellers typically being less conservative than their rural counterparts.

S-shaped curve
Probably the best-supported empirical regularity is the S-shaped curve. When the cumulative percentage of adopters of an innovation is plotted against time, an S-shaped curve typically results (see Figure 2.8). This curve describes a process that begins gradually, then picks up pace only to slow down again in the final stages.

A simple conceptual framework
A simple framework that conceptualizes the process of innovation diffusion is derived from a variety of sources, but especially from Rogers (1962).

There are four components of an innovation diffusion process: the innovation, a population, the communication process, and the adoption process. The *innovation* is assumed

Box 13.3: Cholera diffusion

In the nineteenth century, North America experienced three principal cholera pandemics in 1832, 1848, and 1866. Cholera is spread especially by water contaminated with human feces; it can also be carried by flies and food. Climatically, its spread is exacerbated by warm, dry weather.

Each of the three pandemics diffused differently as the urban and communications systems changed. The 1832 outbreak diffused in a neighbourhood fashion, while the 1866 outbreak diffused in a hierarchical fashion. The 1848 outbreak demonstrated aspects of both types of diffusion (Pyle 1969).

The 1832 pandemic occurred in a limited urban and communication system. Urban centres were few and small and movement was largely by water — coastal, river, and canal. Data on cholera spread show a neighbourhood process. After initial outbreaks near Montreal and New York City, the disease moved along distance-biased, not city-sized-biased, paths. The 1866 pandemic occurred in a very different urban and communication system. Eastern North America was by then well served by rail and an integrated urban system was evident. These changing circumstances allowed cholera to diffuse hierarchically. From New York City, it travelled along major rail routes to distant second order centres, as well as along a number of other minor paths. Areas close to the major cities often had cholera outbreaks at considerably later dates than did more distant large cities.

This example of two different processes of spread is a useful reminder of the importance of economic and other infrastructures to any diffusion process. In general, a neighbourhood spread is more likely in circumstances of limited technology and on a local scale, while a hierarchical spread is more likely in a developed technological context and in a larger area.

to be a desirable advance upon earlier procedures. For such an innovation to be diffused and either accepted or rejected, a *population* is necessary. At any given time during the process of diffusion, there are two classes of individuals in the population: adopters and non-adopters. There are three explanations for the presence of non-adopters: they have not heard of the innovation; they have heard and are in the process of decision making; they have heard and have made the decision to reject. A third category of individuals is often relevant, namely, change agents. These are people who are promoting the innovation, but who may be regarded as external to the system; for convenience they cannot be either adopters or non-adopters.

The *communication process* is how innovation or knowledge relating to the innovation is diffused. The process may operate effectively when a change agent communicates with a non-adopter or when an adopter communicates with a non-adopter. Communications may take two basic forms; pairwise (individual to individual) or mass communication (individual or organization to group). Adoption of an innovation is rarely immediate upon receipt of information, and the *adoption process* may be divided as follows: awareness — interest — evaluation — trial — adoption. Individuals vary in their innovativeness and may be classified as innovators, early majority, late majority, and laggards (Figure 13.7).

The pattern of innovation diffusion is assumed to be equivalent to the pattern of adopters, and the adoption process is assumed to be a learning process. The entire process, then, is envisaged as a stimulus response situation, with receipt of information as the stimulus and the decision to adopt or reject as the response. Adoption or learning may occur by reinforcement, insight, or both of these.

This framework recognizes processes of communication and adoption that allow an individual to be in one of three states:

No knowledge: This is the state prior to the successful completion of the communication process

Knowledge: This is the state after receipt of the information, but prior to the decision making

Adoption: This is the state after the decision to adopt is made

As is the case with any conceptual framework, this one is intended to be a simplified version of a complex reality and an aid to diffusion research.

DIFFUSION RESEARCH AND POLITICAL ECONOMY

For many researchers, the framework noted in the preceding section represents yet another spatial analytic dehumanization of human geography. Accordingly, a third approach to diffusion developed after about 1970. This work centres on the explicitly human consequences of diffusion, rather than on the mechanics of the diffusion process. There are various

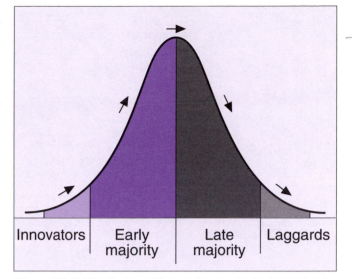

Figure 13.7 Distribution of innovativeness. As the innovation is adopted over time, an S-shaped curve results.

aspects to this third approach, but it is most clearly identified by associating it with the political economy introduced in Chapter 10.

Any diffusion process affects the use of resources; it does not merely add to existing technological capacity. Some innovations result in significant time savings and hence change the daily time budgets of individuals, while others are actually time demanding as in the case of the introduction of formal schooling. Analysing such innovations becomes an analysis of culture change.

Contemporary innovation diffusion research also considers the extent to which there is a spatial pattern of innovativeness related to a wide variety of social variables, such as relative wealth, level of education, gender, age, employment status, and physical ability. These considerations go far beyond the previous reference to such spatial issues as a possible rural/urban dichotomy or the cultural concern with ethnicity. Not surprisingly, there is also a focus on overarching social, economic, and political conditions over which most individuals exercise little or no control.

In a study of the diffusion of agricultural innovations in Kenya since the Second World War, Freeman (1985) discovered that the diffusion process was greatly affected by the preemption of valuable innovations by early adopters. This preemption, which had the effect of producing very different S-shaped curves (Figure 13.8), resulted from the presence of an entrenched élite who, as early adopters, saw the advantages of limiting adoptions by others. Being in some position of social authority, this élite could use political action, such as lobbying for legislation to prevent further spread of the innovation or limiting access to essential agricultural processing facilities. This situation arose in Kenya for three agricultural innovations, coffee, pyrethrum, and processed dairy products.

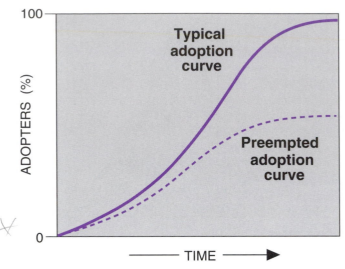

Figure 13.8 Effects of preemption on the adoption curve.

Although diffusion is an important form of movement, it is clearly relevant as much for the consequences of that movement as for the movement itself.

THE DIFFUSION OF DISEASE

Mapping and modelling disease diffusion have employed a variety of procedures prompted by diverse philosophical perspectives. The discussion of cholera diffusion in Box 13.3 is presented as a case-study using Hagerstrand-derived empirical regularities in a spatial analytic tradition to describe the diffusion process. There is also considerable research into the diffusion of infectious diseases that uses mathematical modelling techniques (for example, Cliff, Haggett, and Ord 1986; Cliff and Smallman-Raynor 1992).

In recent years, there has been considerable attention given to the diffusion of AIDS as evidenced by numerous research articles, but especially by the publication of a book (Gould 1993) that demonstrates the value of maps in understanding the spread of the disease and the publication of an AIDS atlas (Smallman-Raynor, Cliff, and Haggett 1992). There is little doubting the value of maps to emphasize the spread and growth of the disease. Together, these various contributions emphasize the complexity of the process by identifying cultural, social, behavioural, economic, political, and transportation factors that have all played roles in the past and present distribution patterns of AIDS.

Many academic disciplines have contributed to the understanding of AIDS; physical and human geographers are making a contribution using their interests and expertise in space, movement, and diffusion.

TRANSPORTATION

Humans have continually striven to facilitate movement across the surface of the earth, reducing distance friction and the costs of interaction. One result is the evolution of transport systems as complex components of the human landscape. A transport system comprises a series of specific modes such as road, rail, water, and air and is constructed to move people and materials, to allow for spatial interaction, and to link centres of supply and demand. Such systems are largely deficient throughout much of the less developed world, thus limiting economic activity such as commercial agriculture and mineral production.

In some countries, a transport system has been constructed to encourage national unity and open up new settlement regions. It is not a coincidence that this has occurred in the two largest countries in the world that have large areas of difficult environment — Canada and Russia. The transcontinental railway in Canada, completed in 1886, proved to be a crucial centripetal factor and encouraged settlement of the prairie region. The trans-Siberian railway between Moscow and Vladivostok has been operating since the beginning of this century.

TRANSPORT GEOGRAPHY: AN OVERVIEW

Geographic interest in transport is not as well developed as the interests in agriculture, settlement, and industry. Nineteenth-century geographers such as Ratzel and Hettner viewed transport routes as landscape features and as factors related to more general landscape changes, and the early twentieth-century French school of human geography regarded transport as a critical component of the geography of circulation. Despite these beginnings, transport geography remained largely unexplored until the 1950s when there were several studies of particular modes of transport. The most significant developments were associated with the rise of spatial analysis and involved quantitative and modelling studies, many of which had a planning orientation. Even today, transport geography continues to have a notable positivistic flavour, having been relatively unaffected by humanist, Marxist, and other social theoretic interests (Hoyle and Knowles 1992).

EVOLUTION OF TRANSPORT SYSTEMS

Three features characterize the evolution of transport systems: intensification, or the filling of space; diffusion, or spread across space; and articulation, or more efficient spatial structures. These three do not, however, change at a steady pace. Indeed, just as agriculture, settlement, and industry have experienced times of revolutionary (as opposed to evolutionary) change, so has transportation. This is because the basic causes of such revolutionary change apply to many aspects of the human geographic landscape. Thus, transport systems change in response to advances in technology and to various social and political factors. A consideration of the evolution of transport networks in Britain will clarify these generalizations.

Britain

The first organized transport system in Britain was created by the Romans between AD 100 and 400 for political and

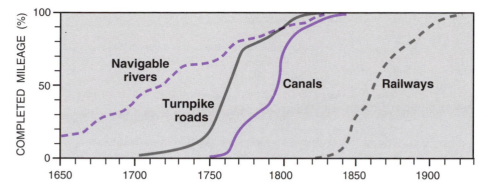

Figure 13.9 Diffusion of transport innovations in Britain, 1650–1930. Source: E. Pawson, *Transport and Economy*. (London: Academic Press, 1977):13.

military reasons. Unlike the earlier system, which permitted social interaction between settlement centres, the Roman system focused on London and allowed for rapid movement between the key Roman centres. Once the Romans withdrew from Britain, their system fell into disuse as it failed to serve the needs of the local population. Only in the seventeenth century did a new national system that reflected an emergent national interest appear in the form of carriers and coach services. By the early eighteenth century, two innovations diffused rapidly across Britain: turnpike roads (an organizational innovation) and navigable waterways (especially canals), a technological innovation. In the nineteenth century, the technological innovation of railways appeared and in the twentieth century, the road system responded to the new technology of automobiles. Finally, air transport added a new dimension to the overall transport system.

Turnpike roads transferred the cost of road maintenance from local residents to actual road users. No new technology was involved and thus, in principle, turnpike roads could have been set up at any time. Beginning about 1700, turnpikes diffused rapidly. During the early industrial period, *canals* played an important role in serving mines and ironworks. After 1790 they were the dominant transport mode and until about 1830 they were an important factor in industrial location decisions and related population distributions. Canals were tied to industry as they were suited for moving heavy raw materials, whereas turnpikes were more suited to the movement of people and information. Thus the two new systems were largely complementary rather than competitive.

Neither of these eighteenth-century developments compared to railways in long-term impact. *Railways* dominated British transport from about 1850 to about 1920. The first railway was completed in 1825. The early railway network was relatively local and related to industry, whereas after 1850 railways linked all urban centres. Railway construction in Britain was organized and financed by small groups of entrepreneurs with minimal government intervention. Elsewhere in Europe and overseas there was typically greater government involvement.

Each of these three innovations — turnpike roads, canals, and railways — displays the characteristic S-shaped growth curve (Figure 13.9): a slow start, rapid expansion, and a slow

completion period. This evolution of transport systems detailed for Britain is not uncharacteristic. Today, most parts of the developed world have an overall transport system that includes navigable waterways, railways, roads, and air traffic, a system that is continually changing in response to technology and demand.

Theories of transport system evolution
Several human geographers have studied transport system evolution from a theoretical perspective. For the less developed world, Taafe, Morrill, and Gould (1963) proposed four stages of development (Figure 13.10):

1. In this first stage there is a series of scattered ports along an entry zone (most likely a coastline); there is no real network.
2. A small number of the early ports prosper and develop networks inland.
3. In this third stage a number of feeder routes and lateral connections arise.
4. In the final stage a number of high-priority linkages emerge.

This generalized model usefully complements detailed empirical studies and has inspired a substantial body of subsequent work (Box 13.4). A second and rather different theoretical focus concerns the complex relationship between transport change and larger issues of economic change. Transport is both cause and effect and must be considered in analyses of, for example, agricultural, settlement, and industrial change (Box 13.5).

NETWORKS

Networks are specific locations linked together by routes to form some interconnected system. The locations that are linked by routes are usually referred to as nodes. *Nodes* are both sources and destinations of all types of movement. They may simply be individuals or aggregates of individuals in the form of specific businesses or settlements. In the latter two cases, nodes are spatially fixed and their distributions can be described as dense, sparse, clustered, dispersed, and so forth. Our consideration of central place theory was very much concerned with settlements as nodes and their relative

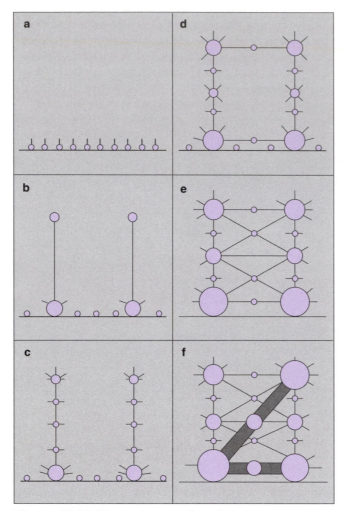

Figure 13.10 Transport network evolution.

times they are more flexible, as in the case of air and sea routes.

Geographers have long shown an interest in the characteristics of routes, especially their locations. Lines of communication are an essential factor in all economic geography: farmers and their products must travel to market and industrialists must move materials and products. In one sense, then, routes are the skeleton on which the flesh of human geography is built. 'Communications are themselves a part of the landscape' (Appleton 1962:xviii). As such, geographers ask their usual questions about where and why.

Graph theory

Networks are most easily analysed by using graph theory. Although they involve simplification, graphs make possible an objective analysis and geographers were introduced to graph theory by one of the pioneers of theoretical geography, William Garrison (1960). Lowe and Moryadas (1975:79) provide an example of the procedure. Figure 13.11a shows Martinique, a mountainous island with a circuitous road network; Figure 13.11b converts the network to a graph. The graph simplifies the network by excluding incidental characteristics, but retaining the essential topology. The elements of the graph are nodes and routes, and these elements can be used to generate a series of measures such as the number of nodes, routes, and subgraphs. Kansky (1963) introduced a large number of other graph theoretic measures such as the beta index, the number of linkages per node:

$$beta = n/r$$

where

n = number of nodes

r = number of routes

locations. In this chapter we are more concerned with the routes that facilitate movement between nodes. *Routes*, the channels along which interaction occurs, are sometimes spatially fixed, as in the case of roads and railways, and at other

A beta value greater than one suggests that there are alternative routings between some pairs of nodes. For the Martinique road network, beta = 1.49. Beta is generally higher

---

Box 13.4: Port system evolution

A study of the ports of Ghana concluded that the system developed from 'a highly unstable scattering of numerous primitive surf-ports with restricted hinterland links' to 'a gradual concentration of traffic and the emergence of a stable system with dependence on two efficient deep-water ports' (Hilling 1977:104).

In this Ghanaian example, transport development and port development are linked. First, there was a long period (1482–1900) of primitive surf-port operations that corresponded with the first scattered ports stage of the Taaffe, Morrill, and Gould (1963) model. Surf-ports were necessary because there were no good natural harbours. They usually involved a pier extending beyond the surf zone. Second, between about 1900

and 1928 the transport system penetrated inland and two ports, Sekondi and Accra, achieved dominance. Both of these locations acquired additional port facilities. Third, after 1928 the interior transport system developed a series of interconnections while deep-water ports were constructed at the two most suitable sites, Tema (near Accra) and Takoradi (near Sekondi). Finally, a high-priority route system developed and was focused on the two deep-water ports, with most of the smaller ports losing trade.

The consequence of these changes is a contemporary port and transport system that is cost-efficient. The case of Ghana does not parallel any of the idealized models.

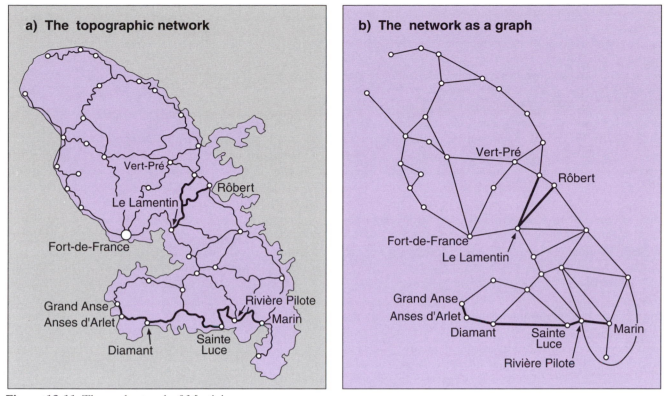

Figure 13.11 The road network of Martinique.
Source: J.C. Lowe and S. Moryadas, *The Geography of Movement* (Boston: Houghton Mifflin, 1975):80.

in more developed countries than in less developed ones. A second measure introduced by Kansky (1963) is the gamma index, the ratio between the actual and the possible number of routes (minimum 0, maximum 1): gamma = $r/3(n-2)$. For the Martinique road network, gamma equals 0.52. This measure is often expressed as a percentage; 0.52 is converted to 52 per cent, meaning that the network has a 52 per cent level of connectivity.

The beta and gamma measures indicate the utility of graph theory and are used extensively. Nevertheless, graph theory shares the drawbacks of all such simplified proce-

dures; information is lost (deliberately) and highly variable nodes and routes are treated as though they were identical. Indeed, many networks have a hierarchical structure with the nodes that are close together being especially well connected.

MODES OF TRANSPORT

In some cases the different modes, which may have evolved at different times and under very different human geographic circumstances, combine to create an integrated system; in other cases route duplication and a general lack of coordination may combine to create a relatively incoherent overall

---

Box 13.5: Railways and economic growth

Traditionally, transport has been viewed as a primary causal variable in economic growth and given a pre-eminent role in several key theories of economic change. Rostow (1960), for example, saw railways as the initiator of economic change.

In the 1960s, this traditional view was seriously questioned by a group of 'new' economic historians. Their analyses relied heavily on the concept of social savings, that is, the difference between actual national income and national income without railways. The usual procedure was to employ an important historical technique known as the counterfactual (recall the argument in Box 9.4). Thus, the following research issue was raised: in a given area at a given date, what would have happened

had there not been a railway system? Central to any such analysis is the argument that we can only understand the importance of railways to economic growth if we describe the area with and without railways. Needless to say, any such work is somewhat controversial in terms of the general approach adopted and the specific use of the approach. Regardless, the general conclusion is that railways were not the primary cause of economic growth — the social savings as defined above were minimal. From our geographic viewpoint, the related conclusion is that the human landscape would have been little different without railways in terms of general content; it would, of course, have involved many differences of detail.

system, although each specific mode may be a perfectly logical system.

Each of the various transport modes has some advantages and disadvantages. *Water* transportation is a particularly inexpensive means of moving people and goods over long distances because of the minimal resistance to movement and because waterways are often available for use at no charge. It is, however, a slow form of movement and may be circuitous. In the case of inland waterways, local relief can be a major obstacle. Moreover, in some areas, problems of freezing and storms may be very serious. In northern Manitoba, the port of Churchill on Hudson Bay is open only from mid-July to mid-November. Canals continue to be a critical component of many transportation systems. In Europe the Rhine-Main-Danube Canal was completed in 1992, allowing for a continuous water link between the North Sea and the Black Sea.

Land transportation includes railways and roads. *Railways* have the second lowest costs to water transportation and are suitable for moving bulk materials when waterways are not available. They are, however, often restricted by relief, or at least they are expensive to build and maintain. The most significant railway construction in recent years has taken place in Europe. The Channel tunnel between England and France, opened in 1994, comprises three tunnels — two for trains and one for services. In addition, several train tunnel projects are being completed in the Alpine region on the routes between Italy and Germany and in the Scandinavian region where a series of rail and road links (including tunnels and bridges) are to link Sweden and Norway through Denmark to Germany (Scanlink).

Roads accommodate a range of vehicles and are typically less expensive to build than are railways and hence are often favoured in regional planning and related schemes. Most of the internal movement within Europe is by road, with inland waterways second and railways third.

Transportation by *air* is most expensive, but is rapid and tends to be favoured for small-bulk, high-value products. Air transport is particularly subject to rapid technological change and especially influenced by political considerations. Even small countries tend to favour having a national airline for status and tourist development. Governments often fund and subsequently maintain airports.

TRADE

Trade can take place if the difference between the cost of production in one area and the market price in another will at least cover the cost of movement (Box 13.6). Domestic trade is identical in principle to international trade; the major difference is that the latter tends to be affected by human-created barriers. International trade is playing an increasingly important role in the world as it has grown much more rapidly in recent years than international production. This confirms that there is increasing international integration in the global economy.

FACTORS AFFECTING TRADE

The movement of goods from one location to another reflects spatial variations in resources, technology, and culture. The single most relevant variable related to trade is distance. As is the case with all movement, trade is highly responsive to the friction of distance, a friction that decreases as technology advances. Other relevant variables include:

- the specific resource bases of areas — needed materials are imported and surplus materials exported
- the size and quality of the labour force — thus a country short of labour but with plentiful resources is likely to produce and export raw materials
- the amount of capital in a country — higher capital prompts export of high-quality, high-value goods

These variables are closely linked to level of development and a process of unequal exchange is common between more and less developed countries. This involves the more developed countries exporting goods above their value and

---

Box 13.6: Explaining commodity flows — Ullman

The study of commodity flows has long been a topic of concern to geographers; the most influential writer in North America has been Edward Ullman. Ullman was especially interested in transportation and, indeed, defined geography as the study of spatial interaction (apparently after hearing a sociologist define sociology as social interaction). On the basis of detailed studies, Ullman (1956) concluded that commodity flows could be reasonably explained by reference to three variables.

First, movements result from complementarity, the situation where a surplus of a product in one area is matched by a deficit of that product in a second area. Flows then occur from the area of surplus to the area of deficit. This logic applies to all spatial scales. Second, complementarity results in product movement only if there is no other area between the two that is either an area of surplus or deficit. If such an intervening opportunity exists, it will either serve the deficit area or receive the surplus because it is closer to both areas than they are to each other. Third, commodity flows are also related to the ease of transferability. A product that is difficult or expensive to move will be less mobile.

Combined, these three factors provide a simplified explanation of commodity flows on a multitude of spatial scales and at various times.

the less developed countries exporting goods below their value. The less developed countries are inclined to excessive specialization in a small number of primary products and may be dependent on a staple. More developed countries may also specialize, but do so in manufactured goods, often as a result of the technologically induced division of labour between countries. The less developed countries are typically dependent on the more developed countries, as trading partners; this is a circumstance that is central to the world systems theory discussed in Chapter 9. The majority of world trade moves between developed countries, although the rise of the newly industrializing countries (NICs) discussed in Chapter 12 is affecting trade flows.

TRADE THEORIES

Trade theories differ from the location theories discussed so far as they have quite separate intellectual origins and evolutions. This is despite the fact that 'trade and location are as the two sides of the same coin' (Isard 1956:107). From our geographic viewpoint, trade theory literature excludes the central concept of location theory, namely distance; quite literally, trade theory typically considers countries as though there was no space between them. Integrating location theory and trade theory seems a difficult task, not least because trade theory is highly abstract.

Classical trade theory has origins in the work of such economists as Ricardo, a contemporary of Thünen, and focuses on the natural or historically created differences between each country specializing in production and related trade. Trade is explained by the *Law of Comparative Advantage*:

> In the absence of barriers to trade, a country will specialize in the production and export of those commodities that it can produce at a comparatively low cost and import those goods that can be produced at a comparatively lower cost in other countries.

Neoclassical trade theory extends classical theory by using geometric techniques and adding further concepts. The notion of external economies comes from this branch of theory and helps to explain why trade can occur between areas that are similar in most respects. External economies are the benefits that an industry enjoys as a result of the expansion of the industry to include related activities.

Modern trade theory extends the range of variables introduced in classical theory, namely labour, to include such variables as production, capital, land, and entrepreneurship. Modern theory thus compares countries in terms of their overall factor endowment. Such theory may focus on the growth of trade associated with the transition from feudalism to capitalism as the new producers seek ever larger markets.

REGULATION OF TRADE

The fact that trade occurs across international boundaries means that it can be regulated and most governments do indeed actively interfere with the import and export of goods. Thus, trade barriers may be set up to protect domestic production against relatively inexpensive imports. The most popular form of regulation involves the imposition of a **tariff.**

Two principal reactions to the widespread imposition of tariffs are the 1948 establishment of the General Agreement on Trade and Tariffs (GATT) and the tendency for countries, as they become more economically secure, to enter into tariff reduction (free trade) agreements with selected partners.

REGIONAL INTEGRATION

One of the most important topics in any discussion of trade is that of regional integration, which involves trade and location. Regional integration, the linking of separate states in at least some form of economic union, has become important since the Second World War. The basic logic behind integration is the opportunity for increased trade and the expansion of potential markets. There are five stages in the process of integration (Conkling and Yeates 1976:237).

The loosest form of integration is the *free trade area*. It involves a group of states that have agreed to remove artificial barriers, such as import and export duties, to allow movement and trade among themselves. Each state retains a separate policy on trade with other countries. Two recent examples of free trade areas are the North American Free Trade Area — NAFTA (comprising Canada, United States, and Mexico) and the ASEAN Free Trade Area (AFTA, comprising Brunei, Indonesia, Malaysia, Philippines, Singapore, and Thailand). (Note that ASEAN is an acronym for Association of Southeast Asian Nations.)

The second stage is the creation of a *customs union*. This type of international economic integration involves a common tariff barrier, in addition to the removal of barriers between member states. Thus there is free trade within the union and a common external tariff towards other states. In addition to the characteristics of the customs union, a *common market* allows capital and labour to flow freely among member states. There is also a common trade policy towards non-member states. At the present time, the best developed common market is the European Community, which established a single European market in 1992. The importance of transportation as a means of encouraging integration is evident in the several ambitious projects recently completed or underway (the Channel tunnel, transalpine routes, and Scanlink).

There are two levels of integration beyond the common market that groups of states may achieve. The *economic union* is a form of international economic integration that includes a common market and harmonization of certain economic policies, such as currency controls and tax policies. Finally, at the level of *economic integration* there are common social policies and some supranational authority with authority over member states. The extent to which individual states may be prepared to sacrifice national identity and independence to achieve the stage of economic integration remains to be seen.

It is clear that the European Community is the front runner in this ongoing process of regional integration and that states left outside of the major groupings may suffer economically (Cleary and Bedford 1993).

GLOBALIZATION AND REGIONAL INTEGRATION

It is common today to refer to the process of **globalization**, which includes the idea of a global economy that is being increasingly integrated as a result of technological changes (especially in communications and information flow) and organizational changes (especially the rise of multinational corporations). But clearly there is also a transformation from a large number of individual states competing on the world market to a smaller number of regional trading blocs. On the one hand, there are moves towards a new global economy, an integrated global market place; on the other hand, there is a trend towards regional integration and hence regional protectionism. The challenge for the human geographer to describe, explain, and anticipate these changes and tensions is a considerable one.

SUMMARY

A discipline in distance

The spatial location of geographic facts is not random, for humans typically choose to minimize the frictional effects of distance, to minimize movement.

Types of distance

Physical distance is the spatial interval between two points usually, but not necessarily, measured with reference to some standard system. Time distance is often more important than physical distance, especially where movement of people is concerned. The concepts of plastic space and time space convergence are valuable and relate to the fact that time distance between any two locations is a function of various changing circumstances. Economic distance is the cost incurred in overcoming physical distance; it is often of paramount importance for industries and other businesses. Cognitive distance refers to our individual perception of distance and is especially difficult to measure. Social distance is related particularly to residential distributions — the creation of distinctive social spaces.

Gravity and potential models

Movement between locations is related to the distance to be covered and the population size of the locations — a human equivalent to the gravity concept in physics. The potential model is similarly derived from Newtonian physics and expresses the potential that a particular location has for interaction with other locations. Both models are attractive simplifications of complex real life situations.

The diffusion process

Following the pioneering work of Hagerstrand, diffusion research in geography changed from a study of landscape features to a study of process. Four empirical regularities are the neighbourhood effect, the hierarchical effect, resistance, and the S-shaped curve. Most recently, diffusion studies have focused less on the process and more on the human consequences of diffusion, especially as it might cause spatial inequalities.

Transport system evolution

Transport systems are both cause and effect of other economic aspects of landscape. Major changes are associated with political expansion and technological advances. In the case of Britain, the most dramatic changes relate to the Industrial Revolution with the introduction of turnpike roads, navigable waterways, railways, and finally an improved road network. Theoretical approaches to transport system evolution focus on the identification of stages that are related to larger geographic issues.

Networks and modes

Transport networks comprise nodes and routes and are often analysed in a simplified form by using graph theory. Each of the principal modes of transport, water, land, and air, has some advantages and disadvantages.

Trade

Trade is clearly related to the distance between locations and yet the three principal branches of trade theory exclude any consideration of distance. Trade is also related to resource base, labour force, and capital. Today, much world trade is affected by various agreements between countries concerning economic integration.

Globalization

On the one hand, there is a new global economy emerging in response to the internationalization of capital, production, and services; on the other hand, there are several major examples of regional integration and associated regional protectionism.

WRITINGS TO PERUSE

CHAPMAN, K. 1979. *People, Pattern and Process: An Introduction to Human Geography*. London: Arnold.

An introductory textbook that is centred on spatial interaction and spatial diffusion and provides a detailed account of geography as a discipline in distance.

CHORLEY, R.J., and P. HAGGETT. 1970. *Network Analysis in Geography*. New York: St Martin's Press.

A detailed account of many of the topics raised in this chapter with a consistent spatial analytic focus.

FULLERTON, B. 1975. *The Development of British Transport Networks*. London: Oxford University Press.

A brief account of transport change from 1750 to 1970 covering waterways, railways, and roads.

GATRELL, A.C. 1983. *Distance and Space: A Geographical Perspective*. Oxford: Clarendon.

A clear account of distance and spatial concepts with many examples of appropriate analyses.

HAGERSTRAND, T. 1967. *Innovation Diffusion as a Spatial Process*. Translated by A. Pred. Chicago: University of Chicago Press.

This is a seminal work integrating the cultural and spatial analytic approaches to diffusion. The postscript by Pred is an excellent summary of diffusion research before 1967.

HANSON, S., ed. 1986. *The Geography of Urban Transportation*. New York: Guilford Press.

This edited volume includes a series of well-written articles on topics such as urban transport planning, transportation and energy, and environmental impacts of transportation.

HAY, A.M. 1973. *Transport for the Space Economy: A Geographical Study*. London: Macmillan.

Although somewhat dated, this is a useful study of transport issues, including the analysis of networks.

LEINBACH, T.R. 1983. 'Transport Evaluation in Rural Development: An Indonesian Case Study'. *Third World Planning Review* 5:23–35.

This article covers a range of issues from the role of transport in development to urban transport problems.

MORRILL, R.L., and J.M. DORMITZER. 1979. *The Spatial Order: An Introduction to Modern Geography*. North Scituate, Mass.: Duxbury.

An introductory textbook with a novel organization and a strong theoretical content that is especially useful for discussion of the whole range of movement-related topics.

PAWSON, E. 1977. *Transport and Economy: The Turnpike Roads of Eighteenth-Century Britain*. London: Academic Press.

A scholarly work that combines detailed analysis with a series of sound generalizations related to diffusion and economic change.

RIDDELL, J.B. 1985. 'Urban Bias in Underdevelopment: Appropriation from the Countryside in Post-Colonial Sierra Leone'. *Tijdschrifte voor Economische en Sociale Geografie* 76:374–83.

A detailed analysis that explains the non-adoption of intensive rice cultivation by peasants in Sierra Leone in terms of urban bias.

VANCE, JR, J.E. 1986. *Capturing the Horizon: The Historical Geography of Transportation*. New York: Harper and Row.

A major conceptual and factual work dealing with the growth of transport networks.

WALLACE, I. 1990. *The Global Economic System*. London: Unwin Hyman.

A clearly written statement of broad spatial economic issues.

Conclusion

The human geography of the future ... and the future of human geography

NOW IS A GOOD TIME to turn back to the introduction where three recurring themes are identified — humans and land, regional studies, and spatial analysis — and where it is stated that the goal of this textbook is to provide a basis for comprehending the human world as it is today and as it has evolved. Hopefully the three themes have been apparent to you throughout this textbook in our numerous discussions of landscapes, regions, and locations. It is also hoped that the goal has been achieved through your exposure to human geography, which is an academic discipline of tremendous breadth and diversity, one with especially close links to the natural sciences, particularly physical geography, and one with a healthy academic pedigree. It has been argued and repeatedly demonstrated that a human geographic perspective on our world is most valuable — later in this conclusion, we call this perspective a geographical imagination. The editor of a major world newspaper, *The Times*, recently wrote about contemporary school geography:

> It seemed a discipline that took a child out of the classroom, into the street and the park and said: "Look what's here! How did it come to be here? What is it made of?" On such empirical enquiry rests all learning. Geography offered such empiricism in the most comprehensible and immediate form. It also offered the basis for argument, for disagreement, for controversy, for the spirit of dialectic that makes for true understanding, not rote learning. (Jenkins 1992:193–2)

It is not possible to do justice to the title of this final chapter, yet we must at least attempt to comment on current changes in our subject matter (human behaviour as it affects the earth's surface) and on current changes in our methods of studying that subject matter. Accordingly, this chapter addresses four closely related questions:

- What are the key processes that have contributed to our present human landscapes?
- How are our human landscapes changing today?
- How is the discipline of human geography changing to address the many important environmental and social problems with which we are confronted?
- How do you now proceed to become a successful practising geographer?

UNDERSTANDING CONTEMPORARY HUMAN LANDSCAPES

THE INDUSTRIAL REVOLUTION

Earlier in this book, we have referred to the Industrial Revolution even when there was no direct interest in industry. This is because the Industrial Revolution involved much more than industrial change. We now summarize some of the principal changes that originated in Europe and that are associated with the period from about the mid-eighteenth to the mid-nineteenth century. Many of these changes were identified earlier in this book and collectively they have contributed much to our contemporary world. Indeed, the social sciences are themselves a product of these changes that emerged in the mid- to late nineteenth century.

The Industrial Revolution was a series of technical changes that involved the large-scale use of new energy sources, especially coal, through inanimate converters. The steam engine was the principal invention and new machines were able to use this energy source to enhance industrial output. The introduction of new machines involved the construction of factories close to the sources of energy, necessary raw materials, or a transport route. The resultant concentrations of factories generated urban growth, including the emergence of distinct industrial landscapes and working-class residential areas, as workers needed to reside near their places of employment. Industrial cities replaced the earlier preindustrial forms of settlement. New transport links for moving raw materials to the factories and finished products to the markets were essential and (in England) initially took the form of improved roads, then canals, and, most critically, railways. Agriculture became increasingly mechanized and there was a demand for new products, especially wool and cotton, from overseas areas. Overall, the percentage of the total labour force that was engaged in agriculture declined rapidly.

Technical changes were accompanied by significant changes in demographic characteristics. As described in the demographic transition model, the first phase of the Industrial Revolution involved a dramatic reduction in death rates, followed by reductions in birth rates. The interval between these two drops was a time of rapid population growth. The world population was estimated at 500 million in 1650 and at 1,600 million by 1900. Industrialization and associated population increase contributed to new areas of high population density, rural to urban migration, rapid urban growth, and considerable movement overseas to the new colonial territories.

A major social and economic change was the collapse of feudal societies and the rise of capitalism. Labour was transformed into a commodity to be sold and there was an associated separation of the producer from the means of production. Politically, this period witnessed the rise of nationalism and the emergence of the modern nation state, as well as the expansion of several countries around the globe to create empires. Although these empires were short-lived, they contributed enormously to the contemporary political map and also to the division into more and less developed worlds. The nation state, as we have already seen, remains as one of the most potent forces in the contemporary world.

The rise of modernism is linked to the Industrial Revolution with the emphasis on the practicality and desirability of scientific knowledge. The transition to an industrial way of life was also a transition from tradition to modernity, from *Gemeinschaft* to *Gesellschaft*. The world in which we now live is very much a product of the numerous changes that occurred during the period from about the mid-eighteenth to the mid-nineteenth century, notwithstanding the fact that new postindustrial forces have emerged since about 1970.

GLOBALIZATION, REGIONAL INTEGRATION, TERRITORIES, AND REGION STATES

'There is widespread acceptance that something is happening, or indeed has already happened, in the way the global economy is organized. However, precisely what these changes are, and how they have been caused, is a matter for considerable debate' (Dicken 1993:31). The four terms included in the heading above are an indication of our current uncertainty about trends in the global economy, political organization, and geographic identities.

Globalization

Globalization refers to the idea that the world is becoming increasingly homogenized economically and culturally. While there is much truth to this suggestion, it is clearly not the sole change occurring and possibly not the most important. The globalization that is occurring is related primarily to our increasing abilities to transmit information almost instantaneously and to the growing significance of multinational corporations that 'slice their way at will through national boundaries and render all state policy-makers redundant' (Dicken 1993:48). One way to identify economic globalization is to focus on the amount and location of international direct investment, which has grown even more rapidly than international trade since the Second World War. Most of this investment is both by and in the more developed countries.

The term 'globalization' only came into widespread use in the 1980s and was first evident in the activities of multinationals striving to increase world sales of their products. Culturally, globalization is criticized by those who do not favour further homogenization (see Box 2.5).

Regional integration

This is the process by which two or more states group together, initially to gain some economic advantage over other areas, although potentially the integration will assume a political character. There are three principal regional blocs today — North America, the European Community, and east and southeast Asia (centred on Japan). The extent to which these regional units will be able to replace the nation state as major actors remains to be seen.

Territories

Despite the evidence for globalization and the undeniable significance of trading blocs, a third trend is playing an important role in our contemporary world. There is compelling evidence for the increasing significance of territories, parts of the world to which specific groups of people lay claim. This has been a common theme in this textbook especially in Chapter 7 during discussions of linguistic, religious, and ethnic groups, and in Chapter 9 during the account of the stability of states. Groups of people with some common identity are emphasizing their distinctiveness, often at the expense of the state or states in which they are included. This process may lead to the creation of new, possibly more culturally meaningful, states. The division of the former Czechoslovakia into two states, the Czech Republic and Slovakia, occurred peacefully, but the collapse of the former Yugoslavia involved much conflict.

Region states

Each of the three trends noted so far represents a weakening of the role played by the nation state; indeed, Ohmae (1993:78) argues that the nation state 'has become an unnatural, even dysfunctional unit for organizing human activity and managing economic endeavor in a borderless world' and proposes that on 'the global economic map the lines that now matter are those defining what may be called "region states".' Region states are regarded as natural economic areas. They may be parts of states, such as north Italy, or they may cross state boundaries, such as Hong Kong and south China. Their principal links are with the global economy and not with their state economies. If this argument is applied in a Canadian context, it might be suggested that British Columbia is increasingly linked to Asia and the western United States, while elsewhere in Canada there are emerging north-south relationships with areas of the United States. The implications for the rise of region states for national unity in Canada, as elsewhere, are enormous.

CHANGING HUMAN LANDSCAPES

It is always exciting to be a human geographer. When reading the first chapters of this book, you might have thought that human geographers today look back nostalgically to the golden days when every journey meant new facts to be discovered, new facts to be catalogued and understood. Certainly, those days of exploration and discovery are no longer with us, but, as has frequently been noted, this does not mean that there are no new facts. As discussed in the preceding section, we are constantly changing our landscapes in a myriad of ways — cultural, social, political, and economic. There may not be a wealth of new facts constantly over the horizon, but we are changing those facts within the horizon. Indeed, we are changing such facts at an ever-increasing pace because there are so many of us to cause change, because technology is always improving, and because we live in a shrinking world in terms of information and other movement. Four examples of changing human landscapes will clarify these generalizations.

DISAPPEARING PEOPLES

Tribal peoples utilize deceptively simple technologies to live in close harmony with their physical environment. All too often, however, they are in the way of what most of us consider to be progress. Such 'progress' has resulted in the destruction of numerous tribal peoples whom the advocates of 'progress' perceive as backward and ignorant. Such perceptions stem from prejudice and cultural arrogance. Although we are becoming increasingly aware of the negative

TABLE CONCLU.1: DISAPPEARING PEOPLES, 1989

Group	Location	Number remaining 1989	Principal threats to survival
Inuit	Alaska/Canada	100,000	Loss of identity
Iroquois	St Lawrence region	22,600	Loss of identity
Hopi	Arizona	10,000	Forced relocation
Hawaiians	Hawaii	9,000	Loss of identity, tourism
Lacandon	Mexico	300	Deforestation, settlement
Kuna	Panama/Colombia	50,000	Deforestation, tourism
Kogi	Colombia	2,000	Settlement, tourism
Yanomami	Venezuela/Brazil	21,000	Mining, epidemics
Emerillion	French Guyana	200	Loss of identity
Waorani	Equador	80	Oil exploration, settlement
Nambikwara	Brazil	300	Settlement, epidemics
Caraja	Brazil	800	Industrialization
Jurana	Brazil	50	Physical extinction
Caingua	Paraguay/Argentina	3,000	Destruction of environment
Taureg	Sahara desert region	300,000	Loss of identity
Pygmy	Zaire/Congo/Gabon	200,000	Deforestation, settlement
Bushmen	Kalahari desert region	60,000	Loss of identity
Sammi	Norway/Sweden/Finland/USSR	58,000	Loss of identity
Gond	India	3,000,000	Industrial activities
Vedda	Sri Lanka	2,000	Destruction of environment
Onge	Andaman Islands (India)	100	Physical extinction
Semang	Malaysia	2,500	Deforestation, forced conversion to Islam
Penan	Sarawak region	10,000	Logging, lack of land rights
Akha	China/Burma	50,000	Lack of land rights
Ainu	Japan	3,500	Loss of identity, tourism
Tasaday	Mindanao (Philippines)	30	Physical extinction
Papuans	West Papua (Indonesia)	1,000,000	Settlement, oil exploration
Aborigines	Australia	250,000	Loss of identity
Maoris	North Island (New Zealand)	300,000	Loss of identity

Source: Adapted from R. Handbury-Tenison and C. Shankey, 'An Uphill Struggle for Survival', *Geographical Magazine* 61 (1989):29–33.

consequences of such 'progress', little is being done to protect the few tribal groups that remain. It is estimated that up to ten such groups disappear each year. In 1969, a group of concerned people formed Survival International to help tribal peoples protect their rights. The group estimated that there remained twenty-nine such peoples in existence in 1989 (Handbury-Tenison and Shankey 1989). (See Table Conclu.1.)

URBAN GROWTH

With up to 60,000 persons per square kilometre, Tokyo is the most crowded city in the world. There is a constant demand for change and more space (Greenlee 1988). Initially this demand assumed the form of urban sprawl, but most recently it has involved internal reorganization. The population of 27 million plus is now making ingenious use of the urban area by building upwards, downwards, and into the sea. Building upwards is a relatively recent alternative for a city prone to earth tremors. In addition to skyscrapers, there are elevated highways and railways and even multistorey golf driving ranges. Building downwards is popular, and some buildings use at least four levels below ground for retail and office purposes. Perhaps most dramatically, Tokyo is reclaiming land from the sea to permit more expansion. Some recent proposals include the idea of building on a platform on Tokyo Bay and perhaps even constructing huge floating cities. Tokyo is an example of an ever-changing urban area with some unusual innovations.

A CHANGING WORLD

If we wish to make sense of change, we must study human geography. The media are constantly reporting items of geographic interest. The Gulf War in early 1991, for example, can be understood by reference to state evolution in the area, the spatial distribution of ethnic groups, the importance of oil in the developed world, and the dependency relations between more and less developed states. Further, the Gulf War included and caused mass population movements, especially of Kurdish peoples, and prompted severe environmental degradation.

At the time of writing in mid-1994, there is continuing conflict in the former Yugoslavia, which we can understand better (as we saw in Chapter 9, Box 9.6) by recognizing the complex cultural and political geography of this region comprising numerous linguistic, religious, and ethnic groups and its history of domination by other empires. Also at the time of writing, the small African state of Rwanda is experiencing ethnic conflict of tragic proportions. Rwanda is a part of one of the highest population density regions in Africa — the highland region surrounding Lake Victoria. Rwanda and neighbouring Burundi were a part of German East Africa and were placed under Belgian administration at the end of the First World War. Rwanda and Burundi became independent as separate states in 1962. About 90 per cent of the population in Rwanda are Hutu, Bantu-speaking herders

and farmers (82 per cent in Burundi). These people were formerly subordinate to the minority Tutsi, who are pastoralists and probably of Hamitic descent. The 1994 violence between these two groups in Rwanda is yet another instance of the conflict that erupted in Rwanda and Burundi on previous occasions.

New geographic facts, new human landscapes, are always coming into being. Indeed, there are still new worlds to be discovered.

CHANGING PEOPLE IN A CHANGING WORLD

Our contemporary world is characterized by the collapse of many spatial barriers. Such collapses have a multitude of impacts. The distinguished human geographer, David Harvey, wrote:

> The more global interrelations become, the more internationalized our dinner ingredients and our money flows, and the more spatial barriers disintegrate, so more rather than less of the world's population clings to place and neighborhood or to nation, region, ethnic grouping, or religious belief as specific marks of identity ... Who are we and to what space/place do we belong? Am I a citizen of the world, the nation, the locality? Not for the first time in capitalist history ... the diminution of spatial barriers has provoked an increasing sense of nationalism and localism, and excessive geopolitical rivalries and tensions, precisely because of the reduction in the power of spatial barriers to separate and defend against others (Harvey 1990:427).

Our uniqueness as human beings resides in our biological attributes and ability to adapt culturally. We humans have managed to avoid extinction because of our culture, which is essentially our ability to develop ideas from experience and to act on the basis of ideas. Once in place, however, culture can actually be an impediment to change as most people are comfortable in adhering to established values, hence our desire to identify with place in a world where, for many of us, places are changing at a bewildering pace — constantly being created and destroyed — recall the discussion in Chapter 8 of the importance of place. In his book *A Question of Place*, Johnston (1991) argued the need for geography to study wholes, places that are milieux, within which ways of life are constructed and reconstructed.

CHANGING HUMAN GEOGRAPHY

Chapters 1 and 2 of this book described the evolution of human geography and identified key philosophies, concepts, and techniques of analysis. One purpose of these accounts was to demonstrate that our discipline is a changing one, largely in response to the changing needs of society. Human geography, in common with other academic disciplines, exists to serve society. There is, then, every reason to believe that the discipline described in this book will be a different discipline tomorrow. Attempting to anticipate the details of such change is hazardous and possibly hardly worthwhile, but we may at least identify what seem to be the most pressing societal demands on human geography today.

SPACE AND PLACE

First, there seems little doubt that there is an ever-increasing need for an academic discipline that focuses on the importance of space and place. Space is important because an understanding of location, distances between locations, and movements between locations is crucial if we are to make sense of the world. Place is important because understanding the value that people assign to locations is crucial if we are to comprehend their attitudes and behaviours.

A SYNTHESIZING DISCIPLINE

Second, the discipline of geography certainly provides our best framework for investigating human impacts on global and local environments. Most ecological research benefits immeasurably from some linking of human and physical geography. Many disciplines may consider the issue of the consequences of global deforestation, but only geographic responses have the benefit of a firm background in both physical and human sciences.

HANDLING DATA

Third, human and physical geography are both serving society today through the development and application of remote sensing and electronic data processing. In an increasingly complex world with ever larger data sets, remote sensing allows us to acquire data rapidly and repeatedly while electronic data processing allows the conversion of data into maps and facilitates data analysis. Perhaps the best-known recent development is that of geographic information systems, integrated computer systems for the input, storage, analysis, and output of spatial data.

UNDERSTANDING AND SOLVING PROBLEMS

Fourth, human geography facilitates an understanding of people and place by explaining how different peoples see and organize space differently. Hence, we are acutely aware that there are no simple or universal solutions to problems. Effecting change can only follow a full analysis of spatial variety. Solutions to a problem in one area may not be solutions to similar problems in another area. Overall, human geographers are sensitive to spatial variations and the significance of place. This sensitivity is evident in different ways in all of our major philosophical traditions — empiricist, positivist, humanist, and Marxist. Understanding space and place is important in itself and because it allows us to propose solutions to social and environmental problems.

VALUES

In addition to the above four contributions that society expects from human geography, there is a key question: what values direct human geographic research? The general answer is that values are socially specific, derived from the concepts that legitimate the society in which we live. Remember from Chapter 2 that the philosophy of positivism was purported to be value free, whereas Marxist and humanistic geography contributed the important insight that values are socially constructed and therefore subject to change. This general answer does not, of course, solve such specific questions as the relative merits of environmental preservation on the one hand and economic gain on the other hand (Buttimer 1974).

BEING A HUMAN GEOGRAPHER: WHERE WE BEGAN

The advantages of a training in human geography are by now self-evident. The advantages of being a human geographer follow logically. A training in human geography allows us to understand our world; being a human geographer allows us to put that training into practice. Most students of human geography do not formally become human geographers; rather, they seek employment in, for example, government or business, but I strongly suggest that those who do not formally become human geographers never actually lose the benefits of their human geographic training. We might usefully summarize the benefits by referring to the concept of a geographical imagination.

GEOGRAPHICAL IMAGINATION

What is a geographical imagination? Simply put, a geographical imagination is an appreciation of the relevance of space and place to all aspects of human endeavour. Peattie (1940:33) recognized this more than fifty years ago when he wrote: 'Geography ... explains ways of living in all their myriad diversity.' More recently, Gregory (1994) presented a detailed commentary on geographical imaginations as sensitivity to space and place in any attempt to understand ways of life.

Regardless of philosophical preference, human geographers characteristically demonstrate possession of a geographical imagination. From an essentially *positivistic* viewpoint, Morrill (1987:535) observed that, 'If there is not a convincing theory of why and how humans create places and imbue them with meaning, then it is time to develop that theory.' From an essentially *humanistic* viewpoint, Prince (1961:231) asserted, 'Good geographical description demands not only respect for truth, but also inspiration and direction from a creative imagination ... it is the imagination that gives [the facts] meaning and purpose through the exercise of judgement and insight.' Finally, from a *Marxist* viewpoint, Harvey (1973:24) observed, 'This imagination enables the individual to recognize the role of space and place in his own biography, to relate to the spaces he sees around him, and to recognize how transactions between individuals and between organizations are affected by the space that separates them.'

Three different geographers, three different theoretical perspectives, one fundamental conclusion. The concept of geographical imagination is a useful way to conclude this introductory human geography textbook; we hope that as human geographers, you have gained an appreciation of and a sensitivity to the role played by space and place in understanding the world we call home.

SUMMARY

Understanding human landscapes

Contemporary human landscapes are a product of past circumstances, especially of the complex period of change that we call the Industrial Revolution. More recently, a series of often contradictory processes, namely globalization, regional integration, cultural identification with territory, and the emergence of region states, are all occurring and contributing to a complex contemporary human geography.

Changing human landscapes

Although the days when new worlds were being discovered are over, we have not entered a period of stable human geographies. As old facts disappear and new facts appear, our world is constantly changing. Four examples of change involve disappearing peoples, urban growth, major political and other events, and links to place.

Changing human geography

Human geography is a changing academic discipline in response to the changing requirements of the larger society of which it is a part. Space and place are constantly being reinforced as our key concepts; our links with physical geography are a major strength; new technical approaches offer ever-improving ways to handle data, and our understanding of issues is leading to more and better solutions.

Being a human geographer

Acquire and use your geographical imagination.

WRITINGS TO PERUSE

Each of the following six items is an excellent, if often somewhat idiosyncratic, commentary on human geography. All are by distinguished geographers and all are well written and highly entertaining. Read and enjoy!

EYLES, J., ed. 1988. *Research in Human Geography: Introduction and Investigations*. Oxford: Blackwell.

GOULD, P. 1985. *The Geographer at Work*. London: Routledge and Kegan Paul.

HAGGETT, P. 1990. *The Geographer's Art*. Oxford: Blackwell.

JOHNSTON, R.J. 1985. 'To the Ends of the Earth'. In *The Future of Geography*, edited by R.J. Johnston. London: Methuen, 3–26.

STODDART, D.R. 1986. *On Geography*. Oxford: Blackwell.

———. 1987. 'To Claim the High Ground: Geography for the End of the Century'. *Transactions, Institute of British Geographers* n.s. 12:327–36.

Because there is such a close relationship between geography and many world events — both human and physical — it is most useful to keep abreast of our changing world. Two excellent factual sources, both published monthly, are available in major reference libraries: *Keesing's Record of World Events* and *Current History*. Finally, one of the most useful ways to explore any aspect of geography in greater detail is to refer to the following publication:

GAILE, G.L., and C.J. WILLMOTT, eds. 1989. *Geography in America*. Toronto: Merrill.

Glossary

accessibility
A variable quality of a location expressing the ease with which it may be reached from other locations.

acculturation
The process by which an ethnic group is absorbed into a larger society but succeeds in retaining aspects of its distinct identity.

acid rain
The deposition on the earth's surface of sulphuric and nitric acids formed in the atmosphere as a result of fossil fuel and biomass burning; causes significant damage to vegetation and built environments.

adaptation
The adjustment that humans make to a particular set of circumstances; changes in behaviour to reduce conflict with the environment.

agglomeration
The spatial grouping of humans or human activities to minimize the distances between them.

agricultural revolution
The slow change in the human way of life, beginning about 12,000 years ago, from foraging to food production by means of plant and animal domestication.

alienation
The circumstance in which a person is indifferent to or estranged from nature or the means of production.

anarchism
A political philosophy that rejects the state and argues that social order is possible without a state.

animism
A group of religions that incorporate beliefs about the spirituality of inanimate objects.

apartheid
The policy of spatially separating four groups of people, as defined by the authorities, in South Africa between 1948 and 1994.

areal differentiation
A term favoured by Hartshorne; a synonym for regional geography.

artefacts
Those elements of culture directly concerned with matters of livelihood.

assimilation
The process by which an ethnic group loses its identity by being absorbed into a larger society.

authority
The right, usually by mutual recognition, to require and receive submission by others.

biomass
The mass of biological material present in an area, including growing and dead plant material.

capitalism
A social and economic system for the production of goods and services that is based on private enterprise, that involves a separation of the producer from the means of production, and that allows rela-tively few individuals to exploit resources.

carrying capacity
The maximum population that can be supported by a given set of resources.

cartography
The conception, production, dissemination, and study of maps.

caste
An exclusive social division based solely on birth to which an individual belongs for life and that limits interaction with members of other castes.

ceiling rent
The maximum rent a potential land user can be charged for use of a given piece of land.

census

The process of collecting, compiling, and publishing demographic and other data relating to all individuals in a given country at a particular time.

centrifugal forces

In political geography, forces that make it difficult to bind an area together as an effective state; in urban geography, forces that favour decentralization of urban land uses.

centripetal forces

In political geography, forces that pull an area together as one unit to create a relatively stable state; in urban geography, forces that favour concentration of urban land uses in a central area.

chain migration

A process of movement from one location to another through time that is sustained by social links of kinship or friendship; often results in the creation of distinct areas of rural or urban ethnic settlement.

chorography

Greek term, revived by nineteenth-century German geographers, that refers to regional descriptions of large areas.

chorology

Greek term, revived by nineteenth-century German geographers, as a synonym for regional geography.

choropleth map

A thematic map utilizing areally based data and colour or shading proportional to density by statistical or administrative area.

civilization

Cultures with agriculture and cities, food and labour surplus, labour specialization, social stratification, and state organization.

class

A large group comprising individuals of similar social status, income, and culture.

collective consumption

The use of services that are produced and managed on a collective basis.

colonialism

The policy of a state or people seeking to establish and maintain their authority over another state or people.

commercial agriculture

An agricultural system in which the production is primarily for sale.

commercial geography

An early term for economic geography; it focused on what was produced where.

conservation

A general term referring to any form of environmental protection, including preservation.

contextual effect

A concept that refers to the fact that individuals may be influenced in their voting decision by their social contacts.

continental drift

The idea that the present continents were originally connected as one or two land masses that separated and drifted apart to their present positions.

conurbation

A continuously built-up area formed by the coalesence of several expanding, originally separate cities.

core/periphery

A concept that acknowledges that states are often divided into unequal areas — powerful cores and dependent peripheries.

cornucopian thesis

The argument that advances in science and technology will continue to create resources sufficient to support the growing world population.

counterfactual

An approach to historical analyses that compares what actually happened with what might have happened had some particular circumstance been different.

counterurbanization

A process of population decentraliza-tion possibly resulting from the high cost of living in cities, improvements in personal spatial mobility, industrial deconcentration, and advances in information technologies.

creole

A pidgin language that assumes the status of a mother tongue for a group.

cultural adaptation

Changes in technology, organization, and ideology that permit sound human/physical environment relationships to develop.

cultural regions

Areas in which there is a degree of homogeneity in cultural character-istics; areas of similar landscape.

culture

Refers to the way of life of the members of a society.

cycle of poverty

The idea that poverty and deprivation are transmitted intergenerationally, reflecting home background and spatial variations in opportunities.

deconstruction

A method of critical interpretation of a text (for example, a landscape) that aims to show how the multiple positioning (for example, in terms of class or gender) of an author or a reader affects the creation or reading of the text.

deglomeration

The spatial separation of humans or human activities so as to maximize the distances between them.

democracy

A form of government involving free and fair elections, openness and accountability, civil and political rights, and a just society.

demography

The study of human populations.

density

A measure of the number of geographic facts (for example, people) per unit area.

dependence

A relationship that involves one state or people being dominated by another state or people.

dependency theory

The idea that European overseas expansion resulted in a symbiotic relationship between the development of dependence and the development of under-development.

desertification

The process by which an area of land becomes a desert; typically involves the impoverishment of an ecosystem because of climatic change and/or human impact.

development

A term to be used cautiously because it often involves an implicit general acceptance of a particular perspective (ethnocentrism); typically interpreted to mean a process of becoming larger, more mature, and better organized; often measured by economic criteria.

developmentalism
Analysis of cultural and economic change that treats each country or region of the world separately in an evolutionary manner; assumes that all areas follow the same stages and that they are autonomous.

devolution
A process of transferring power from central to regional or local levels of government.

dictatorship
A form of government, often by the military, that is oppressive and dominant.

diffusion
The spread of a phenomenon over space and the growth of that phenomenon through time.

disorganized capitalism
In contrast to organized capitalism, refers to a new form characterized by a process of disorganization and industrial restructuring.

distance
The spatial dimension of separation; a fundamental concept in spatial analysis.

distance decay
The declining intensity of any pattern or process with increasing distance from a given location.

distribution
The pattern of geographic facts (for example, people) within an area.

domestication
The process of making plants and/or animals more useful to humans through selective breeding.

dual economy
A region with two distinct types of economy, typically a resource extraction economy and a more traditional economy.

ecology
The study of relationships between organisms and their environments.

economic base theory
A theory that tries to explain the growth or decline of a region in terms of the nature and size of its basic activities for which the demand is beyond the urban area.

economic operator
A model of human behaviour in which each individual is completely rational; economic operators maximize returns and minimize costs.

economic rent
The surplus income that accrues to a unit of land above the minimum income needed to bring a unit of new land into production at the margins of production.

ecosystem
An ecological system; comprises a set of interacting and interdependent organisms and their physical, chemical, and biological environment.

ecotourism
Tourism that is environmentally friendly and allows participants to experience a distinctive ecosystem.

empiricism
A philosophy of science based on the belief that all knowledge results from experience and that therefore gives priority to factual observations over theoretical statements.

energy
The capacity of a physical system for doing work.

entropy
A measure of the disorder or disorganization in a system.

environmental determinism
The view that human activities are controlled by the environment.

equinox
Occurs twice each year when the sun is vertically overhead at the equator; on this day, the periods of daylight and darkness are both twelve hours long all over the earth.

ethnic
A group whose members perceive themselves as different from others because of a common ancestry and shared culture.

ethnocentrism
A form of prejudice or stereotyping that presumes that one's own culture is normal and natural and therefore all other cultures are inferior.

ethnography
The general research approach requiring researcher involvement in the subjects studied; refers to any approach that is based on first-hand observation in the field; qualitative methods such as participant observation are a part of ethnography.

existentialism
A humanistic philosophy that holds that humans are responsible for making their own nature; stresses personal freedom, decision, and commitment in a world without absolute values outside of individuals.

exurbanization
The movement of households from urban areas to locations outside the urban area but within the commuting field.

fascism
The belief that a process of national rebirth is essential to end a period of social and cultural decadence in a state.

fecundity
A biological term; the ability of a woman to conceive; refers to potential rather than actual number of live births produced by a woman.

federalism
A form of government involving a division of authority and power between central and regional governments.

feminism
The advocacy of equal rights for woman and men along with a commitment to improve the position of women in society.

fertility
Generally refers to all aspects of human reproduction that lead to live births, but more specifically to the actual number of live births produced by a woman.

feudalism
A social and economic system prevalent in Europe, prior to the Industrial Revolution, which involved two groups; the land was controlled by lords while the peasants were bound to the land and subject to the lord's authority.

fieldwork
Means of data collection; a general term that includes qualitative (for example observation) and quantitative (for example, questionnaire) methods.

first effective settlement
A concept that refers to the likely importance of the initial occupance of an area as a determinant of later landscapes.

flexible accumulation
Industrial technologies, labour practices, relations between firms, and consumption patterns that are increasingly flexible.

forces of production
A Marxist phrase that refers to the raw materials, tools, and workers that actually produce goods.

Fordism
A group of industrial and broader social practices initiated by Henry Ford and dominant until recently in most industrial countries.

formal region
A region that is identified as such because of the presence of a particular characteristic(s).

friction of distance
A measure of the restraining effect of distance on human movement.

functional region
A region that is identified as such because it comprises a series of linked locations.

garden city
A planned settlement designed to combine the advantages of urban and rural living; an urban centre emphasizing spaciousness and quality of life.

Gemeinschaft
A term introduced by Tonnies; a form of human association based on loyalty, informality, and personal contact; assumed to be characteristic in traditional village communities.

gender
The social aspect of the relations between the sexes.

gentrification
A process of inner city urban neighbourhood social change resulting from the in-movement of higher-income groups.

geographic information system (GIS)
A computer-based tool that combines the storage, display, analysis, and mapping of spatially referenced data.

geographie Vidalienne
French school of geography initiated by Paul Vidal de la Blache at the end of the nineteenth century and that remains influential today. The focus is on the study of human-made (cultural) landscapes.

geopolitics
The study of the importance of space in understanding international relations.

geopolitik
The study of states as organisms that need to expand in territory in order to fulfil their destinies as nation states.

gerrymandering
The creation of electoral boundaries so that one party will benefit.

Gesellschaft
A term introduced by Tonnies; a form of human association based on rationality and depersonalization; assumed to be characteristic in urban areas.

ghetto
A residential district in an urban area with a concentration of a particular ethnic group.

globalization
A process whereby the population of the world is increasingly bonding into a single society.

green belt
An area of planned open, partially rural, land surrounding an urban area.

gross domestic product (GDP)
A monetary measure of the value at market prices of goods and services produced by a country over a given time period (usually one year); provides a better indication of domestic production than the GNP.

gross national product (GNP)
A monetary measure of the value at market prices of goods and services produced by a country, plus net income from abroad, over a given time period (usually one year).

heartland theory
A geopolitical theory of world power based on the assumption that the land-based state controlling the Eurasian heartland held the key to world domination.

historical materialism
A method associated with Marxism that centres on the material basis of society and that attempts to understand social change by referring to historical changes in social relations.

Holocene
Literally 'wholly recent'; the postglacial period that began 10,000 years ago; preceded by Pleistocene.

homeland
A special type of culture region that involves interaction with the physical environment to evoke emotional attachment and bonding.

hominids
The bipedal primate; humans and their ancestors.

humanism
A philosophy that centres on such characteristics of human life as value, quality, meaning, significance, and spirituality.

hypothesis
In positivist philosophy, a general statement deduced from theory but not yet verified.

iconography
The description and interpretation of visual images, including landscape, in order to uncover their symbolic meanings.

ideal type
A term introduced by the sociologist, Max Weber, to refer to a hypothetical norm, comparisons with which can aid understanding of the real world.

idealism
A humanistic philosophy that contends that any understanding of human actions has to be achieved by uncovering the thoughts behind those actions.

idiographic
Concerned with the unique and particular.

image
The perception of reality held by an individual or group.

imperialism
A relationship between states with one dominant over the other.

Industrial Revolution
A process that converted a fundamentally rural society into an industrial society beginning in mid-eighteenth-century England; primarily a technological revolution associated with new energy sources.

informal sector
A part of a national economy involved in productive labour, but without any formal control or remuneration and operating without formal recognition.

infrastructure (base)
A Marxist concept that refers to the economic structure of a society,

especially as it gives rise to political, legal, and social systems.

innovation

The use of an invention or idea to prompt change in human behaviour or production processes.

interaction

The relationship or linkage between locations.

interaction theory

A body of theories that explain movements of goods and people between locations.

irredentism

The argument made by a country, often expressed through war, that a minority living in an adjacent country rightfully belongs to it.

isodapanes

In Weberian least cost industrial location theory, lines of equal additional transport cost drawn about the point of minimum transport cost.

isopleth map

A map utilizing lines to connect locations of equal data value.

isotims

In Weberian least cost industrial location theory, lines of equal transport costs around material sources and markets.

landscape

A principal concern of geographic study; the characteristics of a particular area especially as created through human activity.

landscape school

The American school of geography initiated by Carl Sauer in the 1920s that remains influential today; an alternative approach to environmental determinism; the focus is on human-made (cultural) landscapes.

landschaftskunde

German term, introduced in the late nineteenth century, best translated as landscape science; refers to geography as the study of the landscapes of particular regions.

latitude

Angular distance on the surface of the earth, measured in degrees, minutes and seconds, north and south of the equator (which is the line of 0° latitude); lines of constant latitude are called parallels.

law

In positivist philosophy, a hypothesis that has been proven correct and is taken to be universally true; once formulated, laws can be used to construct theories.

less developed world

All countries not classified as more developed (see more developed world).

life cycle

The process of change experienced by individuals over their lifespans; often divided into stages (such as childhood, adolescence, adulthood, old age) with each stage associated with particular forms of behaviour.

limits to growth

The argument that, in the future, both world population and world economy may collapse because available world resources are inadequate.

lingua franca

An existing language used as a common means of communication between different language groups.

lithosphere

Outer layer of rock on earth; includes crust and upper mantle.

locales

The setting or context for social interaction; a term introduced in structuration theory that has become popular in human geography as an alternative to place.

location

A term that refers to a specific part of the earth's surface; an area where something is situated.

location theory

A body of theories that explain the distribution of economic activities.

locational interdependence

When the selection of a location is related to the selection of locations by others who are competitors for the available market.

longitude

Angular distance on the surface of the earth, measured in degrees, minutes and seconds, east and west of a specific prime meridian (which is the line of 0° longitude); lines of constant longitude are called meridians.

malapportionment

The creation of electoral districts of varying population size so that one party will benefit.

malnutrition

Condition where diet omits some food necessary for health.

Maoism

The revolutionary thought and practice of Mao Zedong involving protracted revolution to achieve power and socialist policies after power is achieved.

material index

An index devised by Weber and used in industrial location theory to show the extent to which the least cost location for a particular industrial firm will be either material or market oriented.

mental map

The individual psychological representation of space.

mentifacts

Those elements of culture that are attitudinal in character; the values held by members of a group.

mercantilism

A school of economic thought dominant in Europe in the seventeenth and early eighteenth centuries that argued for the involvement of the state in economic life so as to increase national wealth and power.

minority language

A language spoken by a minority group in a state in which the majority of the population speaks some other language; may or may not be an official language.

mode of production

A Marxist term that refers to the organized social relations through which a human society organizes productive activity; human societies are seen as passing through a series of such modes.

model

An idealized and structured representation of the real world.

modernism

A view that assumes the existence of a reality characterized by structure, order, pattern, and causality.

monarchy

The institution of rule over a state by the head of a hereditary family; monarchism is the belief that this is best.

more developed world

Using a United Nations classification, this comprises Europe, North America, Australia, Japan, New Zealand, and the former USSR.

multiculturalism
A policy that endorses the right of ethnic groups to remain distinct rather than assimilated into a dominant society.

multilingual state
A state in which the population includes at least one linguistic minority.

multinational corporations
A large business organization that operates in two or more countries; sometimes called transnational companies.

nation
A group of people sharing a common culture and an attachment to some territory; a difficult term to define objectively.

nation state
A political unit that contains one principal national group with which it identifies itself and its territory.

nationalism
The political expression of nationhood or aspiring nationhood; reflects a consciousness of belonging to a nation.

nomothetic
Concerned with the universal and the general.

normative theory
A theory that is concerned with what ought to be rather than with what actually occurs; the aim is to seek what is rational or optimal according to some given criteria.

nuptiality
The extent to which a population marries.

oligarchy
Refers to rule by an élite group of people, typically the wealthy.

organized capitalism
A form of capitalism that developed after the Second World War and that was characterized by increased growth of major (often multinational) corporations and increased involvement by the state (often in the form of public ownership) in the economy.

ozone layer
Layer in the atmosphere 16 to 40 km (10 to 25 miles) above the earth that absorbs dangerous ultraviolet solar radiation; ozone is a gas composed of molecules consisting of three atoms of oxygen (O_3).

participant observation
A qualitative method that requires the researcher to be involved in the people and place being studied.

patriarchy
A social system in which men dominate, oppress, and exploit women.

perception
The process by which humans acquire information about physical and social environments.

phenomenology
A humanistic philosophy that involves the study of the ways in which humans experience everyday life and imbue activities with meaning.

phenotype
Any physical or chemical trait that can be observed or measured.

physiological density
Population per unit of cultivable land.

pidgin
A new language designed to serve the purposes of commerce between different language groups; typically has a limited vocabulary.

place
Location; in humanistic geography, this term has acquired a particular meaning as a context for human action that is rich in human significance and meaning.

place utility
The measure of satisfaction an individual derives from a location relative to his or her goals.

placelessness
Homogeneous and standardized landscapes that lack local variety and character.

Pleistocene
Geological time period, from about 1.5 million years ago to 10,000 years ago, characterized by a series of glacial advances and retreats, succeeded by Holocene.

pollution
The release into the environment of substances that degrade one or more of land, air, or water.

population pyramid
A diagrammatic representation of the age and sex composition of a population.

positivism
A philosophy that contends that science is able to deal only with empirical questions (those with factual content), that scientific observations are repeatable, and that science progresses through the construction of theories and derivation of laws.

possibilism
The view that the environment does not determine either human history or present conditions; rather, humans pursue a course of action that they select from among a number of possibilities.

post-Fordism
A group of industrial and broader social practices evident in industrial countries since about 1970; involves more flexible production methods than those associated with Fordism.

postmodernism
A movement in philosophy, social science, and the arts that argues that reality cannot be studied objectively, stressing that multiple interpretations are possible.

power
In general, the capacity to affect outcomes; more specifically, to dominate others by means of violence, force, manipulation, or authority.

pragmatism
A humanistic philosophy that focuses on the construction of meaning through the practical activities of humans.

primate city
A country's largest city, usually the capital city, that dominates political, economic, and social life.

principle of least effort
A general principle considered as guiding human activities; for human geographers, it refers to minimizing distances and related movements.

projection
Any procedure employed to represent positions of all or a part of the earth's spherical (three-dimensional) surface on to a flat (two-dimensional) surface.

proxemics
The study of personal space, the invisible boundaries that protect and compartmentalize each individual.

public goods
Goods that are freely available to all or that are provided (equally or unequally) to citizens by the state.

qualitative methods
A set of tools to collect and analyse data in order to subjectively understand the phenomena being studied; the methods include passive observation, participation, and active intervention.

quantitative methods
A set of tools to collect and analyse data to achieve a statistical description and scientific explanation of the phenomena being studied; the methods include sampling, models, and statistical testing.

questionnaire
A structured and ordered set of questions designed to collect unambiguous and unbiased data.

race
Subspecies; a physically distinguishable population within a species.

raster
A method used in a GIS to represent spatial data; divides the area into numerous small cells, pixels, and describes the content of each cell.

rational choice theory
The view that social life can be explained by models of rational individual action; an extension of the economic operator concept to other areas of human life.

recycling
The reuse of material and energy resources.

region
A part of the earth's surface that displays internal homogeneity and is relatively distinct from surrounding areas according to some criterion or criteria; regions are intellectual creations.

regionalization
A special case of classification; involves assigning locations on the earth's surface to regions, which must be contiguous spatial units.

relations of production
A Marxist term that refers to the ways in which the production process is organized, specifically to the relationships of ownership and control.

remote sensing
A variety of techniques for acquiring and recording data from points that are not in contact with the phenomena of interest.

renewable resources
Resources that regenerate naturally to provide a new supply within a human life span.

restructuring
In a capitalist economy, changes in or between the various component parts of an economic system resulting from economic change.

rimland theory
A geopolitical theory of world power based on the assumption that the state controlling the area surrounding the Eurasian heartland held the key to world domination.

sacred space
A landscape that is particularly esteemed by an individual or a group, usually (but not necessarily) for religious reasons.

sampling
The selection of a subset from a defined population of individuals to acquire data that are representative of that larger population.

satisficing behaviour
A model of human behaviour that rejects the rationality assumptions of the economic operator, assuming instead that the objective is to reach a level of satisfaction that is acceptable.

scale
The resolution levels used in any human geographic research; most characteristically refers to the size of the area studied, but also to the time period covered and the number of people investigated.

scientific method
The various steps a science takes to obtain knowledge; a phrase most commonly associated with a positivist philosophy.

sense of place
The deep attachments that humans have to specific locations such as home and also to particularly distinctive locations.

simulation
Representing a real process in an abstract form for purposes of experimentation.

site
The location of a geographic fact with reference to the immediate local environment.

situation
The location of a geographic fact with reference to the broad spatial system of which it is a part.

slavery
Labour that is controlled through compulsion and not involving remuneration; in Marxist terminology, it is one example of a mode of production.

social physics
An approach to aggregate human movement and interaction that is based on the assumption that such phenomena are analogous to physical science processes.

socialism
A social and economic system that involves common ownership of the means of production and distribution.

society
Refers to the interrelationships that connect individuals as members of a culture.

sociofacts
Those elements of culture most directly concerned with interpersonal relations; the norms that people are expected to observe.

solstice
Occurs twice each year when the sun is vertically overhead at the farthest distance from the equator, once 23.5°N and once 23.5°S of the equator; affects hours of daylight, for example, when at 23.5°N, there is maximum daylight (longest day) in the Northern hemisphere and minimum daylight (shortest day) in the Southern hemisphere.

space
Areal extent; a term used in both absolute (objective) and relative (perceptual) forms.

spatial monopoly
When a single producer sells the entire output of a particular industrial good or service in a given area.

spatial preferences

Individual (sometimes group) evaluation of the relative attractiveness of different locations.

spatial separatism (or spatial fetishism)

A phrase critical of spatial analysis because it sees space or distance as a cause without reference to humans.

species

A group of organisms that are able to produce fertile offspring among themselves but not with any other group.

spectacle

Places and events that are examples of carefully created mass leisure and consumption.

squatter settlement

A shanty town, a concentration of temporary dwellings, neither owned nor rented, at the city edge; related to large immigration into urban areas that are unable to cope with such population increases.

state

An area with defined and internationally acknowledged boundaries; a political unit.

state apparatus

The institutions and organizations through which the state exercises its power.

stock resources

Minerals and land that take a long time to form and, hence, from a human perspective, are fixed in supply.

structuration

The conditions that govern the continuity and transformation of social structures.

subsistence agriculture

An agricultural system in which the production is not primarily for sale but is consumed by the farmer's household.

suburb

An outer commuting zone of an urban area; implies a degree of social homogeneity and a lifestyle suited to family needs.

superorganic

An interpretation of culture that sees it as above both nature and individuals and therefore as the principal cause of the human world; a form of cultural determinism.

superstructure

A Marxist concept that refers to the political, legal, and social systems of a society.

sustainable development

A term popularized in the 1987 report of the World Commission on Environment and Development that refers to economic development that does not damage the environment.

symbolic interactionism

A group of social theories that see the social world as a social product with meanings resulting from interaction.

system

A set of interrelated components or objects that are linked together to form a unified whole.

tariff

A tax or customs duty charged by a country on imports from other countries.

technology

The ability to convert energy into forms useful to humans.

text

A term that originally referred to the written page, but that has broadened to include such products of culture as maps and landscape; postmodernists recognize that there are any number of realities depending on how a text is read.

theory

In positivist philosophy, an interconnected set of statements, often called assumptions or axioms, that deductively generates testable hypotheses.

time-space convergence

A decrease in the friction of distance between locations as a result of improvements in transport and communications technologies.

topography

A Greek term, revived by nineteenth-century German geographers, that referred to regional descriptions of local areas.

topological map

A diagram that represents a network as a simplified series of straight lines.

topophilia

The affective ties that humans have with particular places; literally, love of place.

undernutrition

When a person who is healthy and able to perform normal activity begins to lose weight steadily or becomes unable to perform normal activity.

urban sprawl

The largely unplanned expansion of an urban area; typically discontinuous, leaving rural enclaves.

urbanism

The urban way of life; involves a decline of community and the rise of complex social and economic organization resulting from size, density, and heterogeneity.

urbanization

The spread and growth of cities.

vector

A method used in a GIS to represent spatial data; describes the data as a collection of points, lines, and areas and describes the location of each of these.

vernacular region

A region that is identified as such on the basis of the perceptions held by those inside and outside the region.

verstehen

A research method, associated primarily with phenomenology, that involves the researcher adopting the perspective of the individual or group under investigation; a German term, best translated as sympathetic or empathetic understanding.

welfare geography

An approach to human geography that maps and explains social and spatial variations.

well-being

The degree to which the needs and wants of a society are satisfied.

References

INTRODUCTION

BONE, R.M. 1992. *The Geography of the Canadian North.* Toronto: Oxford University Press.

FAIRGRIEVE, J. 1926. *Geography in School.* London: University of London Press.

JOHNSTON, R.J. 1985. 'To the Ends of the Earth'. In *The Future of Geography*, edited by R.J. Johnston. New York: Methuen, 326–38.

CHAPTER 1

HART, J.F. 1982. 'The Highest Form of the Geographer's Art. *Annals, Association of American Geographers* 72:1–29.

HARTSHORNE, R. 1939. *The Nature of Geography: A Critical Survey of Current Thought in the Light of the Past.* Lancaster: Association of American Geographers.

MAY, J.A. 1970. *Kant's Concept of Geography and Its Relation to Recent Geographical Thought.* Toronto: University of Toronto Press.

CHAPTER 2

BLAINEY, G. 1968. *The Tyranny of Distance.* Melbourne: Macmillan.

BUNGE, W. 1968. 'Fred K. Schaefer and the Science of Geography.' *Harvard Papers in Theoretical Geography*, Special Papers Series, Paper A (mimeographed).

———. 1979. 'Fred K. Schaefer and the Science of Geography'. *Annals,*

Association of American Geographers 69:128–32 (condensed version of Bunge 1968).

DOHRS, F.E., and L.M. SOMMERS, eds. 1967. *Introduction to Geography: Selected Readings.* New York: Crowell.

FEBVRE, L. 1925. *A Geographical Introduction to History.* London: Routledge and Kegan Paul.

HARTSHORNE, R. 1955. '"Exceptionalism in Geography" Re-examined'. *Annals, Association of American Geographers* 45:205–44.

———. 1959. *Perspective on the Nature of Geography.* Chicago: Rand McNally.

HOLLAND, P., et al. 1991. 'Qualitative Resources in Geography'. *New Zealand Journal of Geography* 92:1–28.

HUNTINGTON, E. 1927. *The Human Habitat.* New York: Van Nostrand.

JACKSON, R.H., and R. HENRIE. 1983. 'Perceptions of Sacred Space'. *Journal of Cultural Geography* 3, no. 2:94–107.

LEWTHWAITE, G.R. 1966. 'Environmentalism and Determinism: A Search for Clarification'. *Annals, Association of American Geographers* 56:1–23.

MARTIN, G.J. 1989. 'The Nature of Geography and the Schaefer-Hartshorne Debate'. In *Reflections on Richard Hartshorne's The Nature of Geography*, edited by J.N. Entrikin and S.D. Brunn. Occasional

Publication, Association of American Geographers:69–88.

MINSHULL, R. 1967. *Regional Geography: Theory and Practice.* London: Hutchinson.

MONMONIER, M. 1991. *How to Lie with Maps.* Chicago: University of Chicago Press.

PEET, R. 1989. 'World Capitalism and the Destruction of Regional Cultures'. In *A World in Crisis: Geographical Perspectives*, 2nd ed., edited by R.J. Johnston and P.J. Taylor. Oxford: Blackwell, 175–99.

RAPER, J.F., and N.P.A. GREEN. 1989. 'The Development of a Tutor for Geographic Information Systems'. *British Journal of Educational Technology* 20:164–72.

RELPH, E. 1976. *Place and Placelessness.* London: Pion.

SAARINEN, T.F. 1974. 'Environmental Perception'. In *Perspectives on Environment*, edited by I.R. Manners and M.W. Mikesell. Washington: Association of American Geographers, Commission on College Geography Publication 13:252–89.

SCHAEFER, F. 1953. 'Exceptionalism in Geography: A Methodological Examination'. *Annals, Association of American Geographers* 43:226–49.

SEMPLE, E. 1911. *Influences of Geographic Environment*. New York: Henry Holt.

SPATE, G.H.K. 1952. 'Toynbee and Huntington: A Study in Determinism'. *Geographical Journal* 118:406–28.

TOBLER, W. 1970. 'A Computer Movie'. *Economic Geography* 46:234–40.

TUAN, YI-FU. 1979. *Landscapes of Fear*. Oxford: Blackwell.

———. 1982. *Segmented Worlds and Self: Group Life and Individual Consciousness*. Minneapolis: University of Minnesota Press.

———. 1983. 'Geographical Theory: Queries from a Cultural Geographer'. *Geographical Analysis* 15:69–72.

———. 1984. *Dominance and Affection*. New Haven: Yale University Press.

WHITTLESEY, D. 1954. 'The Regional Concept and the Regional Method'. In *American Geography: Inventory and Prospect*, edited by P.E. James and C.F. James. Syracuse: Syracuse University Press, 19–68.

WOOD, D. 1993. 'The Power of Maps'. *Scientific American* 268, no. 5:88–93.

YEATES, M.H. 1968. *An Introduction to Quantitative Analysis in Economic Geography*. Toronto: McGraw Hill.

CHAPTER 3

ARKELL, T.C. 1991. 'The Earth's Fault'. *Geographical Magazine* 63, no. 8:19–21.

DEAR, M., and J. WOLCH. 1989. 'How Territory Shapes Social Life'. In *The Power of Geography: How Territory Shapes Social Life*, edited by J. Wolch and M. Dear. Boston: Unwin Hyman, 3–18.

FEDER, K.L. and M.A. PARK. 1993. *Human Antiquity: An Introduction to Physical Anthropology and Archaeology*, 2nd ed. Toronto: Mayfield.

GOULD, S.J. 1981. *The Mismeasure of Man*. New York: W.W. Norton.

———. 1985. 'Human Equality Is a Contingent Fact of History'. In *The Flamingo's Smile*, edited by S.J. Gould. New York: Norton, 185–98.

———. 1987. 'Bushes All the Way Down'. *Natural History* 96, no. 6:12–19.

KENNEDY, K.A.R. 1976. *Human Variation in Space and Time*. Dubuque: Brown.

KIRCHNER, J.W. 1989. 'The Gaia Hypothesis: Can It Be Tested?' *Reviews of Geophysics* 27, no. 2:223–35.

LEWIN, R. 1982. *Thread of Life: The Smithsonian Looks at Evolution*. Washington, DC: Smithsonian Books.

LOVELOCK, J. 1979. *Gaia: A New Look at Life on Earth*. Toronto: Oxford University Press.

———. 1988. *The Ages of Gaia*. New York: W.W. Norton.

MONTAGUE, A., ed. 1964. *The Concept of Race*. New York: Collier.

TRENHAILE, A.S. 1990. *The Geomorphology of Canada*. Toronto: Oxford University Press.

CHAPTER 4

AGNEW, C. 1990. 'Green Belt Around the Sahara'. *Geographical Magazine* 62, no. 4:26–30.

BAHN, P., and J. FLENLEY. 1992. *Easter Island: Earth Island*. New York: Thames and Hudson.

CARSON, R. 1962. *Silent Spring*. Boston: Houghton Mifflin.

DREGNE, H.E. 1977. 'Desertification of Arid Lands'. *Economic Geography* 53:322–31.

GALE, R.J.P. 1992. 'Environment and Economy: The Policy Models of Development'. *Environment and Behavior* 24:723–37.

GLANCE, N.S., and B.A. HUBERMAN. 1994. 'The Dynamics of Social Dilemmas'. *Scientific American* 270, no. 3:76–81.

GOUDIE, A. 1981. *The Human Impact: Man's Role in Environmental Change*. Oxford: Blackwell.

GROVE, R.H. 1992. 'Origins of Western Environmentalism'. *Scientific American* 267, no. 1:42–7.

HARDIN. G. 1968. 'The Tragedy of the Commons'. *Science* 162:1243–8.

JOHNSTON, R.J. 1992. 'Laws, States and Superstates: International Law and the Environment'. *Applied Geography* 12:211–28.

———, and P.J. Taylor. 1986. 'Introduction: A World in Crisis?' In *A World in Crisis: Geographical Perspectives*, edited by R.J. Johnston and P.J. Taylor. Oxford: Blackwell, 1–11.

KAPLAN, R.D. 1994. 'The Coming Anarchy', *Atlantic Monthly* 273, no. 2:44–76.

KELLY, K. 1974. 'The Changing Attitudes of Farmers to Forest in Nineteenth Century Ontario'. *Ontario Geography* 8:67–77.

MARSH, G.P. [1864] 1965. *Man and Nature, or Physical Geography as Modified by Human Action*, edited by D. Lowenthal. Cambridge, Mass.: Harvard University Press, 1965.

PONTING, C. 1991. *A Green History of the World*. New York: St Martin's Press.

POWELL, J.M. 1976. *Environmental Management in Australia, 1788–1814*. New York: Oxford University Press.

REPETTO, R. 1990. 'Deforestation in the Tropics'. *Scientific American* 260:36–42.

RUCKELSHAUS, W.N. 1989. 'Toward a Sustainable World'. *Scientific American* 261, no. 3:166–75.

SIMON, J., and H. KAHN, eds. 1984. *The Resourceful Earth*. Oxford: Blackwell.

SMIL, V. 1987. *Energy, Food and Environment*. New York: Oxford University Press.

———. 1989. 'Our Changing Environment'. *Current History* 88:9–12, 46–8.

———. 1993. *Global Ecology: Environmental Changes and Social Flexibility*. New York: Routledge.

Supply and Services Canada. 1991. *The State of Canada's Environment*. Ottawa: Supply and Services Canada.

THOMAS, W.L., et al., eds. 1956. *Man's Role in Changing the Face of the Earth*. Chicago: University of Chicago Press.

TUAN, Y.F. 1971. *Man and Nature*. Washington, DC: Association of American Geographers, Commission on College Geography, Resource Paper No. 10.

WILSON, E.O. 1989. 'Threats to Biodiversity'. *Scientific American* 261, no. 3:108–16.

World Commission on Environment and Development. 1987. *Our Common Future*. Oxford: Oxford University Press.

CHAPTER 5

BOSERUP, E. 1965. *The Conditions of Agricultural Change*. London: Allen and Unwin.

DEMENY, P. 1974. 'The Populations of the Underdeveloped Countries'. In *The Human Population*. San Francisco: W.H. Freeman, 105–15.

DOENGES, C.E., and J.L. NEWMAN. 1989. 'Impaired Fertility in Tropical Africa'. *Geographic Review* 79:101–11.

DWYER, D.J. 1987. 'New Population Policies in Malaysia and Singapore'. *Geography* 72:248–50.

Economist, The. 1993. 'Eastern Germany: Living and Dying in a Barren Land'. *The Economist* 331, no. 23 (April):54.

EHRLICH, P. 1968. *The Population Bomb*. New York: Ballantine Books.

HALL, R. 1993. 'Europe's Changing Population'. *Geography* 78:3–15.

JOWETT, J. 1993. 'China's Population: 1,133,709,738 and Still Counting'. *Geography* 78:401–19.

KING, R. 1993. 'Italy Reaches Zero Population Growth'. *Geography* 78:63–9.

MEADOWS, D.H., et al. *The Limits to Growth*. New York: Universe Books.

OLSHANSKY, S.J., B.A. CARNES, and C.K. CASSEL. 1993. 'The Aging of the Human Species'. *Scientific American* 268, no. 4:46–52.

RIDKER, R.G., and E.W. CECELSKI. 1979. 'Resources, Environment and Population: The Nature of Future Limits'. *Population Bulletin* 34, no. 3:3–4.

ROBEY, B., S.O. RUTSTEIN, and L. MORRIS. 1993. 'The Fertility Decline in Developing Countries'. *Scientific American* 269, no. 6:60–7.

VANDERPOST, C. 1992. 'Regional Patterns of Fertility Transition in Botswana'. *Geography* 77:109–22.

CHAPTER 6

BARBERIS, M. 1994. 'Haiti'. *Population Today* 22, no. 1:7.

BONGAARTS, J. 1994. 'Can the Growing Human Population Feed Itself'. *Scientific American* 271, no. 3:36–42.

CHAMPION, A. 1992. 'Migration in Britain'. In *Migration Processes and Patterns 1*, edited by A. Champion and A. Fielding. London: Belhaven, 215–26.

COLVILLE, R. 1993. 'Resettlement: Still Vital After All These Years'. *Refugees* 94:4–8.

DEG, M. 1992. Natural Disasters: Recent Trends and Future Prospects'. *Geography* 77:198–209.

GOULD, P., and R. WHITE. 1986. *Mental Maps*, 2nd ed. Boston: Allen and Unwin.

GREEN, A., and D. OWEN. 1991. 'Local Labour Supply and Demand Interactions in Britain During the 1980's'. *Regional Studies* 25:295–314.

GRIGG, D. 1977. 'E.G. Ravenstein and the "Laws" of Migration'. *Journal of Historical Geography* 3:41–54.

KATES, R.W., and R.S. CHEN. 1993. 'Poverty and Global Environmental Change'. *International Geographical Union Bulletin* 73, no. 1–2:5–14.

KLOOS, H., and A. ADUGNA. 1989. 'The Ethiopian Population: Growth and

Distribution'. *Geographical Journal* 155:35–51.

LEE, E.S. 1966. 'A Theory of Migration'. *Demography* 3, no. 1:47–57.

MAHMUD, A. 1989. 'Grameen Bank Bangladesh: A Workable Solution'. *Geographical Magazine* 61, no. 10:14–16.

PETERSEN, W. 1958. 'A General Typology of Migration'. *American Sociological Review* 23:256–65.

PORTER, G. 1992. 'The Nigerian Census Surprise'. *Geography* 77:371–4.

RAVENSTEIN, E.G. 1876. 'Census of the British Isles, 1871; Birthplaces and Migration'. *Geographical Magazine* 3:173–7, 201–6, 229–33.

———. 1885. 'The Laws of Migration'. *Journal of the Statistical Society* 48:167–227.

———. 1889. 'The Laws of Migration'. *Journal of the Statistical Society* 52:214–301.

SIMPSON, E.S. 1989. 'An Island Divided'. *Geographical Journal* 61, no. 10:1–5.

SMIL, V. 1987. *Energy, Food, Environment: Realities, Myths and Options*. Oxford: Clarendon.

SOWDEN, C. 1993. 'Debt Swaps — For or Against Development'. *Geographical Magazine* 25, no. 12:56–9.

TAYLOR, P.J. 1989. 'The Error of Developmentalism in Human Geography'. In *Horizons in Human Geography*, edited by D. Gregory and R. Walford. London: Macmillan, 303–19.

United Nations Development Programme. 1992. *Human Development Report, 1992*. Toronto: Oxford University Press.

United States Committee for Refugees. 1992. *World Refugee Survey: 1992*. Washington, DC: US Committee for Refugees.

United States Department of State. 1992. *World Refugee Report, June 1992*. Washington, DC: Bureau for Refugee Affairs, Department of State Publication 9998.

WALLERSTEIN, I. 1979. *The Capitalist World Economy*. Cambridge: Cambridge University Press.

WOLPERT, J. 1965. 'Behavioural Aspects of the Decision to Migrate'. *Papers of the Regional Science Association* 15:159–69.

World Bank. 1992. *World Tables, 1992*. Baltimore: Johns Hopkins University Press.

———. 1993. *World Development Report*. Toronto: Oxford University Press.

ZELINSKY, W. 1971. 'The Hypothesis of the Mobility Transition'. *Geographical Review* 61:219–49.

CHAPTER 7

BISWAS, L. 1984. 'Evolution of Hindu Temples in Calcutta. *Journal of Cultural Geography* 4:73–84.

CARNEIRO, R. 1970. 'A Theory of the Origin of the State'. *Science* 169:733–8.

CARTWRIGHT, D. 1987. 'Linguistic Territorialization: Is Canada Approaching the Belgian Model?' *Journal of Cultural Geography* 8:115–34.

CHILDE, V.G. 1951. *Man Makes Himself*. New York: Mentor Books.

FEDER, K.L., and M.A. PARK. 1993. *Human Antiquity: An Introduction to Physical Anthropology and Archaeology*, 2nd ed. Toronto: Mayfield.

FRANCAVIGLIA, R.V. 1978. *The Mormon Landscape*. New York: AMS Press.

FRIEDL, J., and J. PFEIFFER. 1977. *Anthropology: The Study of People*. New York: Harper and Row.

GALE, D.T., and P.M. KOROSCIL. 1977. 'Doukhobor Settlements: Experiments in Idealism'. *Canadian Ethnic Studies* 9:53–71.

GIDDENS, A. 1991. *Introduction to Sociology*. New York: Norton.

HARRIS, M. 1974. *Cows, Pigs, Wars and Witches: The Riddles of Cultures*. New York: Vintage Books.

HUNTINGTON, E. 1924. *Civilization and Climate*. New Haven: Yale University Press.

HUXLEY, J.S. 1966. 'Evolution, Cultural and Biological'. *Current Anthropology* 7:16–20.

JAMES, P.E. 1964. *One World Divided*, 2nd ed. Toronto: Xerox College Publishing.

JORDAN, T.G. 1988. *The European Culture Area: A Systematic Geography*, 2nd ed. New York: Harper and Row.

———, and M. Kaups. 1989. *The American Backwoods Frontier: An Ethnic and Ecological Interpretation*. Baltimore: Johns Hopkins University Press.

KARAN, P. 1984. 'Landscape, Religion and Folk Art in Mithila: An Indian Cultural Region'. *Journal of Cultural Geography* 5:85–102.

KAUPS, M. 1966. 'Finnish Places Names in Minnesota: A Study in Cultural Transfer'. *Geographical Review* 56:377–97.

KEARNS, K.C. 1974. 'Resuscitation of the Irish Gaeltacht'. *Geographical Review* 64:82–110.

KROEBER, A.L., and T. PARSONS. 1958. 'The Concepts of Culture and of Social System'. *American Sociological Review* 23:582–3.

MCCRUM, R., W. CRAN, and R. MACNEIL, eds. 1986. *The Story of English*. London: BBC.

MEINIG, D.W. 1965. 'The Mormon Culture Region: Strategies and Patterns in the Geography of the American West, 1847–1964'. *Annals, Association of American Geographers* 55:191–220.

MITCHELL, R.D. 1978. 'The Formation of Early American Cultural Regions'. In *European Settlement and Development in North America*, edited by J.R. Gibson. Toronto: University of Toronto Press, 66–90.

RUSSELL, R.J., and F.B. KNIFFEN. 1951. *Culture Worlds*. New York: Macmillan.

SAUER, C.O. 1925. 'The Morphology of Landscape'. *University of California Publications in Geography* 2:19–53.

SMIL. V. 1987. *Energy, Food and Environment*. New York: Oxford University Press.

SPENCER, J.E., and R.J. HORVATH. 1963. 'How Does an Agricultural Region Originate?' *Annals, Association of American Geographers* 53:74–92.

TOYNBEE, A.J. 1935–61. *A Study of History*, 12 volumes. New York: Oxford University Press.

WITTFOGEL, K. 1957. *Oriental Despotism: A Comparative Study of Total Power*. New Haven: Yale University Press.

ZELINSKY, W. 1973. *The Cultural Geography of the United States*. Englewood Cliffs: Prentice-Hall.

CHAPTER 8

ABLER, R.F., M.G. MARCUS, and J.M. OLSON, eds. 1992. *Geography's Inner Worlds: Pervasive Themes in Contemporary American Geography*. New Brunswick: Rutgers University Press.

BERRY, B.J.L. 1992. Review of *Geography's Inner Worlds: Pervasive Themes in Contemporary American Geography*. *Urban Geography* 13:490–4.

BLAUT, J.M. 1980. 'A Radical Critique of Cultural Geography'. *Antipode* 12:25–9.

DEAR, M. 1988. 'The Postmodern Challenge: Reconstructing Human Geography'. *Transactions, Institute of British Geographers NS* 13:262–74.

EVANS, D.J., and D.T. HERBERT. 1989. *The Geography of Crime*. London: Routledge.

EVANS, R. 1989. 'Consigned to the Shadows'. *Geographical Magazine* 61, no. 12:23–5.

EYLES, J., and W. PEACE. 1990. 'Signs and Symbols in Hamilton: An Iconology of Steeltown'. *Geografiska Annaler* 72B:73–88.

FRANCAVIGLIA, R.V. 1978. *The Mormon Landscape*. New York: AMS Press.

GIDDENS, A. 1984. *The Constitution of Society*. Cambridge: Polity Press.

GREGORY, D. 1981. 'Human Agency and Human Geography'. *Transactions, Institute of British Geographers NS* 6:1–18.

GREGSON, N. 1986. 'On Duality and Dualism: The Case of Time Geography and Structuration'. *Progress in Human Geography* 10:184–205.

HALE, R.F. 1984. 'Vernacular Regions of America'. *Journal of Cultural Geography* 5:131–40.

HARVEY, D.W. 1989. *The Condition of Postmodernity*. Oxford: Blackwell.

HERBERT, D.T., and D.M. SMITH, eds. 1989. *Social Problems and the City: New Perspectives*. Oxford: Oxford University Press.

ISAJIW, W. 1974. 'Definitions of Ethnicity'. *Ethnicity* 1:111–24.

JACKSON, E.L., and D.B. JOHNSON. 1991. 'Geographic Implications of Mega-Malls, with Special Reference to West Edmonton Mall'. *Canadian Geographer* 35:226–32.

JACKSON, P. 1989. *Maps of Meaning: An Introduction to Cultural Geography*. London: Unwin Hyman.

———, and S.J. Smith. 1984. *Exploring Social Geography*. London: Allen and Unwin.

JOHNSTON, R.J. 1986. 'Individual Freedom and the World Economy'. In *A World in Crisis: Geographical Perspectives*, edited by R.J. Johnston and P.J. Taylor. Oxford: Blackwell, 173–95.

———. 1991. *A Question of Place: Exploring the Practice of Human Geography*. Oxford: Blackwell.

JONES, K., and G. MOON. 1987. *Health, Disease and Society: An Introduction to Medical Geography*. London: Routledge.

KNOX, P.L. 1975. *Social Well-Being: A Spatial Perspective*. Oxford: Clarendon Press.

KOBAYASHI, A. 1993. 'Multiculturalism: Representing a Canadian Institution'. In *Place/Culture/Representation*, edited by J. Duncan and D. Ley. London: Routledge, 205–31.

MCQUILLAN, A. 1993. 'Historical Geography and Ethnic Communities in North America'. *Progress in Human Geography* 17:355–66.

MEAD, G.H. 1934. *Mind, Self and Society*. Chicago: University of Chicago Press.

MONK, J. 1992. 'Gender in the Landscape: Expressions of Power and Meaning'. In *Inventing Places: Studies in Cultural Geography*, edited by K. Anderson and F. Gale. Melbourne: Longman Cheshire, 123–38.

NOSTRAND, R.L., and L.E. ESTAVILLE JR. 1993. 'Introduction: The Homeland Concept'. *Journal of Cultural Geography* 13, no. 2:1–4.

OSBORNE, B.S. 1988. 'The Iconography of Nationhood in Canadian Art'. In *The Iconography of Landscape: Essays on the Symbolic Representation, Design and Use of Past Environments*. New York: Cambridge University Press, 162–78.

PAIN, R. 1992. 'Space, Sexual Violence and Social Control: Integrating Geographical and Feminist Analyses of Women's Fear of Crime'. *Progress in Human Geography* 15:415–31.

PAWSON, E., and G. BANKS. 1993. 'Rape and Fear in a New Zealand City'. *Area* 25:55–63.

PRATT, G. 1989. 'Reproduction, Class and the Spatial Structure of the City'. In *New Models in Geography, Volume 2*, edited by N. Thrift and R. Peet. London: Unwin Hyman, 84–108.

PRED, A. 1985. 'The Social Becomes the Spatial, the Spatial Becomes the Social: Enclosures, Social Change and the Becoming of Place in the Swedish Province of Skåne'. In *Social Relations and Spatial Structures*, edited by D. Gregory and J. Urry. London: Macmillan, 336–75.

———, and M.J. Watts. 1992. *Reworking Modernity: Capitalisms and Symbolic Discontent*. New Brunswick: Rutgers University Press.

RAITZ, K.B. 1979. 'Themes in the Cultural Geography of European Ethnic Groups in the United States'. *Geographical Review* 69:79–94.

ROONEY JR, J.F. 1974. *A Geography of American Sport*. Reading, Mass.: Addison Wesley.

SMITH, D.M. 1973. *A Geography of Social Well-Being in the United States.* New York: McGraw Hill.

SMITH, S.J. 1987. 'Fear of Crime: Beyond a Geography of Deviance'. *Progress in Human Geography* 11:1–23.

STEIN, H.F., and G.L. THOMPSON. 1992. 'The Sense of Oklahomaness: Contributions of Psychogeography to the Study of American Culture'. *Journal of Cultural Geography* 11, no. 2:63–91.

TRÉPANIER, C. 1991. 'The Cajunization of French Louisiana: Forging a Regional Identity'. *Geographical Journal* 157:161–71.

WAGNER, P.L. 1975. 'The Themes of Cultural Geography Rethought'. *Yearbook: Association of Pacific Coast Geographers* 37:7–14.

ZELINSKY, W. 1980. 'North America's Vernacular Regions'. *Annals, Association of American Geographers* 70:1–16.

CHAPTER 9

ARKELL, T., and M. DAVENPORT. 1989. 'Africa's Fractious Five'. *Geographical Magazine* 61, no. 12:16–27.

BUNGE, W. 1988. *Nuclear War Atlas.* Oxford: Blackwell.

CLARK, A.H. 1975. 'The Conceptions of Empires of the St Lawrence and the Mississippi: An Historico-Geographical View with Some Quizzical Comments on Environmental Determinism'. *American Review of Canadian Studies* 5:4–27.

CLARK, G.L., and M.J. DEAR. 1984. *State Apparatus: Structures and Language of Legitimacy.* Boston: Allen and Unwin.

DOUGLAS, J.N.H. 1985. 'Conflict Between States'. In *Progress in Political Geography,* edited by M. Pacione. London: Croom Helm.

EAST, W.G., and J.R.V. PRESCOTT. 1975. *Our Fragmented World: Introduction to Political Geography.* London: Macmillan.

ELSOM, D. 1985. 'Climatological Effects of a Nuclear Exchange: A Review'. In *The Geography of Peace and War,* edited by D. Pepper and A. Jenkins. Oxford: Blackwell, 126–47.

EVANS, R. 1991. 'Legacy of Woe'. *Geographical Magazine* 63, no. 6:34–8.

HARTSHORNE, R. 1950. 'The Functional Approach in Political Geography'. *Annals, Association of American Geographers* 40:95–130.

HIRSCH, P. 1993. 'The Socialist Developing World in the 1990s'. *Geography Review* 7, no. 2:35–7.

HUNTINGTON, S.P. 1993. 'The Clash of Civilizations'. *Foreign Affairs* 72, no. 3:22–49.

JOHNSTON, R.J. 1982. *Geography and the State: An Essay in Political Geography.* New York: St Martin's Press.

———. 1985. *The Geography of English Politics.* London: Croom Helm.

———. 1993. 'Tackling Global Environmental Problems'. *Geography Review* 6, no. 2:27–30.

JONES, S.B. 1954. 'A Unified Field Theory of Political Geography'. *Annals, Association of American Geographers* 44:111–23.

KASPERSON, R.E., and J.V. MINGHI. 1969. *The Structure of Political Geography.* Chicago: Aldine.

KOHR, L. 1957. *The Breakdown of Nations.* Swansea: Christopher Davies.

KREBHEIL, E. 1916. 'Geographic Influences in British Elections'. *Geographical Review* 2:419–32.

LARKINS, J., and S. PARISH. 1982. *Australia's Greatest River.* Melbourne: Rigby.

LORENZ, K. 1967. *On Aggression.* London: Methuen.

MACKINDER, H.J. 1919. *Democratic Ideals and Reality.* New York: Henry Holt.

MONTAGUE, A. 1976. *The Nature of Human Aggression.* New York: Oxford University Press.

MORRIS, D. 1967. *The Naked Ape.* New York: McGraw Hill.

OPENSHAW, S., P. STEADMAN, and O. GREENE. 1983. *Doomsday: Britain After Nuclear Attack.* Oxford: Blackwell.

OVERTON, J.D. 1981. 'A Theory of Exploration'. *Journal of Historical Geography* 7:53–70.

PADDISON, R. 1983. *The Fragmented State: The Political Geography of Power.* New York: St Martin's Press.

ROKKAN, S. 1980. 'Territories, Centres and Peripheries'. In *Centre and Periphery,* edited by J. Gottman. London: Sage, 163–204.

ROYLE, S. 1991. 'St Helena: A Geographical Summary'. *Geography* 76:266–8.

TAYLOR, P.J. 1989. *Political Geography: World Economy, Nation State and Locality,* 2nd ed. London: Longman.

WOLFE, R.I. 1962. 'Transportation and Politics: The Example of Canada'. *Annals, Association of American Geographers* 52:176–90.

CHAPTER 10

ARKELL, T. 1991. 'The Decline of Pastoral Nomadism in the Western Sahara'. *Geography* 76:162–6.

BARRETT, H., and A. BROWNE. 1991. 'Environmental and Economic Sustainability: Woman's Horticultural Production in the Gambia'. *Geography* 76:241–8.

BASSETT, T.J. 1988. 'The Political Ecology of Peasant-Herder Conflicts in the Northern Ivory Coast'. *Annals, Association of American Geographers* 78:453–72.

BINFORD, L. 1968. 'Post-Pleistocene Adaptations'. In *New Perspectives in Archaeology,* edited by L. Binford and S. Binford. Chicago: Aldine, 313–41.

BOSERUP, E. 1965. *The Conditions of Agricultural Change.* London: Allen and Unwin.

BOWLER, I.R. 1985. *Agriculture under the Common Agriculture Policy.* Manchester: Manchester University Press.

CHISHOLM, M. 1962. *Rural Settlement and Land Use: An Essay in Location.* London: Hutchinson.

CONZEN, M.P. 1971. *Frontier Farming in an Urban Shadow.* Madison: State Historical Society of Wisconsin.

ERICKSON, R.A. 1989. 'The Influence of Economics in Geographic Inquiry'. *Progress in Human Geography* 13:223–49.

FRIEDMANN, H. 1991. 'Changes in the International Division of Labour: Agri-Food Complexes and Export Agriculture'. In *Towards a New Political Economy of Agriculture,* edited by W. Friedland et al. Boulder: Westview Press, 65–93.

GRIFFIN, E. 1973. 'Testing the Von Thünen Theory in Uruguay'. *Geographical Review* 63:500–16.

GRIGG, D.B. 1974. *The Agricultural Systems of the World: An Evolutionary Approach.* New York: Cambridge University Press.

———. 1992. 'Agriculture in the World Economy: An Historical Geography of Decline'. *Geography* 77:210–22.

GROSSMAN, L.S. 1993. 'The Political Ecology of Banana Exports and Local Food Production in St. Vincent, Eastern Caribbean'. *Annals, Association of American Geographers* 83:347–67.

HALL, P.G., ed. 1966. *Von Thünen's Isolated State.* Oxford: Pergamon.

HORVATH, R.J. 1969. 'Von Thünen's Isolated State and the Area Around Addis Ababa, Ethiopia'. *Annals, Association of American Geographers* 59:308–23.

JOHNSON, H.B. 1962. 'A Note on Thünen's Circles'. *Annals, Association of American Geographers* 52:213–20.

JONASSON, O. 1925. 'Agricultural Regions of Europe'. *Economic Geography* 1:277–314.

KING, L.J. 1969. *Statistical Analysis in Geography*. Englewood Cliffs: Prentice-Hall.

LEAMAN, J.H., and E.C. CONKLING. 1975. 'Transport Change and Agricultural Specialization'. *Annals, Association of American Geographers* 65:425–37.

MCCALLUM, J. 1980. *Unequal Beginnings: Agriculture and Economic Development in Quebec and Ontario until 1870*. Toronto: University of Toronto Press.

MORRISH, M. 1994. 'China Takes the Road to Market'. *Geographical Magazine* 67, no. 4:43–5.

MULLER, P.O. 1973. 'Trend Surfaces of American Agricultural Patterns: A Macro-Thünen Analysis'. *Economic Geography* 49:228–42.

NORTON, W., and E.C. CONKLING. 1974. 'Land Use Theory and the Pioneering Economy'. *Geografiska Annaler* 56B:44–56.

PEET, J.R. 1969. 'The Spatial Expansion of Commercial Agriculture in the Nineteenth Century: A Von Thünen Interpretation'. *Economic Geography* 45:283–301.

RAITZ, K.B. 1973. 'Ethnicity and the Diffusion and Distribution of Cigar Tobacco Production in Wisconsin and Ohio'. *Tijdschrifte voor Economische en Sociale Geografie* 64:293–306.

REITSMA, H.J. 1971. 'Crop and Livestock Production in the Vicinity of the United States–Canada Border'. *Professional Geographer* 23:216–23.

SALAMON, S. 1985. 'Ethnic Communities and the Structure of Agriculture'. *Rural Sociology* 50:323–40.

SAUER, C.O. 1952. *Agricultural Origins and Dispersals*. New York: American Geographical Society.

SCHLEBECKER, J.T. 1960. 'The World Metropolis and the History of American Agriculture'. *Journal of Economic History* 20:187–208.

SINCLAIR, R. 1967. 'Von Thünen and Urban Sprawl'. *Annals, Association of American Geographers* 57:72–87.

SMITH, J.R. 1925. *North America*. New York: Harcourt Brace and Co.

SMITH, W. 1984. 'The "Vortex" Model and the Changing Agricultural Landscape of Quebec'. *Canadian Geographer* 28:358–72.

TICKELL, O. 1992. 'Going to Seed'. *Geographical Magazine* 64, no. 12:29–32.

WOLPERT, J. 1964. 'The Decision Process in Spatial Context'. *Annals, Association of American Geographers* 54:537–58.

YOUNG, L.J. 1991. 'Agriculture Changes in Bhutan: Some Environmental Questions'. *Geographical Journal* 157:172–8.

ZIMMERER, K.S. 1991. 'Wetland Production and Smallholder Persistence: Agriculture Change in a Highland Peruvian Region'. *Annals, Association of American Geographers* 81:443–63.

CHAPTER 11

BASSETT, K., and J. SHORT. 1989. 'Development and Diversity in Urban Geography'. In *Horizons in Human Geography*, edited by D. Gregory and R. Walford. London: Macmillan, 175–93.

BERRY, B.J.L. 1967. *Geography of Market Centres and Retail Distribution*. Englewood Cliffs: Prentice-Hall.

BORCHERT, J.R. 1967. 'American Metropolitan Evolution'. *Geographical Review* 57:301–32.

BRUSH, J.E. 1953. 'The Hierarchy of Central Places in South West Wisconsin'. *Geographical Review* 43:350–402.

BYLUND, E. 1960. 'Theoretical Considerations Regarding the Distribution of Settlement in Inner North Sweden'. *Geografiska Annaler* 42B:225–31.

CATER, J., and T. JONES. *Social Geography: An Introduction to Contemporary Issues*. New York: Arnold.

CHRISTALLER, W. 1966. *Central Places in Southern Germany*. Translated by C.W. Baskin. Englewood Cliffs: Prentice-Hall.

CHURCH, R.L., and T.L. BELL. 1988. 'An Analysis of Ancient Egyptian Settlement Patterns Using Location-Allocation Covering Models'. *Annals, Association of American Geographers* 78:701–14.

COOKE, P. 1990. 'Modern Urban Theory in Question'. *Transactions, Institute of British Geographers* 15:331–43.

COSGROVE, D. 1984. *Social Formation and Symbolic Landscape*. London: Croom Helm.

DAVIES, S., and M. YEATES. 'Exurbanization as a Component of Migration: A Case Study in Oxford County, Ontario'. *Canadian Geographer* 35:177–86.

DEAR, M.J., and J.R. WOLCH. 1987. *Landscapes of Despair*. Princeton: Princeton University Press.

DUNCAN, J.S. 1990. *The City as Text: The Politics of Landscape Interpretation in the Kandyman Kingdom*. New York: Cambridge University Press.

ENGLISH, P.W., and R.C. MAYFIELD, eds. 1972. *Man, Space and Environment*. New York: Oxford University Press, 601–10.

EARLE, C.V. 1977. 'The First English Towns of North America'. *Geographical Review* 67:34–50.

GODFREY, B.J. 1991. 'Modernizing the Brazilian City'. *Geographical Review* 81:18–34.

GREENBIE, B.B. 1981. *Spaces: Dimensions of the Human Landscape*. New Haven: Yale University Press.

GROSSMAN, D. 1971. 'Do We Have a Theory for Rural Settlement?' *Professional Geographer* 23:197–203.

HALL, P., et al. 1973. *The Containment of Urban England*. London: Allen and Unwin.

HARRIS, C.D., and E.L. ULLMAN. 1945. 'The Nature of Cities'. *Annals, American Academy of Political and Social Science* 37:7–17.

HARVEY, D.W. 1969. *Explanation in Geography*. London: Arnold.

———. 1973. *Social Justice and the City*. London: Arnold.

———. 1982. *The Limits to Capital*. Oxford: Blackwell.

———. 1989. *The Urban Experience*. Oxford: Blackwell.

HOSKINS, W.G. 1955. *The Making of the English Landscape*. London: Hodder and Stoughton.

HOYT, H. 1939. *The Structure and Growth of Residential Neighbourhoods in American Cities*. Washington, DC: Federal Housing Administration.

HUDSON, J.C. 1969. 'A Location Theory for Rural Settlement'. *Annals, Association of American Geographers* 59:461–95.

KEDDIE, P.D., and A.E. JOSEPH. 1991. 'The Turnaround of the Turnaround? Rural Population Change in Canada,

1976 to 1986'. *Canadian Geographer* 35:367–79.

KNOX, P. 1992. 'Suburbia by Stealth'. *Geographical Magazine* 44, no. 8:26–9.

———. 1994. *Urbanization: An Introduction to Urban Geography*. Englewood Cliffs: Prentice-Hall.

LEY, D. 1992. 'Gentrification in Recession: Social Change in Six Canadian Inner Cities, 1981–1986'. *Urban Geography* 13:230–56.

———. 1993. *The New Middle Class and the Remaking of the Central City*. Oxford: Oxford University Press.

LÖSCH, A. 1954. *The Economics of Location*. Translated by W.H. Woglom. New Haven: Yale University Press.

LOWDER, S. 1994. 'Urban Development in Ecuador'. *Geography Review* 7, no. 4:2–6.

LYNCH, K. 1960. *The Image of the City*. Cambridge, Mass.: MIT Press.

MEYER, D.R. 1980. 'A Dynamic Model of the Integration of Frontier Urban Places into the United States System of Cities'. *Economic Geography* 56:120–40.

MIDDLETON, N. 1994. 'Environmental Problems in Mexico City'. *Geography Review* 7, no. 4:16–18.

MULLER, E.K. 1976. 'Selective Urban Growth in the Middle Ohio Valley, 1800–1860'. *Geographical Review* 66:178–99.

PAHL, R.E. 1966. 'The Rural/Urban Continuum'. *Sociologia Ruralis* 6:299–327.

PARKE, R.E., and E.W. BURGESS. 1921. *Introduction to the Science of Sociology*. Chicago: University of Chicago Press.

ROZMAN, G. 1978. 'Urban Networks and Historical Stages'. *Journal of Interdisciplinary History* 9:65–92.

SARGENT, C.G. 1975. 'Towns of the Salt River Valley, 1870–1930'. *Historical Geography Newsletter* 5, no. 2:1–9.

SASSEN, S. 1991. *The Global City: New York, London, Tokyo*. Princeton: Princeton University Press.

SJOBERG, G. 1960. *The Pre-Industrial City: Past and Present*. Glencoe: Free Press.

SMAILES, A.E. 1957. *The Geography of Towns*. London: Hutchinson.

SMITH, R.L. 1985. 'Activism and Social Status as Determinants of Neighbourhood Identity'. *Professional Geographer* 37:421–32.

VANCE, J.E. 1970. *The Merchant's World: The Geography of Wholesaling*. Englewood Cliffs: Prentice-Hall.

———. 1971. 'Land Assignment in Pre-Capitalist, Capitalist and Post-Capitalist Societies'. *Economic Geography* 47:101–20.

———. 1990. *The Continuing City: Urban Morphology in Western Civilization*. Baltimore: Johns Hopkins University Press.

WADE, R.C. 1959. *The Urban Frontier: 1790–1830*. Cambridge, Mass.: Harvard University Press.

WHEATLEY, P. 1971. *The Pivot of the Four Quarters: A Preliminary Enquiry into the Origins and Character of the Ancient Chinese City*. Chicago: Aldine.

WIRTH, C. 1938. 'Urbanism as a Way of Life'. *American Journal of Sociology* 44:46–63.

CHAPTER 12

BAND, G.C. 1991. 'Fifty Years of UK Offshore Oil and Gas'. *Geographical Journal* 157:179–89.

BONE, R.M. 1992. *The Geography of the Canadian North*. Toronto: Oxford University Press.

CROMAR, P. 1979. 'Spatial Change and Economic Organization: The Tyneside Coal Industry (1751–1770)'. *Geoforum* 10:45–57.

DANIELS, P.W. 1985. *Service Industries: A Geographical Appraisal*. New York: Methuen.

DICKEN, P. 1992. *Global Shift: The Internationalization of Productive Activity*. London: Chapman.

FREEBERNE, M. 1993. 'The Northeast Asia Regional Development Area: Land of Metal, Wood, Water, Fire and Earth'. *Geography* 78:420–32.

FRIEDMANN, J. 1972. 'A General Theory of Polarized Development'. In *Growth Centers in Regional Economic Development*, edited by N.M. Hansen. New York: Free Press, 82–102.

FRIEDRICH, C., trans. 1926. *Alfred Weber's Theory of the Location of Industries*. Cambridge, Mass.: Harvard University Press.

GRAHAM, J., et al. 1988. 'Restructuring in U.S. Manufacturing: The Decline of Monopoly Capitalism'. *Annals, Association of American Geographers* 78:473–90.

GRUBEL, H.G., and K. WALKER. 1989. *Service Industry Growth: Causes and Effects*. Vancouver: The Fraser Institute.

HOOVER, E.M. 1948. *The Location of Economic Activity*. Toronto: McGraw Hill.

HOPKINS, J.S.P. 1990. 'West Edmonton Mall: Landscape of Myths and Elsewhereness'. *Canadian Geographer* 34:2–17.

HOTELLING, H. 1929. 'Stability in Competition'. *Economic Journal* 39:40–57.

HUDSON, R., and D. SADLER. 1989. *The International Steel Industry: Restructuring, State Policies and Localities*. London: Routledge.

JACKSON, E.L., and D.B. JOHNSON, eds. 1991. 'The West Edmonton Mall and Mega Malls'. *Canadian Geographer* 35, no. 3.

JAKLE, J.A. 1985. *The Tourist: Travel in Twentieth-Century North America*. Lincoln: University of Nebraska Press.

KENNELLY, R.A. 1968. 'The Location of the Mexican Steel Industry'. In *Readings in Economic Geography: The Location of Economic Activity*, edited by R.H.T. Smith, E.J. Taafe, and L.J. King. Chicago: Rand McNally, 126–57.

LEY, D., and K. OLDS. 1988. 'Landscape as Spectacle: World's Fairs and the Culture of Heroic Consumption'. *Environment and Planning D; Society and Space* 6:191–212.

MASSEY, D. 1984. *Spatial Divisions of Labour: Social Structures and the Geography of Production*. New York: Methuen.

MORRISH, M. 1994. 'China Takes the Road to Market'. *Geographical Magazine* 67, no. 4:43–5.

MURGATROYD, T. 1993. 'The New Zealand Power Crisis'. *Geography* 78:433–7.

O'HARE, G., and H. BARRETT. 1993. 'The Fall and Rise of the Sri Lankan Tourist Industry'. *Geography* 78:438–42.

POLLARD, J.S. 1989. 'Gender and Manufacturing Employment: The Case of Hamilton'. *Area* 21:377–84.

PRED, A. 1966. *The Spatial Dynamics of U.S. Urban-Industrial Growth, 1800–1914*. Cambridge, Mass.: MIT Press.

ROSTOW, W.W. 1960. *The Stages of Economic Growth*. Cambridge: Cambridge University Press.

SHIELDS, R. 1989. 'Social Spatialisation and the Built Environment: The Example of West Edmonton Mall'. *Environment and Planning D: Society and Space* 7:147–64.

SQUIRE, S.J. 1994. 'Accounting for Cultural Meanings: The Interface Between Geography and Tourism Studies Re-examined'. *Progress in Human Geography* 18:1–16.

STAFFORD, H.A. 1985. 'Environmental Protection and Industrial Location'. *Annals, Association of American Geographers* 75:227–40.

WEBBER, M.J. 1984. *Industrial Location*. Beverly Hills: Sage.

WHEAT, S. 1994. 'Taming Tourism'. *Geographical Magazine* 67:16–19.

CHAPTER 13

APPLETON, J.H. 1962. *The Geography of Communications in Great Britain*. Toronto: Oxford University Press.

BLAINEY, G. 1968. *The Tyranny of Distance*. London: Macmillan.

CARROTHERS, G.A.P. 1956. 'An Historical Review of the Gravity and Potential Concepts of Human Interaction'. *Journal of the American Institute of Planners* 22:94–102.

CLEARY, M., and R. BEDFORD. 1993. 'Globalisation and the New Regionalism: Some Implications for New Zealand Trade'. *New Zealand Journal of Geography* 20:19–22.

CLIFF, A.D., P. HAGGETT, and J.K. ORD. 1986. *Spatial Aspects of Influenza Epidemics*. London: Pion.

———, and M.R. Smallman-Raynor. 1992. 'The AIDS Pandemic: Global Geographical Patterns and Local Spatial Processes'. *Geographical Journal* 158:182–98.

CONKLING, E.C., and M.H. YEATES. 1976. *Man's Economic Environment*. Toronto: McGraw Hill.

FORER, P. 1978. 'A Place for Plastic Space?' *Progress in Human Geography* 3:230–67.

FREEMAN, D.B. 1985. 'The Importance of Being First: Preemption by Early Adopters of Farming Innovations in Kenya'. *Annals, Association of American Geographers* 75:17–28.

GARRISON, W.L. 1960. 'Connectivity of the Interstate Highway System'. *Regional Science Association, Papers and Proceedings* 6:121–37.

GOULD, P. 1993. *The Slow Plague: A Geography of the AIDS Pandemic*. Oxford: Blackwell.

HAGERSTRAND, T. 1951. 'Migration and the Growth of Culture Regions'. *Lund Studies in Geography* B, 3.

———. 1967. *Innovation Diffusion as a Spatial Process*. Translated by A. Pred. Chicago: University of Chicago Press.

HALL, E.T. 1966. *The Hidden Dimension*. Garden City: Doubleday.

HARVEY, D. *Explanation in Geography*. London: Arnold.

HILLING, D. 1977. 'The Evolution of a Port System: The Case of Ghana'. *Geography* 62:97–105.

HOYLE, B.S., and R.D. KNOWLES. 1992. *Modern Transport Geography*. London: Belhaven.

ISARD, W. 1956. *Location and Space Economy*. New York: Wiley.

———, et al. 1960. *Methods of Regional Analysis: An Introduction to Regional Science*. Cambridge, Mass.: MIT Press.

JANELLE, D.G. 1968. 'Central Place Development in a Time-Space Framework'. *Professional Geographer* 20:5–10.

———. 1969. 'Spatial Reorganization: A Model and Concept'. *Annals, Association of American Geographers* 59:348–64.

JORDAN, T.G., M. DOMOSH, and L. ROWNTREE. 1994. *The Human Mosaic: A Thematic Introduction to Cultural Geography*, 6th ed. New York: Harper Collins.

KANSKY, K. 1963. *The Structure of Transportation Networks*. Chicago: University of Chicago, Department of Geography, Research Paper No. 84.

KNIFFEN, F. 1951. 'The American Covered Bridge'. *Geographical Review* 41:114–23.

KNOX, P. 1994. *Urbanization: An Introduction to Urban Geography*. Englewood Cliffs: Prentice-Hall.

LOWE, J.C., and S. MORYADAS. 1975. *The Geography of Movement*. Boston: Houghton Mifflin.

PYLE, G. 1969. 'The Diffusion of Cholera in the United States in the Nineteenth Century'. *Geographical Analysis* 1:59–75.

RAITZ, K.B. 1973. 'Ethnicity and the Diffusion and Distribution of Cigar Tobacco Production in Wisconsin and Ohio'. *Tijdschrifte voor Economische en Sociale Geografie* 64:293–306.

REILLY, W.J. 1931. *The Law of Retail Gravitation*. New York: Free Press.

ROGERS, E.M. 1962. *Diffusion of Innovations*. New York: Free Press.

ROSTOW, W.W. 1960. *The Stages of Economic Growth*. Cambridge: Cambridge University Press.

SMALLMAN-RAYNOR, M.R., A.D. CLIFF, and P. HAGGETT. 1992. *Atlas of AIDS*. Oxford: Blackwell.

TAAFE, E.J., R.L. MORRILL, and P. GOULD. 1963. 'Transport Expansion in Underdeveloped Countries: A Comparative Analysis'. *Geographical Review* 53:503–29.

ULLMAN, E.L. 1956. 'The Role of Transportation and the Bases for Interaction'. In *Man's Role in Changing the Face of the Earth*, edited by W.L. Thomas Jr. Chicago: University of Chicago Press, 862–80.

WARNTZ, W. 1957. 'Transportation, Social Physics and the Law of Refraction'. *Professional Geographer* 9:2–7.

WATSON, J.W. 1955. 'Geography: A Discipline in Distance'. *Scottish Geographical Magazine* 71:1–13.

WILSON, A.G., and R.J. BENNETT, eds. 1986. *Mathematical Methods in Human Geography and Planning*. Chichester: Wiley.

ZIPF, G.K. 1949. *Human Behavior and the Principle of Least Effort*. Cambridge, Mass.: Addison Wesley.

CONCLUSION

BUTTIMER, A. 1974. *Values in Geography*. Washington, DC: Association of American Geographers, Commission on College Geography, Resource Paper No. 24.

DICKEN, P. 1993. 'The Changing Organization of the Global Economy'. In *The Challenge for Geography: A Changing World, A Changing Discipline*, edited by R.J. Johnston. Oxford: Blackwell, 31–53.

GREENLEE, J. 1988. 'Infinite Accretion'. *Geographical Magazine* 60, no. 11:24–6.

GREGORY, D. 1994. *Geographical Imaginations*. Oxford: Blackwell.

HANDBURY-TENISON, R., and C. SHANKEY. 1989. 'An Uphill Struggle for Survival'. *Geographical Magazine* 61, no. 9:29–33.

HARVEY, D.W. 1973. *Social Justice and the City*. London: Arnold.

———. 1990. 'Between Space and Time: Reflections on the Geographical Imagination'. *Annals, Association of American Geographers* 80:418–34.

JENKINS, S. 1992. 'Four Cheers for Geography'. *Geography* 77:193–7.

JOHNSTON, R.J. 1991. *A Question of Place: Exploring the Practice of Human Geography*. Oxford: Blackwell.

MORRILL, R.L. 1987. 'A Theoretical Imperative'. *Annals, Association of American Geographers* 77:535–41.

OHMAE, K. 1993. 'The Rise of the Region State'. *Foreign Affairs* 76, no. 2:78–87.

PEATTIE, R. 1940. *Geography as Human Destiny*. Port Washington, NY: Kennikat Press.

PRINCE, H.C. 1961. 'The Geographical Imagination'. *Landscape* 11, no. 2:22–5.

Index

(Boldface indicates illustrations)

Abler, 178
aboriginal populations: and
    colonialism, 195; mortality rate,
    106
abortion, 102–3
absolute location, 37
absolute space, 36
accessibility, 41
acculturation, 187
acid rain, 80, 91; and dead fish, **78**
Addis Ababa (Ethiopia), 229, 262
administrative: centres, 248;
    principles, 255–6
adoption, of innovation, 297, **298**
aerial photography, 27, 51
Afghanistan refugees, 133
Africa, 141, **201**, 202–3, **202**;
    decolonization of, 4;
    refugees, 133
age-sex structure: Brazil, **112**;
    Canada, **113**; China, **110**
age structure of populations, **111**
agglomeration, 41, 270, 281
aggression, 213
aging, of world population, 111, 114
Agra (India), 168
agriculturalists, impact on
    ecosystems, 82
agricultural landscape, 219–23
agricultural revolution, 79, 148–9;
    and population growth, 114
agriculture: core areas, 231; and
    ethnicity, 222–3; industry of,
    236–7; and institutional change,

233; and location, 218–27;
    political ecology, 237–9; and
    technology, 233, 246; and
    world economy, 236; world
    regions, **232**
AIDS, diffusion of, 298
air transport, 299, 302
Alexander the Great, 193
alienation, 180
Alsace, 203
Amazon rain forest, **86**
Amritsar (India), 168
anarchism, 209, 250
Andalusia, 204
animal domestication, 79, 86, 88, 220
animism, 165
Annales de Géographie (Vidal), 21
*Annals of the Association of American
    Geographers*, 33
Antarctica, 81, 93, 95
Antarctic ice sheets, 69
anthropogeography, 29
Anthropographie (Ratzel), 20
antinatalist policies, 109–10
apartheid, 181, 182, **182**
apes, 70
Apian (map maker), 16
Appalachian-Acadian mountain
    system, 63
applied geography, 26–7
Arabia, 13
Arab Maghreb union, 207, 208, 234
Aral Sea, 90–1, 108
ARC/INFO software, 49
areal differentiation, 23
Aristotle, 11, 31, 193, 197, 207
artefacts, 146

artificial selection of plants and
    animals, 231
Asby (Sweden), **292**
ASEAN Free Trade Area, 303
Asia: decolonization of, 4; and
    immigration, 131; refugees, 133
assimilation, 187
Association of Southeast Asian
    Nations *See* ASEAN
atmosphere, 56, 77
Australia, 22, 209, 233, 252, 294
Australoid, 72
Australopithecus afarensis, 70, 71, 72
authority and power, 181
auto-air-amenity, 252
Azerbaijan, 204

Bacon, Francis, 57
Baghdad, 248
Bahn, 78
Bahrain, 90
Baikal, Lake, 90
Balfour Declaration (1917), 197, 198
bands, 148
Banff mountain landscape, **58**
Bangladesh, 139, 140, 206; sea-level
    change, **94**
Banks, 183
Basque country, 201, 203
Bassett, 258
behavioural: approach to industrial
    location, 271–2; change, 155
Belgium, 161, **161**, 162, 194, 201,
    272
Belize, 279–80
Berlin Wall, **4**, 132, 200

Berry, Brian, 178, 257
beta index, 300
Bhopal, 81
Bhutan, 237, 238
Bible Belt, 40
binational states, 194
Binford, 231
biodiversity crisis, 85
biogeochemical cycles, 77
biomass, 79
biosphere, 77
biotechnology, 239
birth control policies, 109; *See also*
    contraception
birth rate, Finland, 6
Biswas, 168
black blizzards, 90
Blainey, Geoffrey, 294
Blaut, 180
Boas, Franz, 145
Bodin, Jean, 17, 30, 197
Bone, Robert M., 7
Borchert, J.R., 252
Boserup, Ester, 117, 231; theory of
    population growth, 117
Botswana, 101, 105
boundaries, 206
branch plant economy, 277
Brasilia, 251
Brazil, 260; age-sex structure, **112**;
    rain forest, 85; urban slum, **184**
Britain, 207, 211; agriculture after
    1750, 232–3; colonies and,
    196–7; energy supplies, 275;
    transport systems, 298–9, **299**;
    urban planning, 251; *See also*
    United Kingdom

British Antarctic Survey, 93
British empire, 194, **195**
British Land Utilization Survey, 26
Brittany, 203
Brundlandt Report (1987), 82
Brundtland Commission, 81
Brunhes, Jean, 21
Brush, 257
Buddhism, 164–5, **165**
Buffon (naturalist), 17, 80
Bulgaria, 204
Bunge, W., 33, 214
bureaucratic systems, 150
Burgess, 256
Burundi, 309
Büsching (geographer), 17
Bylund, E., 244

Cairo, 262
Cajun group, 172–3
Calcutta, 168
Cameroon, 101
Canada, 194, 199; age-sex structure,
    113; French and English in, **162**;
    geomorphology of, 63; industry
    in, 276–7; internal divisions in,
    206; as multilingual state, 161,
    162, **162**; population data, 106;
    prairies, **40**; refugees in, 135;
    regional inequalities, 284;
    regions, 153, **153**; and staple
    exports, 294; trans-continental
    railway, 298
Canadian Shield, **63**, 284
canals, 299, 302, 307
Canberra, **250**
cane toad, 88
capitalism, 35, 36, 116, 141, 207,
    249, 263, 268, 307; impact on
    environment, 80; and modes of
    production, 154; and
    restructuring, 237; and state
    power, 210; and surplus
    population, 117; *See also* Industrial
    Revolution
Carey, H.C., 294
Carneiro, 150
Carrothers, G.A.P., 294
carrying capacity, 109–10, 231
Carson, Rachel, 80
cartography, 47–8
caste systems, 292
Catalonia, 203
Catholic religion, 96
cattle: herding, 67; ranching, 85
Caucasoid, 72
CBR (crude birth rate), 100, 105–6,
    118; world distribution, **104**
CDR (crude death rate), 104–6, **107**,
    118
ceiling rent, 223
Celtic languages, 159, 163
Central African Republic, 101
central Asia, life expectancy, 108

centralization, 281
central place: administration
    principle, **257**; marketing
    principle, **256**; theory, 253–6,
    257, **257**; transportation
    principle, **257**
Central Places in Southern Germany
    (Christaller), 253
centrifugal forces, 200–1
centripetal forces, 200–1
ceremonial centres, 248
CFCs *See* chloro-fluorocarbons
chain migration, 187
Chamberlain, Houston Stewart, 73
Champlain, Samuel de, 199; map of
    New France, **17**
chance factors, 295
Chandigarh (India), 251
Chang Chi'en, 13
Channel tunnel, 302, 303
chemical: cycling and energy flows,
    77; weathering, 59
Chernobyl, 81
Childe, 149
Chile, 209
chimpanzees, 70
China, 116, 195, 209; age-sex
    structure, **110**; agricultural
    change, 238; fertility rate, 110;
    geography in, 12–13; Great Leap
    Forward, 284; and industry, 278;
    map making, 13; population in,
    111
Chisholm, 227
chloro-fluorocarbons (CFCs), 81, 92–3
cholera, 91, 296; epidemic North
    America, 42
chorography, 12
chorology, 20, 21, 23–4, 36
choropleth map, 47, **48**
Christaller, Walter, 253, 254, 271
Christchurch (New Zealand), 183
Christianity, 13, 14, 164–5, **165**,
    166, 167–8
Churchill, Winston, 248
circulation technologies, 280
cities, 248; industrial, 249; in less
    developed world, 248;
    postindustrial, 260–1;
    postmodern, 261; preindustrial,
    248–9; visual quality, 261
city states, 12, 193
civilization, 149–50, **150**, 214–16
class, 180, 292
classical economics, 219
classical geography, 11–12
classical Greece, 10–12, 129, 193,
    207, 248
classical trade theory, 303
climate, 63, 65, 92; and agriculture,
    220; global distribution of, 65,
    65; Mediterranean, 234
Club of Rome, 115
clustered pattern, 41
coal, 79, 273, 274, 275

Coalbrookdale, 249
coercive theory, 149–50
cognitive distance, 41, 290–2
cold war, 200
collective consumption, 280
colonialism, 141, 194–7, 203; and
    Africa, 202–3; and environmental
    degradation, 80; and
    industrialization, 278; and
    plantation agriculture, 235
Columbus, Christopher, 14
Comecon (Council for Mutual
    Economic Assistance), 207
commercial: agriculture, 223, 232–4;
    geography, 219
Commission of the European
    Communities, 206
commodity flows, 302
common market, 303
Commonwealth, 207
communal living, 169
communism, 180–1, 200
Company of Hostmen, 273
compound unitary government, 209
computer-assisted cartography, 46, 49
Comte, Auguste, 32
concentric zone model, 256, 258
Condition of Postmodernity, The
    (Harvey), 179
conflicts, **212**, 213–14
conformal projection, 49
Confucianism, 13, 165
conservation, 95
Conservative party: Britain, 211;
    Canada, 211
constitutional monarchy, 207
contextual effect, 213
continental drift, 58, **58**
contraception, 101–2, **103**, 118
conurbations, 260
convergence rate, 290
Cook, James, 14, 15
Cooke, 258
cooperative public-private projects,
    280
Copernicus, 15
Cordilleran: belts, **62**, 63; mountain
    system, 63
core/periphery, 155, **156**, 252, 277,
    283; economic spatial structure,
    204
cornucopian thesis, 115
correlation and regression analysis,
    228
Corsica, 203
Cosgrove, 261
Cosmography (Munster), 16
Cosmos (Humboldt), 18
cost: distance, 290; of war, 214
cotton industry, 274
Council for the Development of
    French in Louisiana (CODOFIL),
    172
Council for Mutual Economic
    Assistance (Comecon), 207

counterfactual method, 198, 199
counterurbanization, 246
Cousin, Victor, 31
cows, as sacred to Hindus, 169
creole, 163, 172, 173
crime, 181
cycle of poverty, 263

da Gama, Vasco, 14
dairying, 235
Dalrymple, Alexander, 15–16
Daniels, P.W., 281, 371
Darby, Abraham, 249
Darwin, Charles, 20, 30, 73, 88
Davidson Glacier, **60**
Davies, 247
Davis, William Morris, 21, 23, 31
dead: use of land for, 169; fish, **78**
Dead Sea, 56
Dear, M.J., 264
death control policies, 109
death rate, Finland, 6
debt, as world problem, 138–9
decentralization, 7, 281
deep ecology, 82, 96
defence centres, 248
deforestation, 78, 84
deglomeration, 41, 270
deindustrialization, 281
Demangeon, Albert, 21
Demeny, 105
democracy, 207
Democratic party, 211
demographic data (Finland and
    France), 6
demographic transition: model, 307;
    theory, 117–18, **117**, 120, 126
demography, 99
Deng Xiaoping, 284
density, 122–5
dependency theory, 197
depopulation in rural areas, 246
deregulation, 280
desertification, 84, 86
deserts, 68
de Seversky, 200
determinism, 30
Deutsch (political scientist), 197, 198
development: concept, 45–6;
    measures, 46
developmentalism, 127
devolution, 203
dewesternization, 215
Dias, Bartolomeu, 14
dictatorship, 209
Die Erdkunde (Ritter), 19
diffusion, 295–8; and agriculture,
    295; concept of, 42; of disease,
    298; human consequences, 297;
    of industry, 274; innovation, 295,
    297; research, 295–8
digital mapping, 49; *See also*
    computer-assisted cartography
disappearing peoples, 308–9, **308**

disasters, 139
discrimination by employers, 282
disease, diffusion of, 298
Disney World, 279
dispersion of rural settlements,
    243–4, 247
distance, 294; concept of, 40–2,
    288–95; decay curve, **41**; and
    international relations, 198;
    measuring systems, 41; and time,
    41; and transport cost, 227
distemic space, 262
distribution, 122–5
Distribution of Industry Act (1945),
    251
domain, 155, **156**
domestication of plants and animals,
    230–1
domestic space, 185
dot maps, 47, **48**
doubling time, 108
Doukhobors, 169, **169**; and farm
    villages, 244
Doxiadis, 289
Drake, Francis, 15
drought, 81
dry season horticultural production,
    238
Durkheim, Emile, 21, 31, 145, 146
Dust Bowl, 90, **90**, 174

Earle, C.V., 252
earth, 56–69; crust of, 57–9;
    evolution, 56; life on, 69–70;
    revolution around sun, **57**
Earth Day, 80
Earth Observing System, 52
earth orbital satellites, 51
earthquakes, 58–9
Earth Resources Technology Satellite
    (ERTS), 51
Easter Island, 129; deforestation, 78
East India Company, 80
Ebstorf, 13
ecology: concept of, 76; shallow vs
    deep, 82
economic: base theory, 250;
    colonialism, 235; determinism,
    36; distance, 290; inequalities,
    283–4; operator, 220;
    regionalism, 215; union, 303
Economic Community of West
    African States, 207
economic factors, and fertility, 100–1
economic growth, 283; and
    employment, **281**; and railways,
    301
economic integration, 303; and
    European Community, 284
economic rent, 223, **225**, 226, 227,
    256, 258
economics, and human geography,
    219
Economics of Location, The (Lösch), 253

ecosphere, 77
ecosystems, 76–7; human impacts on,
    82
ecotourism, 279–80
Edmonton, **290**
Edrisi (geographer), 14
education, 141; and geography, 181
Ehrlich, 115
ekistics, 289
elections and geography, 210–13
electricity, 79
Electricorp, 275
electromagnetic radiation, 51
élite, 150, 207; in less developed
    world, 215; in preindustrial cities,
    249
élitist landscapes, 183
empires and exploration, 194, **195**
empiricism, 30, 119
employment, 178; and economic
    growth, **281**; geography of, 7
enclosure, 233, 243
energy, 77, 79; crisis in New Zealand,
    275; sources, 274–5
Engels, Friedrich, 36
English language, development,
    159–60
entropy, 295; -maximizing models,
    295
environment, 80–2; and cultural
    landscape, 31; and economy,
    81–2; and global problems, 210;
    human impact on, 310; industrial
    impacts, 282; politics of, 80–1;
    and population density, 125; and
    religion, 96; and trade policies, 82
environmental degradation: and
    colonialism, 80; prevention, 95
environmental determinism, 22–3,
    29–31, 33, 80, 145, 199
environmental ethics, 79–82
epeiric seas, 63
epidemics, 114
equal area projection, 49
equinox, 57
Eratosthenes, 11; map, **11**
Erdkunde, Die (Ritter), 19
Erie, Lake, 91
Esperanto, 163
Essay on the Inequality of Human Races
    (Gobineau), 73
Essay on the Principle of Population, An
    (Malthus), 116
Estaville, 175
Estonia, 204
Ethiopia, 136, 229; fertility in, 101
ethnic groups: former USSR, **205**;
    former Yugoslavia, **204**
ethnicity, 186–7; and agriculture,
    221–2; and religion, 165
ethnic regions: Africa, **201**; Europe,
    208
ethnocentrism, 53
ethnography, 52
Euclidean geometry, 288

Europe: areas of dissent, **202**; as
    cultural region, 154, **154**; ethnic
    regions, **208**; expansion overseas,
    14, 195–6; and free migration,
    129–30; integration, 206–7;
    medieval, 12–13; and nationalism,
    193; separatist movements,
    203–4; urbanization, 249
European Atomic Energy
    Community, 206
European Coal and Steel
    Community, 206
European Community, 303, 304; and
    economic integration, 284
European Economic Community
    (EEC), 4, 206–7, 223
European Free Trade Association,
    206
Everest, Mount, 56
evolution: human, 71; of primates,
    69–70
*Exceptionalism in Geography: A
    Methodological Examination*
    (Schaefer), 33
existentialism, 34
exploration, 14; and development,
    195, **195**; or invasion, 14
export processing zones, 278, 280
exurbanization, 247
Exxon Valdez, 91

factories, 273
factory towns, 274
famine, 136
farming: Canadian prairies, **68**;
    commercial, 234–6; in Illinois,
    222; mixed, 234
fascism, 209
fear, geography of, 183, 185
fecundity, 100
federalism, 209
feedlot system, 235
feminism, 178–9
feminist geography, 119, 176
fertility, 99–114, 100; and cultural
    factors, 101–2; rate, 104–6, 110;
    transition, 118; urban vs rural
    areas, 104
feudalism, 179, 193, 249, 307
field patterns, 243
fieldwork, 52
Finnish place names, 163–4
fire, 71, 79, 114, 147; and vegetation,
    83, **83**
fish killed by acid rain, **78**
Flanders, 203
Flenley, 78
flexible accumulation, 280
flooding (Bangladesh), 140
folk culture, 188–9
food: aid, 136, **137**; as commodity,
    142; global trade, 237; less
    developed world, 80–1; as world
    problem, 137–8, 141–2

forced migration, 129
forces of production, 35
Fordism, 179, 280; modes of
    production, 258
forests: boreal, 68–9; monitoring, 84
formal region, 40
fossil fuels, 83, 274
Foster, George, 18
Fra Mauro (map maker), 14
France, geography in, 21
Franco-Prussian War (1870–1), 20
Freeman, D.B., 297
free migration, 129–31
free trade area, 303
free will, 30, 40
friction of distance, 288
Friedmann, John, 236, 283
frontier agriculture, 232
Frontier Wage, The (Thünen), 223
functional region, 40
fundamentalism, 214

Gabon, fertility in, 101
Gaelic language, 159, 160, 163
Gaia hypothesis, 70
Galicia, 204
Gambia, 238
gamma index, 301
garden city, 250
Garrison, William, 300
Gatterer (geographer), 17
gazetteer (Ptolemy), 11
Geddes, Patrick, 250
Gemeinschaft (community), 245–6, 307
gender, 119, 178, 185–6; relations
    and agriculture, 238–9
gendered employment, 282
gender-specific theory, 119
General Agreement on Trade and
    Tariffs (GATT), 303
general fertility rate (GFR), 100
General Foods, 239
gentrification, 260–1
geodesics, 289
Geographia (Strabo), 11
Geographia Generalis (Varenius), 16–17
geographic: analysis, 14; description,
    15–16; knowledge, 36; literacy, 36
geographical imagination, 310–11
geographical societies, 19
geographic information system (GIS),
    46, 49–51, 310
geographie Vidalienne, 22
geography: classical, 11–12, 27;
    contemporary, 23, 23–8;
    institutionalization, 19–22, 27,
    29; Middle Ages, 12–14, 27;
    preclassical, 10, 27; universal,
    18–19, 27
*Geography's Inner Worlds: Pervasive
    Themes in Contemporary American
    Geography*, 178
geophagy, 189
geopolitics (and *Geopolitik*), 198–200

German Catholics, and farm villages, 244
Germany, 106, 200; geography in, 20–1; refugees, 131, 133, 135; unification, 4
Gerry, Elbridge, 211
gerrymandering, 211, **211**
Gesellschaft (mass society), 245–6, 296, 307
GFR *See* general fertility rate (GFR)
Ghana, 203, 300
ghettos, 186, 262
Giddens, Anthony, 176
global: city, 261; climate, **65**; climate and human evolution, 71; cordilleran belts, **62**; diffusion, 283; environments, 66–9, **66**; ocean currents, **62**; soils, **64**; vegetation, **64**; warming, 85, 93; warming and sea levels, 94
globalization, 304, 307
Gobineau, Joseph Arthur de, 73, 149
Golden Temple, 168
Gondwanaland, **58**
Gonneville (explorer), 15
gorillas, 70
Gottman, 197
Gould, Steven Jay, 73, 127, 299
government: and agricultural landscape, **224**; forms of, 207–10
gradation, 59
Graham, J., 282
Grameen Bank, 87, 139
Grand Allies, 273
Grand Canyon, **61**
Grapes of Wrath, The (Steinbeck), 174
graph theory, 300–1
grasslands, temperate, 68
gravity model, 293–4
grazing economy, 236
Great American Desert, 164
Great Britain *See* Britain
Great Leap Forward, 284
green belt, 250
Greenbie, B.B., 262
Greene, 214
greenhouse: effect, 70, 86, 92–3; gases, 93
Greenland ice sheets, 69
green parties, 81, 96
Green Plan, 81
green revolution, 234, 237
Gregory, 176, 310
grid system: in China, 13; of mapping, 11–12
Griffen, 229
Grigg, 231, 233
Grossman, D., 238, 245
gross national product (GNP), 140–1
groundwater depletion, 90
group engineering, 189–90
group formation, 148
Group of Seven, 188
group unity, and language, 155–6
growth curves, S-shaped, **43**

growth pole concept, 283
Guide to Geography (Ptolemy), 11
guild system, 249
Gulf War, 309
Guthrie, Woody, 174

Hagerstrand, T., 42, 292, 295, 298
Haiti, 138
Hale, 173
Hamilton, **44**, 188
Han dynasty, 13
Hardin, Garrett, 81
Harris, 169, 258
Hartshorne, Richard, 24, 25, 33, 35, 197, 200
Hartshorne-Schaefer debate, 33
Harvey, David, 179, 258, 263, 309, 310
Haushofer (geographer), 200
health care, geography of, 181, 183
heartland: and hinterland, 277; theory, 199, **199**
Henry, Prince (Portugal), 14
Herbert, 181
Herbertson (geographer), 23
Hereford map, 13
Herodotus, 11
Hettner, Alfred, 18, 20, 24, 28, 298
hexagons, 254–5, **255**
hierarchies, 42, 255, 296
high-technology, 278, 281
Hinduism, 165, **165**, 168, 169
Hipparchus, 11
historical materialism, 35
Hitler, 33
Holocene period, 72
homelands, 174–5
homeless people, 185
hominids, 70
Homo erectus, 71, 72
Homo habilis, 71
Homo sapiens sapiens, 71–3
Hong Kong, 194
Hoover, 270
Horn of Africa refugees, 134
Horvath, 229
Hotelling, Harold, 272
households, 178
housing, geography of, 7
Howard, Ebenezer, 250
How to Lie with Maps (Monmonier), 50
Hoyt, 258
Hudson, J.C., 276; theory, 245
human determinism, 145
human development index (HDI), **140**, 141, 186
Human Development Report, 141
human disasters, 139
human geography, 1–2, 10, 26–7; concepts, 36–46; evolution of, 309–10; historical perspective, 7; humanistic, 34; Marxist, 35; and philosophy, 35–6; and physical geography, 23, 24

human impacts, 82–93
humanism, 30, 32–4, 119, 175–6, 258, 310
humanistic geography, 32–4
human landscape, and cities, 249
humans: and ecosystems, 77–9, 82–3; and land, 1, 22
human species, 56
Humboldt, Alexander von, 18–19, 22, 24, 27, 30, 42, 80, 145
hunter gatherers, and ecosystems, 82
Huntington, Ellsworth, 22, 30, 31, 149, 214
Hutterite colony, **221**
hydraulic hypothesis, 149–50
hydrosphere, 77

ibn-Batata, 14
ibn-Khaldun, 14, 197
ice erosion, 63
iconography, 176; of cities, 261–2; of landscape, 188
idealism, 34
ideal type, 227
idiographic method, 35
IDRISI software, 49
illegal migration, 131
images, 42–3
immigrants, Eastern European, **130**
immigration policies, 130
impelled migration, 129
imperialism, 197
import system, 230
INR (infant mortality rate), 104
India, 81, 162, 206; fertility rate, 110; and industry, 278; refugees, 131, 133
Indian-Pakistan wars, 213
indigenization, 215
Indo-European languages, 159–60, **160**, **161**
industrial change, 284
industrialists, and ecosystems, 83
industrial landscapes, 273–4, 278, 307
industrial location: behavioural approach, 271–2; factors involved, 268, 280–1; and pollution regulations, 282; theory, 267–72; *See also* least cost theory
industrial regions, major centres, 275–6
Industrial Revolution, 7, 176, 248, 249, 279; and capitalism, 179–80, 272–4; effect on demographics, 307; and energy sources, 79; and modern world, 306–7; and organized sports, 189; and population growth, 114; social and economic changes, 307
industrial waste, 91
infant mortality rate *See* IMR
information exchange, 280–1
information service industry, 281

information technology, 7; and decentralization, 282
infrastructure, 34, 35
inner-city landscape, 185
Innis, Harold, 294
innovation diffusion, 295, 297; in Kenya, 297
Innuitian mountain system, 63
institutionalization, 19–22, 29
intensity theory, 226
interaction, 41; theory, 220
interglacial period, 92
international: cooperation, 210; direct investment, 307; relations, 198
interregional factors, 269
intraregional factors, 269
Inuit civilization, 151
invasion, or exploration, 14
Iran, 200; and refugees, 133
Iraq, 10, 200
iron horse, 252
iron industry, 273–4
Iroquois longhouse, **148**
irredentism, 202
irrigation, 149
Isard, Walter, 295
Islam, 164–5, **165**, 195; and geography, 12–14; and physical environment, 168
isochromes, 290
isodapane, 270; map, **270**
isolated state, 226, 253
Isolated State, The (Thünen), 223
isoline map, **48**
isopleth map, 47
isotims, 270; map, **270**

Jackson, P., 175, 188
James, Preston, 144–5, 189, 214
Janelle, 290
Japan, 124, 274; industry in, 276–7
Jerusalem, 198
Jewish state, 198
Jews: and farm villages, 244; and forced migration, 129; *See also* Judaism
Johnston, R.J., 187, 212
Jonasson, 230
Jones, 197–8; field theory, **197**
Jordan, 133, 155, 164
Joseph, 246
J-shaped curve of death rates and age, **106**
Judaism, 165, 167–8; *See also* Jews
Jura region, 204
justice and geography, 181

Kansky, K., 300, 301
Kant, Immanuel, 18, 27, 36, 214
Kaplan, 93
Karan, P., 168
Kaups, 155, 163

Keddie, 246
Kennelly, 271
Kenya, 297; and refugees, 134
Kenyan Green Belt Movement, 87
Kerguelen (explorer), 15
Kim Il Sung, 209
Kim Jong Il, 209
Kjellen (political scientist), 198, 199
Kniffen, F.B., 151, 295
Knox, Thomas, 181, 209–10, 261, 263
Kohr, L., 207
Krebheil, E., 210
Kroeber, Alfred, 145, 146, 149
Kropotkin, Peter, 209, 250
Kurds, 200, 202, 309
Kurgan culture, 159

labour: as commodity, 179, 307; costs, 269–70; and forced migration, 129; in industrial location, 280–1
Labour party, 211
Lagos, 262
land: competition for, 223; fertility, 223; human impacts on, 82–3; and humans, 1–2; private ownership, 150; productivity and population, 124; transportation See railways; roads
land-forms, 59
Landsat, 51, 84, 85
landscape, 1; of consumption, 188; élitist, 183; geography, 22–6; and human meaning, 187; in language, 164; physical, 59–65; as place, 187; and power relations, 181–6; school, 23, 145–6; of stigma, 183–5
Landscapes of Despair (Dear and Wolch), 264
Landschaftskunde, 23
land survey, 243, 244
land use, rent-paying abilities, 225
language, 155–64, 157–9; evolution, 147; families, 156–60; groups, 156, 203–4; in landscape, 156, 163–4; and nationalism, 160–1; world distribution of, 160
Laos, 209
large-scale grain production, 236
laterite, 90
Latin American Integration Association, 207
latitude, 14; calculation, 11; establishing, 14
Latvia, 204
Laurasia, 58
Laurentianism, 294
Laurentian Shield, 63, 284
Law of Comparative Advantage, 303
Law of Retail Gravitation, 294
least cost theory, 268–70, 273, 275; locational triangle, 269; testing, 271

Lebanon, 133
lebensraum, 200
Le Corbusier, 251
Lee Kuan Yew, 135
LE (life expectancy), 104–5, 107; France, 6
Le Play (sociologist), 21
Léros, 185
less developed world, 5, 80–1, 135–9; agricultural modernization, 237–8; and cities, 248; contraceptive use, 101; energy use in, 79; and environmental concerns, 210; fertility rate, 105; industry in, 278–9; and population, 99; and refugees, 133; and socialism, 209; Thünen theory, 229–30; tourism in, 279–80; trade, 303; urbanization, 260
Lewthwaite, 31
liberal feminism, 179
Liberal party, 211
life cycle, and migration, 128
life on earth, 69–70
life expectancy See LE
lifestyle: irrational, 169; rural vs urban, 146
Limitation of the Vend, The, 273
limits to growth, 115
lingua franca, 162
lithosphere, 57, 77
Lithuania, 204
livestock-fattening economy, 236
locales, 176
location: concept of, 37; and distance, 288; patterns of, 2
location-allocation models of movement, 295
locational interdependence, 271, 272
locational triangle, 269
location theory, 220, 223
logarithmic transformation: of distances, 292
logging, effect on landscape, 83
London underground system, 292–3, 293
longitude, 14; calculation, 11; establishing, 14
Lorenz, Konrad, 213
Los Angeles, 262
Lösch, August, 253, 256, 271
Louisiana, 172–3
Love Canal, 50, 91
Lovelock, James, 70
Lowe, J.C., 300
Lusaka, 6
Lynch, K., 261

Macao, 194
McCarthyism, 33
McCarthy, Joseph, 33
McKenzie, 256
Mackinder, Halford J., 21, 198–200; heartland theory, 199, 199

Magellan (explorer), 14, 15
malapportionment, 211
Malaysia, 109
Mali, 203
malnutrition, 135–6
Malte-Brun, Conrad, 18
Malthus, Thomas Robert, 17, 80, 115–16; theory of population growth, 116
Man and Nature, or Physical Geography as Modified by Human Action (Marsh), 80
Mandeville, John, 16
manifest destiny, 196
Man's Role in Changing the Face of the Earth (Thomas), 80
Maoism, 209
Mao Zedong, 209
map making, 19, 36; early, 15
Mappemonde (Ebstorf), 13
mapping, 46, 47
map projections, 49
maps: early, 10, 11, 15; limitations, 50; medieval, 12–13; political divisions, 3–6, 3
map scale, 47
maquiladoras, 278
Marcus, 178
market area analysis, 270–1, 275
marketing, principle, 255, 256
market places, 248
Marsh, G.P., 21
Martin, G.J., 33
Martinique, road network, 300–1, 301
Marxism, 30, 34–6, 119, 175–6, 258, 282, 310; and early civilization, 150; economics, 219; human geography, 35, 147; and population growth, 116–17; and postmodernism, 179; and urban areas, 263
Marx, Karl, 36, 175, 283; and capitalism, 116–17, 154, 180; and industrial location, 268
Massey, D., 282
material index, 269
Mauritius, 80
Maydan-e-Naghsh-e-Jahan (Iran), 39
Mead (philosopher), 175
Mediterranean: agriculture, 234; areas, 68
Mediterranean Sea, 92
Meinig, D.W., 155
Mennonites, and farm villages, 244
mental images, 42–3
mentally ill, 183, 185
mental maps, 43, 45, 127; Bancroft (Ontario), 128; and cities, 261; and decision makers, 272; Pointe Claire (Quebec), 128; and safety, 185
mentifacts, 146
mercantile model, 251–2
mercantilism, 249

Mercator (map maker), 15, 38; projection, 15, 48, 49
Mesopotamia, 10
metatheory, 263
metropolitan evolution, 252
Mexican steel industry, 271
Mexico City, 259, 262
Meyer, D.R., 252
middle ages, 12–14
middle class, changes, 7
Middle East, 133
migration, 125–31; laws, 126; Ravenstein laws, 126; selectivity of, 127–8; types of, 128–9
military: centres, 248; expenditures, 214
Mill, J.S., 268
minimum cost distances, 41
minorities, 186; languages of, 162
Mithila region, 168
mixed farming, 234
moa species, 88
mobility transition, 126; and demographic transition theory of population growth, 126–7; model, 283
mode of production, 35
modernism, 176, 251, 307
modernization, and demographic change, 6
modern trade theory, 303
modes of transport, 301
Mollweide projection, 48, 49
monarchy, 207
monetarist economic policy, 138
Mongoloid, 72
Monmonier (cartographer), 49, 50
monogamy, 71, 114
monsoon areas, 66
Montague, Ashley, 213
Monte Carlo simulation, 295
Montesquieu (philosopher), 17, 30, 31, 197
Montreal protocol (1987), 81, 82, 92, 93
more developed world, 5–6; and agricultural activity, 237; energy use in, 79; fertility rate, 106; industry in, 275–6; and loans, 138–9; and population, 99; and trade, 303
Morgan (anthropologist), 149
Mormons, 37–8, 155, 169, 172, 187; and farm villages, 244
Morrill, R.L., 299, 310
Morris, Desmond, 213
Morris, William, 272
mortality, 99–114; rate, 104–7, 106
Moryadas, S., 300
Mozambique, 133
Muhammad, 13
Muller, E.K., 230, 252
multiculturalism, 187
multilingual states, 161

multinational: corporations, 235, 275, 278, 307; states, 194
multiple nuclei model, 258
Munster, Sebastian, 16

Naess, Arne, 82
Naked Ape, The (Morris), 213
Namibia, fertility in, 101
Napoleonic wars, 193
nation, 192
National Geographic Society, 49
national identity, 192, 193, 203; and classical Greece, 12
nationalism, 193, 214, 307; and language, 160–1
National Topographic System, 47
National Wool Act (1954), 224
nation state, 12, 192–4, 308
Native Canadians, 7
natural disasters, 139
natural gas, as fuel, 274
natural resources, cultural appraisal, 79
Nature of Geography, The (Hartshorne), 23, 24
Nature of Human Aggression, The (Montague), 213
Nazi Germany, 200
Neanderthals, 71, 72
nearest neighbour analysis, 255
near-infrared spectral region, 51
Negroid, 72
neighbourhoods, 42, 262–4, 296; geography of, 7
neighbourhood watch schemes, 185
neoclassical: economics, 219; trade theory, 303; urban land rent, 258
neo-Ricardian economics, 219
Netherlands, 91, 94
networks, 299–301, **300**; and graph theory, 300–1
New England farming village, **173**
newly industrializing countries (NICs), 276–8; and trade, 303
New Towns Act (1958), 251
New York state, 228
New Zealand, 275
Niagara Falls, **39**
Niagara Peninsula, 247
Nigerian census (1991), 124
nodes, 299–300
Nomad civilization, 151
nomothetic method, 35–6
normative theory, 220
North America, 275; cultural regions, **152**; industrial regions, 275; regions, 153–4; vernacular regions, 172–4, 173
North American Free Trade Agreement (NAFTA), 207, 303
North Atlantic Treaty Organization (NATO), 200
Northern Ireland, 203
North Korea, 209

North Sea, 82, 275
Nostrand, 175
nuclear power, 79, 83, 275; station, **78**
nuclear war, geography of, 214
Nuclear War Atlas (Bunge), 214
nucleated rural settlements, 243–4; patterns, **245**
nuptiality, 101

Oaxaca, 59
objectivism, 34
ocean: currents, **62**; pollution, 91
Ohmae, 308
oil, 79, 83, 274; sources, 274; spills, 91
Oklahoma City, 40
Oklahoma State, and psychogeography, 174
oligarchy, 207
Olson, 178
On Aggression (Lorenz), 213
One World Divided (James), 144, 189, 214
Ontario, 228–9
Openshaw, 214
Ordnance Survey, 19
Organization of African Unity, 207
Organization of American States (OAS), 207
Organization of Petroleum Exporting Countries (OPEC), 207
Origin of Species, The (Darwin), 20, 30, 88
Ortelius (map maker), 15
Osborne, 188
overpopulation: and food problem, 137–8; and Malthus, 117
Owen, Robert, 250
oxygen depletion, 91
ozone: and chloro-fluorocarbons (CFCs), 92; layer, 56, 69, 81, 93

Pacific Rim countries, 276–8
Paddison, R., 209
Pahl, 246
Pakistan, 132, 133, 206
Palestine, 133, 198
Pangea, **58**, 70
pariah landscapes, 183
Park, 256
Parsons, 145
participant observation, 52
pastoral: frontier, 233; nomadism, 234
patriarchy, 178, 185
Pawson, 183
peace and war, 213–16
Pearl (scientist), 115, 116
peasant-herder conflict, 239
Peattie, R., 310
Peet, 230
perception, 42–5

permafrost, 69
Persian empire, 193
Perspective on the Nature of Geography (Hartshorne), 24
Peru, 238
Petersen (sociologist), 128
petroglyphs, 147
phenomenology, 33–4, 37, 175
phenotypes, 73
physical distance, 289, **289**, 292
physical geography, 21; as cause, 22; and human geography, 24; landscape as, 1
physiography, 21
physiological density, and agriculture, 123
Physische Geografie (Kant), 18
pidgin languages, 163
Pinatubo, Mount, 59
pioneer periphery, 252
pipe open-pit mine, **89**
pixels, 51
place: concept of, 37–8; and space, 310
placelessness, 38
place names, 163–4
place utility, 127
planning in socialist countries, 284
plantation agriculture, 235
plant domestication, 79; and ecosystems, 83–4
Plato, 31, 79, 193, 197, 207
Pleistocene period, 71
point patterns, **41**
polar regions, 69, 93
political economy, 219, 258
Pollard, J.S., 282
pollution, 79; and industrial locations, 282; water, 90–2
Polo, Marco, 13
Polynesian civilization, 151
Pompeii, 59
popular culture, 188–9
population: Canada, 106; density, 122–5, **123**, **138**; distribution, 122, **123**; growth, 114–18, **115**; mapping, 124–5; ten most populous countries, **123**, **124**; world, 6, 123–4, **123**
Population Bomb, The (Ehrlich), 115
population pressures, and desertification, 86
population pyramid, 111, 112–13
populism, 214
Portolano chart, **13**
port system, evolution in Ghana, 300
positive checks, 116
positivism, 30–2, 119, 175, 219, 310; and postmodernism, 179
possibilism, 23, 30, 145, 146
postindustrial society, 7, 281
postmodernism, 175–6, 178–9, 258, 261
postsuburban expansion, 260–1
potential model, 294–5

power: and authority, 181; of state, 210; and structuration theory, 177
pragmatism, 33
prairies, **40**
preclassical geography, 10
Pred, A., 177, 283
premodern urban development, 252
preventive checks, 116
primate city, 256
primitive migration, 128–9
Prince, H.C., 310
principle of least effort, 288
printing and map making, 14, 15
production technologies, 280
profit maximization *See* economic operator
projection, 47
pronatalist policies, 109
Protestantism, 96; and capitalism, 180
proxemic space, 262, 291
psychogeography, 174
Ptolemy, **12**, 14; map of world, **12**; works of, 14
public goods, 210
push-pull factors, 125–6, **129**, 247

qualitative data collection methods, 52–3
quantitative data collection methods, 53
Quebec, 45, 198, 201, 236
Queensland, 88
Question of Place, A (Johnston), 309

rabbit-proof fence (Australia), 88, **88**
races, 72, 73
racism: history of, 73; and immigration, 130
Radarsat satellite, 51
radical feminism, 179
railways, 274, 298–9, 302, 307; and economic growth, 301
rainfall: variability, 220; and wheat yield, **221**
Raitz, K.B., 186
ranching, 235–6
random: pattern, 41; sampling, 53
range of a good, 253
rank size rule, 256
rate of natural increase *See* RNI
rational choice theory, 220
Ratzel, Friedrich, 20, 22, 28–30, 197–9, 298; laws, 197
Ravenstein, E.G., 126, 294
Réclus, Elisée, 21, 209, 250
recreation and tourism industry, 279–81
recycling, 95
Red River, **26**
refugees, 131–5; by region, **132**; in need of protection, **132**; numbers, 132, **132**, 133; sources of, **133**

region: concept of, 38–40; definition of, 1, 187; states, 308
regional: geography, 22–3, 25–6, 38–40; inequalities, 283–4; integration, 307; science, 295
regionalization, 40
reindustrialization, 281
relations of production, 35
relative location, 37
relative space, 36
religion, 164–70; and agriculture, 221; and civilization, 151; conflicts, 167; distribution of, 165, **165**, **166**, **167**; and group identity, 165, 167, 215; and landscape, 167–70; and use of land, 168–9
remote sensing, 46, 50–2, 84, 85
renewable resources, 79
reproductive revolution, 118
Republican party, 211
resettlement, 133, 135
residential areas, 262–3
resistance to innovations, 296
restrictive immigration, 131
restructuring, 237, 280–4
revolution of earth, 57, **57**
Rhine Action Plan, 91–2
Rhine River, 91
Ricardo, David, 219, 223, 268, 303
rice: cultivation by women, 238; paddies (China), **67**
Richthofen, Ferdinand von, 20
Riemann geometry, 288
rimland theory, 200
Rio Earth Summit (1992), 81, 86
Ritter, Carl, 18–19, 22, 30, 197
RNI (rate of natural increase), 106–8, **108**, **109**, 118
roads, 302, 307
Robinson (geographer), 49; projection, **48**, 49
Rogers, 296
Rokkan, S., 203
Roman civilization, 12
Romania, 204; fertility rate, 110
Rome, 248
Rooney, 189
Rostow, Walt W., 283, 301
rotation of earth, 57
Rotestein, Frank, 115
routes, 300
Royal Geographical Society, 21
Rozman, G., 252
rural landscape (Vermont), **25**
rural settlement: Canadian prairies, 244, **244**; patterns, 243–4; theory, 244–5; in transition, 245–6
rural society, contemporary, 246
rural-urban fringe, 246–8
Russell, R.J., 151
Russia: life expectancy, 108; trans-Siberian railway, 298
Rwanda, 309

sacred: space, 37; structures, 168–9
Sadler, 276
Sahel region, 81, 86
Sahrawis, 234
sailing craft, 79
sail wagon, 252
St Helen, Mount, 59
St Helena, 194
salt, 70
sampling method, 53
San Andreas fault, 58
Sardinia, 204
Sargent, C.G., 252
satellite remote sensing, 51
satisficing behaviour, 220, 271
Sauer, Carl, 23–4, 28, 30, 33, 145–6, 189, 219, 231, 295
savannas, 66, 68
scale, 147; concept of, 42; human, 147
Scanlink, 303
Schaefer, Fred W., 23
Schaefer-Hartshorne debate, 33
Schlebecker, 230
Schlüter, Otto, 22–3, 28, 30
Schutz, 175
scientific method, 31–2
Scotland, 203
sea level: change, **94**; and global warming, 94
secessionist movements, 201–2
Second World War, 4, 249; and capitalism, 180
sector model of urban land use, 258
Semple, Ellen Churchill, 21, 22, 30, 31
separation, political, 4–5
separatism, in Europe, 203–4
service industries, 281–2
settlement: concept, 155; formation, 242; nodes, 299; in northern Sweden, 244; science of, 289; See also rural settlement
sexist hiring practices, 282
sexuality, and landscape, 188
shanty towns, 259, 260
sheep farming, 233
shifting agriculture, 233
Shinto, 165
shopping malls, 279; as popular culture, 188
Short, 258
Silent Spring (Carson), 80
Silicon Valley, 278
Sinclair, 230
Singapore, 109
site, concept of, 37
situation, concept of, 37
Sjoberg, 249
slavery, 129
slums and poverty, 263
small-scale maps, 47
Smil, V., 76, 79, 89, 93, 94, 148
Smith, 175, 181, 236

Smith, Adam, 219, 223, 268
smog, 93
social: democracy, 180; determinism, 145; distance, 41, 291–2; physics, 294–5; scale, 42; status, 292; theory, 175–9
social geography See landscape, geography
socialism, 35, 180–1, 207, 209; regional development, 284
socialist feminism, 179
social life, and industrialization, 249
society: and culture, 145–7; and industry, 282
sociofacts, 146
soils, 63; abuse of, 89–90; erosion, 90; global distribution of, **64**; and population density, 125; and relief, 220–1
solstice, 57
South Africa, 181, 182, 232, 252
South East Asia Treaty Organization, 207
South Korea, 277
space: concept of, 36–7, 288–95; and international relations, 198; and place, 310
spatial: barriers, 309; cognition, 290; competition, 253; concepts (North America 1763), **45**; model, 53; monopoly, 271; patterns, 295; preferences, 127; separatism, 37
spatial analysis, 2, 22–3, 25–6, 32, 36, 40; and distance, 41; -oriented diffusion research, 295–6; and quantitative methods, 53
spatial scale, 42; Canada, 122; point patterns, **43**
spatial variations, in fertility, 102, 104
specialized periphery, 252
species, 72
spectacle, 279
sphere, 155, **156**
sport as popular culture, 189
Spykman, 200
squatter settlements, **6**, 258–9
Sri Lanka, 137; tourism in, 280
S-shaped curve, 42, **43**; of innovativeness, 296, **297**; population growth, 115–16
Stafford, H.A., 282
Stamp (geographer), 26
staple: crop, 142, 236; economies, 277; model, 252; theory, 283
state, 192; and agriculture, 223; apparatus, 210; creation, 192–8
states: and internal divisions, 206; stability, 200–7; world distribution, 210
statistical sampling theory, 53
Steadman, 214
steam engine, 79, 306
steel: industry, 273–4, 276; mills, **44**; rail, 252

Stein, 174
Steinbeck, John, 174
stock resources, 79
Strabo, 11, 197
structuralism, 34
structuration, 175–7
subjectivism, 34
subsidized prices, 224
subsistence agriculture, 223
subsistence societies, and population growth, 117
suburban expansion, 185, 247, 260–1
superorganic, 146
superstructure, 34, 35
supply and demand curves, **225**
Surtsey, 59
survey questionnaire, 53
Survival International, 309
sustainable development, 82, 95–6
Swahili, 162
Switzerland as multilingual state, 161, **161**
symbolic interactionism, 175, 213
Syria, 133
systems, 76

Taafe, E.J., 299
taiga, 68–9
Taj Mahal, 168
Tangshan, 59
Tansley (botanist), 76
Tanzania, 209
Taoism, 165; and physical environment, 168
tariffs, 224, 303
Taylor, Griffith, 22, 30
techniques of analysis, 46–53
technological change: and agriculture, 233; and environment, 82
technology, 79
temperature and population density, 125
temporal scale, 42
Tennessee Valley, **87**
territorial expansion, 198
territories, 307
textile industry, 274
TFR See total fertility rate (TFR)
Thailand, 105, 133
Third World See less developed world
Thomas, W.L., 80
Thompson, 174
Thomson, Tom, 188
threshold populations, 253
Thünen, Johann Heinrich von, 223, 242; theory, 223, 226–30, **226**, **227**, 236, 254
Tickell, 239
time: dimension, 2; distance, 41, 290
Times (London), 306
Tokyo, 262
T-O maps, **12**

Tonnies, 245

tool making, 147

topographic maps, early, 19

topography, 12

topological map, 292–3

toponyms, 163–4

topophilia, 38

topophobia, 38

Toronto, **5**, 22, 188, 228–9; in physical space and time space, **291**

total fertility rate (TFR), 100

tourism, 279

Town and Country Planning Act (1947), 251

Toynbee, Arnold, 151, 192, 214

trade: factors affecting, 302–3; regional integration, 303–4; regulation, 303; theories, 303

Tragedy of the Commons, 81, 210

transaction technologies, 280

trans-continental railway, 298

transformation of distance, 292–3

transitional periphery, 252

transport: cost, 269, **271**; geography, 298; networks *See* networks; systems, 298–302

transportation principle, 255

trans-Siberian railway, 298

tree planting, against desertification, 87

Trépanier, 172

tribal people, decline of, 308–9

tropical rain forest, 66, 81, 85; effect of removal, 84–6; location, **85**

Trudeau, P.E., 187

Tuan Yi-Fu, 34, 38

tundra, 69

Turkey, 200

turnpike roads, 299

Tyneside, 273

typhoid, 91

tyranny of distance, 42

Tyranny of Distance, The (Blainey), 294

Tyrol, 203

Ullman, Edward, 258, 302

undernutrition, 135–6

unification, political, 4–5

uniform pattern, 41

Union of Soviet Socialist Republics (USSR), 205, **205**

unitary government, 209

United Kingdom, 201, 284; *See also* Britain

United Nations, 139, 213; disputes and conflicts, 212; population projections, 115

United Nations Conference on Environment and Development (Rio Earth Summit), 81

United Nations Conference on Human Environment, 80

United Nations Environmental Program, 86, 262

United Nations Food and Agricultural Organization, 85

United Nations High Commissioner for Refugees (UNHCR), 133

United Nations Plan of Action (1977), 86

United States, 274; and coal, 274; cultural regions, **152**, 153; expansion, **196**; geography in, 21–2; refugees in, 135; urban planning, 251

United States Environmental Protection Agency, 80

United States National Environmental Policy Act, 282

universal: geography, 18–19; language, 163

Unzen, Mount, 59

urban: climates, 93; decline and crime, 181; ecology, 258; growth, 246, 249–51, 309; planning, 250–1; population relative to world, **243**; slum, **184**; sprawl, 246–7; theory, 258

urban centres, 5; definitions, 242; internal structure, 256–8, **259**; in Manitoba, **43**; as places, 262–4; ten largest, 248

urban expansion: less developed world, 258–60; and profits to landowners, 230

urbanism, 248

urbanization, 248, 249

urban land values, 258, **259**

urban location, 251–6; central place theory, 253–6, **256–7**

urban managerialism, 258

urban street scene: Peshawar, **102**; Toronto, **5**

Uruguay, land use, **229**

values: cultural variations, 145; and human geographic research, 310; and more developed world, 190

Vance, James, 248, 249; mercantile model, 251–2

VanderPost, 105

Varenius, Bernhardus, 17

vegetation: global distribution of, **64**; human impacts on, 83–6, **84**

Venice, 94, 248

Vermont rural landscape, **25**

vernacular regions, 40, 172–5; North America, **173**; and sport, 189

verstehen, 33

vertical integration, 235–7

Vesuvius, 59

Vidal, Paul de la Blache, 21, 22, 28, 30, 31, 145–6, 242

Vietnam, 132, 209

Vietnamese: boat people, **131**; refugees, 132, **137**

Vietnam War, 132, 213

violence, 185

viticulture, 221–2

volcanoes, 58–9

voluntary repatriation, 133

voting: and class, 211; and place, 212–13

Wade, 252

Wagner, 175

Wales, 203

Wallerstein, 141

Wallonia, 203

war, cost of, 214

Warntz, 295

Warsaw Pact, 200

water, 57, 59; global cycle, 90; human impacts on, 90–2; industrial use of, 90; and population density, 125; transport, 302

water-mill, 79

Watson, Wreford, 288

weather, and agricultural yield, 65

weathering, 59

Webber, M.J., 269

Weber, Alfred, 253, 254, 268–70, 280

Weber, Max, 258; and Protestantism, 180

weeds, 83

Wegener, A.W. (meteorologist), 57

welfare geography, 181

well-being, 181

West Edmonton Mall, 188, 279

Westminster Abbey, 168

wet rice farming, 233–4

wheat, 230

White, R., 127

Whittlesey, 233

Wilson, 295

wind erosion, 63

windmill, 79

Winnipeg, 37, **37**, **38**

Wirth, 246, 249

Wisconsin, 228

Wittfogel (historian), 149

Wolch, J.R., 264

Wolfe, R., 198

Wolpert, J., 127

women, 178; and farming, 238; fear of crime, 183; status of, 118

Woodstock (Ontario), 247

woollen industry, 274

working class: decline, 7; popular culture, 188; residential areas, 307

world: agriculture, 230–6, **232**; debt *See* debt; food problem *See* food; industrial patterns, 274–80; political divisions (1938), **3**; political divisions (1994), **3**; population relative to urban, **243**; systems theory, 141–2, 197

World Bank, 85, 136, 140, 236

World Development Report (World Bank), 140

World Health Organization, 105, 185, 262

World Travel and Tourism Corporation, 279

Wright, Frank Lloyd, 251

Yeates, 247

Yellowstone National Park, 95

yojana, 289

Yosemite National Park, **95**

Yugoslavia, 204, **204**, 309

Zaire, fertility in, 101

Zelinsky, W., 126–7, 173

Zimmerer, 238

Zionism, 197

Zipf, G.K., 288

zoning, 251